Gyorgy Szekely
Sustainable Process Engineering

Gyorgy Szekely

Sustainable Process Engineering

2nd Edition

DE GRUYTER

Author
Prof. Gyorgy Szekely
Advanced Membranes and
Porous Materials Center (AMPM)
King Abdullah University of Science
and Technology (KAUST)
4700 Building 4/4222
6900 Thuwal 23955
Saudi Arabia
gyorgy.szekely@kaust.edu.sa

ISBN 978-3-11-102815-6
e-ISBN (PDF) 978-3-11-102816-3
e-ISBN (EPUB) 978-3-11-103032-6

Library of Congress Control Number: 2024934456

Bibliographic information published by the Deutsche Nationalbibliothek
The Deutsche Nationalbibliothek lists this publication in the Deutsche Nationalbibliografie;
detailed bibliographic data are available on the internet at http://dnb.dnb.de.

© 2024 Walter de Gruyter GmbH, Berlin/Boston
Cover image: Sakorn Sukkasemsakorn/iStock/Getty Images Plus
Typesetting: Integra Software Services Pvt. Ltd.
Printing and binding: CPI books GmbH, Leck

www.degruyter.com

This book is dedicated to the current and former members of my research group, alongside whom I have been maturing and growing a very stimulating and enthusiastic environment.

Preface

Welcome to the second edition of my book *Sustainable Process Engineering*, the inspiration for which arose from my university teachings on sustainability and green engineering. In this textbook, you will find an engaging and interesting reference guide on sustainability that will provide you with a solid foundation in this exciting field of science. This textbook is intended for chemists, materials scientists, and chemical engineers to provide them with the necessary tools to develop sustainable processes and implement the concepts of sustainability in designing synthetic routes and chemical processes. The book describes the basic principles of sustainable process design using simple, practical examples from various fields. An extensive reference list is also provided so that interested readers can quickly find the most relevant up-to-date literature for a more detailed description of special aspects. The book also includes QR codes to direct readers to animations, short videos, magazines, and blogs on the specific topics covered to enhance the learning experience.

One of the primary aims of sustainability is to manufacture products in the most environmentally, economically, and socially beneficial way by conserving materials, energy, and natural resources. Scientists, engineers, and industries have made great efforts to design greener chemicals and materials and develop more efficient synthetic routes and chemical processes. A set of green principles were initially drawn up to facilitate sustainable design through proactive thinking and industry engagement. These green principles have since matured over time into different green metrics. Throughout this book, the principles of green chemistry and engineering are referenced to guide the reader to the best sustainable practices. However, even with the considerable progress in greenness and sustainability in recent years, significant difficulties remain in assessing the greenness and sustainability of processes. Hence, the critical aspects and shortcomings of green technologies and methods are also identified and discussed.

This textbook is divided into 15 chapters, each covering a fundamental topic: Chapter 1 introduces sustainable process engineering with definitions and comparisons of green chemistry and green engineering. Chapter 2 discusses the increasing number of green metrics used to quantify and compare the sustainability of reactions and chemical processes. Chapter 3 provides an overview of green solvents and their role in sustainable manufacturing and includes solvent selection guides to green solvent manufacturing processes. Chapter 4 describes an example of the step-by-step development of a sustainable process and compares the original patented route with a process-intensified alternative. Chapter 5 presents a survey of the different process intensification methods and equipment employed. This chapter is followed by detailed accounts of continuous flow reactions and continuous separation processes in Chapters 6 and 7, respectively. Continuing on from the different methods and equipment described in the previous chapters, Chapter 8 focuses on the recovery and recycling of solvents. Chapter 9 describes the toolbox of process analytical technologies

https://doi.org/10.1515/9783111028163-202

and gives examples of how they can achieve more efficient and safer manufacturing processes. Chapter 10 introduces the basics of nuclear fuels and their future in terms of sustainability. Chapter 11 provides an overview of biofuels and their sustainable production. The manufacturing and use of green polymers and green building blocks are summarized in Chapter 12, followed by an introduction to solar powered engineering in Chapter 13. Chapter 14 describes data-driven optimization of chemical processes and practical applications of machine learning. Finally, the book's different topics are summarized in Chapter 15 as worked examples that aim to help readers deepen their understanding of the sustainable assessment of chemical manufacturing processes.

Let me conclude by gratefully acknowledging my students, who helped me to craft and write this textbook. They are all members of the KAUST community and are either members of my research group or attended my course Sustainable Process Development. I have been working in process development for twelve years now, and all my colleagues and students have contributed both directly or indirectly to the development of this book. I am particularly indebted and grateful to my distinguished students, who have made a tremendous effort in both the logistics and writing of this book; many thanks go to them for all their hard work.

Gyorgy Szekely
Advanced Membranes and Porous Materials Center
King Abdullah University of Science and Technology (KAUST)
Thuwal, Saudi Arabia

About the author

Prof. Gyorgy Szekely received his MSc degree in Chemical Engineering from the Technical University of Budapest, Hungary. He subsequently earned his PhD degree in Chemistry from the Technical University of Dortmund, Germany, under Marie Curie Actions. Gyorgy worked as an Early-Stage Researcher in the pharmaceutical research and development center of Hovione PharmaScience Ltd. in Portugal and as an IAESTE fellow at the University of Tokyo, Japan. He was a visiting researcher at Biotage MIP Technologies AB in Sweden. Gyorgy was a Postdoctoral Research Associate working with Prof. Andrew Livingston in Imperial College London, UK. He was appointed a Lecturer in Chemical Engineering at The University of Manchester, UK, between 2014 and 2019, followed by a Visiting Academic position until 2022. During his time in the UK he received the Distinguished Visiting Fellowship of the Royal Academy of Engineering. He is currently an Associate Professor in Chemical Engineering at the Advanced Membranes and Porous Materials Center at King Abdullah University of Science and Technology (KAUST), Saudi Arabia. His multidisciplinary professional background covers green process engineering, green solvents and materials, continuous reactions, and membrane separations. He serves as an Academic Editor for the journals *Sustainability Science and Technology*, *Sustainability & Circularity NOW*, *Journal of Membrane Science*, and *Advanced Membranes*; as an Associate Editor for the Separation Processes section of *Frontiers in Chemical Engineering*; and he is a member of the Editorial Advisory Board of *ACS Applied Polymer Materials*. He is a Fellow of the Royal Society of Chemistry and a Fellow of the Higher Education Academy. Gyorgy has been designing novel materials and processes for molecular level separations, which has resulted in more than 150 articles, industrial collaborations and consultancy works, books, patents, and invited keynote lectures. He received the ACS Sustainable Chemistry & Engineering Lectureship Award in Long Beach, 2023, and the ACS Class of Influential Researchers Award 2022. He is listed among the World's Top 2% Scientists by Stanford University annually since 2020. To learn more about the author, follow the QR code to his group's website (www.SzekelyGroup.com).

https://doi.org/10.1515/9783111028163-203

Contents

Preface —— VII

About the author —— IX

1   Introduction to sustainable processing —— 1
1.1   Sustainable development —— 2
1.2   Introduction to green chemistry —— 4
1.3   The 24 principles of green chemistry and green engineering —— 5
             Bibliography —— 11

2   Green process metrics —— 13
2.1   Atom economy —— 13
2.2   Reaction mass efficiency —— 15
2.3   Carbon efficiency —— 15
2.4   Effective mass yield —— 16
2.5   Environmental factor —— 16
2.6   Mass intensity —— 19
2.7   Process mass intensity —— 20
2.8   Mass productivity —— 20
2.9   Wastewater intensity —— 21
2.10   Solvent intensity —— 21
2.11   Carbon footprint, carbon emission factor, and carbon intensity —— 22
2.11.1   Methodology for carbon footprint industrial standards —— 23
2.11.2   Carbon footprint in the pharmaceutical industry —— 24
2.11.3   Carbon footprint in the petrochemical industry —— 25
2.12   Health and safety hazards —— 27
2.13   Defining a good chemical process —— 28
             Bibliography —— 30

3   The role of solvents in sustainable processes —— 32
3.1   Classification of solvents —— 33
3.2   Solvent usage and safety concerns —— 34
3.3   Green solvents —— 37
3.4   Solvent selection guides —— 42
             Bibliography —— 45

4   Sustainable process development from alpha to omega —— 48
4.1   PolarClean: a green polar aprotic solvent —— 48
4.2   The patented production of PolarClean —— 49
4.3   Toward the design of greener synthetic routes —— 51

4.4	Quality assessment	**54**
4.5	Green metrics analysis	**56**
4.5.1	Complexity and Ideality	**56**
4.5.2	Carbon intensity	**57**
4.5.3	Atom economy	**60**
4.5.4	Yield	**63**
4.5.5	E-factors	**65**
4.5.6	Health and safety risks	**66**
4.5.7	Solvent intensity	**68**
4.6	Room for improvement: Further optimization potential	**69**
	Bibliography	**70**
5	**Process intensification: methods and equipment**	**72**
5.1	Evolution of chemical processes	**77**
5.2	Process-intensifying equipment	**78**
5.2.1	Microreactors	**78**
5.2.2	Rotating devices	**80**
5.3	Process-intensifying methods	**83**
5.3.1	Membrane reactors	**84**
5.3.2	Hybrid separations	**86**
5.3.3	Use of alternative energy: ultrasound and microwave	**87**
5.3.4	Other methods	**92**
	Bibliography	**93**
6	**Continuous microflow processes**	**98**
6.1	Introduction	**98**
6.2	The advantages and disadvantages of continuous microfluidic systems	**100**
6.3	The green attributes of continuous flow processes	**102**
6.3.1	Principle 1: Prevention	**103**
6.3.2	Principle 2: Atom economy	**106**
6.3.3	Principle 6: Design for energy efficiency	**109**
6.3.4	Principle 9: Catalysis	**109**
6.3.5	Principle 11: Real-time analysis for pollution prevention	**111**
6.3.6	Principle 12: Safer chemistry for accident prevention	**112**
6.4	Microflow reactor systems	**114**
6.5	Lab-of-the-future and automated robotic platforms	**116**
	Bibliography	**117**
7	**Continuous separation processes**	**121**
7.1	Downstream processing in organic synthesis	**121**
7.2	Batch versus continuous separations	**121**

7.3	Continuous processing with supercritical fluids —— **123**	
7.4	Continuous membrane separations —— **125**	
7.5	Continuous crystallization processes —— **132**	
7.6	Centrifugal partition chromatography —— **135**	
7.7	Pressure and temperature swing adsorption —— **137**	
7.8	Artificial intelligence in chemical and separation technologies —— **141**	
	Bibliography —— **143**	

8	**Solvent recovery and recycling —— 147**	
8.1	Distillation processes —— **148**	
8.2	Adsorption processes —— **152**	
8.3	Membrane-based solvent recovery processes and their comparison with distillation and adsorption —— **154**	
8.4	Tools for solvent recovery process design —— **161**	
	Bibliography —— **162**	

9	**Process analytical technology —— 165**	
9.1	Introduction —— **165**	
9.2	PAT for green chemistry and engineering —— **167**	
9.3	Development of PAT systems —— **170**	
9.4	Industry outlook —— **172**	
9.5	PAT methods —— **173**	
9.5.1	Infrared spectroscopy —— **173**	
9.5.2	Raman spectroscopy —— **175**	
9.5.3	Nuclear magnetic resonance spectroscopy —— **175**	
9.5.4	Ultraviolet-visible spectroscopy —— **176**	
9.6	Case studies —— **177**	
9.6.1	Control of ammonia content and reaction monitoring with FTIR —— **177**	
9.6.2	FTIR spectroscopy-enabled control strategy for brivanib alaninate manufacturing —— **180**	
9.6.3	Implementation of Raman spectroscopy in reaction monitoring —— **182**	
9.6.4	Process control of continuous synthesis and solid drug formulation by IR and Raman spectroscopy —— **183**	
	Bibliography —— **185**	

10	**Sustainable nuclear fuels —— 188**	
10.1	Benefits of nuclear energy —— **192**	
10.2	Disadvantages of nuclear energy —— **193**	
10.3	Uranium as a nuclear fuel —— **195**	
10.3.1	Availability of uranium —— **195**	

10.3.2   Current methods for uranium sourcing —— **195**
10.3.3   Sustainable extraction of uranium from seawater —— **197**
10.4   Waste management —— **201**
           Bibliography —— **204**

11   **Toward sustainable biofuel production processes** —— **207**
11.1   Production of alcohols as fuels —— **208**
11.1.1   Biochemical conversion of lignocellulosic biomass —— **210**
11.1.2   Grinding —— **210**
11.1.3   Pretreatment —— **211**
11.1.4   Hydrolysis/saccharification —— **212**
11.1.5   Fermentation —— **213**
11.1.6   Distillation/dehydration —— **214**
11.1.7   Case study of a membrane integrated bioreactor system for the
           continuous production of bioethanol —— **215**
11.2   Biodiesel and its conventional production —— **217**
11.2.1   Alternative routes for biodiesel production —— **220**
           Bibliography —— **225**

12   **Green polymers and green building blocks** —— **229**
12.1   Introduction —— **229**
12.2   Polymers and the environment —— **231**
12.3   Plastic waste management: methods and limitations —— **235**
12.4   Bioplastics —— **236**
12.5   Green polymers —— **238**
12.6   Green monomers and building blocks —— **241**
12.7   Extraction methods —— **248**
12.7.1   Mechano-catalytic depolymerization —— **248**
12.7.2   Integrated conversion —— **249**
12.7.3   Ultrasound-assisted radical depolymerization —— **251**
12.7.4   Fermentation —— **252**
12.7.5   Segmented continuous flow fractionation —— **253**
12.8   New design technology concepts for advanced polymer materials —— **254**
12.8.1   Reactor design configuration —— **255**
12.8.2   Online monitoring —— **255**
12.8.3   Automation —— **256**
12.8.4   Membranes —— **256**
12.8.5   Membranes from chitosan and PLA —— **258**
12.8.6   Production of bio-based polyethylene (bio-PE) —— **260**
12.8.7   Bio-based 1,4-butanediol —— **262**
12.8.8   BioFoam —— **263**
12.8.9   Desmodur eco N —— **263**

12.8.10   Rilsan HT and Rilsan Invent —— **264**
12.8.11   Polycarbonates —— **264**
           Bibliography —— **265**

13   Solar powered engineering —— 269
13.1   Water harvesting from air —— **269**
13.2   Solar-driven membrane processes —— **272**
13.3   Concentrated solar power —— **275**
13.4   Photochemistry and photocatalysis —— **279**
13.4.1   Heterogeneous photocatalysis —— **280**
13.4.2   Solar-driven advanced oxidation processes —— **283**
13.4.3   Hybrid advanced oxidation processes —— **284**
13.4.4   Homogeneous photocatalysis —— **286**
13.4.5   Luminescent solar concentrator reactors —— **287**
13.4.6   Cloud-inspired photochemical reactor —— **288**
13.4.7   Chiral separation using light —— **290**
           Bibliography —— **291**

14   Data-driven optimization of chemical processes —— 294
14.1   Self-optimizing systems —— **295**
14.1.1   Autonomous experimentation platforms —— **296**
14.2   Fault detection and diagnosis systems in industrial processes —— **301**
14.2.1   Shallow machine learning algorithms —— **303**
14.2.2   Deep learning —— **305**
14.2.3   Transfer learning —— **307**
14.2.4   Unsupervised machine learning algorithms —— **309**
14.3   Refinery production scheduling —— **310**
14.3.1   Optimizing production scheduling: industry 3.0 vs. industry 4.0 in oil
           refinery operations —— **311**
14.3.2   Challenges —— **315**
14.3.3   Real case: Abqaiq Plants, a digital transformation success story —— **317**
14.3.4   Enhancing heating control to increase refinery throughput —— **319**
14.3.5   Model predictive control in scheduling a refinery —— **320**
           Bibliography —— **323**

15   Worked examples —— 327
15.1   Example 1 – Green metrics analysis for hazardous chemistry scale-up
           and decision-making —— **327**
15.1.1   Part A problem statements —— **327**
15.1.2   Part B problem statements —— **327**
15.1.3   Part A solutions —— **328**
15.1.4   Part B solutions —— **329**

15.2	Example 2 – Green metric analysis of catalytic synthesis and purification of a pharmaceutical building block —— **332**	
15.2.1	Part A problem statements —— **332**	
15.2.2	Part B problem statements —— **332**	
15.2.3	Part A solutions —— **333**	
15.2.4	Part B solutions —— **334**	
15.3	Example 3 – Comparison of batch and microflow processes in diazomethane-based chemistry —— **336**	
15.3.1	Part A problem statements —— **336**	
15.3.2	Part B problem statements —— **337**	
15.3.3	Part A solutions —— **338**	
15.3.4	Part B solutions —— **342**	
15.4	Example 4 – Bioethanol production: conventional batch fermentation versus continuous membrane bioreactor —— **344**	
15.4.1	Part A problem statements —— **344**	
15.4.2	Part A solutions —— **345**	
15.4.3	Part B problem statements —— **348**	
15.4.4	Part B solutions —— **349**	
15.5	Example 5 – Application of process analytical technologies in continuous catalytic hydrogenation —— **354**	
15.5.1	Problem statements —— **354**	
15.5.2	Solutions —— **355**	
15.6	Example 6 – Green metrics analysis for hazardous chemistry and purification optimization —— **356**	
15.6.1	Part A problem statements —— **357**	
15.6.2	Part A solutions —— **357**	
15.6.3	Part B problem statements —— **358**	
15.6.4	Part B solutions —— **359**	
15.7	Example 7 – Green metrics analysis and reaction optimization —— **360**	
15.7.1	Part A problem statements —— **361**	
15.7.2	Part A solutions —— **361**	
15.7.3	Part B problem statements —— **362**	
15.7.4	Part B solutions —— **362**	
15.7.5	Part C problem statements —— **364**	
15.7.6	Part C solutions —— **364**	
15.7.7	Part D problem statements —— **365**	
15.7.8	Part D solutions —— **365**	
	Bibliography —— **366**	

Index —— **367**

1 Introduction to sustainable processing

Diana Gulyas Oldal, Gyorgy Szekely

In 2015, the United Nations released a blueprint titled *Transforming our world: The 2030 agenda for sustainable development*, which describes 17 sustainable development goals to achieve a better and more sustainable future world for all [1]. Most of these goals are directly or indirectly connected with the work of chemists and chemical engineers. *Chemistry* involves the knowledge of chemical structures, reactions, properties, and the underlying theories. Chemists are tasked with making novel materials, examining their fundamental characteristics and reaction mechanisms, determining their structures, and implementing analyses (Figure 1.1a).

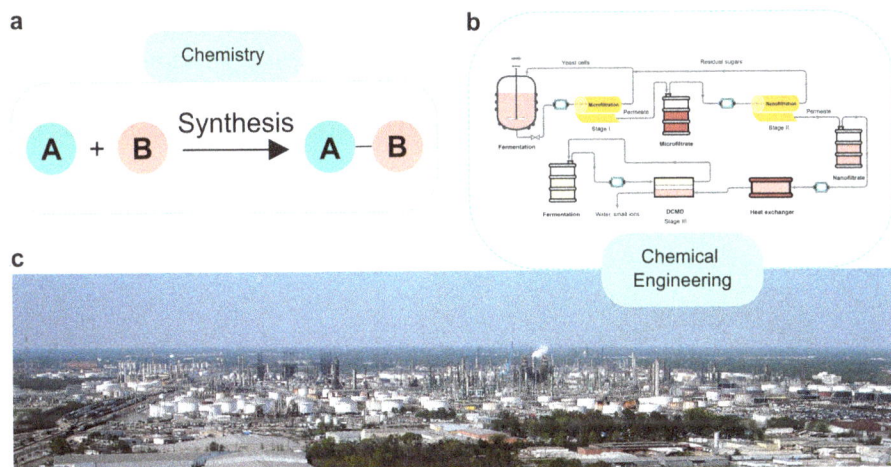

Figure 1.1: Schematic comparison between chemistry (a) and chemical engineering (b–c). The photo of the ExxonMobil oil refinery in Baton Rouge is reproduced under the Creative Commons license.

The field of *chemical engineering* requires a general knowledge of chemistry; however, the primary emphasis is on a detailed knowledge of heat and mass flow and the underlying thermodynamics and mathematics in these processes. Chemical engineers scale up methods for synthesis and design systems for heating, cooling, and transporting large amounts of material. Furthermore, they work on improving the efficiency and economics of industrial processes (Figure 1.1b).

https://doi.org/10.1515/9783111028163-001

1.1 Sustainable development

Sustainable development "meets the needs of the present without compromising the ability of future generations to meet their own needs" as per the definition by Brundtland [2]. Sustainability depends on the extent of utilized *resources* and *generated waste* [3]. Sustainable development requires two essential criteria: (i) natural resources need to be utilized sustainably to prevent depletion of supplies in the long run; (ii) any residues or waste should be generated at lower rates than the natural environment can readily assimilate them.

To preserve current global resources and provide future generations the opportunity to have at least the same standards that we currently enjoy, supply depletion should be controlled. The gold standard for the three crucial aspects of sustainability, i.e., *social*, *environmental*, and *economic* development, should be met, as Figure 1.2 illustrates [4]. These aspects can be described as people, planet, and profit – the three Ps – as represented by overlapping circles [5]. The 2D metrics are shown as the interactions of two aspects: socio-economic, eco-efficiency, and socio-ecological metrics. Genuinely sustainable solutions will fulfill all three aspects. Follow the QR code on this page to learn about the three pillars of sustainability.

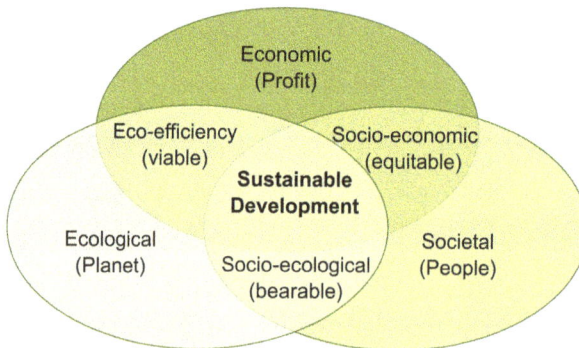

Figure 1.2: The importance of sustainability among social, environmental, and economic factors. Eco-efficiency, socio-economic, and socio-ecological metrics are 2D, while sustainability is 3D.

The broad scope of sustainability requires strong cooperation between social, economic, and environmental professionals. Sustainable engineering aims to develop technologically and economically viable systems with a particular emphasis on protecting ecosystems and human health. In addition to the environmental aspects, to achieve a fully sustainable technology, socio-economic metrics must be considered.

Isoni and co-workers described sustainability as "an anthropocentric concept based on human judgment of the delicate balance of social, environmental and economic factors, and as such, it is not uncommon to observe trade-offs in borderline situations" [6]. Sustainability is driven by human activities, since it relies on engineers, policymakers, healthcare experts, social scientists, and economists, among others. It is a global intersection of addressing societal needs while reducing the negative impact on the environment and considering the business aspects of a system; therefore, sustainability requires significant compromise. Two of the three pillars of sustainability are highly subjective (the economic and social aspects). The environmental factor is also subjective to some extent because there is no clear borderline when a system/process/product can be considered detrimental to the environment. Thus, these metrics change considerably depending on the field and geographical area, and even from company to company [5].

Transferring processes from one scale to another implies considerable temporal and spatial interdependence (Figure 1.3), starting from the molecular scale and ending at the macroeconomic level [7]. In the vast majority of cases, this represents a challenging task, since many processes, materials, and aspects do not behave in the same way at different scales.

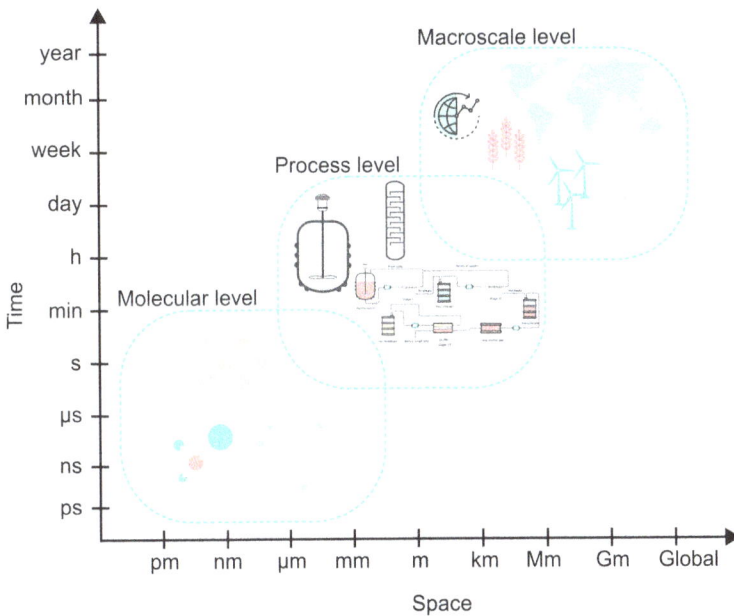

Figure 1.3: Temporal versus spatial scales in sustainable development for modeling and optimization. The smallest level considers molecules, which consume the least time and space, with the primary aim to make new substances (green chemistry). The middle level deals with processes and their development and optimization (green engineering). The largest level describes complete systems at the macroscale, including global network optimization (sustainability). Inspired by reference [7].

Sustainability is equally important in both the corporate and the scientific world. BASF is making significant efforts toward obtaining a leading sustainability position, including the issuance of green bonds [8]. According to a recent statement, they successfully issued corporate bonds of 2.28 billion USD to finance sustainable projects. They will fund instruments based on sustainability requirements to finance environmentally and socially friendly products and projects. This is a fine example of how to attract investors' attention and capital to achieve sustainable business management, and how to satisfy the increasing demand for cleaner production.

1.2 Introduction to green chemistry

Green chemistry deals with the fundamental aspects of chemistry without regard for industrial processes or profitability [9]. Figure 1.4a illustrates the importance and position of green chemistry in the hierarchy of safety controls. It is the most effective control measure because it eliminates or substitutes the hazard before it occurs. Consequently, green chemistry is a powerful and useful tool for eliminating the use of hazardous substances. Follow the QR code on this page to learn more about the hierarchy of safety controls.

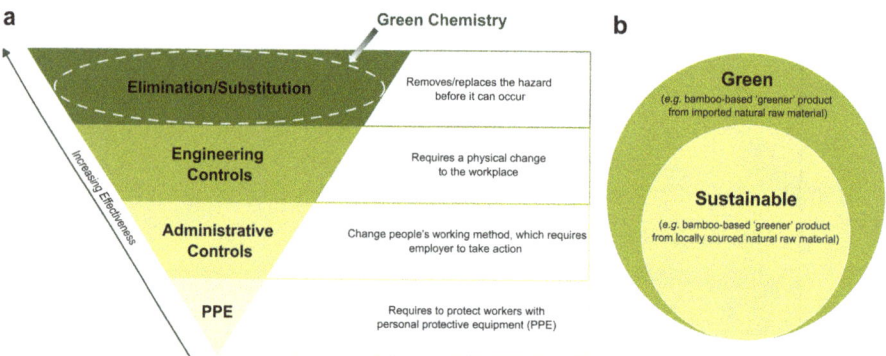

Figure 1.4: The importance and position of green chemistry. (a) The role of green chemistry in safety controls, and the steps illustrating the hierarchical structure of different controls in safety measures [10]. (b) Conglomeration of green and sustainable chemistry.

Green chemistry pursues to unify the academic, governmental, and industrial sectors by paying more attention to the environmental effect during the initial invention and innovation phase [11]. Paul Anastas [12] provided an ultimate academic definition of green chemistry: "Green chemistry is the utilization of a set of principles that reduces or eliminates the use or generation of hazardous substances in the design, manufacture, and application of chemical products." The ultimate governmental definition of green chemistry was formulated by the Environmental Protection Agency (EPA) [13]: "Green chemistry is the design of chemical products and processes that reduce or eliminate the use or generation of hazardous substances. Green chemistry applies across the life cycle of a chemical product, including its design, manufacture, use, and ultimate disposal."

Green chemistry aims to design and develop processes and materials that are environmentally benign, resulting in safer initial and final products with diminished energy consumption and expense [14]. The application of green chemistry results in facile separation processes, high yields, and minimum waste generation.

Sustainable chemistry is a subset of green chemistry (Figure 1.4b), which incorporates the design, manufacture, and use of efficient, effective, safe, and more environmentally benign chemical products and processes. Moreover, implementing the concept of sustainability in the production and use of chemicals and chemical products is crucial for enabling sustainable development. Another essential role of this field is to enable industrial processes to produce better products, mitigate pollution, and increase profit margins.

1.3 The 24 principles of green chemistry and green engineering

Paul Anastas, together with J. C. Warner [12] and J. B. Zimmerman [15], introduced the 24 principles of green chemistry and green engineering (Table 1.1). The term *green engineering* has the following definition according to the EPA [16]: "Green engineering is the design, commercialization, and use of processes and products in a way that reduces pollution, promotes sustainability and minimizes risk to human health and the environment without sacrificing economic viability and efficiency."

Table 1.1: Summary of the 24 principles of green chemistry [12] and green engineering [15].

The 12 principles of green chemistry	The 12 principles of green engineering
Prevention	Inherent rather than circumstantial
Atom economy	Prevention instead of treatment
Less hazardous chemical syntheses	Design for separation
Designing safer chemical	Maximize efficiency
Safer solvents and auxiliaries	Output-pulled versus input-pushed
Design for energy efficiency	Conserve complexity
Use of renewable feedstocks	Durability rather than immortality
Reduce derivatives	Meet need, minimize excess
Catalysis	Minimize material diversity
Design for degradation	Integrate material and energy flows
Real-time analysis for pollution prevention	Design for commercial "afterlife"
Inherently safer chemistry for accident prevention	Renewable rather than depleting

The principles of green chemistry outline a real effort to preserve the surrounding environment and commit to clean chemical technology. Anastas pioneered the definition of the principles [12] which have become the most widely used list of definitions in the field. The 12 principles are explained below:

1. *Prevention*: It is better to prevent waste generation than treat or clean up waste after it has been generated. Waste consumes time, energy, money, and resources when generated, handled, and disposed of, particularly hazardous waste [17].

2. *Atom economy*: Synthetic methods should be designed to maximize all materials used to produce the final product. The atom economy (AE) measures the efficacy of different chemical reactions and facilitates the comparison of diverse synthesis routes [18].

3. *Less hazardous chemical synthesis*: Wherever practicable, synthetic methods should be designed to use and generate substances that are non-toxic to human health and the environment. While reducing the amount of material and energy consumed represents a crucial element of green chemistry, considering resources and energy also play an essential role. Highly reactive chemical substances are frequently used for chemical reactions due to their favorable thermodynamic and kinetic properties [10]. These compounds may react with unintended biological cells of living organisms, leading to detrimental consequences.

4. *Designing safer chemicals*: Chemical products should be designed to preserve a function's efficacy while reducing toxicity. One of the most notorious examples for the need for safer chemical design is the case of the drug thalidomide, which was used to treat anxiety, insomnia, and morning sickness. This racemic sedative was removed from the market in the 1960s because it was discovered to cause severe teratogenic effects that resulted in congenital malformations, such as the lack of arms and legs or extraordinary short limbs in newborns. Further research revealed that the (R)-(+)-isomer was therapeutically active, while the (S)-(–)-isomer was responsible for the teratogenic effect [19]. Follow the QR codes on this page for more examples of safer chemical design.

5. *Safer solvents and auxiliaries*: The use of auxiliary substances[1] should be avoided wherever possible and innocuous when their use is unavoidable. An example of this principle is the substitution of the toxic and low-boiling point benzene with its less toxic and volatile alternative, toluene. Another example is the replacement of pentane and hexane with the greener alternative, heptane, which possesses a higher boiling point and lower toxicity. For more examples of solvent replacement, please refer to [20].

---

1 Auxiliary substances are solvents, separation agents, drying agents, surfactants, stabilizers, colorants, preservatives, emulsifiers, and every substance that is not directly incorporated into the product, but it is still needed for the reaction to occur.

6. *Design for energy efficiency*: Energy requirements for the intended scale of production should be estimated and minimized. Synthetic methods and separations should be performed at ambient temperature and pressure, if possible.
7. *Use of renewable feedstocks*: A raw material or feedstock should be renewable rather than depleting whenever technically and economically practicable.
8. *Reduce derivatives*: Unnecessary derivatization (use of protecting groups, protection/deprotection, temporary modification of physical/chemical properties) should be minimized or avoided if possible because such steps require additional time, reagents, and energy and generate waste.
9. *Catalysis*: Catalysts (as selective as possible) are superior to stoichiometric or excess use of reagents. Catalysts are usually easier to recover and reuse than excess reagents because they often have a similar chemical nature to the product.
10. *Design for degradation*: Chemical products should be designed so that they break down into innocuous degradation products at the end of their function and do not persist in the environment.
11. *Real-time analysis for pollution prevention*: Analytical methodologies need to be in place to allow for real-time, in-process monitoring and control before forming hazardous substances. Refer to Chapter 9 for process analytical technologies.
12. *Inherently safer chemistry for accident prevention*: Substances and the form of a substance used in a chemical process should be chosen to minimize the potential for chemical accidents, including accidental release, explosions, and fires.

Green chemistry can be considered as a tool for attaining sustainability. Sheldon reported the correlations between green chemistry principles and green metrics [5]. Mass-based metrics are related to five of the principles of green chemistry (1, 2, 5, 8, and 9). They represent the backbone of green product design. Nevertheless, they are not sufficient for obtaining a comprehensive picture of the greenness of a process. Accordingly, more factors should be taken into account, such as energy efficiency (principle 6), health and safety risks (principles 3, 11, and 12), the renewability of resources (principle 7), and the environmental effect of input and output materials (principle 3) [5].

Energy security appears as one of the main factors impacting sustainable and green chemical design, along with environmental and human health aspects. Persistence may be an advantageous property in terms of invested energy and complexity for obtaining the desired final product. To ensure that this investment ultimately ends up as a benefit, the decision pathway in Figure 1.5 should be followed [21]. The decisions are related to the molecular complexity, embedded energy, and distribution in the environment.

Figure 1.6 represents the optimal factors for green chemical syntheses. These contributors can best succeed if they are taken into account during the design phase of a particular process.

Redesign + using less or non-toxic materials

Redesign + using renewable sources

Redesign

| Desired function | No |

Yes

| Finite fossil resources | Yes |

No

| Toxic | Yes |

No

| High molecular complexity | Yes |

No

| High invested energy | Yes | Distributed into environment |

No | Yes | No

| Degradable in the environment | | Utilizing in a recycle loop |

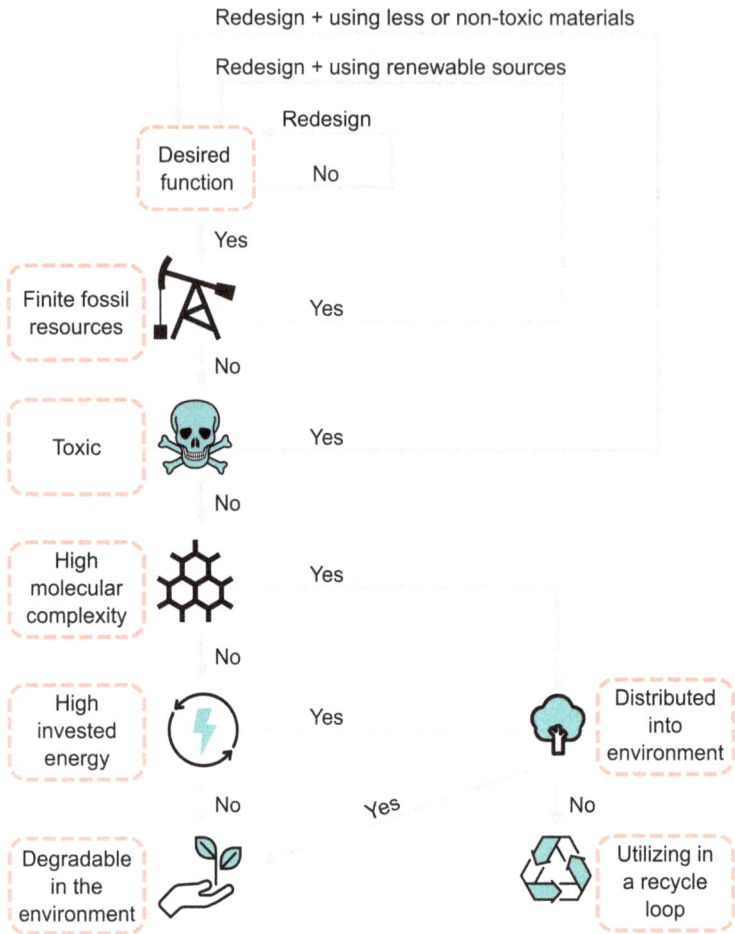

Figure 1.5: The decision pathway for chemical design. Adapted from [21].

Material

Hazards

Expenses

Reduction

Energy

Waste

Finite resources

Figure 1.6: Elements of reduction in chemical syntheses.

Paul Anastas and Julie Zimmerman developed the 12 principles of green engineering [22]. These principles emphasize the actions that will contribute to establishing a greener chemical process or product. The definitions of these principles are provided below.

1. *Inherent rather than circumstantial*: Process engineers should strive to ensure that all material and energy inputs and outputs are as inherently non-hazardous as possible.
2. *Prevention instead of treatment*: It is better to prevent waste generation than treat or clean up waste after formation.
3. *Design for separation*: Separation and purification unit operations should be designed to minimize energy consumption and materials use.
4. *Maximize efficiency*: Products, processes, and systems should be designed to maximize mass, energy, space, and time efficiency.
5. *Output-pulled versus input-pushed*: Products, processes, and systems should be *output-pulled* rather than *input-pushed* through the use of energy and materials.
6. *Conserve complexity*: Embedded entropy and complexity must be viewed as an investment when making design choices on recycling, reuse, or beneficial disposition.
7. *Durability rather than immortality*: Targeted durability, not immortality, should be a design goal. A successful design for durability rather than immortality is demonstrated by developing biodegradable starch-based pellets and foams for packaging materials. These products can be readily dissolved in water systems at the end of their lifetime, making them sustainable alternatives to conventional polystyrene packaging [23].
8. *Meet need, minimize excess*: Design solutions for unnecessary capacity or capability (e.g., "one size fits all") should be considered a design flaw. The reasonable anticipation of the necessary process swiftness and product adaptability is a significant factor at the design stage to avoid consequences arising from overdesign or unusable capacities. If these requirements are met, there is no need to dispose of and treat components, which would not be accomplished in the vast majority of processing conditions.
9. *Minimize material diversity*: Material diversity in multicomponent products should be minimized to promote disassembly and value retention.
10. *Integrate material and energy flows*: The design of products, processes, and systems must include integration and interconnectivity with available energy and materials flows.
11. *Design for commercial afterlife*: Products, processes, and systems should be designed for performance in a commercial afterlife; i.e., they can be reused or transformed to preserve their usefulness for other new products. For example, the raw material of polyethylene terephthalate (PET) can be recovered, or new material can be produced through chemical recycling via methanolysis, resulting in the formation of dimethyl terephthalate and ethylene glycol. Another example of the

successful implementation of this principle is that approx. 90% of Xerox equipment is designed for remanufacturing [15].

12. *Renewable rather than depleting*: Material and energy inputs should be renewable rather than depleting. Consuming finite substances leads to depletion, and depleting virgin resources demand repetitive processes, which have environmentally alarming consequences. An input can be considered renewable if the generated waste may be reused as recycled material or as an alternative feedstock that does not lose its value. Thus, considering and applying renewable inputs make an essential contribution to sustainable development.

The green engineering principles were developed from the green chemistry principles. Following both sets of principles, sustainable process design can be achieved by paying particular attention to the environment, our society, and the economy (refer to Figure 1.2). Green chemistry mainly focuses on designing a product to minimize its hazard, whereas the focal point of green engineering is the manufacturing process. The interconnectivity of green chemistry and green engineering is shown in Figure 1.7. There is a complex relationship between these two fields, which implies that, in reality, chemists and chemical engineers need to cooperate closely to achieve sustainable development.

Figure 1.7: Interconnectivity between the 12 principles of green chemistry and the 12 principles of green chemical engineering.

The main aim of sustainable development is to address our present needs without jeopardizing future generations from meeting their own needs. Consequently, the chemical sector needs to develop sustainable products for the present and subsequent generations. Figure 1.8 shows the status of today's chemical sector and what transformations are needed to make the sector more sustainable. A transition is necessary from linear to circular processes [21]. The inherent nature of the products must be changed to meet the future needs of the chemical sector. Any waste generated should be designed to be utilized as a raw material for other reactions. Follow the QR code on this page to learn more about the future of the chemical industries.

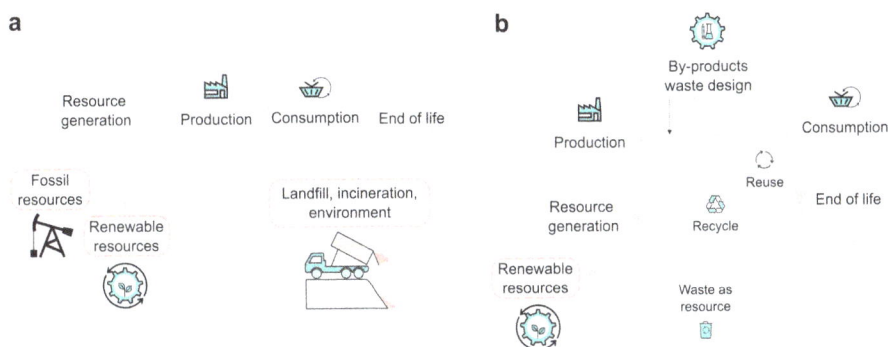

Figure 1.8: Comparison of the sustainability aspects of the present (a) and future (b) chemical sectors, from obtaining the resources to the end of a product's lifetime. Adapted from [21].

Bibliography

[1] Sustainable Development Goals www.un.org/sustainabledevelopment/sustainable-development-goals (accessed Jul 26, 2020).

[2] Keeble, B. R. The Brundtland Report: "Our Common Future". *Med. War.* **1988**, *4* (1), 17–25. 10.1080/07488008808408783.

[3] Graedel, T. E. *Em Handbook of Green Chemistry and Technology*. Clark, J. H., Ed.; 2002.

[4] Marmolejo-Correa, D.; Bean, J.; El-Halwagi, M. M. *Sustain. Metrics* **2009**, *49* (8), 1928–1932.

[5] Sheldon, R. A. Metrics of Green Chemistry and Sustainability: Past, Present, and Future. *ACS Sustain. Chem. Eng.* **2018**, *6* (1), 32–48. 10.1021/acssuschemeng.7b03505.

[6] Isoni, V.; Wong, L. L.; Khoo, H. H.; Halim, I.; Sharratt, P. Q-SAESS: A Methodology to Help Solvent Selection for Pharmaceutical Manufacture at the Early Process Development Stage. *Green. Chem.* **2016**, *18* (24), 6564–6572. 10.1039/c6gc02440h.

[7] Guillén-Gosálbez, G.; You, F.; Galán-Martín, Á.; Pozo, C.; Grossmann, I. E. Process Systems Engineering Thinking and Tools Applied to Sustainability Problems: Current Landscape and Future Opportunities. *Curr. Opin. Chem. Eng.* **2019**, *26*, 170–179. 10.1016/j.coche.2019.11.002.

[8] BASF places chemical industry first green bond, anchors sustainability in its financing strategy | Market Report Company – analytics, Prices, polyethylene, polypropylene, polyvinylchloride, polystyrene, Russia, Ukraine, Europe, Asia, reports http://www.mrcplast.com/news-news_open-371344.html (accessed Jun 26, 2020).

[9] Chemistry International – Newsmagazine for IUPAC https://old.iupac.org/publications/ci/2008/3002/bw1_bull.html (accessed Sep 1, 2020).

[10] 12 Principles of Green Chemistry – American Chemical Society https://www.acs.org/content/acs/en/greenchemistry/principles/12-principles-of-green-chemistry.html (accessed May 10, 2020).

[11] Höfer, R. History of the Sustainability Concept – Renaissance of Renewable Resources. *RSC Green Chem.* **2009**, *4*, 1–11. 10.1039/9781847552686-00001.

[12] Anastas, P. T.; Warner, J. C. Principles of Green Chemistry. *Green Chem. Theory Pract.* **1998**, 29–56.

[13] Basics of Green Chemistry | Green Chemistry | US EPA https://www.epa.gov/greenchemistry/basics-green-chemistry (accessed Mar 26, 2020).

[14] Boodhoo, K.; Harvey, A. *Process Intensification Technologies for Green Chemistry: Engineering Solutions for Sustainable Chemical Processing*; John Wiley & Sons, 2013.

[15] Anastas, P. T.; Zimmerman, J. B. Peer Reviewed: Design Through the 12 Principles of Green Engineering. *Environ. Sci. Technol.* **2003**, *37* (5), 94A–101A. 10.1021/es032373g.

[16] About Green Engineering | Green Engineering | US EPA https://www.epa.gov/green-engineering/about-green-engineering#principles (accessed Mar 26, 2020).

[17] Zimmerman, J. B.; Anastas, P. T. When Is a Waste Not a Waste. *Sustain. Sci. Eng. Defin. Princ.* **2005**, 201–221.

[18] Anastas, P. T.; Zimmerman, J. B. The Periodic Table of the Elements of Green and Sustainable Chemistry. *Green. Chem.* **2019**, *21*, 6545–6566. 10.1039/c9gc01293a.

[19] Nguyen, L. A.; He, H.; Pham-Huy, C. Chiral Drugs: An Overview. *Int. J. Biomed. Sci.* **2006**, *2* (2), 85–100.

[20] Alfonsi, K.; Colberg, J.; Dunn, P. J.; Fevig, T.; Jennings, S.; Johnson, T. A.; Kleine, H. P.; Knight, C.; Nagy, M. A.; Perry, D. A.; Stefaniak, M. Green Chemistry Tools to Influence a Medicinal Chemistry and Research Chemistry Based Organisation. *Green. Chem.* **2008**, *10* (1), 31–36. 10.1039/b711717e.

[21] Zimmerman, J. B.; Anastas, P. T.; Erythropel, H. C.; Leitner, W. Designing for a Green Chemistry Future. *Science (80).* **2020**, *367* (6476), 397–400. 10.1126/science.aay3060.

[22] Anastas, P. T.; Zimmerman, J. B. Design through the 12 Principles of Green Engineering. *IEEE Eng. Manag. Rev.* **2007**, *35* (3), 16. 10.1109/EMR.2007.4296421.

[23] Anastas, P. T.; Zimmerman, J. B. Chapter 2 the Twelve Principles of Green Engineering as a Foundation for Sustainability. *Sustain. Sci. Eng.* **2006**, *1*, 11–32. 10.1016/S1871-2711(06)80009-7.

2 Green process metrics

Diana Gulyas Oldal, Gergo Ignacz, Gyorgy Szekely

Improvement in sustainability should represent a measurable change that can be used for comparison, the communication of the work, and the transmission to manufacturing [1]. The evaluation of sustainability through comparison with conventional methods is feasible through green metrics analysis. Several metrics have been defined for *sustainability assessment* purposes [2]. An appropriate metric needs to be simple, clearly defined, measurable, and objective [1]. Various metrics are used during the design stage of synthetic routes, which are presented in the following sections.

Figure 2.1 illustrates the synthesis route of aspirin (acetylsalicylic acid), which is used to treat pain and inflammation. Aspirin is derived from the reaction of salicylic acid and acetic anhydride in the presence of an acid catalyst. Acetic anhydride is in excess, and therefore, the reaction needs quenching by the addition of water. The product precipitates upon the addition of water, and acetic anhydride undergoes hydrolysis to yield two molecules of acetic acid. The mixture is left to cool to room temperature. The solid product is isolated via vacuum filtration, followed by recrystallization using ethanol and water, and a subsequent filtration step. The green metric analysis in the following sections is based on the synthesis of aspirin in Figure 2.1.

2.1 Atom economy

Barry Trost first proposed the atom economy (AE) as a tool for organic chemists to pursue greener chemistry [3]. AE measures the amount of substrates and reagents incorporated into a final product, assuming exact stoichiometric quantities and 100% chemical yield. AE is defined as the ratio of the final product's molecular weight and the sum of the molecular weights of all reagents described in eq. (2.1), where MW is the molecular weight and n is the stoichiometric number of a reagent. Note that, in reality, the actual yield is always less than 100%. Furthermore, AE does not take into account solvent use. Calculation of AE for a generic multistep process with I as the final product is presented in eq. (2.2).

$$\text{AE}(\%) = \frac{\text{MW}_{\text{product}}}{\sum n \times \text{MW}_{\text{reagents}}} \times 100\%, \tag{2.1}$$

$$A + B \rightarrow C + D,$$
$$C + E \rightarrow F + G,$$
$$F + H \rightarrow I + J,$$

$$\text{AE}(\%) = \frac{\text{MW}_I}{\text{MW}_A + \text{MW}_B + \text{MW}_E + \text{MW}_H} \times 100\% \tag{2.2}$$

https://doi.org/10.1515/9783111028163-002

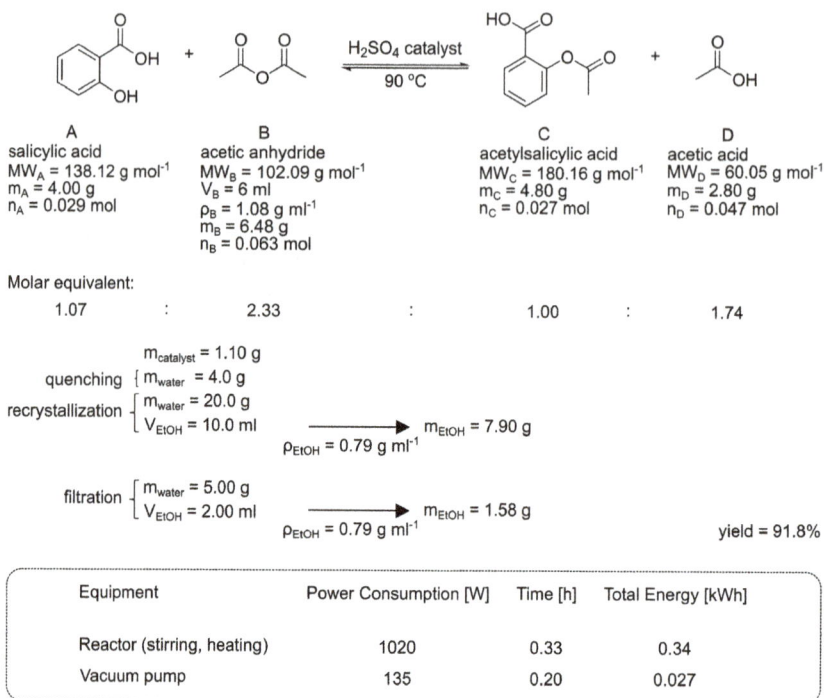

A
salicylic acid
$MW_A = 138.12$ g mol^{-1}
$m_A = 4.00$ g
$n_A = 0.029$ mol

B
acetic anhydride
$MW_B = 102.09$ g mol^{-1}
$V_B = 6$ ml
$\rho_B = 1.08$ g ml^{-1}
$m_B = 6.48$ g
$n_B = 0.063$ mol

C
acetylsalicylic acid
$MW_C = 180.16$ g mol^{-1}
$m_C = 4.80$ g
$n_C = 0.027$ mol

D
acetic acid
$MW_D = 60.05$ g mol^{-1}
$m_D = 2.80$ g
$n_D = 0.047$ mol

Molar equivalent:

| 1.07 | : | 2.33 | : | 1.00 | : | 1.74 |

$m_{catalyst} = 1.10$ g

quenching { m_{water} = 4.0 g

recrystallization { m_{water} = 20.0 g ; V_{EtOH} = 10.0 ml → m_{EtOH} = 7.90 g, ρ_{EtOH} = 0.79 g ml^{-1}

filtration { m_{water} = 5.00 g ; V_{EtOH} = 2.00 ml → m_{EtOH} = 1.58 g, ρ_{EtOH} = 0.79 g ml^{-1}

yield = 91.8%

Equipment	Power Consumption [W]	Time [h]	Total Energy [kWh]
Reactor (stirring, heating)	1020	0.33	0.34
Vacuum pump	135	0.20	0.027

Figure 2.1: The synthesis of aspirin (acetylsalicylic acid) from salicylic acid and acetic anhydride. This example is used to calculate the green metrics in the following sections.

The higher the AE, the fewer by-products and co-products formed, and consequently, the more sustainable the synthetic route. Simple addition and isomerization reactions have 100% AE because all of the starting materials are incorporated into the final product. There is no *co-product* or *by-product* (the by-product is structurally related to the final product, whereas the co-product is not) [4]. On the other hand, substitution and elimination reactions usually result in lower AE values. The calculation of AE does not require any experimental work. It can already be used during reaction design to indicate and assess waste generation among diverse alternative routes [5]. Equation (2.3) presents an example calculation for AE based on the synthesis of aspirin (Figure 2.1). The equation includes the final product (acetylsalicylic acid, MW: 180.16 g mol^{-1}) and the starting materials (salicylic acid, MW: 138.12 g mol^{-1}; acetic anhydride, MW: 102.09 g mol^{-1}).

$$AE = \frac{180.16}{138.12 + 102.09} \times 100\% = 75\%. \tag{2.3}$$

2.2 Reaction mass efficiency

Reaction mass efficiency (RME) is a mass-based metric developed by Constable and co-workers at GlaxoSmithKline (GSK) [2]. RME represents an improvement to AE because it takes yields and excess reactants into account. RME is defined as the percentage of the mass of the reactants that remains in the isolated product. For the generic reaction of $A + B \rightarrow C$, the RME is mathematically defined in eq. (2.4), and simplified in eq. (2.5). The RME describes reactions only, and not the whole process, and it does not consider the generated waste. Equations (2.6) and (2.7) present example calculations for RME based on aspirin synthesis (Figure 2.1) following eqs. (2.4) and (2.5), respectively.

$$\text{RME}\,(\%) = \left(\frac{\text{MW}_C}{\text{MW}_A + (\text{MW}_B \times \text{molar ratio } B/A)}\right) \times \text{yield}, \qquad (2.4)$$

$$\text{RME}\,(\%) = \frac{\text{mass}_C}{\text{mass}_A + \text{mass}_B} \times 100\%, \qquad (2.5)$$

$$\text{RME} = \frac{180.16}{138.12 + 102.09 \times \frac{2.33}{1.07}} \times 91.8\% = 46\%, \qquad (2.6)$$

$$\text{RME} = \frac{4.80}{4.00 + 6.48} \times 100\% = 46\%. \qquad (2.7)$$

2.3 Carbon efficiency

Carbon efficiency (CE) is a metric also introduced by the pharmaceutical industry[2]. CE was developed to distinguish the carbon mass between reactants and products.[1] CE is defined as the percentage of carbon in the reactants that remains in the final isolated product. The mathematical formula for the reaction of $A + B \rightarrow C$ is represented in eq. (2.8). Equation (2.9) shows the relation for CE calculation when the reaction results in product C and co-product or by-product formation. The CE takes into account both the yield and the stoichiometry of reagents and products.

$$\text{CE}\,(\%) = \frac{\text{Amount of carbon in product}}{\text{Total carbon present in reactants}} \times 100\%, \qquad (2.8)$$

$$\text{CE}\,(\%) = \frac{\text{moles of } C \times \text{carbons in } C}{(\text{moles of } A \times \text{carbons in } A) + (\text{moles of } B \times \text{carbons in } B)} \times 100\%. \qquad (2.9)$$

---

1 With the exception of a few inorganic salts (potassium chloride or lithium chloride), pharmaceutical products are exclusively carbon-based drugs.

Equation (2.10) presents an example calculation for CE based on aspirin synthesis (Figure 2.1). Equation (2.9) was applied because, in this example, co-product formation occurs (D).

$$CE = \frac{0.027 \times 9\,\text{carbons}}{0.029 \times 7\,\text{carbons} + 0.063 \times 4\,\text{carbons}} \times 100\% = 53.4\%. \tag{2.10}$$

2.4 Effective mass yield

The effective mass yield (EMY) is defined as the percentage of the desired product's mass relative to the mass of all *non-benign* materials used in its synthesis (eq. (2.11)) [6]. However, there is no clear definition for non-benign reagents. *Benign* substances are chemicals that pose no or negligible environmental risk and health concerns, such as pure or saline water, cellulose, or dilute ethanol. A more exact definition is required because all materials (except water) are environmentally hazardous to some extent. Moreover, the definition does not consider the level of toxicity based on objective data [1].

$$EMY\,(\%) = \frac{\text{Mass of product}}{\text{Mass of non-benign reagents}} \times 100\%. \tag{2.11}$$

Note that the definition of *non-benign reagents* is somewhat vague and includes those by- and co-products, reagents, and solvents that pose a negligible environmental risk, e.g., water, low-concentration saline, dilute ethanol, and autoclaved cell mass [6]. This definition does not consider the level of toxicity or any objective data regarding the reagents' non-benign nature. The EMY value for the example in Figure 2.1 is given in eq. (2.12). Salicylic acid is an irritant and corrosive substance, and acetic anhydride is classified as a flammable, irritant, and corrosive material. Thus, both will contribute to the EMY of the example reaction because of their non-benign nature.

$$EMY = \frac{4.80}{4.00 + 6.48} \times 100\% = 45.8\%. \tag{2.12}$$

2.5 Environmental factor

The environmental factor (EF), often shortened as E-factor, was one of the first green metrics proposed by Roger Sheldon in the early 1990s [7, 8]. EF is defined as the mass of waste per unit of product (eq. (2.13)). The EF is simple to use, and therefore it is one of the most common green metrics [4]. EF considers the waste generated from all *auxiliary components* during a process (e.g., solvent loss and chemicals applied for purification), not only what is generated during a reaction [5]. Lower values are preferable, as a high E-factor refers to generation of a large amount of waste. There is no difference between the types of waste regarding their hazardous nature. A large amount of benign waste is considered worse than a small amount of hazardous waste. The ideal E-factor value would be zero, which would ultimately address the first principle of green chemistry: prevention of waste generation [9].

$$EF = \frac{Total\ mass\ of\ waste\ (excl.\ H_2O)}{Mass\ of\ final\ product} = \left[\frac{kg}{kg}\right]. \tag{2.13}$$

The contributors to the EF originate from three significant sources (eq. (2.14)): the *core chemical reaction* (such as by-products and co-products), called EF_{kernel}; the *excess reagents*, which include all the unreacted materials, called EF_{excess}; and the *auxiliary materials* from the work-up and purification processes, called EF_{aux}.

$$EF_{total} = EF_{kernel} + EF_{excess} + EF_{aux}. \tag{2.14}$$

The general concept of dividing the EF into subgroups is a powerful tool for comparing different reactions or processes. For example, EF_{aux} provides the total auxiliary materials irrespective of the individual contributors. Consider the final product of aspirin (Figure 2.1), which needs recrystallization and filtration. New sub-EFs can be generated under EF_{aux}, namely, $EF_{crystallization}$ and $EF_{filtration}$, for the respective processes. Comparing the values of $EF_{crystallization}$ and $EF_{filtration}$, the effectiveness of EF_{aux} can be further evaluated. Even though sub-EFs are useful, their derivation can be cumbersome or even impossible in some cases.

Table 2.1 highlights the need for waste reduction, particularly in the fine chemical and pharmaceutical sectors [7]. The petrochemical industry generates less waste per product than other chemical sectors due to (i) the net margins with the demand of reducing the amount of waste, (ii) utilizing the materials instead of disposal, and (iii) relatively few process steps [1]. The pharmaceutical industry is at the opposite end of the spectrum. Pharmaceutical compounds are produced with high purities through multiple reaction and purification steps. Moreover, higher net profit margins do not necessitate waste minimization. Nonetheless, the pharmaceutical sector produces considerably less waste than many other industries [1]. However, bear in mind that petrochemicals are usually produced using significantly fewer steps than active pharmaceutical ingredients.

Table 2.1: E-factors for different chemical industries.

Industry sector	Annual product tonnage	E-factor
Oil refining	10^6–10^8	Less than 0.1
Bulk chemicals	10^4–10^6	1–5
Fine chemicals	10^2–10^4	5–50
Pharmaceuticals	10–10^3	25–100

The E-factor can also be divided into the simple E-factor (sEF, eq. (2.15)) and the complete E-factor (cEF, eq. (2.16)), depending on the process development phase [10]. The sEF does not consider solvents and water, and thus, it is much more suitable during the early development stage of synthetic routes. The cEF considers all process materials, such as raw materials, reagents, solvents, and water, without considering recycling. Thus, the cEF is adequate for *total waste stream analysis* (eq. (2.16)).

$$\text{sEF} = \frac{\sum m_{\text{raw materials}} + \sum m_{\text{reagents}} - m_{\text{product}}}{m_{\text{product}}} = \left[\frac{\text{kg}}{\text{kg}}\right], \tag{2.15}$$

$$\text{cEF} = \frac{\sum m_{\text{raw materials}} + \sum m_{\text{reagents}} + \sum m_{\text{solvents}} + \sum m_{\text{water}} - m_{\text{product}}}{m_{\text{product}}} = \left[\frac{\text{kg}}{\text{kg}}\right]. \tag{2.16}$$

One of the limitations of the original E-factor is the lack of consideration for a *reaction's energy consumption* or, more precisely, the energy required for heating and cooling. Energy generation represents one of the significant sources of greenhouse gas (GHG) emission in the atmosphere (refer to Figure 2.4). Thus, this factor should not be ignored, and for this, a new type of E-factor, called E^+-factor (eq. (2.17)), has been introduced [11].

$$E^+ = \frac{\sum m_{\text{waste}}}{m_{\text{product}}} \left[\text{kg kg}^{-1}\right] + \frac{W[\text{kWh}] \times CO_2EF\left[\text{kg}CO_2\,\text{kWh}^{-1}\right]}{m_{\text{product}}\,[\text{kg}]}. \tag{2.17}$$

To better understand CO_2EF, please refer to eq. (2.36) and Section 2.11 in general. Consider the example of aspirin synthesis in Figure 2.1. Equation (2.18) can be used to calculate the original E-factor based on the definition in eq. (2.13). The filtration step is done twice, before and after the recrystallization, and thus it needs to be considered twice.

$$\text{EF} = \frac{4.00 + 6.48 + 1.10 + 1.58 + 7.90 + 1.58 - 4.80}{4.80} = 3.7\,\text{g g}^{-1} = 3.7\,\text{kg kg}^{-1}. \tag{2.18}$$

According to eq. (2.14), the original E-factor of the aspirin example can be divided into different parts. We can differentiate between the core reaction ($\text{EF}_{\text{kernel}}$), the recrystallization step ($\text{EF}_{\text{cryst.}}$), and the filtration step ($\text{EF}_{\text{filtr.}}$). Thus, the total E-factor is the sum of the previous E-factors (eq. (2.19)). Note that the filtration step was performed twice and that the materials from the kernel reaction to recrystallization, i.e., the starting materials, should only be counted in the kernel step. The mass of the product should be excluded from EF_{total}; hence the subtraction at the end of eq. (2.19).

$$EF_{total} = EF_{kernel} + EF_{cryst.} + 2 \times EF_{filtr.}$$

$$= \frac{4.00 + 6.48 + 1.10}{4.8} + \frac{7.90}{4.8} + 2 \times \frac{1.58}{4.8} - \frac{4.80}{4.80}. \tag{2.19}$$

The definition of sEF in eq. (2.15) does not consider the solvents, and thus, they should be excluded from the example calculation on aspirin synthesis (eq. (2.20)).

$$sEF_{total} = sEF_{kernel} + sEF_{cryst.} + 2 \times EF_{filtr.} = \frac{4.00 + 6.48 + 1.10}{4.8} + 0 + 0 - \frac{4.80}{4.80}$$

$$= 1.41 \, g\,g^{-1} = 1.41 \, kg\,kg^{-1}. \tag{2.20}$$

The cEF (eq. (2.16)) considers all the inputs, such as the starting materials, the solvents, and water. Accordingly, the example calculation for aspirin synthesis includes all process materials (eq. (2.21)).

$$cEF_{total} = cEF_{kernel} + cEF_{cryst.} + 2 \times cEF_{filtr.}$$

$$= \frac{4.00 + 6.48 + 1.10 + 4.00}{4.8} + \frac{20.00 + 7.90}{4.8} + 2 \times \frac{5.00 + 1.58}{4.8} - \frac{4.80}{4.80}$$

$$= 10.8 \, g\,g^{-1} = 10.8 \, kg\,kg^{-1}. \tag{2.21}$$

The E⁺-factor for the aspirin example is calculated in eq. (2.22). The *emission conversion factor* is taken from a 2018 UK government report [12]. The CO_2EF is calculated by multiplying the total consumed energy with the emission factor for the used electricity (0.30720 kg CO_2 eq.kWh^{-1}). The energy consumption plays a significant part in the waste contribution, contributing 77.3% of the total E⁺-factor value. Follow the QR code on this page to watch a webinar on green metrics by Andrew P. Dicks from the University of Toronto.

$$E^{+} = 3.7 \, kg\,kg^{-1} + \frac{0.34 \, kWh \times 0.30720 \, kg\,CO_2 \, eq.\,kWh^{-1}}{0.0048 \, kg}$$

$$+ \frac{2 \times 0.027 \, kWh \times 0.30720 \, kg\,CO_2 \, eq.\,kWh^{-1}}{0.0048 \, kg} = 28.9 \, kg\,kg^{-1}. \tag{2.22}$$

2.6 Mass intensity

The mass intensity (MI) is defined as the mass ratio of the total input of materials (excluding water) to the final product (eq. (2.23)) [2]. Optimally, the value of MI should approach 1. This metric considers the yield, the stoichiometry, the solvents, and the reagents utilized in a chemical process. The total mass includes everything needed for a process such as reagents, solvents, catalysts, material mass used in the work-up, and purification, except water. MI can be expressed alongside the E-factor (eq. (2.24)). Equation (2.25) presents an example calculation for MI based on aspirin synthesis (Figure 2.1). The same value of MI is obtained using eq. (2.24) by adding 1 to the result of EF in eq. (2.18).

$$MI = \frac{\text{Total mass in process (excl. } H_2O)}{\text{Mass of product}} = \left[\frac{kg}{kg}\right], \tag{2.23}$$

$$MI = EF \text{ (excl. } H_2O) + 1, \tag{2.24}$$

$$MI = \frac{4.00 + 6.48 + 1.10 + 1.58 + 7.90 + 1.58}{4.80} = 4.7 \text{ g g}^{-1} = 4.7 \text{ kg kg}^{-1}. \tag{2.25}$$

2.7 Process mass intensity

The Green Chemistry Institute Pharmaceutical Round Table reinvented the MI term into a new term called process mass intensity (PMI, eq. (2.26)), which includes water in the solvent input materials and the inputs already included in MI (eq. (2.24)) [13]. This metric expresses the overall sustainability of the process because it does not ignore water consumption. The same value of PMI is obtained using eq. (2.27) by adding 1 to the result of cEF in eq. (2.21). Equation (2.28) presents an example calculation for PMI based on the synthesis of aspirin (Figure 2.1).

$$PMI = \frac{\text{Total mass in process (incl. } H_2O)}{\text{Mass of product}} = \left[\frac{kg}{kg}\right], \tag{2.26}$$

$$PMI = EF \text{ (incl. } H_2O) + 1, \tag{2.27}$$

$$PMI = \frac{4.00 + 6.48 + 1.10 + 4.00 + 5.00 + 1.58 + 20.00 + 7.90 + 5.00 + 1.58}{4.80}$$
$$= 11.8 \text{ g g}^{-1} = 11.8 \text{ kg kg}^{-1}. \tag{2.28}$$

2.8 Mass productivity

Mass productivity (MP) is defined as the percentage value of MI's reciprocal (eq. (2.29)). The introduction of this metric was suggested to improve the understanding of MI in the *business world* [14]. The terms "productivity" and "the percentage as unit" are more tangible and understandable for non-experts who do not work in green chemistry. Equation (2.30) presents an example calculation for MP based on aspirin synthesis (Figure 2.1).

$$MP = MI^{-1} \times 100\% = \frac{\text{Mass of product}}{\text{Total mass (excl. } H_2O)} \times 100\%, \tag{2.29}$$

$$MP = 4.7^{-1} \times 100\% = 21.3\%. \tag{2.30}$$

2.9 Wastewater intensity

Wastewater intensity (WWI) expresses the mass ratio of the *total generated wastewa-ter* to the product, as described in eq. (2.31). WWI takes into account wastewater but not the stoichiometry, the yield, and other solvents [10]. This green metric aims to provide information on the water consumption of a process instead of the E-factor, MI, and the MP, which excludes water.

$$\text{WWI} = \frac{\text{Mass of total wastewater generated}}{\text{Mass of product}} = \left[\frac{\text{kg}}{\text{kg}}\right]. \tag{2.31}$$

Equation (2.32) presents an example calculation for WWI based on aspirin synthesis (Figure 2.1). Every water molecule that is not incorporated into the final product is considered wastewater, such as the 4.00 g of water used for quenching and 25 g used in the purification steps.

$$\text{WWI} = \frac{4.00 + 5.00 + 20.00 + 5.00}{4.80} = 7.1\,\text{g}\,\text{g}^{-1} = 7.1\,\text{kg}\,\text{kg}^{-1}. \tag{2.32}$$

2.10 Solvent intensity

Solvent intensity (SI) is defined as the mass ratio of used solvents to the final product (eq. (2.33)). This metric accounts for solvent usage; however, the yield, stoichiometry, and water are not considered [10]. SI varies significantly among different industrial sectors. The higher the desired purity, the more purification steps are usually needed, which increases the SI. The SI does not take into account the used water in the reaction. Equation (2.34) presents an example calculation for SI based on aspirin synthesis (Figure 2.1).

$$\text{SI} = \frac{\text{Mass of solvents (excl. water)}}{\text{Mass of product}} = \left[\frac{\text{kg}}{\text{kg}}\right]. \tag{2.33}$$

$$\text{SI} = \frac{1.58 + 7.90 + 1.58}{4.80} = 2.3\,\text{g}\,\text{g}^{-1} = 2.3\,\text{kg}\,\text{kg}^{-1}. \tag{2.34}$$

2.11 Carbon footprint, carbon emission factor, and carbon intensity

The carbon footprint refers to the amount of GHG[2] emitted by a particular source per year (eq. (2.35)) [15]. Carbon dioxide comprises approx. 80% of the total GHGs [16]. In practice, carbon dioxide is regarded as the only GHG, and to simplify the calculations, the CO_2 footprint is calculated instead of the GHG footprint. However, other GHGs such as methane or nitrous oxide are also included in the carbon footprint by converting them to kg CO_2 equivalents. With the proper conversion factors, various data, including energy and waste, can be converted to kg eq. CO_2. The carbon footprint represents the total amount of carbon dioxide emitted during material utilization or energy consumed by an individual, a car, a process, a company, a country, or a system. The CO_2 emission factor (CO_2EF, eq. (2.36)) of a particular source describes the ratio of the mass of carbon dioxide released to the quantity of activity.

$$\text{Carbon footprint} = \text{GHG emission} \times \text{year}^{-1}, \tag{2.35}$$

$$CO_2EF = \frac{\text{Mass of } CO_2 \text{ emitted}}{\text{Quantity of activity}}. \tag{2.36}$$

The phrase *quantity of activity* can be defined in a variety of ways. If the emission source is public transport, then the quantity of activity can be expressed as the amount of burned fuel. If the source is cooking, then the quantity of activity can be described as kWh of used electricity. If the CO_2 quantity is expressed as a function of the product mass (kg CO_2 per kg product), it is termed as gravimetric emission factor [15]. On the other hand, when the amount of CO_2 is expressed for a given volume (kg CO_2 per liter), the correct term is the volumetric emission factor [15]. Carbon intensity (CI) is the mass emission of CO_2 over the product, which can also be expressed as a mass (kg) and energy (kWh or MJ) unit (eq. (2.37)). In this book, the kg kg^{-1} form is used throughout the calculations; however, numerous reports use both units inconsistently. Carbon intensity is useful for calculating and comparing the emission intensity and the greenness related to different processes and is simple to use. For example, CI can be calculated for various systems by taking the power consumption and converting it into CO_2 emission via emission conversion factors. The usefulness of CI for comparing different processes is demonstrated in Section 8.3.

$$CI = \frac{m_{CO_2 \text{ emission}}}{\text{Product}}. \tag{2.37}$$

---

2 GHGs are gases that absorb and emit infrared radiation. They include all emitted gases that contribute to the warming of our planet.

2.11.1 Methodology for carbon footprint industrial standards

The classification of different contributors to carbon footprint plays a crucial role in sustainability assessment. The Greenhouse Gas Protocol: A Corporative Accounting and Reporting Standard was developed by the Greenhouse Gas Protocol Initiative [17]. It consists of a partnership of businesses and non-governmental organizations whose primary aim is to develop a globally recognized greenhouse gas accounting and reporting standard and promote its application. An essential part of this document is the setting of operational boundaries by proposing the concept of "scope." Three scopes are introduced to facilitate distinction between direct and indirect emission sources [18], providing feasibility for organizations by helping them to determine what is relevant regarding emissions, and to prepare GHG standards that are effective in different situations (Figure 2.2):

1. *scope 1*: for direct GHG emissions, which originate from sources owned or controlled by the company;
2. *scope 2*: for electricity-based indirect GHG emissions, i.e., emissions from the generation of purchased electricity;
3. *scope 3*: for other indirect emissions, which are the implications of the activities of the company, but from sources not owned or controlled by the company.

Reporting scope 1 and scope 2 emissions is compulsory in line with the protocol; however, scope 3 is optional. Consequently, some companies do not disclose their indirect emissions, mainly due to double-counting of emissions between and within the scopes. For example, double-counting can happen if a manufacturer and an independent retailer use a third-party transportation company. In this case, both the manufacturer and the retailer could incorporate the emissions from transportation.

Evonik Industries, one of the largest specialty chemical companies, disclosed a compilation of data based on direct and indirect emissions [19]. They classified emissions from different sources and provided a complete evaluation of emissions starting from purchased energy and raw resources, across transports and waste generation, ending with the disposal of marketed products. The company applied the Greenhouse Gas Protocol Corporate Standard [17] for accounting and reporting of emissions.

Figure 2.3a illustrates the increase in greenhouse emissions in Evonik Industries, which is likely due to increased productivity. Figure 2.3b shows the data for 2018 divided into different categories based on their contribution to the overall emissions. The largest contributor to the carbon footprint is the procurement of raw materials, with 42% share (11.5 million metric tons), followed by the recycling of sold products (24%) and direct emissions (21%) with approx. the same proportion. In summary, more than half of the carbon footprint of a major company in the chemical industry falls in *scope 3*, with carbon emissions originating from outside the company.

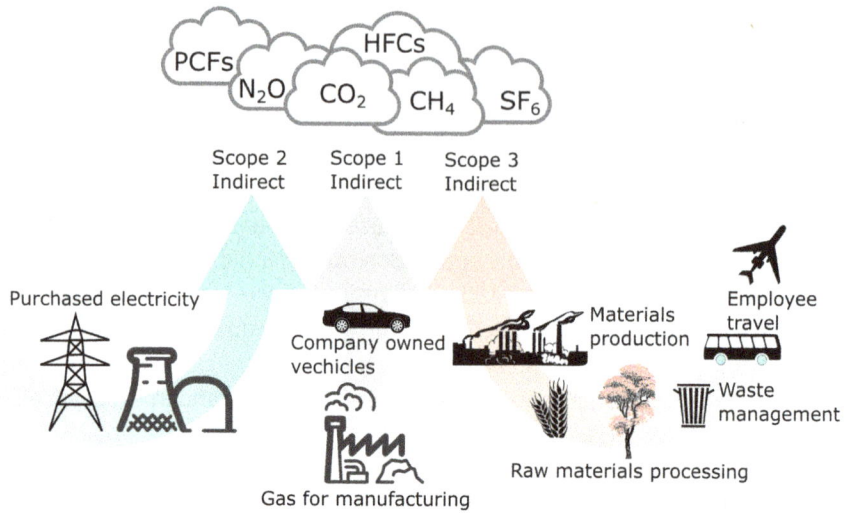

Figure 2.2: Summary of scopes to set operational boundaries. Scope 1 includes direct emissions controlled and owned by the company, such as company-owned cars. Scope 2 includes indirect emissions, which originate from sources owned or produced by other companies, such as purchased electricity. Scope 3 involves other indirect emissions related to the company's activity, utilizing sources from other companies, such as raw material production.

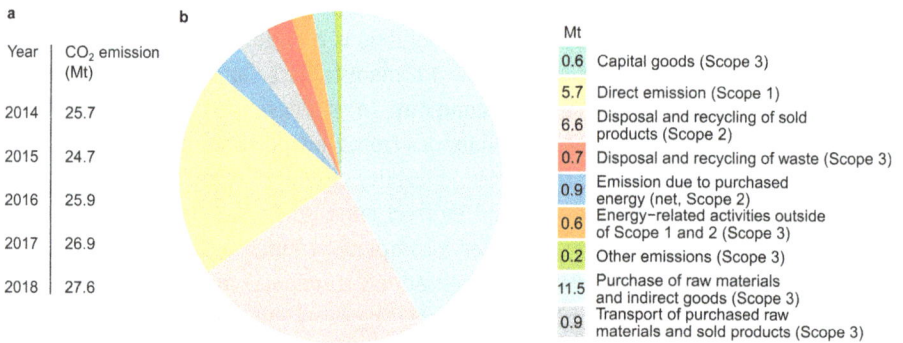

a

Year	CO_2 emission (Mt)
2014	25.7
2015	24.7
2016	25.9
2017	26.9
2018	27.6

b

Mt

- 0.6 Capital goods (Scope 3)
- 5.7 Direct emission (Scope 1)
- 6.6 Disposal and recycling of sold products (Scope 2)
- 0.7 Disposal and recycling of waste (Scope 3)
- 0.9 Emission due to purchased energy (net, Scope 2)
- 0.6 Energy–related activities outside of Scope 1 and 2 (Scope 3)
- 0.2 Other emissions (Scope 3)
- 11.5 Purchase of raw materials and indirect goods (Scope 3)
- 0.9 Transport of purchased raw materials and sold products (Scope 3)

Figure 2.3: Rising tendency of greenhouse gas emissions in Evonik Industries (a) and their carbon footprint in 2018 (b) (unit in millions of metric tons of CO_2). Adapted from [19].

2.11.2 Carbon footprint in the pharmaceutical industry

The importance of the carbon footprint is demonstrated by the fact that GSK, one of the world's largest pharmaceutical companies, recently added carbon footprint into its three primary long-term goals regarding sustainability, in addition to waste and water [20]. GSK examined their industrial processes and found that direct energy usage, i.e., heating,

ventilation, and air conditioning (HVAC), is the major contributor to their carbon footprint. Figure 2.4 illustrates the assessed contributors, showing that more than half of the
carbon footprint is caused following Good Manufacturing Practice (GMP) [20]. Reaching a
certain amount of purity with specified chemicals or solvents requires a significant
amount of energy and results in increased waste generation. The green metric analysis
helped GSK make the necessary decisions and focus the company on optimizing the
manufacturing facilities, particularly by developing energy-efficient HVAC systems.

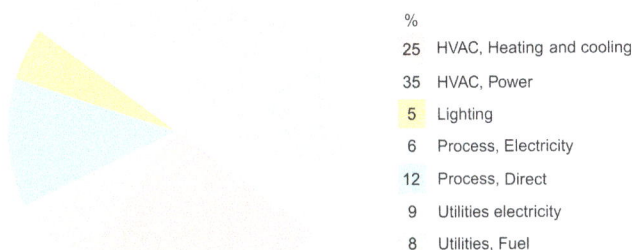

%

25	HVAC, Heating and cooling
35	HVAC, Power
5	Lighting
6	Process, Electricity
12	Process, Direct
9	Utilities electricity
8	Utilities, Fuel

Figure 2.4: Breakdown of GSK's process carbon footprint. Adapted from [20].

2.11.3 Carbon footprint in the petrochemical industry

The overall energy landscape is steadily evolving, with a decrease in conventional energy resources and a gradual progression to renewable energy [21]. Oil and gas companies are in the early phases of this transition to establish low-carbon emission
energy development. The oil field carbon intensity study carried out by Stanford University in 2018 reported global values of crude oil production for countries with at
least 0.1% of global oil production [22]. In total, 8966 on-stream oil fields in 90 different countries were analyzed and the report covered approx. 98% of global crude oil
and condensate production. The report found that GHGs produced in the considered
oilfields are equivalent to 1.7 Gt of carbon dioxide, which amounts to approx. 5% of
all emissions from fuel combustion in 2015. The data illustrated in Figure 2.5 is ranging from Denmark (3.3 g CO_2 MJ^{-1}) to Algeria (20.3 g CO_2 MJ^{-1}) showed that oil in Saudi
Arabia has the second-lowest CI value, 4.6 g CO_2 MJ^{-1}, after Denmark.

Saudi Aramco is the largest company in the sector, with enormous productivity,
even though they have relatively few reservoirs. The gas flaring rates per barrel and
the generation of produced water in their oil production process are low. Consequently,
less mass needs to be lifted to gain a given amount of oil. The energy consumption for
fluid separation, handling, treatment, and reinjection is relatively low compared to
other countries.

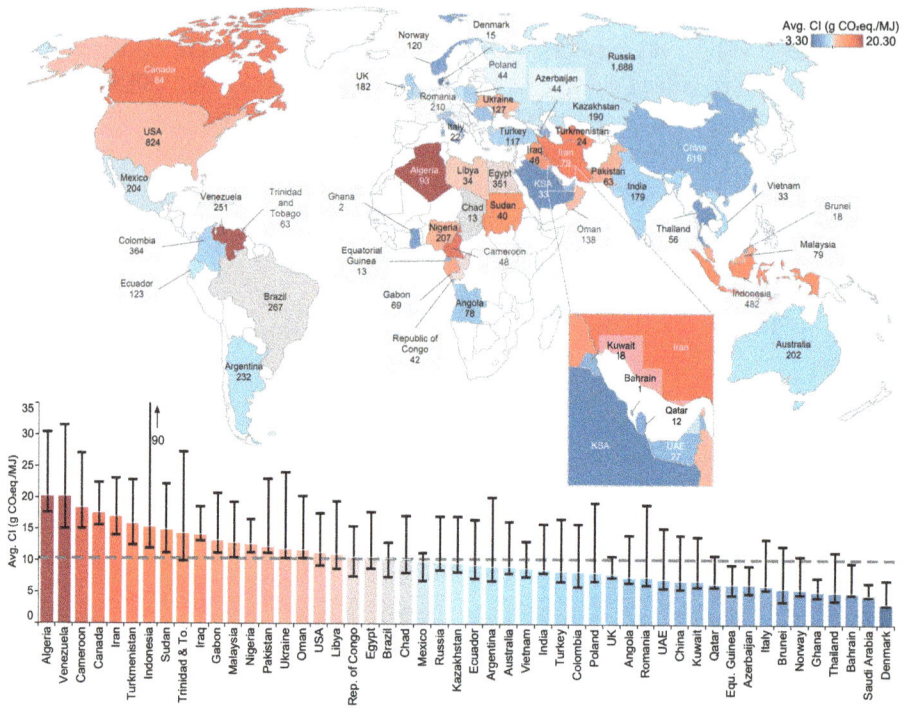

Figure 2.5: Estimated global crude oil carbon intensity (numbers below each country represent the analyzed fields). Reprinted with permission from [22]. Bar chart: national volume-weighted average upstream GHG intensities in g CO_2 eq./MJ crude oil delivered to the refinery. Dark blue: lowest; dark red: highest. The dashed line represents the global average.

A reduction in GHG emissions in the oil industry could be achieved by carefully managing crude oils. Producing, transporting, and refining crude oil into fuels such as gasoline or diesel contributes to approx. 15%–40% of the transport fuels' GHG emission [22]. The report highlighted that flaring has a considerable impact on CI values. Nevertheless, the study suggests that new and present oil fields should consider conservation methods, such as CO_2 capture, and the elimination of routine flaring and reducing methane emissions (e.g., through capturing). Algeria produces the world's lightest crude oil and possesses the highest CI values, which is a consequence of regularly burning large amounts of gas by the oilfield operators [23].

Achieving lower carbon intensity requires thoughtful reservoir management by minimizing flare and GHG emissions, energy efficiency, methane leak detection, and immediate repair [24]. The implementation of careful reservoir management includes practices that can be managed with balanced production, such as using seawater injection as a recovery technique to maintain the pressure on the reservoir with maximizing reservoir sweep. For example, Saudi Aramco tries to mitigate the amount of gas flared, which results in a flaring intensity of less than 1%. In contrast, 15% of all associated gas is flared globally [25]. However, according to the new industrial perspective called "Zero Routine Flaring by 2030," routine flaring will be eliminated globally by 2030. Lowering the flaring rate can be achieved by applying mobility geosteering, multilateral wells, and peripheral water flooding [26]. Methane leak detection is crucial in terms of sustainability because methane is 84 times more powerful than CO_2 as a GHG [27]. Saudi Aramco uses modern solutions such as drone-mounted and thermal cameras, laser detection, and quantification sensors for mitigating methane leaks, which has resulted in a substantially lower methane intensity of 0.06% (the average methane intensity in the industry was 1.34% in 2012) [24, 26].

2.12 Health and safety hazards

The fourth principle of green chemistry states that chemical and materials design should focus on function and mitigate health and safety hazards [28]. This principle is one of the most complex and interdisciplinary requirements of designing chemicals and materials. The mitigation of organic pollutants and toxic, bioaccumulative chemicals has gained increased attention in green chemistry in recent years [29]. Measuring and quantifying the different hazardous properties of chemicals is notoriously difficult. One available tool for identifying hazard and toxicity of materials are the labels provided by the Globally Harmonized System of Classification and Labelling of Chemicals (GHS, Figure 2.6). This labeling system aims to protect health and the environment during the handling, transportation, and use of chemicals; [30] however, GHS was not originally intended to quantify the hazards.

Figure 2.6: Hazard pictograms by the Globally Harmonized System of Classification and Labelling of Chemicals [32].

Creating a quantifiable method to measure the effective and overall hazard of a process has been proposed [31]. Follow the QR code on this page for the corresponding GHS pictograms and their descriptions. The pictograms show the common GHS symbols, and the areas of the pictograms are based on the sum of those materials that have the specific hazard symbol. For example, the flammable materials pictogram would have the largest area if the largest percentage of chemicals are flammable. The other symbols, such as irritants, specific toxicity hazards, and corrosives, gradually decrease as fewer chemicals with these traits are present. The quantifying method introduces an easy method to compare different processes and to help with decision-making. Refer to Section 4.5.6 for more details and an example of the use of this system.

However, the above-mentioned quantifying method does not differentiate between the severities of hazards. For example, both dimethyl sulfoxide and diethyl ether are labeled as flammable; however, diethyl ether is extremely flammable, whereas dimethyl sulfoxide is only slightly flammable. H-symbols (NFPA Rating in the USA) have been introduced to overcome this problem [33].

2.13 Defining a good chemical process

Table 2.2: The eight criteria needed to achieve a robust chemical manufacturing process [34].

Primary factor	Subfactor	Criterion	Description	Weighting [%]
Material cost		Material cost	Cost of all purchased chemicals	15
Conversion cost	Process efficiency	Atom economy	The efficiency of synthesis in terms of raw material strategy	5
		Yield	The efficiency of synthesis in terms of productivity	5
		Volume-time output	Efficiency in terms of reactor capacity and cycle time	40
		E-factor/process mass intensity	Efficiency in terms of process waste and environmental impact	10
	Process reproducibility	Quality service level	Reproducibility in terms of product quality	5

Table 2.2 (continued)

Primary factor	Subfactor	Criterion	Description	Weighting [%]
		Process excellence index	Reproducibility in terms of yield and cycle time	10
Modified ecoscale		14-point multicriterion analysis system including environment, health, and safety considerations	Ballpark lab procedure analysis tool acting as a reward system (and not a penalty)	10

Besides conventional green chemistry metrics, there are other directives for defining GMP [34]. GMP describes eight applicable assessment criteria to define a good chemical manufacturing process. These criteria are classified into three categories: material cost factors, conversion cost factors (process efficiency and process reproducibility factors), and the modified ecoscale. Table 2.2 summarizes all the GMP criteria and their contribution to the GMP. Process efficiency comprises five criteria. However, the largest contributor is the volume-time output (VTO) with a 40% weighting. VTO is a nominal volume of all unit operations (V_{UO}) multiplied by the time spent in these unit operations per batch (t_{UO}), divided by the product output mass per batch (m_p) (eq. (2.38)), and refers to the whole production process. VTO is also a powerful metric to compare different processes in batch and flow configuration. For example, a process with large VTO is economically and financially less preferable than processes with small VTO. VTOs close to or below 1 are favored. Flow processes have substantially lower VTOs than their batch counterparts. Refer to Chapter 6 for a detailed batch and flow comparison. VTO is calculated as follows:

$$VTO = \frac{V_{UO} \ [m^3] \times t_{UO} \ [h]}{m_p \ [kg]}.$$
(2.38)

These criteria are suitable for research and development in the chemical processing industry. Volume and time efficiency play significant roles, which can be expressed as the VTO parameter. Material costs and yield, along with environmental factors, are also crucial for chemical process assessment.

Another current green chemistry metric is the *green motion* metric [30]. This metric is based on a *penalty point system* using a scale of 1–100, with a higher rating ranking better in sustainability and green chemistry. However, this metric lacks a clear definition and breakdown of the scaling system. Precise and traceable information about the calculations for the energy and renewable E-factors are not provided, and therefore, the obtained results cannot be validated.

Bibliography

[1]   Boodhoo, K.; Harvey, A. *Process Intensification Technologies for Green Chemistry: Engineering Solutions for Sustainable Chemical Processing*; John Wiley & Sons, 2013.

[2]   Curzons, A. D.; Constable, D. J. C.; Mortimer, D. N.; Cunningham, V. L. So You Think Your Process Is Green, How Do You Know? – Using Principles of Sustainability to Determine What Is Green – A Corporate Perspective. *Green Chem.* **2001**, *3* (1), 1–6. 10.1039/b007871i.

[3]   Trost, B. M. The Atom Economy – A Search for Synthetic Efficiency. *Science (80)* **1991**, *254* (5037), 1471–1477. 10.1126/science.1962206.

[4]   Welton, T. Solvents and Sustainable Chemistry. *Proc. R. Soc. A: Math Phys. Eng. Sci.* **2015**, *471* (2183). 10.1098/rspa.2015.0502.

[5]   Sheldon, R. A. Metrics of Green Chemistry and Sustainability: Past, Present, and Future. *ACS Sustain. Chem. Eng.* **2018**, *6* (1), 32–48. 10.1021/acssuschemeng.7b03505.

[6]   Hudlicky, T.; Frey, D. A.; Koroniak, L.; Claeboe, C. D.; Larry, E. Toward a 'Reagent-free' Synthesis. **1999**, No. April, 57–59.

[7]   Sheldon, R. A. Organic Synthesis-Past, Present and Future. *Chem. Ind.* **1992**, *23*, 903–906.

[8]   Sheldon, R. A. The E Factor 25 Years On: The Rise of Green Chemistry and Sustainability. *Green Chem.* **2017**, *19* (1), 18–43. 10.1039/c6gc02157c.

[9]   Sheldon, R. A. The E Factor: Fifteen Years On. *Green Chem.* **2007**, *9* (12), 1273–1283. 10.1039/b713736m.

[10] Roschangar, F.; Sheldon, R. A.; Senanayake, C. H. Overcoming Barriers to Green Chemistry in the Pharmaceutical Industry-the Green Aspiration Level™ Concept. *Green Chem.* **2015**, *17* (2), 752–768. 10.1039/c4gc01563k.

[11] Tieves, F.; Tonin, F.; Fernández-Fueyo, E.; Robbins, J. M.; Bommarius, B.; Bommarius, A. S.; Alcalde, M.; Hollmann, F. Energising the E-Factor: The E + -Factor. *Tetrahedron* **2019**, *75* (10), 1311–1314. 10.1016/j.tet.2019.01.065.

[12] Greenhouse gas reporting: conversion factors 2018 – GOV.UK https://www.gov.uk/government/publications/greenhouse-gas-reporting-conversion-factors-2018 (accessed Sep 8, 2020).

[13] Tools for Innovation in Chemistry »ACS GCI Pharmaceutical Roundtable Portal https://www.acsgcipr.org/tools-for-innovation-in-chemistry/ (accessed Sep 6, 2020).

[14] Constable, D. J. C. Green Chemistry Metrics. In *Handbook of Green Chemistry*; Wiley-VCH Verlag GmbH & Co. KGaA: Weinheim, Germany, 2018; pp 1–28. 10.1002/9783527628698.hgc124.

[15] Treptow, R. S. Carbon Footprint Calculations: An Application of Chemical Principles. *J. Chem. Educ.* **2010**, *87* (2), 168–171. 10.1021/ed8000528.

[16] Overview of Greenhouse Gases | Greenhouse Gas (GHG) Emissions | US EPA https://www.epa.gov/ghgemissions/overview-greenhouse-gases (accessed Sep 8, 2020).

[17] WBCSD; WRI. A Corporate Accounting and Reporting Standard. *Greenh. Gas Protoc.* **2012**, 116.

[18] Feist, B. T. The Three Scopes of Greenhouse Gas Emissions. *J. Propery. Manag.* **2018**, *83* (3), 24–25.

[19] Evonik Carbon Footprint. 2018 Evonik Industries Contents https://corporate.evonik.com.

[20] Benchmarking Biopharma's Carbon Footprint https://www.pharmamanufacturing.com/articles/2017/benchmarking-biopharmas-carbon-footprint/ (accessed Apr 15, 2020).

[21] Lu, H.; Guo, L.; Zhang, Y. Oil and Gas Companies' Low-Carbon Emission Transition to Integrated Energy Companies. *Sci. Total Environ.* **2019**, *686*, 1202–1209. 10.1016/j.scitotenv.2019.06.014.

[22] Masnadi, M. S.; El-Houjeiri, H. M.; Schunack, D.; Li, Y.; Englander, J. G.; Badahdah, A.; Monfort, J. C.; Anderson, J. E.; Wallington, T. J.; Bergerson, J. A.; Gordon, D.; Koomey, J.; Przesmitzki, S.; Azevedo, I. L.; Bi, X. T.; Duffy, J. E.; Heath, G. A.; Keoleian, G. A.; McGlade, C.; Nathan Meehan, D.; Yeh, S.; You, F.; Wang, M.; Brandt, A. R. Global Carbon Intensity of Crude Oil Production. *Science (80)* **2018**, *361* (6405), 851–853. 10.1126/science.aar6859.

[23] Measuring crude oil's carbon footprint | Stanford News https://news.stanford.edu/2018/08/30/mea suring-crude-oils-carbon-footprint/ (accessed Apr 22, 2020).

[24] Meet the excellence behind Saudi Aramco's low carbon intensity | Saudi Aramco https://www.sau diaramco.com/en/magazine/elements/2020/low-carbon-intensity# (accessed Apr 16, 2020).

[25] Stohl, A.; Klimont, Z.; Eckhardt, S.; Kupiainen, K.; Shevchenko, V. P.; Kopeikin, V. M.; Novigatsky, A. N. Black Carbon in the Arctic: The Underestimated Role of Gas Flaring and Residential Combustion Emissions. *Atmos. Chem. Phys*. **2013**, *13* (17), 8833–8855. 10.5194/acp-13-8833-2013.

[26] Study shows record low carbon intensity of Saudi crude oil | Saudi Aramco https://www.saudiar amco.com/en/news-media/news/2018/study-shows-record-low-carbon-intensity-of-saudi-crude-oil (accessed Apr 19, 2020).

[27] Huang, J.; Mendoza, B.; Daniel, J. S.; Nielsen, C. J.; Rotstayn, L.; Wild, O. Anthropogenic and Natural Radiative Forcing. *Clim. Chang. 2013 Phys. Sci. Basis Work. Gr. I Contrib. To Fifth Assess. Rep. Intergov. Panel Clim. Chang*. **2013**, *9781107057*, 659–740. 10.1017/CBO9781107415324.018.

[28] Anastas, P. T.; Warner, J. C. Principles of Green Chemistry. *Green Chem. Theory Pract. 1998*, 29–56.

[29] Brooks, B. W. Greening Chemistry and Ecotoxicology Towards Sustainable Environmental Quality. *Green Chem*. **2019**, *21* (10), 2575–2582. 10.1039/c8gc03893g.

[30] Phan, T. V. T.; Gallardo, C.; Mane, J. GREEN MOTION: A New and Easy to Use Green Chemistry Metric from Laboratories to Industry. *Green Chem*. **2015**, *17* (5), 2846–2852. 10.1039/c4gc02169j.

[31] Cseri, L.; Szekely, G.; Towards Cleaner PolarClean: Efficient Synthesis and Extended Applications of the Polar Aprotic Solvent Methyl 5-(Dimethylamino)-2-Methyl-5-Oxopentanoate. *Green Chem*. **2019**, *21* (15), 4178–4188. 10.1039/c9gc01958h.

[32] GHS pictograms – Transport – UNECE http://www.unece.org/trans/danger/publi/ghs/pictograms. html (accessed Apr 6, 2020).

[33] Health and safety: H/P phrases – SAMANCTA https://ec.europa.eu/taxation_customs/dds2/SA MANCTA/EN/Safety/HP_EN.htm (accessed Sep 8, 2020).

[34] Dach, R.; Song, J. J.; Roschangar, F.; Samstag, W.; Senanayake, C. H. The Eight Criteria Defining a Good Chemical Manufacturing Process. *Org. Process Res. Dev*. **2012**, *16* (11), 1697–1706. 10.1021/op300144g.

3 The role of solvents in sustainable processes

Diana Gulyas Oldal, Gyorgy Szekely

Solvents and dissolution have a long history because they have always formed an integral part of scientific curiosity. In the fifteenth century, alchemists sought new solvents to be used in dissolution processes, particularly to dissolve gold (Figure 3.1). The ultimate goal was to find *menstruum universale*, a mystical solvent with the power to eradicate disease from the human body [1].

Figure 3.1: The universal solvent sought by alchemists was called alkahest. The picture depicts Alchimia, the spirit of alchemy.

The word "solvent" is derived from the Latin word *solvo*, meaning to loosen something. The main characteristic of solvents is to loosen the intermolecular bonds between the molecules of a solute and separate them [2]. According to IUPAC's classification, the definition of a solution involving a solvent and a solute is the following: [3] "A liquid or solid phase containing more than one substance, when for convenience one (or more)

https://doi.org/10.1515/9783111028163-003

substance, which is called the solvent, is treated differently from the other substances, which are called solutes." Solvents represent an important part of every industrial sector, from food and beverage, through pharmaceuticals to metallurgy. The size of the global solvent market is worth approx. 26–28 billion USD, which is increasing steadily on an annual basis (Table 3.1).

Table 3.1: Global market for petrochemical-based and green solvents per raw material in 2017 and 2018 and predictions for 2023 (expressed in billions of USD) [4]. CAGR, compound annual growth rate.

Raw material	2017	2018	2023	CAGR% 2018–2023
Fossil-derived solvents	25.9	26.3	27.9	1.2
Green solvents	3.9	4.2	5.9	7.0
Total	29.8	30.5	33.8	2.1

The following sections discuss the classification of solvents, present guidelines on choosing appropriate solvents for sustainable processes and how to recover them, explain the difference between conventional and green solvents, and draw attention to hazards associated with using organic solvents.

3.1 Classification of solvents

Solvents can be classified into two main categories: *polar* and *non-polar*. The dielectric constant of solvents provides a measure of polarity; i.e., it is a measure of the reduction of the strength of the electric field surrounding a charged particle immersed in a substance relative to the field strength around the same particle in a vacuum [5]. Small electrostatic attractions and repulsions between ionic species occur when the dielectric constant of a solvent is high because ions of opposite charge have a greater tendency to dissociate. Solvents with a dielectric constant less than 15 are considered to be non-polar [5]. On the contrary, solvents with a dielectric constant above 15 fall into the category of polar or dipolar solvents.

Polar solvents can be further categorized into two subgroups: *protic* and *aprotic* solvents. Protic solvents are good hydrogen donors; i.e., they have hydrogen bound to oxygen or nitrogen and can solvate cations and anions. Aprotic solvents have large dipole moments and do not contain hydrogen atoms directly connected to an electronegative atom; thus, they cannot form strong hydrogen bonds. Polar protic solvents possess acidic hydrogen with the capability for strong hydrogen bonding. They usually have high dielectric constants. Some of the examples include water, most alcohols, and ammonia. On the other hand, polar aprotic solvents are not good hydrogen donors and therefore have medium dielectric constant values. Solvents included in this category are ethyl acetate (EtOAc), *N,N*-dimethylformamide (DMF), acetonitrile (MeCN), and dimethyl sulfoxide (DMSO), among others.

Non-polar solvents contain atoms with similar electronegativities, and therefore the molecules lack partial charges; i.e., the electric charge is evenly distributed, resulting in a low dielectric constant. Some representatives of this group include hydrocarbons, diethyl ether (Et$_2$O), and dichloromethane (DCM). In summary, three general distinctive groups for solvents exist based on their molecular structure. Figure 3.2 provides general information about these groups, including examples for each category. Follow the QR code on this page to learn more about the classification of solvents.

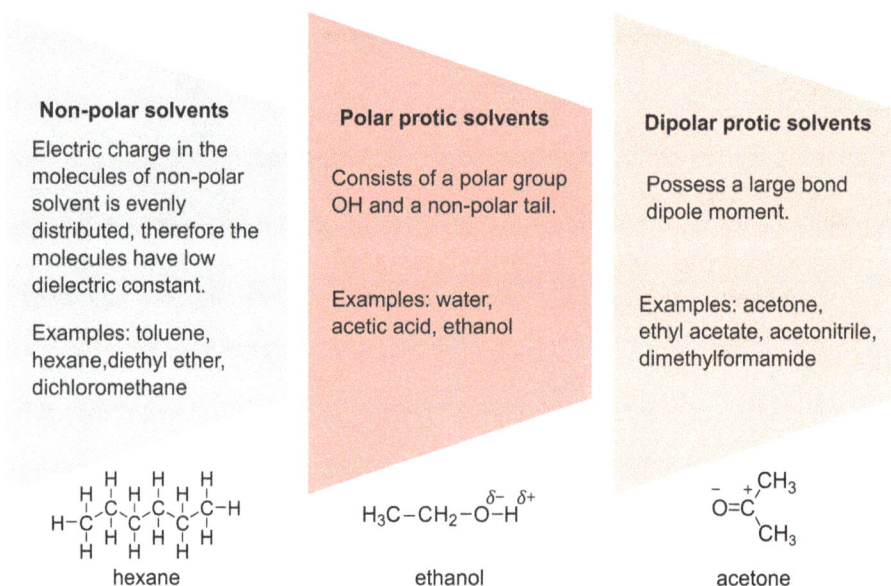

Non-polar solvents

Electric charge in the molecules of non-polar solvent is evenly distributed, therefore the molecules have low dielectric constant.

Examples: toluene, hexane, diethyl ether, dichloromethane

Polar protic solvents

Consists of a polar group OH and a non-polar tail.

Examples: water, acetic acid, ethanol

Dipolar protic solvents

Possess a large bond dipole moment.

Examples: acetone, ethyl acetate, acetonitrile, dimethylformamide

hexane

$H_3C-CH_2-\overset{\delta-}{O}-\overset{\delta+}{H}$

ethanol

acetone

Figure 3.2: General classification and characterization of solvents.

3.2 Solvent usage and safety concerns

Solvents can play many different roles. They can be used to dissolve or disperse materials in a given medium or be applied as carriers. Cleaning and washing processes are further applications where solvents are in high demand. Solvents are also widely used as a reaction and mixing medium.

Organic solvents are broadly used in manufacturing and versatile sectors such as paint, agricultural, food, pharmaceutical, and oil industries (Figure 3.3). The paint industry accounts for almost half of the total solvent consumption globally, followed by the pharmaceutical sector, with nearly one-tenth of the overall consumption [2]. The paint industry mainly uses aromatic and aliphatic hydrocarbons, which are produced through petrochemical production. The second-largest organic solvent user, the pharmaceutical industry, requires large quantities of high-purity solvents to produce high-purity active pharmaceutical ingredients (APIs). APIs are manufactured through multistep synthesis consisting of several reactions and purification steps.

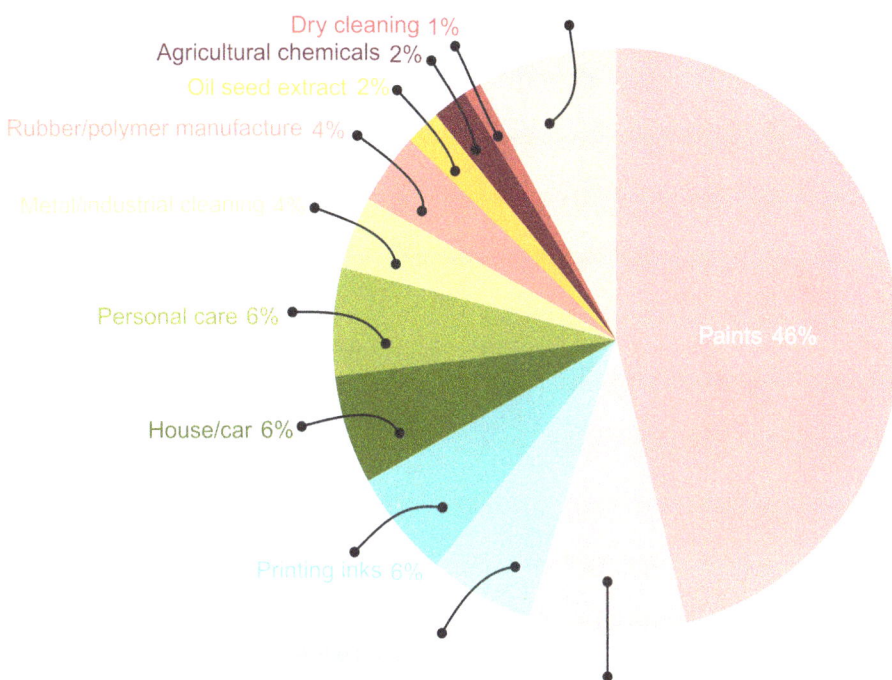

Figure 3.3: The breakdown of organic solvent usage among diverse industrial sectors [2].

The breakdown of solvents used by solvent classes reveals that the main contributors include alcohols with 30%, aromatics with 20%, and ether with 13%, followed by esters and alkanes, each with 9% of overall usage. Chlorinated and polar aprotic solvents account for 8% and 6%, respectively [6]. The solvents market is forecasted to grow continuously and reach approx. 57 billion USD by 2023, which corresponds to a 10 billion USD increase and a 4% annual growth rate since 2018 [7]. The car manufacturing and construction sectors drive the demand for petrochemical-based, conventional organic solvents. Sustainable development and strict regulations aim to mitigate the use of volatile organic compounds (VOCs) and are thus driving the demand for green solvents.

In many chemical productions, batch processes are still widely used, which implies multiple reaction and purification steps with the isolation of intermediates [2]. In particular, the pharmaceutical industry utilizes a large amount of solvents due to liquid phase reactions. A report by GlaxoSmithKline (GSK) revealed that solvents typically account for 80–90% of the total mass utilization in a pharmaceutical batch process [8]. Consequently, solvents significantly affect the waste generation of industrial processes (pharmaceutical sector) and products (paint industry). Since most solvents are hazardous and detrimental to the environment, solvent usage and disposal must be carefully managed. Solvents represent an important element of a process because they affect yield, energy consumption, process efficiency, kinetics, toxicity, health, and safety hazards (Figure 3.4) [2].

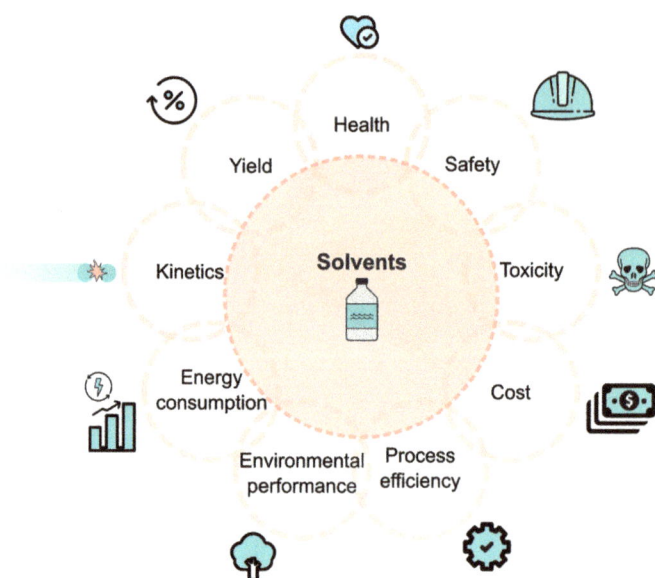

Figure 3.4: Solvents can make or break a chemical process because they affect various parameters in a process and beyond. Adapted from [2].

A study carried out by the American Chemical Society Green Chemistry Institute Pharmaceutical Roundtable (ACS GCI-PR) revealed that 46 kg material is used to produce 1 kg of API in a commercial process. Given that 56% of the mass used is solvent, it can be derived that 22 kg solvent is required to produce 1 kg of API (Figure 3.5) [9]. Such a large amount of solvent will inevitably have a significant effect on an industry's overall sustainability. Therefore, solvent management, particularly in terms of disposal and reuse, is key to improving the API sector's sustainability.

Primary health hazards of solvents can be classified as short-term, reversible, acute effects and chronic, long-term, irreversible effects. Most organic solvents are neurotoxic, and many can cause liver and kidney toxicity. Chlorinated and brominated organic solvents are potential carcinogens, while some organic solvents may be harmful to the skin, causing contact dermatitis or defatting of skin, for instance. VOCs, which partly originate from organic solvents, contribute to ground-level ozone pollution. Siloxanes and chlorinated solvents are persistent in the environment, and

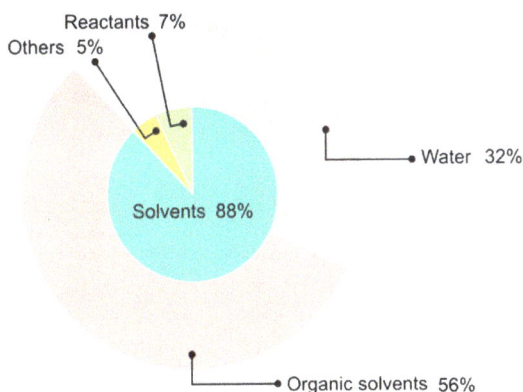

Figure 3.5: Composition by mass of the types of materials used for the production of APIs. Adapted from [9].

most organic solvents are flammable. Solvent spills can be a potential groundwater contamination source, endangering our waterways and food chains.

In the past, solvents were used without appropriate awareness of their detrimental effects on the environment and human health following their disposal [2]. In the 1970s, regulations were introduced to mitigate the hazards of solvents in the workplace and beyond. Some conventional solvents can be replaced by green solvents that possess lower toxicity or lower exposure. The next section discusses the potential of green solvents to mitigate the hazards associated with conventional solvents.

3.3 Green solvents

As discussed in Section 3.2, conventional organic solvents are petrochemical-based and pose health and environmental risks. Green solvent alternatives are a significant part of the solution for achieving more sustainable solvent use. Hallett et al. summarized the most important green and sustainable solvents in terms of their reaction performance with respect to their environmental and economic aspects [10]. They assessed each class of green solvent and provided a comprehensive guide on the present-day status of these solvents. We recommend readers to read this review for detailed information on the comparison of green solvents. The global green solvents market size was estimated to be worth approx. 5.5 billion USD in 2015, and has been growing considerably since [11]. A comprehensive market analysis forecasted the value of the green solvents market to double by 2022. The sector is mainly driven by soaring demand from the application segments dominating the breakdown outlined in Figure 3.3. Follow the QR code on this page to watch a webinar on the role of green solvents in a more sustainable future.

The use of green solvents instead of conventional solvents can facilitate meeting the requirements of green chemistry and green engineering. Figure 3.6 illustrates the way in which green solvents bridge the gap between green chemistry and green engineering. Their renewable nature and safer use due to their lower volatility are the key features that make them sustainable.

Green Chemistry
Safer Chemistry for Accident Prevention
Real-time Analysis for Pollution Prevention
Design for Degradation
Catalysis
Reduce Derivatives
Use of Renewable Feedstocks
Design for Energy Efficiency
Safer Solvents and Auxiliaries
Design Safer Chemicals
Less Hazardous Chemical Synthesis
Atom Economy
Prevention

Green Engineering
Inherent Rather Than Circumstantial
Prevention Instead of Treatment
Design for Separation
Maximize Efficiency
Output-pulled vs Input-pushed
Conserve Complexity
Durability Rather Than Immortality
Meet Need, Minimize Excess
Minimize Material Diversity
Integrate Material and Energy Flows
Design for Commercial Afterlife
Renewable Rather Than Depleting

Figure 3.6: Green chemistry and engineering principles applying for green solvents.

Figure 3.7 illustrates the most important green solvent classes and their beneficial green characteristics. The most common property among them is their low volatility, which facilitates the mitigation of VOC emissions into the atmosphere. Water is the most preferred green solvent according to most solvent selection guides [12, 13]. Section 3.4 introduces the solvent selection guides in detail. Other promising green solvents include supercritical fluids [14] such as supercritical CO_2, which is the most famous example [15, 16], ionic liquids [17–20], liquid polymers (mostly polyethylene glycol and polypropylene glycol) [21–23], switchable solvents [24], and renewable-based solvents [25, 26].

Supercritical fluids exist in a state above their critical temperature and critical pressure, which is an intermediate phase between the liquid and gaseous phase; i.e., the boundary between these two phases disappears when a liquid is in equilibrium above a critical temperature and pressure. These fluids have a liquid-like density and a gas-like viscosity, and consequently have gas-like transport properties (diffusivity) with good compressibility. A small change in pressure will cause a large change in volume and density; therefore, changing their solubility is simple. Follow the QR code on this page to watch an animation about phase diagrams and supercritical fluids. Some examples of often-used supercritical fluids include CO_2, water, ethylene, propane, ethane, and ammonia. Carbon dioxide is the most common due to its environmentally friendly nature, low flammability and toxicity, and low cost. *Ionic liquids* are molten at room temperature with extremely low volatility and flammability and high thermal and chemical stability. These materials are also tunable because the combinations of cations and anions are quasi-infinite. Ionic liquids are promising alternatives for reducing the environmental impact of conventional solvents. However, their implementation in industrial processes is limited because of biodegradability and toxicity issues and high cost. The manufacturing of common ionic liquids, namely, [TEA][HSO$_4$], was compared to acetone production from fossil-based resources and glycerol from renewable sources [27]. The total cost, including direct production and indirect costs from externalities (such as monetized resource damage, human health, and ecosystem quality), was 3.33, 2.22, and 1.87 USD kg^{-1} for glycerol, acetone, and [TEA][HSO$_4$], respectively. This case study demonstrated that the real costs of ionic liquids are not necessarily higher than those of conventional solvents.

Liquid polymers are fluids with very low volatility. The most common examples are polyethylene glycol (PEG) and polypropylene glycol (PPG), although poly(tetrahydrofuran) is also a known liquid polymer. Both PEG and PPG are biodegradable; however, PPG is slightly less biodegradable than PEG. *Switchable solvents* are liquids that can be reversibly changed from one form to another. Their properties can be changed by introducing triggers, such as gas, heat, or light. CO_2 can act as a trigger to induce a change in switchable solvents, for instance, by switching their polarity or hydrophilicity [28]. Switchable-polarity solvents can switch between a low-polarity and high-polarity form by adding an atmosphere of CO_2 to switch to higher polarity. The reaction can be reversed to switch to a lower polarity form by removing the CO_2, which is usually carried out by heating or purging the solution with N_2. Secondary amines such as dipropylamine or *N*-ethyl-*N*-propylamine are switchable non-ionic liquids with a low polarity that can switch to carbamate ionic liquids with higher polarity triggered by CO_2 [29]. Switchable hydrophilicity solvents are typically hydrophobic and immiscible with water, thus forming a biphasic mixture with water. The solvents turn into a hydrophilic form triggered by CO_2 and thus become miscible with water. The reversible reaction occurs by removing the CO_2 through an N_2 purge. Tertiary amines (e.g., *N,N,N*-trimethylamine) and amidines (e.g., *N,N,N*-tripropylbutanamidine) can be used as switchable hydrophilicity solvents due to the miscibility change caused by the acid-base reaction between the solvent and CO_2 (or carbonic acid in carbonated water) [30]. The reaction results in a hydrophilic bicarbonate salt. The phase separation, while using switchable hydrophilicity solvents, can eliminate the need for distillation and volatile solvents.

Deep eutectic solvents (DES) are composed of mixtures of Brønsted or Lewis acids and bases. These mixtures exhibit a melting point that is lower than that of each individual component, which is the basis for their classification as eutectic solvents. Most of DES contain a hydrogen-bond acceptor (HBA), which are mostly quaternary ammonium and phosphonium salts and hydrogen-bond donors (HBD) that can be metal halides, carbohydrates, amides, alcohols, or carboxylic acids [31]. DES have been used in numerous applications such as extraction of metals, separation and gas capture, biocatalysis, pharmaceuticals, batteries or biomass processing [32]. The physicochemical properties of DES can be tuned by altering the chemical structure and/or the relative molar ratio of HBA to the HBD, and their hydrophobicity and hydrophilic-

Figure 3.7: Classification of green solvents and their sustainable and beneficial characteristics. sc, supercritical; PEG, polyethylene glycol; PPG, polypropylene glycol.

ity can be tailored by modifying the functional group substituents of the HBA and HBD [31]. Natural DES (NADES) are DES constituting of primary metabolites, such as aminoacids, organic acids, sugars or choline derivates, which are fully biocompatible

Figure 3.8: Application areas of green solvents. sc, supercritical; PEG, polyethylene glycol; PPG, polypropylene glycol.

and sustainable [33]. NADES possess low vapor pressure, and they are biodegradable meaning that their disposal is straightforward and inexpensive.

Renewable- or bio-based solvents are obtained from biodegradable biomass sources (e.g., crops, forest waste) as an alternative to fossil-based solvents. A representative example of this group is 2-methyl tetrahydrofuran (2-MeTHF), which is an alternative to conventional tetrahydrofuran (THF). 2-MeTHF is obtained from sugars in biomass and has higher stability and is less volatile than its corresponding conventional solvent.

Figure 3.8 represents the application areas of the green solvents introduced in Figure 3.7. Most of them are used as reaction and separation media, but their use in washing and lubrication is also common. The versatile applicability of green solvents is shown by the various industrial sectors that use them, from the chemical industry, materials science, and mechanical engineering, to construction.

3.4 Solvent selection guides

Solvent selection represents an important step in every process design involving the dissolution of a compound or material in a particular media. Solvents have a wide range of physicochemical properties that determine their potential applications. However, hazards, health and safety, and sustainability are also important aspects of solvent selection. To make an informed decision on selecting a green solvent, it is necessary to gather a plethora of information, such as the carbon footprint of the solvent, biodegradability, solubility in water, VOC-forming potential, the solvent's potential for recycling, waste disposal, and production through renewable-based routes, among others [34]. Therefore, solvent selection guides have been crafted by pharmaceutical manufacturers, including GSK, Sanofi, Pfizer, and ACS GCI-PR, to gather information on solvents and categorize them [2].

GSK published the first guide in 1999 [35], where 35 solvents were color-coded depending on their impact on the environment, health and safety risks, and waste. This guide was further expanded in 2004 with a life cycle assessment (LCA) [36]. The subsequent guideline in 2011 took into account regulatory concerns along with process safety factors [9]. During the evolution of the solvent selection guides, the number of classified solvents rose from 47 initially to 154 in the guideline published in 2016 [13]. The listed guidelines show a chronological development in solvent sustainability. Follow the QR code on this page to learn how GSK scientists discover new medicines, while reducing the environmental impact of their manufacture.

Pfizer proposed a traffic light-based color-coding system that ranked solvents as green, yellow, or red, which classified them as preferred, usable, and undesirable categories, respectively [37]. Sanofi's guide compares solvents based on their chemical functionality (alcohols, ketones, esters, ethers, hydrocarbons, halogenated solvents, aprotic polar solvents, bifunctional and mixed solvents). It ranks them in four different categories: recommended, substitution advisable, substitution requested, and banned [38].

The Innovative Medicines Initiative (IMI)-CHEM21, which includes various manufacturing and academic collaborators, unified the solvent selection guides of GSK, GCI-PR, Pfizer, AstraZeneca, and Sanofi [39]. They compared these different solvent selection guides to find the overlaps and divergences between them and thus compiled a general guide. The issues with previously published solvent selection guides were the differences between ranking structures and criteria; i.e., each company invented their own ranking system. However, CHEM-21 tended to overcome this obstacle by ranking the solvents across three categories: environment, safety, and health. Each category is ranked from 1 to 10 (1 as most recommended, and 10 as most hazardous). CHEM-21 ranked solvents from previously published selection guides and incorporated the bio-based solvents in the ranking system for the first time to compare their properties with conventional solvents [40].

Table 3.2 provides a summary of some of the most important specifications that should be considered in a solvent selection process. The solvents are ranked from 1 and 10, from worst to best, in different classes, as proposed in GSK's guide from 2011 [9]. The waste category refers to the recycling, incineration, volatility, and biotreatment aspects, while the environmental impact covers the fate and effects of solvents on the environment. The following category is health, which ranks solvents based on their acute and chronic effects on human health and exposure potential. Issues linked with solvent storage and handling are assessed under the flammability and explosion class. The stability (reactivity) category covers factors affecting the stability of the solvent. An LCA ranking covers the entire lifetime of the solvent, from production to disposal; however, in Table 3.2, LCA ranking refers to environmental impacts related to the solvent's production. The boiling points of the solvents are indicated as an essential physical property for solvent selection. The solvents' reaction types are presented for each solvent, where the letters refer to different reaction types. The flashpoint and the acute-toxicity LD_{50} numerical values can be used to compare the risks originating from different solvents. The short-term exposure limit values cover the tolerable average exposure during a short time period (a few minutes). On the other hand, the International Council for Harmonization of Technical Requirements for Registration of Pharmaceuticals for Human Use (ICH) guide categorizes solvents into four classes based on their hazardous nature. Class 1 solvents should not be used to produce drug substances due to their unacceptably toxic nature or their detrimental environmental effects. Nevertheless, if their use cannot be avoided for producing an API with a significant therapeutic benefit, then their levels should be limited according to the regulations [41]. The solvents in Class 2 need to be limited in pharmaceutical products due to their inherent toxicity. Class 3 solvents have low toxic potential; i.e., there are no solvents in this category hazardous to human health at levels normally accepted in APIs.

Table 3.2: Solvent selection guide example. Properties and classification of some common solvents. Adapted from [2]. EI, Environmental impact; F&E, Flammability and explosion. Reaction types: A, SNi; B, oxidation; C, ozonization; D, epoxidation; E, catalytic hydrogenation; F, hydride reduction; G, aldol reaction; H, Wittig reaction; I, Diels–Alder cycloaddition; J, Friedel–Crafts reaction; K, diazotization; L, diazo coupling.

Solvent	Waste	EI	Health	F&E	Stability	LCA ranking	BP [°C]	Reaction types	FP [°C]	LD$_{50}$ [mg kg^{-1} rat]	STEL [mg m^{-3}]	EMA ICH
Benzene	5	6	1	3	10	7	80	B, D, F, G, H, I, J, K	−11	5,960	9	Class 1
Toluene	6	3	4	4	10	7	110	A, B, E, G, I, K, L	4	5,580	384	Class 2
Water	4	10	10	10	10	10	100	A, B, E, G, I, K, L	–	–	–	–
scCO$_2$	10	5	6	6	10	N/A	N/A	B, D, E, I, J	–	N/A	27,400	–
Ethanol	3	8	8	6	9	9	78	A, C, E, G, H, I, K, L	14	10,470	5,760	Class 3
Ethyl acetate	4	8	8	4	8	6	77	B, C, E	−3	5,620	400	Class 3

General principles for solvent selection based on scientific facts and experience are gathered into a collection of the following points [42]:

- Use a single solvent in the entire process, if possible.
- Recycle the solvent if economically feasible (critical on an industrial scale due to high volumes).
- Do everything to reduce the environmental footprint.
- Prioritize renewable-based solvents.
- Avoid solvents that are tainted with legal restrictions.
- Refrain from intrinsically reactive solvents.
- Bring to the fore solvents with low viscosity to enhance mixing.
- Use solvents with high specific heat capacity due to their beneficial capability of acting as a heat sink. However, do not neglect that more energy is needed for heating.
- Choose solvents with a higher flash point and auto-ignition temperatures because they minimize sparking hazards.

Bibliography

[1]   DeSimone, J. M. Practical Approaches to Green Solvents. *Science (80).* **2002**, *297* (5582), 799–803. 10.1126/science.1069622.

[2]   Cseri, L.; Razali, M.; Pogany, P.; Szekely, G. *Organic Solvents in Sustainable Synthesis and Engineering*; Elsevier Inc., 2018. 10.1016/B978-0-12-809270-5.00020-0.

[3]   Iupac. IUPAC Compendium of Chemical Terminology. In Nič, M., Jirát, J., Košata, B., Jenkins, A., McNaught, A., Eds.; *IUPAC: Research Triangle Park, NC*; 2009; Vol. 2167. https://doi.org/10.1351/goldbook.

[4]   Chen J. CHM087A Green Solvents: Technologies, Emerging Opportunities and Markets https://www.bccresearch.com/market-research/chemicals/green-solvents.html (accessed Aug 29, 2020).

[5]   Richardson, K. S.; Lowry, T. H. *Mechanism And Theory In Organic Chemistry – Ihomas H. Lowry*; 1976.

[6]   Curzons, A. D.; Constable, D. J. C.; Mortimer, D. N.; Cunningham, V. L. So You Think Your Process Is Green, How Do You Know? – Using Principles of Sustainability to Determine What Is Green – A Corporate Perspective. *Green Chem.* **2001**, *3* (1), 1–6. 10.1039/b007871i.

[7]   MarketsandMarkets™. *CH 2218, Market Research Report*; 2018.

[8]   Constable, D. J. C.; Jimenez-Gonzalez, C.; Henderson, R. K. Perspective on Solvent Use in the Pharmaceutical Industry. *Org. Process Res. Dev.* **2007**, *11* (1), 133–137. 10.1021/op060170h.

[9]   Henderson, R. K.; Jiménez-González, C.; Constable, D. J. C.; Alston, S. R.; Inglis, G. G. A.; Fisher, G.; Sherwood, J.; Binks, S. P.; Curzons, A. D. Expanding GSK's Solvent Selection Guide – Embedding Sustainability into Solvent Selection Starting at Medicinal Chemistry. *Green Chem.* **2011**, *13* (4), 854–862. 10.1039/c0gc00918k.

[10] Clarke, C. J.; Tu, W. C.; Levers, O.; Bröhl, A.; Hallett, J. P. Green and Sustainable Solvents in Chemical Processes. *Chem. Rev.* **2018**, *118* (2), 747–800. 10.1021/acs.chemrev.7b00571.

[11] Grand View Research. *Green & Bio-Based Solvents Market Size Forecast 2016–2024*; 2016.

[12] Byrne, F. P.; Jin, S.; Paggiola, G.; Petchey, T. H. M.; Clark, J. H.; Farmer, T. J.; Hunt, A. J.; Robert McElroy, C.; Sherwood, J. Tools and Techniques for Solvent Selection: Green Solvent Selection Guides. *Sustain. Chem. Process.* **2016**, *4* (1), 1–24. 10.1186/s40508-016-0051-z.

[13] Alder, C. M.; Hayler, J. D.; Henderson, R. K.; Redman, A. M.; Shukla, L.; Shuster, L. E.; Sneddon, H. F. Updating and Further Expanding GSK's Solvent Sustainability Guide. *Green Chem.* **2016**, *18* (13), 3879–3890. 10.1039/c6gc00611f.
[14] Eckert, C. A.; Knutson, B. L.; Debenedetti, P. G. Supercritical Fluids as Solvents for Chemical and Materials Processing. *Nature.* **1996**, *383* (6598), 313–318. 10.1038/383313a0.
[15] Han, X.; Poliakoff, M. Continuous Reactions in Supercritical Carbon Dioxide: Problems, Solutions and Possible Ways Forward. *Chem. Soc. Rev.* **2012**, *41* (4), 1428–1436. 10.1039/c2cs15314a.
[16] Beckman, E. J. Supercritical and Near-Critical CO2 in Green Chemical Synthesis and Processing. *J. Supercrit. Fluids.* **2004**, *28* (2–3), 121–191. 10.1016/S0896-8446(03)00029-9.
[17] Inamuddin; Rangreez, T.A.; Asiri, A. M. In Green Solvents II. In Mohammad, A., Inamuddin, D., Eds.; Springer Netherlands: Dordrecht, 2012. 10.1007/978-94-007-2891-2.
[18] Mallakpour, S.; Dinari, M. *Ionic Liquids as Green Solvents: Progress and Prospects*; 2012. 10.1007/978-94-007-2891-2_1.
[19] Yoo, C. G.; Pu, Y.; Ragauskas, A. J. Ionic Liquids: Promising Green Solvents for Lignocellulosic Biomass Utilization. *Curr. Opin. Green Sustain. Chem.* **2017**, *5*, 5–11. 10.1016/j.cogsc.2017.03.003.
[20] Earle, M. J.; Seddon, K. R. Ionic Liquids: Green Solvents for the Future. *ACS Symp. Ser.* **2002**, *819* (1), 10–25. 10.1021/bk-2002-0819.ch002.
[21] Feu, K. S.; De La Torre, A. F.; Silva, S.; De Moraes Junior, M. A. F.; Corrêa, A. G.; Paixão, M. W. Polyethylene Glycol (PEG) as a Reusable Solvent Medium for an Asymmetric Organocatalytic Michael Addition. Application to the Synthesis of Bioactive Compounds. *Green Chem.* **2014**, *16* (6), 3169–3174. 10.1039/c4gc00098f.
[22] Chen, J.; Spear, S. K.; Huddleston, J. G.; Rogers, R. D. Polyethylene Glycol and Solutions of Polyethylene Glycol as Green Reaction Media. *Green Chem.* **2005**, *7* (2), 64–82. 10.1039/b413546f.
[23] Heldebrant, D. J.; Witt, H. N.; Walsh, S. M.; Ellis, T.; Rauscher, J.; Jessop, P. G. Liquid Polymers as Solvents for Catalytic Reductions. *Green Chem.* **2006**, *8* (9), 807–815. 10.1039/b605405f.
[24] Pollet, P.; Davey, E. A.; Ureña-Benavides, E. E.; Eckert, C. A.; Liotta, C. L. Solvents for Sustainable Chemical Processes. *Green Chem.* **2014**, *16* (3), 1034–1055. 10.1039/c3gc42302f.
[25] Horváth, I. T. Solvents from Nature. *Green Chem.* **2008**, *10* (10), 1024–1028. 10.1039/b812804a.
[26] Lomba, L.; Giner, B.; Bandrés, I.; Lafuente, C.; Pino, M. R. Physicochemical Properties of Green Solvents Derived from Biomass. *Green Chem.* **2011**, *13* (8), 2062–2070. 10.1039/c0gc00853b.
[27] Baaqel, H.; Díaz, I.; Tulus, V.; Chachuat, B.; Guillén-Gosálbez, G.; Hallett, J. P. Role of Life-Cycle Externalities in the Valuation of Protic Ionic Liquids-a Case Study in Biomass Pretreatment Solvents. *Green Chem.* **2020**, *22* (10), 3132–3140. 10.1039/d0gc00058b.
[28] Jessop, P. G.; Mercer, S. M.; Heldebrant, D. J. CO2-Triggered Switchable Solvents, Surfactants, and Other Materials. *Energy Environ. Sci.* **2012**, *5* (6), 7240–7253. 10.1039/c2ee02912j.
[29] Heldebrant, D. J.; Koech, P. K.; Ang, M. T. C.; Liang, C.; Rainbolt, J. E.; Yonker, C. R.; Jessop, P. G. Reversible Zwitterionic Liquids, the Reaction of Alkanol Guanidines, Alkanol Amidines, and Diamines with CO2. *Green Chem.* **2010**, *12* (4), 713–772. 10.1039/b924790d.
[30] Vanderveen, J. R.; Durelle, J.; Jessop, P. G. Design and Evaluation of Switchable-Hydrophilicity Solvents. *Green Chem.* **2014**, *16* (3), 1187–1197. 10.1039/c3gc42164c.
[31] Farooq, M. Q.; Odugbesi, G. A.; Abbasi, N. M.; Anderson, J. L. Elucidating the Role of Hydrogen Bond Donor and Acceptor on Solvation in Deep Eutectic Solvents Formed by Ammonium/Phosphonium Salts and Carboxylic Acids. *ACS Sustain. Chem. Eng.* **2020**, *8* (49), 18286–18296. 10.1021/acssuschemeng.0c06926.
[32] Hansen, B. B.; Spittle, S.; Chen, B.; Poe, D.; Zhang, Y.; Klein, J. M.; Horton, A.; Adhikari, L.; Zelovich, T.; Doherty, B. W.; Gurkan, B.; Maginn, E. J.; Ragauskas, A.; Dadmun, M.; Zawodzinski, T. A.; Baker, G. A.; Tuckerman, M. E.; Savinell, R. F.; Sangoro, J. R. Deep Eutectic Solvents: A Review of Fundamentals and Applications. *Chem. Rev.* **2021**, *121* (3), 1232–1285. 10.1021/acs.chemrev.0c00385.

[33] Paiva, A.; Craveiro, R.; Aroso, I.; Martins, M.; Reis, R. L.; Duarte, A. R. C. Natural Deep Eutectic
      Solvents – Solvents for the 21st Century. *ACS Sustain. Chem. Eng*. **2014**, *2* (5), 1063–1071. 10.1021/
      sc500096j.
[34] Criteria for solvent selection: Solvent selection guides: CHEM21 learning platform http://learning.
      chem21.eu/methods-of-facilitating-change/tools-and-guides/criteria-for-solvent-selection/
      (accessed May 18, 2020).
[35] Curzons, A. D.; Constable, D. C.; Cunningham, V. L. Solvent Selection Guide: A Guide to the
      Integration of Environmental, Health and Safety Criteria into the Selection of Solvents. *Clean Technol.
      Environ. Policy*. **1999**, *1* (2), 82–90. 10.1007/s100980050014.
[36] Jiménez-Gonzalez, C.; Curzons, A. D.; Constable, D. J. C.; Cunningham, V. L. Expanding GSK's Solvent
      Selection Guide – Application of Life Cycle Assessment to Enhance Solvent Selections. *Clean Technol.
      Environ. Policy*. **2004**, *7* (1), 42–50. 10.1007/s10098-004-0245-z.
[37] Alfonsi, K.; Colberg, J.; Dunn, P. J.; Fevig, T.; Jennings, S.; Johnson, T. A.; Kleine, H. P.; Knight, C.; Nagy,
      M. A.; Perry, D. A.; Stefaniak, M. Green Chemistry Tools to Influence a Medicinal Chemistry and
      Research Chemistry Based Organisation. *Green Chem*. **2008**, *10* (1), 31–36. 10.1039/b711717e.
[38] Prat, D.; Pardigon, O.; Flemming, H. W.; Letestu, S.; Ducandas, V.; Isnard, P.; Guntrum, E.; Senac, T.;
      Ruisseau, S.; Cruciani, P.; Hosek, P. P. Sanofi's Solvent Selection Guide: A Step Toward More
      Sustainable Processes. *Org. Process Res. Dev*. **2013**, *17* (12), 1517–1525. 10.1021/op4002565.
[39] Prat, D.; Hayler, J.; Wells, A. A Survey of Solvent Selection Guides. *Green Chem*. **2014**, *16* (10),
      4546–4551. 10.1039/c4gc01149j.
[40] Prat, D.; Wells, A.; Hayler, J.; Sneddon, H.; McElroy, C. R.; Abou-Shehada, S.; Dunn, P. J. CHEM21
      Selection Guide of Classical- and Less Classical-Solvents. *Green Chem*. **2015**, *18* (1), 288–296. 10.1039/
      c5gc01008j.
[41] European Medicines Agency. ICH Guideline Q3C (R5) on Impurities: Guideline for Residual Solvents.
      Int. Conf. Harmon. Tech. Requir. Regist. Pharm. Hum. Use. **2019**, *44* (August 2019), 24.
[42] Federsel, H. J. En Route to Full Implementation: Driving the Green Chemistry Agenda in the
      Pharmaceutical Industry. *Green Chem*. **2013**, *15* (11), 3105–3115. 10.1039/c3gc41629a.

4 Sustainable process development from alpha to omega

Diana Gulyas Oldal, Gyorgy Szekely

Improving the sustainability of production processes is increasingly important. Considerable efforts focus on designing and developing greener synthetic routes for green solvents such as Cyrene [1] and PolarClean [2]. In this chapter, the conventional *production process* of PolarClean is presented, followed by the design and realization of a more sustainable process. The role of PolarClean as a green solvent and its patented process are discussed, and the retrosynthetic approach and the green metrics analysis of the various possible manufacturing routes are explained. Health and safety considerations are also presented and taken into account for decision-making. This chapter provides a comprehensive case study, following the process from problem identification, through design and assessment, to the final decision-making.

4.1 PolarClean: a green polar aprotic solvent

The demand for sustainable solutions is increasing due to environmental concerns, which implies new efforts and research plans [3]. Substantial research on obtaining and applying green solvents has mainly focused on alcohols and low-polarity esters (Figure 4.1). However, there are significantly fewer sustainable alternatives available for polar aprotic solvents. The importance of polar aprotic solvents is related to their solvation characteristics and the promotion of various chemical reactions. Conventional solvents are problematic because of their toxicity, and therefore, the demand for green counterparts is growing (Table 4.1) [2].

Methyl 5-(dimethylamino)-2-methyl-5-oxopentanoate (**1**) (Figure 4.1), sold under the commercial name Rhodiasolv PolarClean by Solvay, is a non-toxic alternative to conventional polar aprotic solvents. PolarClean has low acute toxicity and is considered non-carcinogenic, non-genotoxic, and non-mutagenic [4]. Due to its high solvency and eco-friendly characteristics, its popularity as a solvent or co-solvent in crop formulation [5], membrane technology [6–10], chemical synthesis [11–13], and solubilization of agrochemicals [14] is increasing. Aside from its application as a solvent, it acts as a crystal growth inhibitor and facilitates the cold stability improvement of final formulations. It is a non-flammable compound with vapor pressure as low as 0.01 Pa at 20 °C [9] and a boiling point of approx. 280 °C, and it is miscible with water. This green solvent is produced from methyleneglutarodinitrile, a by-product of Nylon 66 manufacturing [13]. Such an approach is termed waste utilization or waste upcycling, which contributes to the circular economy.

https://doi.org/10.1515/9783111028163-004

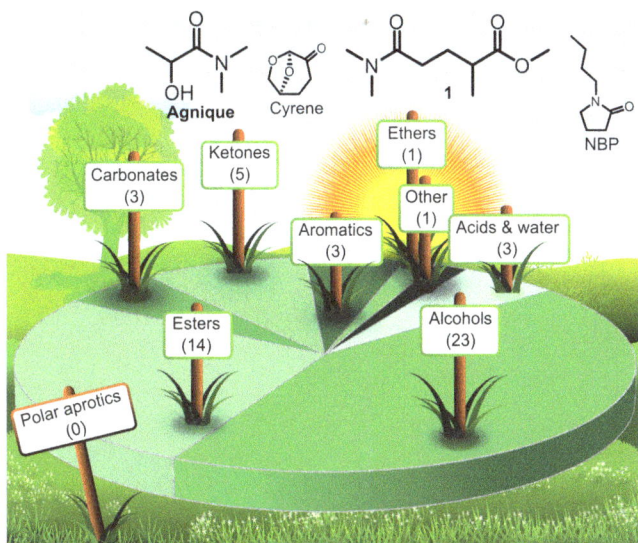

Figure 4.1: Distribution of different green solvents based on solvent type. Polar aprotic solvents are currently underrepresented compared to other types of solvents. Adapted from [2] under the CC-BY-NC 3.0 license.

Table 4.1: Examples and characteristics of conventional and green polar aprotic solvents. Based on [2]. GVL, γ-valerolactone; PC, propylene carbonate; NBP, N-butylpyrrolidone; DMF, dimethylformamide; DMAc, dimethylacetamide; NMP, N-methyl-2-pyrrolidone; THF, tetrahydrofuran; DCM, dichloromethane.

Green polar aprotic solvents	Conventional polar aprotic solvents
GVL, Cyrene, PC, NBP, PolarClean	DMF, DMAc, NMP, THF, DCM
Limited commercial availability	High reproductive toxicity
Low toxicity	
Low environmental impact	
Some are renewable-based	

4.2 The patented production of PolarClean

The patented production of PolarClean entails three synthetic routes (Figure 4.2a) [15]. All three routes start with the hydrocyanation of butadiene, a readily available and inexpensive material, leading to 2-methylglutaronitrile. The hydrolysis of 2-methylglutaronitrile results in 2-methylglutaric acid. Routes A_1 and A_2 proceed with a cyclization step leading to 2-methylglutaric anhydride. The subsequent esterification and amidation of the two carbonyl groups take place in different orders in the different routes, resulting in the formation of PolarClean (**1**). The third route, A_3, reaches the final product formation from 2-methylglutaric acid through consecutive diesterification and amidation steps.

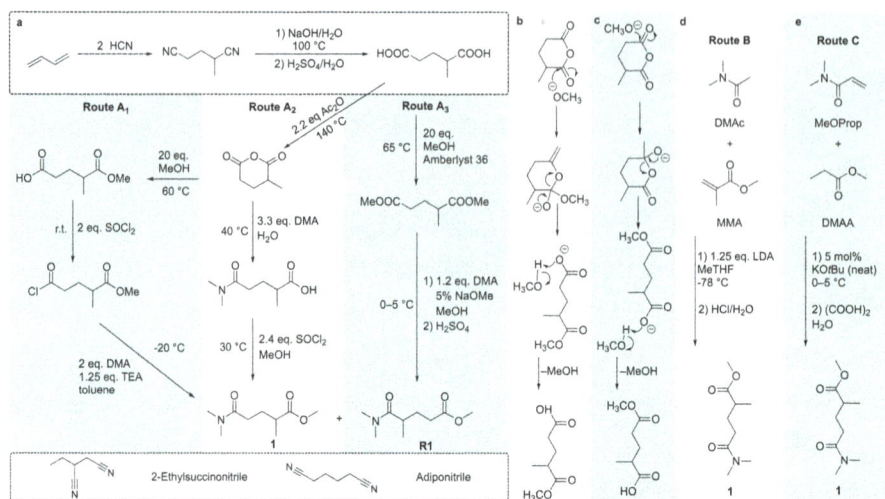

Figure 4.2: Commercial synthetic routes for the patented production of PolarClean (1) (a). Ring-opening mechanism ultimately yielding the desired product (b) and the regioisomer by-product (c). The improved reaction routes (d and e) are presented as Route B and Route C.

Each patented synthetic route of PolarClean consists of multiple steps, and they all result in a yellowish, multicomponent mixture. 2-Methylglutaronitrile is a common intermediate in all three routes, which is a product of the hydrocyanation of butadiene and consists of approx. 11% dinitrile isomers (Figure 4.2). The dinitrile isomer impurities react in the same way as the main compound, which results in the carryover of the isomers' derivatives with the desired product. This carryover can ultimately lead to a multicomponent mixture of the final product.

Furthermore, the cyclic anhydride's ring-opening can occur in two positions due to the similar reactivity of the two carbonyl groups (Figure 4.2b and Figure 4.2c). This reaction results in the formation of the desired final product (1) and its regioisomer (R1), methyl 5-(dimethylamino)-4-methyl-5-oxopentanoate. Due to the ring-opening in different positions, the final product is a mixture of 1 and R1, and its exact composition can be determined using gas chromatography analysis (Table 4.4 in Section 4.4). The content of the desired product from Routes A_1 (37%) and A_2 (69%) was relatively low based on the analysis. The isomer composition for Route A_3 has not been reported. More detailed inline analytics could reveal how the isomer composition changes over time, ultimately resulting in a better process. Refer to Chapter 9 for the various analytical technologies and their use in sustainable process development.

Overall, the patented synthesis of PolarClean consists of multiple synthetic and purification steps, and there are several sources of impurities that are structurally related to the final product. Consequently, the E-factor is relatively high and estimated to be in the range of 13–26 kg kg^{-1}.

Route B and Route C represent the improved reaction routes for the synthesis of PolarClean. The subsequent chapters present the process of designing and evaluating these new synthetic methods from a green and sustainable point of view and deciding whether the improved processes represent a better route than the patented one.

4.3 Toward the design of greener synthetic routes

Life cycle assessment considers the environmental impact of products from the cradle to the grave. Therefore, mitigation of the drawbacks of the patented synthesis of PolarClean will have an overall positive impact on this green solvent's sustainability. New synthetic routes have been designed using a retrosynthetic approach as a practical tool for solving problems in the design of organic reactions.

Before discussing the retrosynthetic analysis applied to PolarClean, the definitions and terminology related to retrosynthesis are reviewed. Retrosynthetic analysis is often used to identify a synthetic strategy. This method transforms the desired molecule into simpler precursors by literally chopping it up and creating so-called synthons. Synthons are non-existing (or extremely unstable) molecules related to stable reagents, called synthetic equivalents. The goal is to reduce a molecule into its building blocks of lower complexity until commercially available molecules are obtained. To achieve a green synthesis, these molecules and the reactions that form the desired product need to be environmentally friendly. Thus, their synthetic equivalents must be obtained from a sustainable source and meet circular economy standards.

Disconnection describes the process of breaking a bond to form possible new starting materials. Disconnection should occur during known reactions, such as the Wittig reaction, oxidation, or reduction. If the compound contains a heteroatom such as an ester, amide, ether, or sulfide group, then the disconnection should be carried out along the heteroatom. In the case that there are many heteroatoms present in the compound, chemoselectivity should be considered. A *synthon* is a fragment or species generated from bond disconnection. A *synthetic equivalent* is an existing molecule that can be ascribed to synthon.

The retrosynthetic analysis applied to PolarClean is shown in Figure 4.3 [2]. In the patented synthesis (Route A), the product's carbon backbone is created using a $C_4 + C_1 + C_1$ pattern. Two new strategies, $C_4 + C_2$ (Route B, Figure 4.3) and $C_3 + C_3$ (Route C, Figure 4.3), can be considered alternative methods to obtain the same product. Michael addition may be an option for carbon–carbon bond formation in both alternative routes due to the CH acidic character of the α position in carbonyl derivatives. Michael addition involves the nucleophilic addition of a carbon nucleophile to an unsaturated carbonyl compound.

The first alternative $C_4 + C_2$ strategy, Route B, is constructed from methyl methacrylate (MMA) and N,N-dimethylacetamide (DMAc) building blocks, which are readily available and inexpensive chemicals. Although they are currently derived mainly from fossil feedstock, their renewable-based synthesis from itaconic acid and acetic

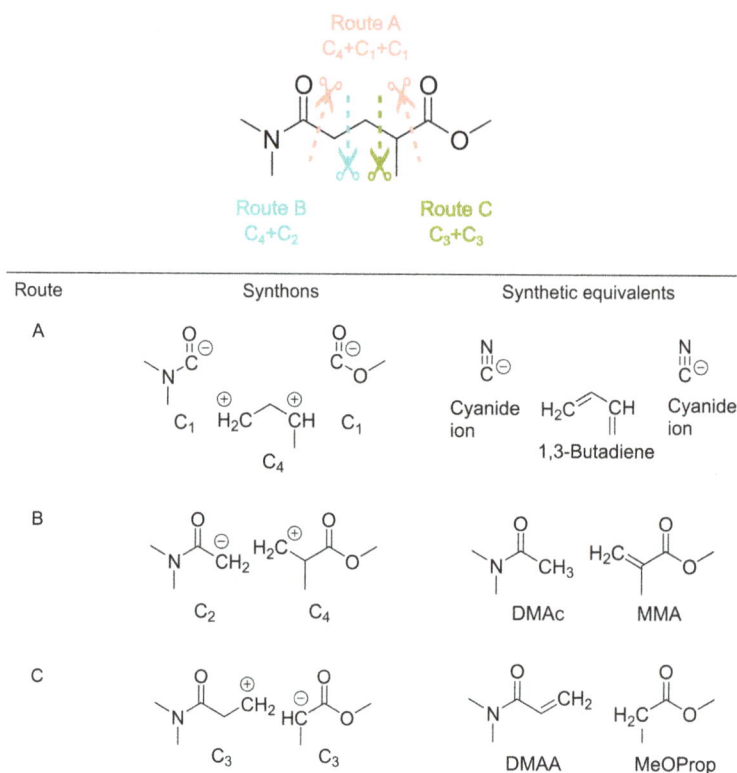

Figure 4.3: Retrosynthetic approach for methyl 5-(dimethylamino)-2-methyl-5-oxopentanoate (PolarClean, **1**). The generated synthons represent the building blocks of the molecules for the different routes. Synthons can be *translated* into synthetic equivalents, which represent real, existing molecules. Adapted from [2] under the CC BY-NC 3.0 license.

acid is growing in popularity [16, 17]. Routes B and C represent a single-step synthesis, unlike the original multistep Route A, which addresses the 8th principle of green chemistry, i.e., by minimizing derivatives. Refer to Section 1.3 for the 12 principles of green chemistry.

The Michael reaction requires a base to deprotonate the CH acidic reagent. Therefore, the pK_a value of the molecules needs to be considered during material selection. pK_a is the negative logarithm of the acidity constant (K_a), commonly used to characterize a molecule's acidity. A list of green acids and bases are presented in Table 4.2. All the listed acids and bases are labeled as green by the GlaxoSmithKline (GSK) standard [18]. The lower the pK_a, the stronger the acid, and therefore the weaker the conjugate base. For example, hydrochloric acid is a strong acid with a pK_a of -6.12 [19], while ammonia has weak acidity owing to its pK_a of 32.5 [20]. As mentioned,

$$pK_a = -\log_{10} K_a. \tag{4.1}$$

Table 4.2: Different green acids and bases, according to GSK [18]. Only those acids and bases considered green by the GSK standard are listed.

Acid	pK_a	Base	pK_b
Hydrochloric acid	−8	Sodium bicarbonate	5.95
Hydrobromic acid	−9	Potassium bicarbonate	5.95
Phosphoric acid	1.9, 6.74, 11.74	Sodium carbonate	9.1
Sulfuric acid, dilute	3.19, 1,98	Potassium carbonate	9.1
Glutaric acid	4.41, 5.52	Cesium carbonate	9.1
Citric acid	2.93	Trisodium phosphate	11.74
Ascorbic acid	4.09	Tripotassium phosphate	11.74
p-TsOH (monohydrate)	−6.57	Potassium hydroxide	15.74
Benzoic acid	4.2	Sodium hydroxide	15.74
Oxalic acid	1.25, 4.23	Calcium hydroxide	15.74
Pivalic acid	4.94	Barium hydroxide	15.74
Succinic acid	4.24	Potassium acetate	4.76
Acetic acid	4.76	Sodium acetate	4.76
Propionic acid	4.79	2-Methylpyridine	5.97
Formic acid	3.75	2,6-Lutidine	6.75
Methanesulfonic acid	−1.92	DBN	13.5
		Pyridine	5.17
		4-Methylpyridine	6.02
		Morpholine	8.49
		Diethylaminopropylamine	10.48
		Tetramethylguanidine	13.6
		DBU	12.5
		2,2,6,6-Tetramethylpiperidine	11.1

The Michael addition starts with the deprotonation of the CH acidic reagent, which is, in this case, the DMAc. It has a weak CH acidic character with a pK_a value of approx. 29.4 [21]. Consequently, a strong base such as lithium diisopropylamide (LDA) with a pK_a value of approx. 35.7 for the conjugated acid [22] is needed to deprotonate the DMAc. LDA is a pyrophoric and highly reactive material that poses a safety hazard. Moreover, Route B's reaction proceeds rapidly, even at temperatures as low as −78 °C, which requires a significant amount of energy for cooling. The acidic work-up employs hydrochloric acid, a highly corrosive material that poses an additional safety hazard. Route B also generates a large amount of waste, at approx. 62 kg kg^{-1}. This aqueous and organic waste originates from the reaction, acidic work-up, extraction, and distillation processes.

The second alternative, Route C, is constructed by the $C_3 + C_3$ strategy using methyl propionate and N,N-dimethylacrylamide as building blocks, which are also derived from renewable resources [16, 17]. This reaction route exploits the higher pK_a (approx. 25) of the ester. The base used for the Michael addition is potassium tert-butoxide (t-BuOK). Its basicity heavily depends on the solvent used. t-BuOK has higher base strength in polar solvents, which facilitates the reversible deprotonation of weak

CH acids. DMAA is not only a building block in the reaction but also a polar solvent, which reduces waste generation and the number of components. The neat reaction addresses the 5th principle of green chemistry, i.e., by avoiding auxiliary substances.

The reaction carries on swiftly using only a catalytic amount of the base, which addresses the 9th principle of green chemistry. Virtually 100% conversion is reached within only 2.5 min. Because this is an exothermic reaction, cooling is necessary, but, in contrast to the −78 °C used in Route B, a milder temperature (0–5 °C) is adequate, which allows for substantial energy saving. Another advantage of Route C is the application of saturated aqueous oxalic acid for the work-up, which is less corrosive than the hydrochloric acid employed previously [18].

4.4 Quality assessment

The impurity profiles provide an assessment of the quality of the products obtained through Routes A, B, and C (Figure 4.2). Two different batches, labeled X and Y and obtained through Route A, are analyzed [23] and compared with the products from Routes B and C (Table 4.3). Quality analysis is assessed based on the different batches' colors using UV-Vis spectroscopy and the platinum-cobalt (Pt-Co) scale [24]. PolarClean X and Y have medium yellow coloration above 200, while the reference value of another commercial batch (PolarClean R) shows a weaker discoloration with a value less than 100. Product 1B possesses a strong yellow coloration, higher than 500, which exceeds the upper limit of the used Pt-Co scale, while 1C is transparent with a coloration value of 22. The coloration of high-purity products is an important quality parameter. Discoloration usually indicates the presence of impurities, and they can lead to a failed batch [25]. From a coloration point of view, Route C outperformed the other routes. Figure 4.4 shows the identified by-products during the different synthetic routes.

Table 4.3: The coloration of different PolarClean batches obtained from Routes A, B, and C. The values refer to the Pt-Co coloration scale.

	PolarClean X	PolarClean Y	PolarClean R	PolarClean 1B	PolarClean 1C
Pt-Co scale	233 ± 1	203 ± 1	< 100	> 500	22 ± 1

Quantitative analysis was performed to determine the composition of the reaction products using gas chromatography-mass spectrometry (GC-MS), and the results are summarized in Table 4.4. PolarClean X and PolarClean Y possess similar composition and purity. Their purity is 85.7% and 83.8%, respectively. In both cases, the main impurity is regioisomer 2. The ratio of regioisomers 1 and 2 is 9:1, which is strikingly similar to the reported composition of 2-methylglutaronitrile (Figure 4.2a) and 2-ethylsuccinonitrile (Figure 4.2a) starting materials used in Route A (84% and 11%, re-

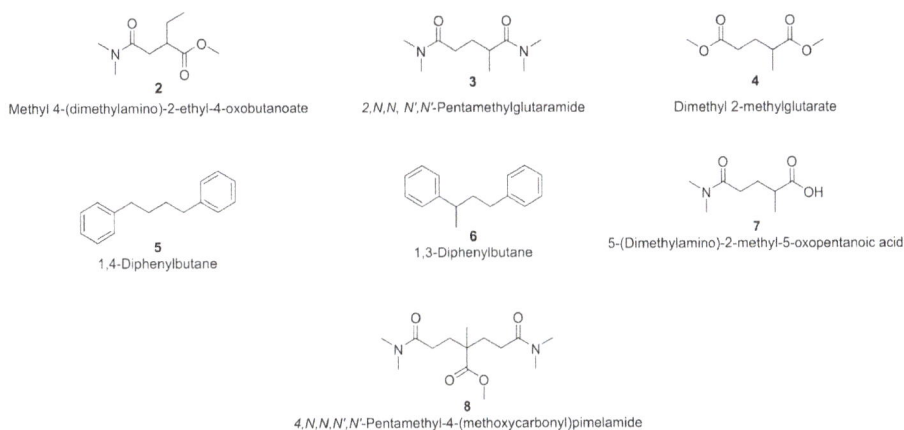

2
Methyl 4-(dimethylamino)-2-ethyl-4-oxobutanoate

3
2,N,N, N',N'-Pentamethylglutaramide

4
Dimethyl 2-methylglutarate

5
1,4-Diphenylbutane

6
1,3-Diphenylbutane

7
5-(Dimethylamino)-2-methyl-5-oxopentanoic acid

8
4,N,N,N',N'-Pentamethyl-4-(methoxycarbonyl)pimelamide

Figure 4.4: By-products originating from different synthesis routes of PolarClean.

spectively) [26]. Since neither of the X and Y products contains R1 as an impurity originating from the ring-opening (Figure 4.2c), we can postulate that these batches are produced through Route A₃. A previously reported impurity profile for commercial PolarClean R suggests a substantially different composition [23], as it contains 4.6% **7** as the main impurity.

Table 4.4: GC-MS impurity profiles of PolarClean (1) obtained through different synthetic pathways. Refer to Figure 4.3 for the synthetic Routes A, B, and C. The numbers in bold indicate the impurities, whose name and structure are provided in Figure 4.4.

Batch	1	2	3	4	5	6	7	8
PolarClean **X**	88.7	8.9	4.3	1.1	–	–	–	–
PolarClean **Y**	83.8	10	5.0	1.2	–	–	4.6	–
PolarClean **R**	95.1	–	Trace	Trace	–	–	–	–
1B	90.1	–	–	–	8.9	1.0	–	–
1C	>99	–	–	–	–	–	–	–

Route B gives the product with high purity (90.1%) and only two impurities (5 and 6), which are aromatic compounds from the LDA solution. The impurities' boiling point is near that of **1**, and thus, they became enriched in the final product. Furthermore, the previously discussed strong yellow coloration of 1B could be caused by the **5** and **6** aromatic hydrocarbons. These impurities result in the formation of an oil-in-water emulsion during dilution of 1B with water, thus limiting the product's application. This issue can be avoided by using the LDA that is produced without a styrene-containing process. However, most commercial LDA products contain these impurities, which require custom orders at a higher price.

Batch 1C has the highest purity among all the examined batches (>99%), and **8** is the only observed by-product produced from Route C. The boiling point of **8** is significantly higher than that of **1**, which facilitates removing this impurity. It forms due to the Michael reaction between the desired product **1** and the DMAA starting material.

4.5 Green metrics analysis

Green metrics analysis was performed to facilitate the comparison of the different synthesis routes. First, the following metrics were considered: number of steps, complexity, ideality, carbon intensity (CI), atom economy (AE), yield, and complete E-factor (Figure 4.5). The health and safety risks were also assessed at a later stage. The key process performance indicators were the number of steps, complexity, and ideality. The definitions of complexity and ideality were the same as those proposed by Roschangar et al [27].

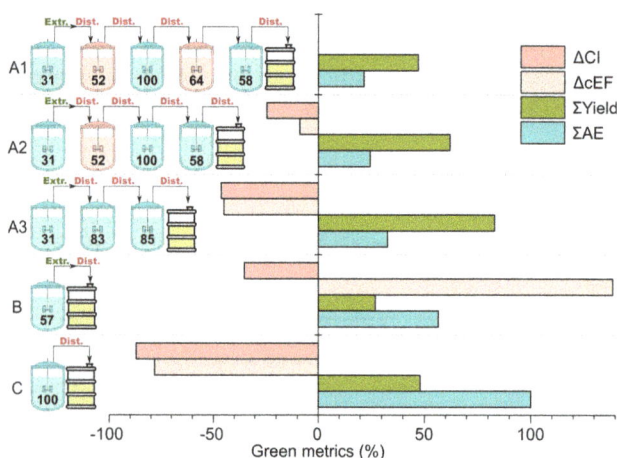

Figure 4.5: Green metrics comparison of the synthetic routes leading to **1**. The blue reactors refer to construction steps, while the pink reactors represent concession steps. Adapted from [2] under the CC BY-NC 3.0 license. Extr., extraction; Dist., distillation; ΔCI, carbon intensity change; ΔcEF, complete E-factor change; ΣYield, overall yield; ΣAE, overall atom economy.

4.5.1 Complexity and Ideality

Before discussing the terms of *complexity* (eq. (4.2)) and *ideality* (eq. (4.3)), the construction and concession steps should be defined. *Construction steps* are chemical transformations that form skeletal C–C, C–X, C–H, and X–H bonds (X = heteroatom) present in the product [27]. *Concession steps* are all non-constructive reactions and do not form skeletal bonds, e.g., carbonyl activation steps. For example, the use of protective groups is a concession step.

$$\text{Complexity} = \sum \text{construction steps,} \qquad (4.2)$$

$$\text{Ideality} = \frac{\text{complexity}}{\sum \text{steps}} \cdot 100\%. \qquad (4.3)$$

Figure 4.5 reveals that the different synthesis routes consist of a different number of construction steps (Table 4.5). There are three construction steps during Route A, the patented synthesis of the product. On the contrary, the proposed Routes B and C have a complexity of 1, indicating an improvement in simplicity.

Table 4.5: Complexity and ideality of the examined synthesis routes yielding 1.

Route	Complexity [%]	Ideality [%]
A_1	3	60
A_2	3	75
A_3	3	100
B	1	100
C	1	100

Ideality demonstrates a tendency in the way molecules are synthesized. The ideal synthesis creates complex structures from available simple starting materials by connecting them using construction steps, leading to the target compound without intermediary concession steps [28].

Table 4.5 illustrates the calculated ideality values for each route. There is a progressive decrease across the patented routes ($A_1 \rightarrow A_2 \rightarrow A_3$) due to the elimination of concession steps. The proposed new Routes B and C have ideality values of 100%, demonstrating their beneficial nature. Note that the three different Routes A all have a complexity of 3, although A_3 has an ideality of 100%. Therefore, of the five examined, Routes B and C are preferred because they have the lowest achievable complexity and highest ideality at the same time.

4.5.2 Carbon intensity

The pharmaceutical industry has also introduced carbon efficiency (CE) [29] to distinguish the carbon mass between reactants and products. CE is defined as the percentage of carbon in the reactants remaining in the final isolated product:

$$CE(\%) = \frac{\text{Amount of carbon in product}}{\text{Total carbon present in reactants}} \times 100\%. \qquad (4.4)$$

To attain a comprehensive picture of the different synthetic routes' sustainability, CI is estimated to compare the different reaction routes. CI is the mass emission of CO_2 associated with producing a unit mass of the product (eq. (2.8)). CI values are presented relative to the longest route (A_1, CI = 223 kg kg^{-1}) in Figure 4.5 and Table 4.6. Compared to the reference Route A_1, A_2 and A_3 have lower CI values due to the lower number of synthesis and purification steps, which results in less overall waste generation. Route B has a higher CI value than A_3; however, it only comprises one synthesis step with a single-stage vacuum distillation. The overall best route is C, with 29 kg kg^{-1}, which is more than 87% less than for Route A_1.

Table 4.6: Comparison of CI values of different synthetic routes.

Route	CI [kg kg^{-1}]	ΔCI compared to A_1 [%]	Reason
A_1	223	–	–
A_2	169	−24	Fewer synthesis and purification steps
A_3	112	−50	
B	145	−35	One-step synthesis; single vacuum distillation
C	29	−87	One step; neat; no extraction

Each synthesis was simplified to the following components: reactions, extractions, filtrations, atmospheric distillations, and vacuum distillations. CI consists of three components, namely, chemical waste (CI_{chem}), energy consumption (CI_{en}), and cooling water consumption (CI_{cw}), and it is mathematically described in eq. (4.5). For the energy consumption, four activities were considered: stirring, heating, chilling, and applying vacuum (eq. (4.6)).

$$CI = CI_{chem} + CI_{en} + CI_{cw}, \qquad (4.5)$$

$$CI_{en} = CI_{stirring} + CI_{heating} + CI_{chilling} + CI_{vacuum}. \qquad (4.6)$$

The energy consumptions were calculated using estimations based on the catalog power consumption of the equipment used and the activity duration. The cooling water and energy consumption were converted to CI values using emission factors [30]. Emission factors are scalar coefficients that allow converting activity into overall emission. For example, a modern car emits around 0.011 g CO_2 per km, and burning one ton of anthracite coal emits around 1.6 g of N_2O while generating 7353 kWh energy. The CI values related to the chemical waste were calculated by assuming that all the organic waste is disposed of via incineration. Ideal combustion was hypothesized for this process; more precisely, the chemicals' total carbon content was converted to CO_2. The CI_{chem} was calculated as shown in eq. (4.7), where $n_{C, raw materials}$, $n_{C, reagents}$, $n_{C, solvents}$, and $n_{C, product}$ are the amount of carbon in raw materials, reagents, sol-

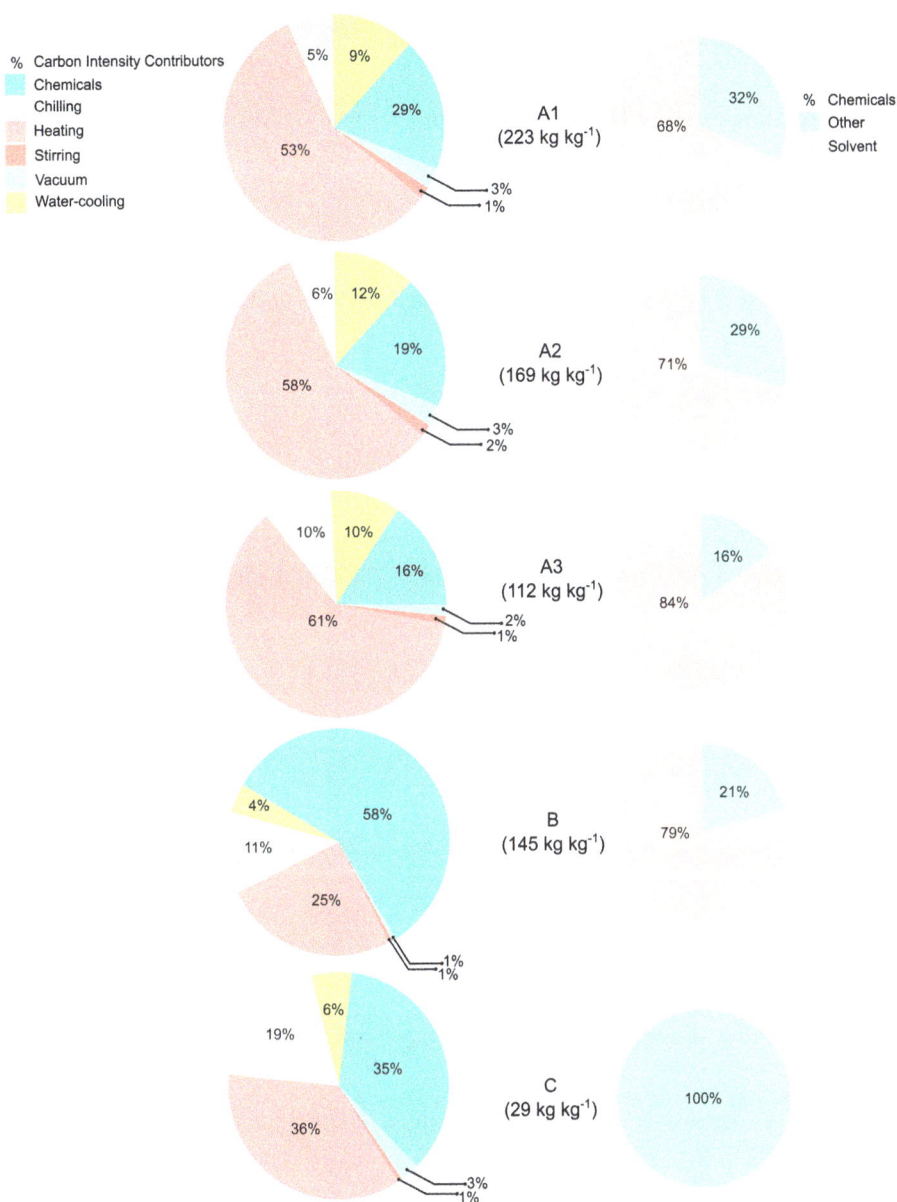

% Carbon Intensity Contributors
- Chemicals
- Chilling
- Heating
- Stirring
- Vacuum
- Water-cooling

A1 (223 kg kg^{-1})

A2 (169 kg kg^{-1})

A3 (112 kg kg^{-1})

B (145 kg kg^{-1})

C (29 kg kg^{-1})

% Chemicals
Other
Solvent

Figure 4.6: Overall CI and its breakdown for the different synthesis routes.

vents, and product in the synthesis, respectively. M_{CO_2} represents the molar weight of CO_2, and $m_{product}$ is the mass of the product.

$$CI_{chem} = \frac{\left(\sum n_{C,raw\ materials} + \sum n_{C,reagents} + \sum n_{C\ solvents} - n_{C,product}\right) \cdot M_{CO_2}}{m_{product}}. \tag{4.7}$$

The breakdown of the individual contributors to CI is presented in Figure 4.6. In every case, the two main CI constituents are energy consumption, specifically the heating, and the chemical waste, accounting for 71–83% of the total CI. Heating accounts for more than 50% for each patented synthetic route, while it is substantially reduced in the proposed new Routes B and C, representing only 25% and 35%, respectively. Solvents generally represent the most considerable portion of chemical manufacturing chemicals, at 84% in Route A_3.

4.5.3 Atom economy

The AE expresses the amount of starting material that ends up in the desired product. AE is defined in eq. (2.1), where MW is the corresponding molecular weight and n is the stoichiometric number of reagent. Refer to Chapter 2 for more details on this green metric. The AE calculations for Routes A_1–A_3 are shown in Figures 4.7–4.9 and

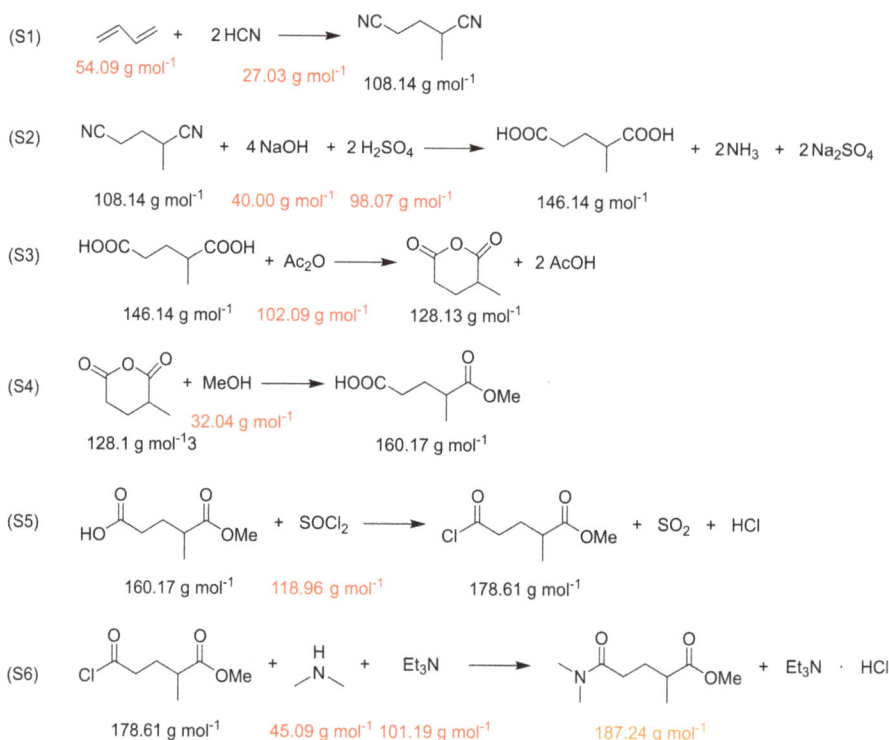

Figure 4.7: The derivation of the AE for synthesis Route A_1.

eqs. (4.8)–(4.10). The overall AE is calculated by considering all starting materials and reagents but neglecting the intermediate compounds.

$$\sum AE = \frac{187.24}{54.09 + 2 \cdot 27.03 + 4 \cdot 40.00 + 2 \cdot 98.07 + 102.09 + 32.04 + 118.96 + 45.09 + 101.19}$$

$$\cdot 100\% = 22\%,$$

$$(4.8)$$

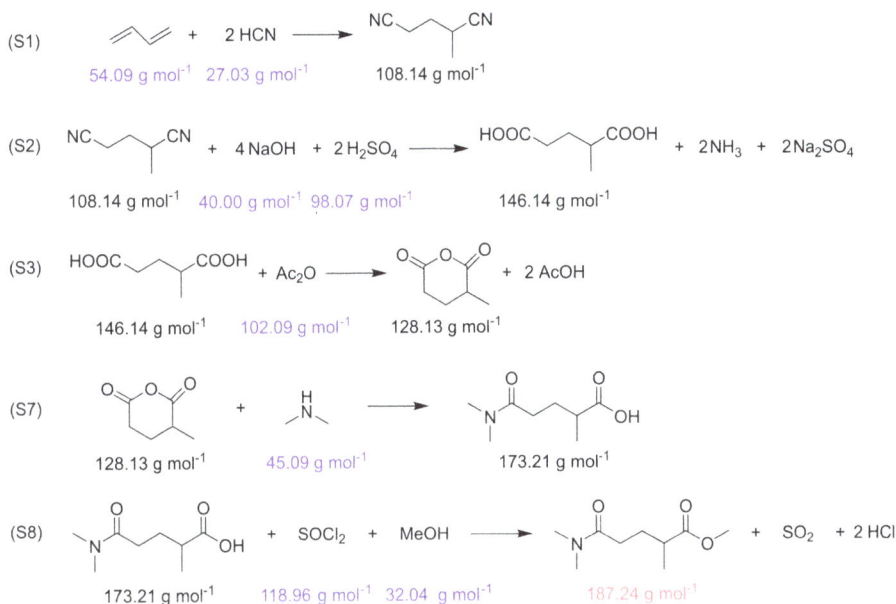

Figure 4.8: AE calculation for Route A$_2$.

$$\sum AE = \frac{187.24}{54.09 + 2 \cdot 27.03 + 4 \cdot 40.00 + 2 \cdot 98.07 + 102.09 + 45.09 + 118.96 + 32.04} \quad (4.9)$$

$$\cdot 100\% = 25\%,$$

(S1)

54.09 g mol⁻¹ 27.03 g mol⁻¹   108.14 g mol⁻¹

(S2)

108.14 g mol⁻¹   40.01 g mol⁻¹ 98.07 g mol⁻¹   146.14 g mol⁻¹

(S9)

146.14 g mol⁻¹   32.04 g mol⁻¹   174.20 g mol⁻¹

(S10)

174.20 g mol⁻¹   45.09 g mol⁻¹   187.24 g mol⁻¹

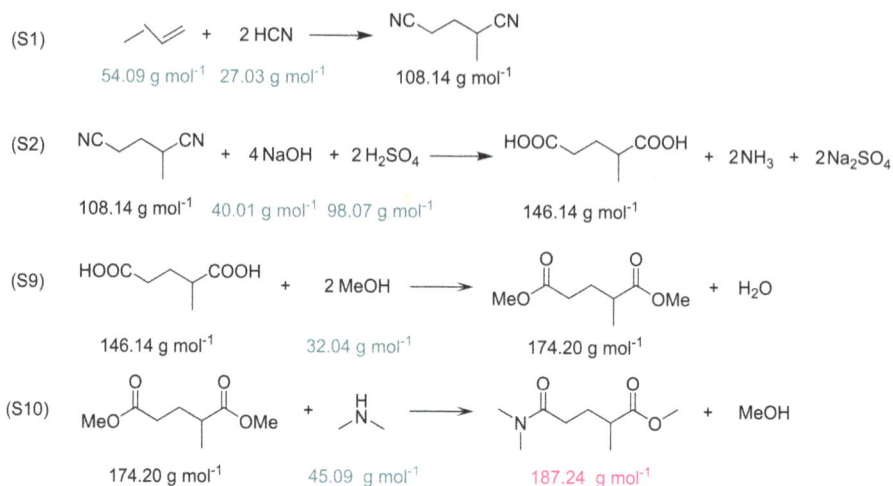

Figure 4.9: AE calculation for Route A₃.

$$\sum \text{AE}_A = \frac{187.24}{54.09 + 2 \cdot 27.03 + 4 \cdot 40.00 + 2 \cdot 98.07 + 2 \cdot 32.04 + 45.09} \cdot 100\% = 33\%. \quad (4.10)$$

The calculation of the AE for Route B is shown in Figure 4.10 and eq. (4.11).

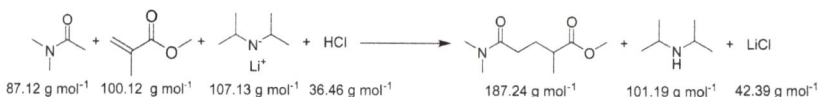

87.12 g mol⁻¹ 100.12 g mol⁻¹ 107.13 g mol⁻¹ 36.46 g mol⁻¹   187.24 g mol⁻¹   101.19 g mol⁻¹ 42.39 g mol⁻¹

Figure 4.10: AE calculation for Route B.

$$\sum \text{AE}_B = \frac{187.24}{87.12 + 100.12 + 107.13 + 36.46} \cdot 100\% = 57\%. \quad (4.11)$$

The AE calculation for Route C is presented in Figure 4.11 and eq. (4.12).

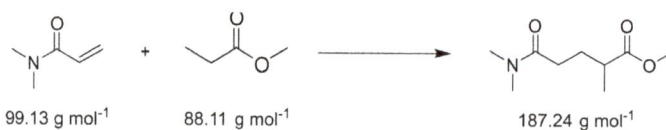

99.13 g mol⁻¹   88.11 g mol⁻¹   187.24 g mol⁻¹

Figure 4.11: AE calculation for Route C.

$$\sum \text{AE}_C = \frac{187.24}{99.13 + 88.11} \cdot 100\% = 100\%. \quad (4.12)$$

The numerical data calculated for each step and the overall synthesis routes are presented in Table 4.7. The AE values strongly depend on each step of the synthesis, varying between 31% and 100% among Routes A, B, and C. In the case of Route A, the AE of 31% means that less than a third of the mass of the starting materials is incorporated into the final product via the patented synthesis. The new single-step synthesis, Route B, has a considerably higher AE value of 57%, whereas Route C demonstrated a significant improvement with an AE as high as 100%.

Table 4.7: The AE values for each reaction step in synthetic Routes A, B, and C.

Steps	AE (%)			
(S1)	100	$\sum A_1 = 22$	B = 87	C = 100
(S2)	31			
(S3)	52			
(S4)	100	$\sum A_2 = 25$		
(S5)	64			
(S6)	58			
(S7)	100	$\sum A_3 = 33$		
(S8)	58			
(S9)	83			
(S10)	85			

4.5.4 Yield

To calculate mass-related performance metrics, we need a comprehensive table comprising each route's actual values. Table 4.8 lists all the chemicals used in the different routes. The following calculations use the data from Table 4.8.

Table 4.8: Comprehensive list of the used chemicals in Routes A_1, A_2, A_3, B, and C.

	A_1	A_2	A_3	B	C	GHS symbols
S1 (kg)	1.23	0.94	0.69	–	–	
S2 (kg)	1.61	1.74	0.91	–	–	
S3 (kg)	1.3	0.99	–	–	–	
S4 (kg)	1.38	–	–	–	–	
S5 (kg)	1.7	–	–	–	–	

Table 4.8 (continued)

	A_1	A_2	A_3	B	C	GHS symbols
S6 (kg)	–	1.38	–	–	–	
S7 (kg)	–	–	1.01	–	–	–
Water (kg)	6.32	7.07	3.57	6.94	0.48	–
Methanol (kg)	6.24	10.93	7.4	–	–	
Toluene (kg)	10.14	–	–	–	–	
Ethyl acetate (kg)	–	–	–	13.89	–	
tert-Butyl methyl ether (MTBE, kg)	5.05	3.86	2.85	–	–	
Tetrahydrofurane (kg)	–	–	–	6.62	–	
Methyl tetrahydrofurane (kg)	–	–	–	9.49	–	
Sodium hydroxide (kg)	0.96	0.73	0.54	–	–	
Sulfuric acid (kg)	1.19	0.91	0.68	–	–	
Dimethyl amine (kg)	0.85	1.1	0.32	–	–	
Thionyl chloride (kg)	2.3	2.26	–	–	–	
Triethyl amine (kg)	1.19	–	–	–	–	
Acetic anhydride (kg)	2.66	2.87	–	–	–	
Sodium methoxide (kg)	–	–	0.01	–	–	
Amberlyst 36 (kg)	–	–	0.74	–	–	
Lithium dimethyl amine (LDA, kg)	–	–	–	2.38	–	
Aq. Hydrochloric acid (kg)	–	–	–	5.56	–	
Aq. Sodium hydrogen carbonate (kg)	–	–	–	6.94	–	–
Brine (kg)	–	–	–	6.94	–	–
Dimethyl acetamide (DMAc, kg)	–	–	–	1.83	–	

Table 4.8 (continued)

	A₁	A₂	A₃	B	C	GHS symbols
Methyl metacrylate (MMA, kg)	–	–	–	2	–	◇! ◇
Methyl propionate (MeOProp, kg)	–	–	–	–	4.87	◇! ◇
Dimethylacrylamide (DMAA, kg)	–	–	–	–	1.1	◇ ◇
Potassium *tert*-butoxide (KO*t*Bu, kg)	–	–	–	–	0.06	◇ ◇
Oxalic acid (kg)	–	–	–	–	0.05	◇! ◇

A chemical process's effectiveness can be described, for example, by the yield of a re-action or a process. The percentage yield shows the ratio of the actual yield to the theoretical yield (eq. (4.13)). The *theoretical yield* is the highest achievable product output, while the *actual yield* is the isolated product after the process. For example, in a non-catalyzed asymmetric reaction, the theoretical yield is 50% (1:1 ratio of R and S isomers), and the actual yield is always lower than that.

$$\text{Percent yield} = \frac{\text{actual yield}}{\text{theoretical yield}} \cdot 100\%. \qquad (4.13)$$

Table 4.9 shows the percent yield values for the different synthesis routes of Polar-Clean. The patented route is synthetically less challenging and therefore has a rela-tively good overall yield in the range of 47–83%. However, the overall yields for Routes B and C are 27% and 48%, respectively, due to the reaction's low selectivity, resulting in a significant number of side-products.

Table 4.9: Overall percent yields for the different reaction routes.

Route	Percent yield (%)
A₁	47
A₂	68
A₃	83
B	27
C	48

4.5.5 E-factors

The E-factors are used as estimates for the waste generated during the production of a particular chemical compound. Thus, E-factors constitute a significant aspect of the overall sustainability of a designed process. The complete E-factor (cEF) includes water as waste, whereas the simple E-factor (sEF) excludes water.

$$sEF = \frac{\sum m_{\text{raw materials}} + \sum m_{\text{reagents}} - m_{\text{product}}}{m_{\text{product}}} = \left[\frac{kg}{kg}\right], \quad (4.14)$$

$$cEF = \frac{\sum m_{\text{raw materials}} + \sum m_{\text{reagents}} + \sum m_{\text{solvents}} + \sum m_{\text{water}} - m_{\text{product}}}{m_{\text{product}}} = \left[\frac{kg}{kg}\right]. \quad (4.15)$$

The E-factor values of the different routes are presented relative to Route A_1 (cEF = 37.1 kg kg^{-1} and sEF = 30.8 kg kg^{-1}) in Figure 4.5 and Table 4.10.

Table 4.10: cEF values of different routes with the corresponding reasons.

Route	cEF	ΔcEF (relative to A₁)	sEF	ΔsEF (relative to A₁)	Explanation
A_1	37.1	–	30.8	–	
A_2	29.6	–9%	22.6	–8%	Fewer synthesis and purification steps
A_3	15.8	–50%	12.2	–60%	
B	61	+138%	55.1	+80%	More solvent in the extraction step
C	5.6	–78%	5.1	–83%	Solvent-free catalytic reaction

Table 4.10 provides information about the E-factor values and their changes related to the different synthetic routes. Routes A_2 and A_3 have lower cEF values than A_1, which means less waste is generated when producing a mass unit of **1**. Meanwhile, Route B possesses a significantly higher cEF value due to extensive solvent usage. The high solvent consumption of B is because the reaction work-up consists of a liquid-liquid extraction step using different solvents such as ethyl acetate, aqueous $NaHCO_3$ solution, brine, and water. Note that without the purification procedure, the cEF$_B$ would only be 27, which is comparable to that of Route A_2. Route C showed the best cEF value of 5.6 kg kg^{-1} due to the applied base's favorable properties, i.e., no solvent was used, and a catalytic amount of t-BuOK proved to be sufficient. Moreover, in Route C, the final product did require fewer purification materials than the other routes, further lowering the cEF values. However, the selectivity (1:3.5 compared to the side-product) and the conversion (62%) could be further improved, for example through microflow implementation of the procedure.

4.5.6 Health and safety risks

The health and safety hazards associated with the compounds used in the synthesis of PolarClean were assessed using GHS pictograms. In Figure 4.12, the area of the illustrated pictograms is proportional to the number of chemicals used per unit mass of the product. The patented synthetic routes (A_{1-3}) have similar hazard profiles. However, A_3

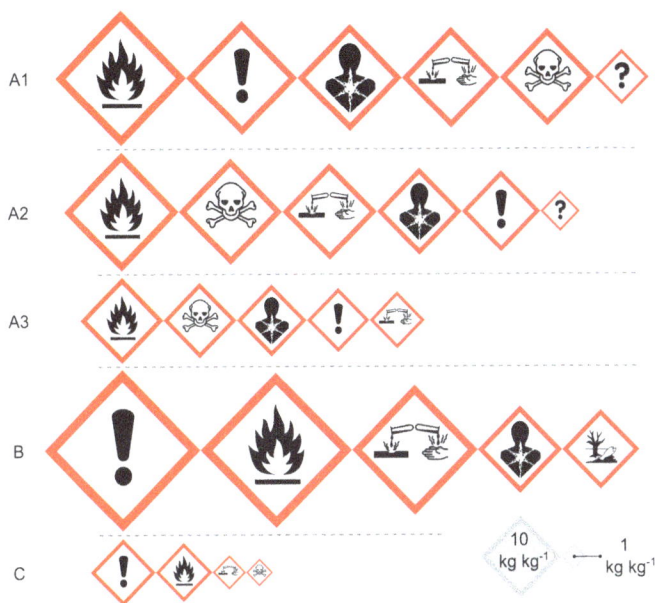

Figure 4.12: Chemical hazards associated with the different synthetic routes. The areas of pictograms are proportional to the amount of chemicals. The question mark refers to the chemicals with unknown hazards. Reproduced from [2] under the CC BY-NC 3.0 license.

does not have unknown hazards from the isolated intermediates. In Route B, more non-toxic chemicals are used. Route C is the most favorable reaction path based on the amount and nature of the chemicals used to obtain the product. Significantly less toxic and corrosive chemicals are used in Route C compared to the other reaction routes.

The effective mass yield (EMY) is also estimated as part of the risk assessment. The EMY is defined as the percentage of the desired product's mass relative to the mass of all *non-benign* materials used in the synthesis (eq. (2.11)) [31]. However, there is no clear definition for non-benign reagents. *Benign* substances are chemicals that pose no or negligible environmental risk and health concerns, such as pure or saline water, cellulose, or dilute ethanol. The EMY is calculated as follows:

$$\text{EMY}(\%) = \frac{\text{Mass of product}}{\text{Mass of non-benign reagents}} \times 100\%.$$

The example below shows the calculation method for Route C. First, the hazardous and toxic reagents need to be identified to estimate the EMY. As discussed previously, this metric lacks a precise definition. Nonetheless, eq. (4.16) attempts to provide information about the amount and hazardous nature of the chemicals used in the reaction (Figure 4.3). Based on this dataset, the EMY can be estimated for the synthesis of Polar-Clean. The EMY for Route C is calculated as follows:

$$EMY_C = \frac{1.00}{4.87 + 1.10 + 0.06 + 0.05} \cdot 100\% = 16.4\%. \tag{4.16}$$

Similarly, the EMY can be derived for the other synthesis routes, presented in Table 4.11. Note that, based on the definition, higher EMY percentage values are preferred. The patented Routes A_1 and A_2 have EMY values of 3.1% and 4.2%, respectively, due to their similar hazardous nature and the slightly higher number of utilized chemicals. Route A_3 shows a substantial improvement in the A-series due to the decrease in the amount of materials used in the synthesis. Route B possesses the least favorable EMY value, which results from a large amount of used chemicals, in particular solvents. In line with the expectations, Route C proved to have the highest EMY value of 16.4% because it is a solvent-free synthesis, and less toxic chemicals are required compared to other reactions.

Table 4.11: Comparison of the effective mass yield (EMY) values for the different synthetic routes.

Route	EMY (%)
A_1	3.1
A_2	4.2
A_3	7.6
B	2.4
C	16.4

4.5.7 Solvent intensity

The solvent intensity (SI) accounts for solvent usage, but yield, stoichiometry, and water are not taken into consideration (eq. (2.33)). The SI is calculated as follows:

$$SI = \frac{\text{Mass of solvents (excl. water)}}{\text{Mass of product}} = \left[\frac{\text{kg}}{\text{kg}}\right]. \tag{4.17}$$

The SI for the five different routes is displayed in Table 4.12.

Table 4.12: Comparison of the SI values for the different synthetic routes.

Route	SI (kg kg^{-1})
A_1	21.43
A_2	14.79
A_3	10.25
B	28.38
C	0

The SI varies between 10 and 28 kg kg^{-1} for Route A$_1$ and B, while that of Route C is technically zero due to the neat conditions. However, there is an important note here: both starting materials in Route C (and B as well) are liquids, and MeOProp is used in 10 molar excess equivalents compared to DMAA. The unreacted MeOProp, in this case, could act as a solvent and, depending on how we define the term "solvent," increase SI to 3.89 kg kg^{-1}.

The overall green metrics analysis is summarized in Table 4.13. In every segment of the analysis (except for the percentage yield), Route C was superior to Routes A$_{1-3}$ and B. In particular, the CI, AE, and cEF significantly improved. Moreover, the most advantageous health and safety profile using less toxic and hazardous materials was also obtained for the solvent-free synthesis Route C. Overall, it can be concluded that Route C is the most sustainable process with regard to both quality and green metrics performance.

Table 4.13: Summary of the green metrics analysis with the best values highlighted in green.

	Complexity	Ideality (%)	CI (%)	AE (%)	Percent yield (%)	cEF (kg kg^{-1})	EMY (kg kg^{-1})	SI (kg kg^{-1})
A$_1$	3	60	223	22	47	25.8	3.1	21.43
A$_2$	3	75	169	25	68	23.5	4.3	14.79
A$_3$	3	100	112	33	83	13.5	7.6	10.25
B	1	100	145	57	27	61.6	2.4	28.38
C	1	100	29	100	48	5.6	16.4	0

4.6 Room for improvement: Further optimization potential

The previous sections demonstrated that Route C is the most promising process, although there is room for improvement, which can be realized through further optimization. Route C's major disadvantage is the formation of a by-product, namely, the double Michael adduct **8** (Figure 4.4). The use of MeOProp in excess could mitigate this by-product formation. However, this comes with a drawback because more base catalyst is required for 100% conversion of DMAA, and excess MeOProp leads to higher CI and cEF values. The by-product is crystalline, and thus, it can be easily isolated with high purity. If **8** could be used as a material for other purposes (i.e., as a useful product), then the cEF and CI values would decrease by 14% from 29 to 25 and from 5.6 to 4.8 kg kg^{-1}, respectively.

However, the formation of **8** may be avoided by efficient and selective Brønsted base catalysts. Strong, commercially available non-ionic alternatives to t-BuOK can be considered from the GSK's base selection guide [18]. The base called P$_4$-phosphazene (pK_a = 42.1) proved to behave similarly to t-BuOK. There was no conversion observed in the case of weaker bases. As mentioned before, selectivity could be significantly improved via precise temperature control. The reactions were only conducted in batch reactors, where the temperature profile was uneven due to the small surface-to-volume ratio. The temperature and residence time could be controlled via microflow chemistry, thus increasing the overall conversion and yield.

Bibliography

[1] Mouterde, L. M. M.; Allais, F.; Stewart, J. D. Enzymatic Reduction of Levoglucosenone by an Alkene Reductase (OYE 2.6): A Sustainable Metal- and Dihydrogen-Free Access to the Bio-Based Solvent Cyrene®. *Green Chem.* **2018**, *20* (24), 5528–5532. 10.1039/c8gc03146k.

[2] Cseri, L.; Szekely, G. Towards Cleaner PolarClean: Efficient Synthesis and Extended Applications of the Polar Aprotic Solvent Methyl 5-(Dimethylamino)-2-Methyl-5-Oxopentanoate. *Green Chem.* **2019**, *21* (15), 4178–4188. 10.1039/c9gc01958h.

[3] Axon, S.; James, D. The UN Sustainable Development Goals: How Can Sustainable Chemistry Contribute? A View from the Chemical Industry. *Curr. Opin. Green Sustain. Chem.* **2018**, *13*, 140–145. 10.1016/j.cogsc.2018.04.010.

[4] Bradbury, S. Exemptions For Pesticide Environmental Protection. **2013**, *78* (103), 91–95.

[5] Vidal, T.; Bramati, V.; Murthy, K.; Abribat, B. A New Environmentally Friendly Solvent of Low Toxicity for Crop Protection Formulations. *J. ASTM Int.* **2011**, *8* (6), 1–8.

[6] Hassankiadeh, N. T.; Cui, Z.; Kim, J. H.; Shin, D. W.; Lee, S. Y.; Sanguineti, A.; Arcella, V.; Lee, Y. M.; Drioli, E. Microporous Poly(Vinylidene Fluoride) Hollow Fiber Membranes Fabricated with PolarClean as Water-Soluble Green Diluent and Additives. *J. Membr. Sci.* **2015**, *479*, 204–212. 10.1016/j.memsci.2015.01.031.

[7] Marino, T.; Blasi, E.; Tornaghi, S.; Di Nicolò, E.; Figoli, A. Polyethersulfone Membranes Prepared with Rhodiasolv®Polarclean as Water Soluble Green Solvent. *J. Membr. Sci.* **2018**, *549*, 192–204. 10.1016/j.memsci.2017.12.007.

[8] Wang, H. H.; Jung, J. T.; Kim, J. F.; Kim, S.; Drioli, E.; Lee, Y. M. A Novel Green Solvent Alternative for Polymeric Membrane Preparation via Nonsolvent-Induced Phase Separation (NIPS). *J. Membr. Sci.* **2019**, *574*, 44–54. 10.1016/j.memsci.2018.12.051.

[9] Dong, X.; Al-Jumaily, A.; Escobar, I. C. Investigation of the Use of a Bio-Derived Solvent for Non-Solvent-Induced Phase Separation (NIPS) Fabrication of Polysulfone Membranes. *Membranes.* **2018**, *8* (2). 10.3390/membranes8020023.

[10] Xie, W.; Tiraferri, A.; Liu, B.; Tang, P.; Wang, F.; Chen, S.; Figoli, A.; Chu, L. Y. First Exploration on a Poly(Vinyl Chloride) Ultrafiltration Membrane Prepared by Using the Sustainable Green Solvent PolarClean. *ACS Sustain. Chem. Eng.* **2020**, *8* (1), 91–101. 10.1021/acssuschemeng.9b04287.

[11] Llevot, A.; Grau, E.; Carlotti, S.; Grelier, S.; Cramail, H. Dimerization of Abietic Acid for the Design of Renewable Polymers by ADMET. *Eur. Polym. J.* **2015**, *67*, 409–417. 10.1016/j.eurpolymj.2014.10.021.

[12] Lebarbé, T.; More, A. S.; Sane, P. S.; Grau, E.; Alfos, C.; Cramail, H. Bio-Based Aliphatic Polyurethanes through ADMET Polymerization in Bulk and Green Solvent. *Macromol. Rapid Commun.* **2014**, *35* (4), 479–483. 10.1002/marc.201300695.

[13] Luciani, L.; Goff, E.; Lanari, D.; Santoro, S.; Vaccaro, L. Waste-Minimised Copper-Catalysed Azide-Alkyne Cycloaddition in Polarclean as a Reusable and Safe Reaction Medium. *Green Chem.* **2018**, *20* (1), 183–187. 10.1039/c7gc03022c.

[14] Rhodiasolv® PolarClean – Green solvent | Solvay https://www.solvay.com/en/brands/rhodiasolv-polarclean (accessed Apr 23, 2020).

[15] Olivier, J.; Massimo, G. Use of Esteramides as Solvents, Novel Esteramides and Process for Preparing Esteramides. *US 2014/0221211 A1,* **2014**, *1* (19).

[16] Harmsen, P. F. H.; Hackmann, M. M.; Bos, H. L. Green Building Blocks for Bio-Based Plastics. *Biofuels Bioprod. Biorefining.* **2014**, *8* (3), 306–324. 10.1002/bbb.1468.

[17]   Vennestrøm, P. N. R.; Osmundsen, C. M.; Christensen, C. H.; Taarning, E. Beyond Petrochemicals: The
        Renewable Chemicals Industry. *Angew. Chem. Int. Ed.* **2011**, *50* (45), 10502–10509. 10.1002/anie.201102117.
[18]   Henderson, R. K.; Hill, A. P.; Redman, A. M.; Sneddon, H. F. Development of GSK's Acid and Base
        Selection Guides. *Green Chem.* **2015**, *17* (2), 945–949. 10.1039/c4gc01481b.
[19]   Trummal, A.; Lipping, L.; Kaljurand, I.; Koppel, I. A.; Leito, I. Acidity of Strong Acids in Water and
        Dimethyl Sulfoxide. *J. Phys. Chem. A.* **2016**, *120* (20), 3663–3669. 10.1021/acs.jpca.6b02253.
[20]   Perrin, D. D. *Ionisation Constants of Inorganic Acids and Bases in Aqueous Solution*; Elsevier, 2016.
[21]   Richard, J. P.; Williams, G.; O'Donoghue, A. M. C.; Amyes, T. L. Formation and Stability of Enolates of
        Acetamide and Acetate Anion: An Eigen Plot for Proton Transfer at α-Carbonyl Carbon. *J. Am. Chem.
        Soc.* **2002**, *124* (12), 2957–2968. 10.1021/ja0125321.
[22]   Fraser, R. R.; Mansour, T. S. Acidity Measurements with Lithiated Amines: Steric Reduction and
        Electronic Enhancement of Acidity. *J. Org. Chem.* **1984**, *49* (18), 3442–3443. 10.1021/jo00192a059.
[23]   Randová, A.; Bartovská, L.; Morávek, P.; Matějka, P.; Novotná, M.; Matějková, S.; Drioli, E.; Figoli, A.;
        Lanč, M.; Friess, K. A Fundamental Study of the Physicochemical Properties of
        Rhodiasolv®Polarclean: A Promising Alternative to Common and Hazardous Solvents. *J Mol Liq.*
        2016, *224*, 1163–1171. 10.1016/j.molliq.2016.10.085.
[24]   Glycols, P. Color of Clear Liquids (Platinum-Cobalt Scale) 1. **2011**, *XIV (Reapproved)*, 1–4. https://doi.
        org/10.1520/D1209-05R11.2.
[25]   Oram, P. D.; Strine, J. Color Measurement of a Solid Active Pharmaceutical Ingredient as an Aid to
        Identifying Key Process Parameters. *J. Pharm. Biomed. Anal.* **2006**, *40* (4), 1021–1024. 10.1016/j.
        jpba.2005.08.006.
[26]   Leconte, P.; Denis, C. Conversion of Nitrile Compounds into Carboxylic Acids and Corresponding
        Esters Thereof, US Pat., 20090326260, **2012**.
[27]   Roschangar, F.; Zhou, Y.; Constable, D. J. C.; Colberg, J.; Dickson, D. P.; Dunn, P. J.; Eastgate, M. D.;
        Gallou, F.; Hayler, J. D.; Koenig, S. G.; Kopach, M. E.; Leahy, D. K.; Mergelsberg, I.; Scholz, U.; Smith,
        A. G.; Henry, M.; Mulder, J.; Brandenburg, J.; Dehli, J. R.; Fandrick, D. R.; Fandrick, K. R.; Gnad-
        Badouin, F.; Zerban, G.; Groll, K.; Anastas, P. T.; Sheldon, R. A.; Senanayake, C. H. Inspiring Process
        Innovation: Via an Improved Green Manufacturing Metric: IGAL. *Green Chem.* **2018**, *20* (10),
        2206–2211. 10.1039/c8gc00616d.
[28]   Hendrickson, J. B. Systematic Synthesis Design. IV. Numerical Codification of Construction Reactions.
        J. Am. Chem. Soc. **1975**, *97* (20), 5784–5800. 10.1021/ja00853a023.
[29]   Curzons, A. D.; Constable, D. J. C.; Mortimer, D. N.; Cunningham, V. L. So You Think Your Process Is
        Green, How Do You Know? – Using Principles of Sustainability to Determine What Is Green – A
        Corporate Perspective. *Green Chem.* **2001**, *3* (1), 1–6. 10.1039/b007871i.
[30]   Government emission conversion factors for greenhouse gas company reporting – GOV.UK
        https://www.gov.uk/government/collections/government-conversion-factors-for-company-reporting
        (accessed May 2, 2020).
[31]   Hudlicky, T.; Frey, D. A.; Koroniak, L.; Claeboe, C. D.; Larry, E. Toward a 'Reagent-Free' Synthesis.
        1999, *No. April*, 57–59.

5 Process intensification: methods and equipment

Diana Gulyas Oldal, Gyorgy Szekely

There is currently a growing trend in environmental awareness and efficiency improvement in different technologies, leading to a more sustainable process design on every scale, particularly in industrial manufacturing. Process intensification (PI) provides an engineering tool for achieving better process efficiency through adequate equipment and novel method design or redesign. PI was first introduced in the 1970s. Since then, there have been many attempts to define it for the chemical and energy industries. PI has several definitions, which are summarized in Table 5.1. Most definitions center on innovation but have different meanings. The term PI was first defined in the 1980s by Ramshaw [1], which focused on decreasing volumes and costs. Cross and Ramshaw [2], a few years later, complemented the definition to achieve a specific goal. Stankiewicz and Moulijn [3] expanded the definition of PI by incorporating the concept of *minimization*, calling for the minimization of energy consumption and waste generation: "PI is any chemical engineering development that leads to a substantially smaller, cleaner, and more energy-efficient technology." In 2003, PI developed a new meaning that embraced the replacement of conventional equipment and processes with more efficient ones [4]. Moulijn et al. proposed that PI is a tool to enhance the efficiency of processes by introducing novel and radical measures [5]. Following that, Becht et al. stated that PI is a method for product and process innovation in the chemical fields, whose main aim should be to maintain *profitability* [6]. Lutze et al. expanded PI's meaning by emphasizing process development and applying integration principles to different unit operations, functions, and phenomena [7]. Portha et al. introduced the concepts of local and global intensification because PI is connected to the sequence of the unit operations due to interactions among all units in the process, thus strongly improving the whole process [8]. In the following sections, we will use the following definition when referring to PI [9]: "any chemical engineering development that leads to a significantly smaller, cleaner, and more energy-efficient technology."

Van Gerven and Stankiewicz, instead of suggesting a new definition, gathered the principles of PI in the following four points [11]. These goals can be achieved through four domains: structure, energy, synergy, and time.

Table 5.1: Various process intensification definitions from the literature.

Definition	Source
"PI is the devising of exceedingly compact plants, which reduces both the 'main plant item' and the installation costs."	Ramshaw [1]
"PI is the strategy of reducing the size of a chemical plant needed to achieve a given production objective."	Cross and Ramshaw [2]

https://doi.org/10.1515/9783111028163-005

Table 5.1 (continued)

Definition	Source
"PI is the development of innovative apparatuses and techniques that offer drastic improvements in chemical manufacturing and processing, substantially decreasing equipment volume, energy consumption, or waste formation, and ultimately leading to cheaper, safer, sustainable technologies."	Stankiewicz and Moulijn [3]
"PI refers to technologies that replace large, expensive, energy-intensive equipment or process with ones that are smaller, less costly, more efficient or that combine multiple operations into fewer devices (or a single apparatus)."	Tsouris and Porcelli [4]
"PI tries to achieve drastic improvements in the efficiency of chemical and biochemical processes by developing innovative, often radically new types of equipment processes and their operation."	Moulijn et al. [5]
"PI stands for an integrated approach for process and product innovation in chemical research and development, and chemical engineering in order to sustain profitability even in the presence of increasing uncertainties."	Becht et al. [6]
"PI is a process development/design option, which focuses on improvements of a whole process by enhancing of phenomena through the integration of unit operations, integration of functions, integration of phenomena and/or targeted enhancement of a phenomenon within an operation."	Lutze et al. [7]
"PI is a holistic overall process based intensification (i.e., global process intensification) in contrast to the classical approach of PI based on the use of techniques and methods for the drastic improvement of the efficiency of a single unit or a device."	Portha et al. [8]
"PI is any chemical engineering development that leads to substantially smaller, cleaner, safer and more energy-efficient technology or that combine[s] multiple operations into fewer devices (or a single apparatus)."	Baldea [10]

1. *Maximize the effectiveness of intra- and intermolecular events*. This point refers to changing the kinetics of a process, leading to a potential solution for low conversions and undesired by-products.
2. *Give each molecule the same processing experience*. Processes allowing all the molecules to go through the same path, i.e., have the same history, would ideally result in uniform products with minimum waste generation. It is important to consider meso- and micro-mixing and temperature gradients besides the macroscopic residence time distribution, dead zones, or bypassing.
3. *Optimize the driving forces at every scale and maximize the specific surface area to which these forces apply*, i.e., transporting rates across interfaces. The effect of driving forces, e.g., concentration difference, should be maximized, which is achieved by maximizing the interfacial area for that driving force.
4. *Maximize the synergistic effects from partial processes*, i.e., synergistic effects should be utilized whenever possible.

Based on the reviewed definitions, *reduction* plays a crucial role in PI. For example, lower mass and heat transfer resistances allow reduced reaction time of the diffusion/conduction interfaces. This advantage, coupled with higher-intensity fluid dynamics in intensified equipment, enables the processes to achieve their inherent rates [12]. Consequently, faster mixing provided by lower reaction volumes allows an increase in selectivity.

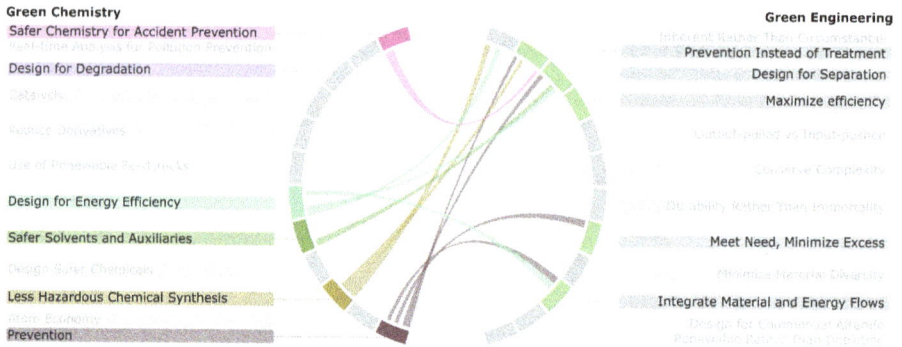

Figure 5.1: The multiple contribution potential of process intensification to address green chemistry and green engineering principles.

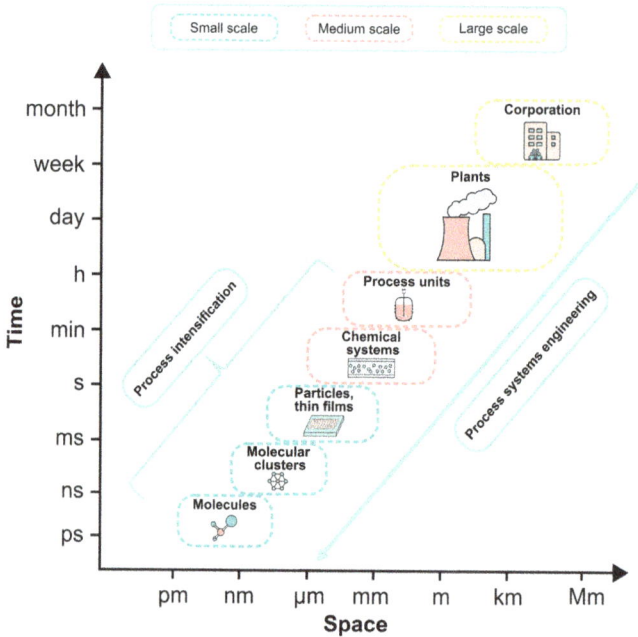

Figure 5.2: Scales of study in process intensification and process systems engineering. On a time-length scale, PI has an overlapping area of activity with process systems engineering. Adapted from [14].

Process Intensification

Equipment

- Equipment for carrying out chemical reactions
- Equipment for operations not involving chemical reactions

Methods

- Multifunctional reactors
- Hybrid separations
- Alternative energy sources
- Other methods

Examples

Equipment for carrying out chemical reactions

- Spinning Disk Reactor
- Static Mixer Reactor (SMR)
- Static Mixing Catalysis (KATAPAKs)
- Monolithic Reactors
- Microreactors
- Heat Exchange Reactors (HEX)
- Supersonic Gas/Liquid Reactor
- Jet-Impingement Reactor
- Rotating Packed-Bed Reactor

Equipment for operations not involving chemical reactions

- Static Mixers
- Compact Heat Exchangers
- Microchannel Heat Exchangers
- Rotor/Stator Mixers
- Rotating Packed Beds
- Centrifugal Adsorber

Multifunctional reactors

- Reverse-Flow Reactors
- Reactive Distillation
- Reactive Extraction
- Reactive Crystallization
- Chromatographic Reactors
- Periodic Separating Reactors
- Membrane Reactors
- Reactive Extrusion
- Reactive Comminution
- Fuel Cells

Hybrid separations

- Membrane Adsorption
- Membrane Distillation
- Adsorptive Distillation

Alternative energy sources

- Centrifugal Fields
- Ultrasound
- Solar Energy
- Microwaves
- Electric Field
- Plasma Technology

Other methods

- Supercritical Fluids
- Dynamic Reactor Operation

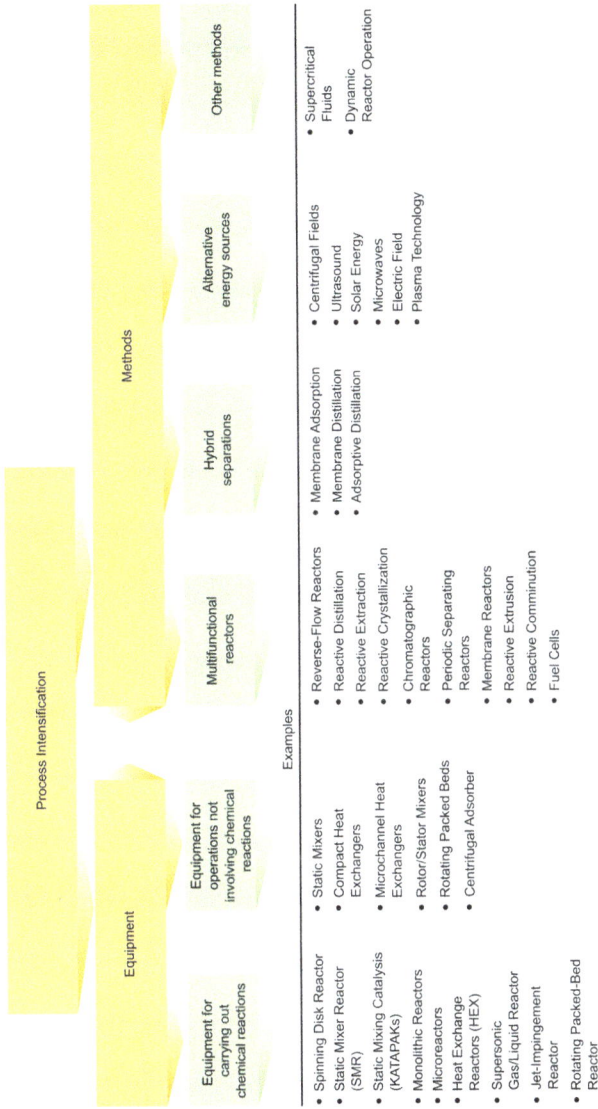

Figure 5.3: Process intensification can be broken down into methods and equipment. Adapted from [3].

PI supports green chemistry and green engineering principles, and the underlying relationships are presented in Figure 5.1. *Prevention* and *safety* are prevalent, along with *energy efficiency* and *integration* concepts. It is important to emphasize that PI contributes to the inherently safe nature of the proposed methods, equipment, and energy-saving. Thus, some of the goals of green chemistry and green engineering can be achieved by applying PI.

PI overlaps with the area of process systems engineering, whose main aim is to enhance decision-making for the creation and operation of the chemical supply chain involved in the discovery, design, manufacturing, and distribution of chemical products [13]. This common area of activity can be presented on a time-length scale (Figure 5.2).

It is important to mention that PI concerns only engineering methods and equipment, which means that the development of a new chemical route or a change in the composition of a catalyst does not qualify as PI. PI concerns novel equipment, processing techniques, and process development methods that significantly improve chemical processing. The *PI toolbox* consists of two dimensions with a broad spectrum of equipment and processing methods (Figure 5.3) [5]. PI equipment includes novel reactors, intensive mixers, and devices for better mass transfer and heat transfer. On the other hand, PI methods consist of integrated processes (integration of reaction and separation, heat exchange, or phase transition) in multifunctional reactors, methodologies utilizing alternative energy resources (e.g., light, microwaves, ultrasound), and hybrid separation operations, as well as other techniques such as dynamic operations.

Figure 5.4 summarizes the main benefits of using different PI methods and equipment. Most of them are characterized by increased yield and better selectivity.

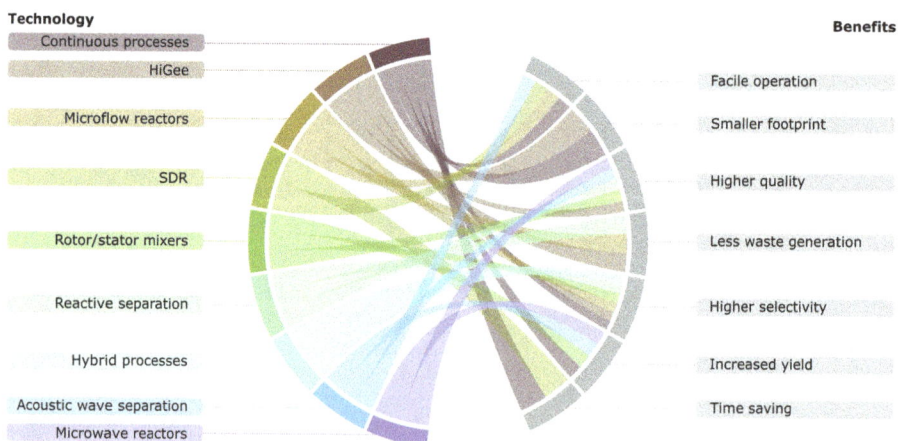

Figure 5.4: The technology benefits of using process intensification methods and equipment.

5.1 Evolution of chemical processes

There are numerous chemical process progression stages, starting from step-by-step operations carried out in different equipment to microreactors which fulfill the same or even more roles as conventional microreactors (Figure 5.5). *Batch processes* are often used in the fine, specialty, and pharmaceutical industries primarily by convention. However, it is also the batch reactors' flexibility (the same reactor can be used for several different types of reaction without modification), affordable price, batch integrity (in the pharmaceutical industry, every process has to be precisely defined, which is easier with distinguished batches), and long residence time for slow processes contribute to their widespread use [12]. However, scale-up and heat and mass transfer issues are the main drawbacks of batch processes.

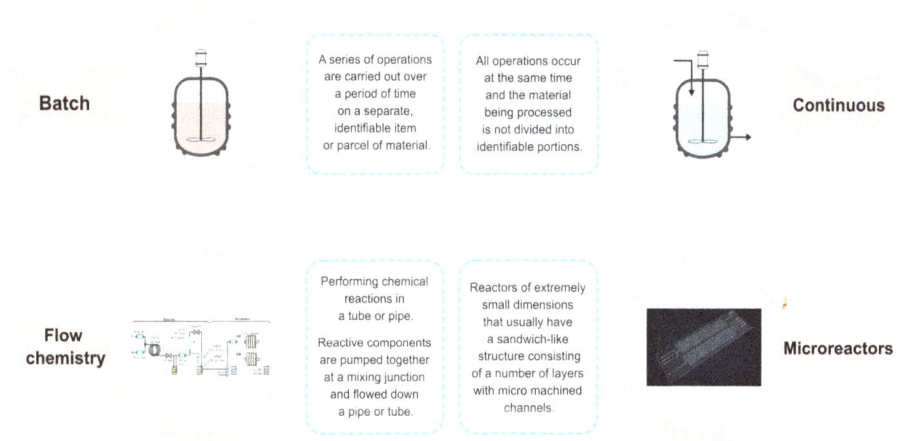

| Batch | A series of operations are carried out over a period of time on a separate, identifiable item or parcel of material. | All operations occur at the same time and the material being processed is not divided into identifiable portions. | Continuous |

| Flow chemistry | Performing chemical reactions in a tube or pipe. Reactive components are pumped together at a mixing junction and flowed down a pipe or tube. | Reactors of extremely small dimensions that usually have a sandwich-like structure consisting of a number of layers with micro machined channels. | Microreactors |

Figure 5.5: The evolution of chemical processes through the development of reactors.

Continuous processes usually have better heat and mass transfer properties than batch operations; thus, they are usually safer and more efficient. A continuous reactor is smaller than the equivalent batch reactors due to two main factors: greater occupancy and more effective mixing (length scales are smaller) [12]. Furthermore, continuous operations are more energy-efficient and easier to control, provide better product quality, and have reduced capital and running costs.

Flow chemistry is a system containing channels and tubings for carrying out a reaction in a continuous stream rather than in batch production. A fluid is pumped into a tube, and, at the connection with another tube, the reactive fluids mix with each other, and the reaction proceeds. These reactions are carried out continuously; therefore, only a small part of the reaction mixture reacts at a given time, implying better reaction time and selectivity [15]. Refer to Chapter 6 for a detailed sustainability overview of continuous microflow reactors and processes.

5.2 Process-intensifying equipment

PI is divided into two approaches: equipment and methods. Process-intensifying equipment includes novel reactors, intensive mixers, and heat transfer and mass transfer devices. The equipment is classified into (i) reactors for chemical reactions and (ii) devices not involving chemical reactions directly (Figure 5.3). Chemical reactors include many versatile devices such as a spinning disk reactor (SDR) which utilizes centrifugal forces or microreactors that are micro-structured or micro-channel reactors [14]. Static mixers are elements inside pipes responsible for the generation of ideal radial mixing and interfacial surface area. A static mixer reactor is another type of equipment with tailored tubings, representing a combination of intensive mixing, mass, and heat transfer for improved radial mixing. Monolithic reactors are devices that allow chemical reaction engineering of segmented flow in micro-channels. Equipment not involving chemical reactions are used as non-reactive systems such as a rotating packed bed or centrifugal absorber. A static mixer is applied for the continuous mixing of fluids without moving parts.

5.2.1 Microreactors

Microreactors are usually a type of continuous flow reactor with a small footprint and small reaction volumes with channel-shaped design. They consist of micron-sized channels, and these channel networks are linked between the reagents and products. Microreactors are usually small, with a sandwich-like structure consisting of several layers and micromachined channels with a diameter range of 10–100 μm (Figure 5.6) [3]. These reactors utilize micrometer-scale reaction spaces that allow more precise control of diffusion, heat exchange, retention time, and flow patterns in chemical reactions [16]. Owing to their microstructured features, a smaller amount of materials can be used, which implies safety and economic benefits and environmental profits. Since reactions performed in microreactors represent continuous techniques, the rapid processing of unstable intermediates can be realized.

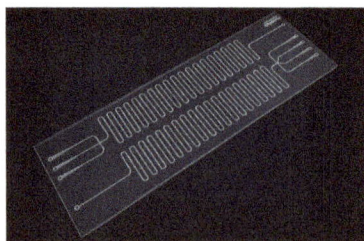

Figure 5.6: A microreactor with channels having a diameter in the range of 10–100 μm. Reprinted with permission from [17].

The microreactors possess an important advantage: mixing, catalytic reaction, heat exchange, or separation can be incorporated within a single unit in various combinations and stages [3]. Moreover, this reactor has an inherently small footprint. Besides, these reactors allow exceptional heat transfer rates [18] that are barely achievable by other equipment. The enhanced heat exchange, together with the increased surface-to-volume ratio, provides great thermal control. Consequently, exothermic reactions can be performed and safely controlled that were previously inaccessible on a larger scale [19]. Moreover, microreactors allow the decrease of reaction time under carefully controlled conditions by increasing the temperature while avoiding the formation of by-products due to the improved temperature distribution [20].

Their very low reaction volume-to-surface area ratio makes them attractive for processes involving toxic or explosive reactants because these materials can be utilized or generated *in situ* [12]. As discussed previously, microreactors' main advantage is the great temperature control of the reaction, which, together with the low internal volume, provides higher process safety. [21] Therefore, this is an inherently safe reactor.

Despite all these advantages, microreactors also suffer from drawbacks. Tubular reactors, and consequently microreactors, can work well with lower-viscosity liquids. However, using high-viscosity fluids can result in undisturbed laminar flow, which can lead to inadequate mixing and a non-homogeneous outcome. The insufficient mixing of the inlets due to the laminar flow is a bottleneck that should be considered. The issue of laminar flow can be solved with the application of static mixers (Figure 5.7) [22]. The tiny channel dimensions, along with the use of static mixers, provide millisecond mixing times [19]. Owing to their small dimensions, hot spots occurring in batch reactors can be prevented, resulting in better selectivity and yield for many reactions [23]. The small channel diameter can result in the formation of insoluble solids in the reaction, i.e., the flow path's clogging, which can stop the reaction route. Another occurring issue is the high pressure drops originating from gas formation during the reaction.

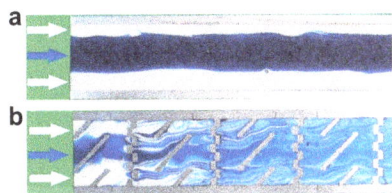

Figure 5.7: Flow profile in microreactors (a) without a mixer and (b) with a static mixer. Reprinted with permission from [22].

Some examples are presented in the next paragraphs to demonstrate how microreactors can lead to more sustainable processes. The first example is the optimization of regioselectivity in a sensitive Grignard reaction (Figure 5.8) [23]. The Grignard reaction is a widely known organometallic chemical reaction for the formation of carbon–carbon bonds.

Process	Yield [%]	Ratio of A & B
Batch	49	65:35
Microreactor	78	95:5

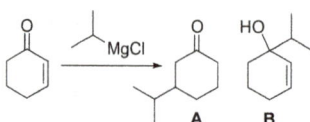

Figure 5.8: Selective Grignard reaction in a batch reactor and a microreactor, and the corresponding yields and product ratios for each process.

The table in Figure 5.8 represents the obtained data after performing the same reaction in two different manners. The initial 49% yield of the batch process was increased to 78%, together with an increase in the regioisomer ratio from 65:35 (A:B) to 95:5 (A:B). Another example showing the advantages of microreactor is the Sonogashira reaction between iodobenzene and phenylacetylene in ([BMIm][PF$_6$]) ionic liquid is given in Figure 5.9 [24]. Sonogashira coupling is a useful reaction to prepare acetylene compounds, usually performed in organic solvents, such as DMF or THF, with a palladium catalyst and copper co-catalyst. In this case, the coupling was carried out in an ionic liquid in a microreactor allowing easy separation of the product from the copper-free catalyst, which was subsequently recovered. The results revealed a negligible drop in yield from 95% in batch to 93% in microreactor processing. However, the reaction time decreased tenfold. In brief, even though the yield declined by 2% when switching from batch to microreactor, the time reduction was nonetheless excellent. A reduction in time is important in manufacturing processes as shorter reaction times mean less energy consumption, less maintenance, less depreciation of the equipment, and often higher selectivity.

Process	Yield [%]	Time [min]
Batch	95	180
Microreactor	93	10

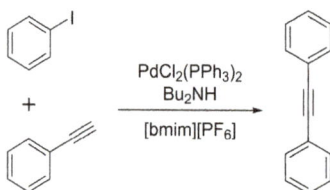

Figure 5.9: Comparison of the results of a Sonogashira reaction carried out in an ionic liquid in a batch reactor and a microreactor.

5.2.2 Rotating devices

5.2.2.1 Spinning disk reactors

SDRs were primarily designed for swift and very fast liquid/liquid reactions with large heat effects such as nitrations, sulfonations, and polymerizations [3]. The working principle of SDRs is shown in Figure 5.10. In this reactor, a thin layer (typically 100 µm) of liquid moves on the surface of a disk spinning at up to 1000 rpm. The thin film containing the reactants is created utilizing centrifugal force. These reactors were invented to enhance heat and mass transfers combined with good processing times. They operate in a continuous mode and are inherently safe due to the small reaction volumes. However, the moving parts require regular maintenance due to the high-speed rotation.

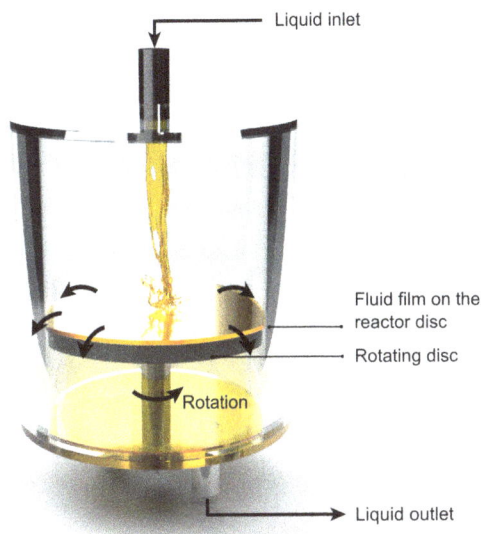

Figure 5.10: Schematic of a spinning disk reactor is designed for very fast liquid/liquid reactions with large heat effects such as nitration, sulfonation, and polymerization.

5.2.2.2 Rotor/stator mixers

Rotor/stator mixers are designed for processes that need very fast mixing on a micro-scale [3]. This equipment consists of a high-speed rotor spinning close to a stationary stator. The fluid goes through the region where the rotor and stator interact, which results in highly pulsating flow and shear. The four stages in Figure 5.11 explain the operation of a high-shear rotor/stator mixer [25]. Follow the QR code on this page for the corresponding animation that shows the entire process.

In stage 1, the rotor blades' high-speed rotation within the precision-machined mixing workhead creates a powerful suction effect. It drags the liquid and solid materials upwards from the bottom of the vessel into the center of the workhead. During stage 2, the centrifugal force drives the fluids and solids towards the edge of the workhead. These materials are milled in the precision-machined clearance between the ends of the rotor blades and the stator's inner wall. Stage 3 is comprised of (i) an intense hydraulic shear as the liquid and solid materials are forced out at high velocity

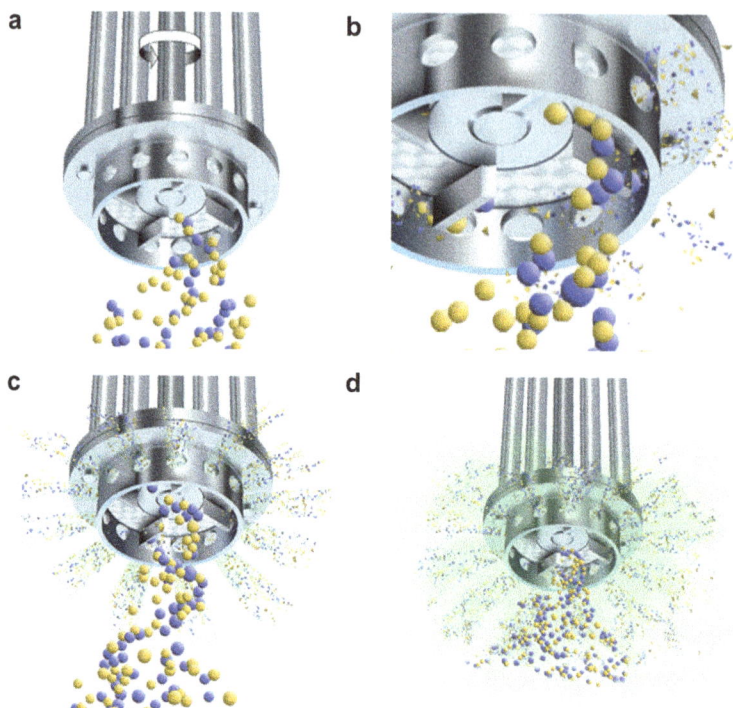

Figure 5.11: Different operation stages of a high-shear rotor/stator mixer. (a) Stage 1. (b) Stage 2. (c) Stage 3. (d) Stage 4. Reprinted with permission from [25].

through the perforations in the stator and (ii) circulation into the main body of the mixture. In the final stage, the fluid and solids discharged from the head are launched radially at high speed towards the mixing vessel's sides. Simultaneously, fresh substances are continuously pulled into the workhead to maintain the mixing cycle. The radial discharge effect and suction into the head create a circulation pattern that minimizes exposure to the air induced by disturbing the fluid's surface.

5.2.2.3 High-gravity technology

High-gravity (Higee) technology intensifies the heat and mass transfer processes by performing the processes in rotating packed beds. The centrifugal force of the equipment induces a high-gravity environment [26]. In the rotating packed bed, thin liquid films and/or tiny droplets are generated. The interface between gas/liquid or liquid/liquid is renewed vigorously, leading to a significant intensification of mass transfer and micromixing, i.e., 1–3 orders of magnitude more mass transfer and micromixing than conventional packed beds [27]. Follow the QR code on this page for the corresponding video that shows how rotating packed bed reactors operate. This equipment originally pertained to separation processes (distillation, extraction, absorption) but has since been applied to reacting systems as well, in particular mass-limited reactions [3]. The technology can be utilized for systems of three phases (gas/liquid/solid).

The following example of biomass torrefaction presents the benefits of using Higee technology [28]. Owing to increasing environmental awareness, alternatives to conventional energy sources are becoming increasingly prominent. Biomass represents a renewable energy resource as a sustainable raw material for biofuel production. Bamboo can be used as a biomass feedstock. In this study, torrefaction was applied as a technology where biomass is thermally pretreated at high temperatures (200–300 °C) in an oxygen-free environment to produce high-quality fuels. Torrefaction is a mild type of pyrolysis that results in a dehydrated product by removing volatiles and moisture from the biomass. The higher heating values (HHVs) were examined under different conditions because they provide measurable data of the calorific value, thus improving fuel quality. The HHV is the maximum potential energy released during complete oxidation of a unit of fuel. HHV includes the thermal energy recaptured by condensing and cooling all products of combustion. The higher the HHV value, the higher the calorific value of the fuel. The highest attainable HHV was 28.389 MJ kg^{-1} with an average centrifugal force in the rotating packed bed at 234 g located at 1800 rpm. The energy yield was approx. 64%, while the HHV showed a 60% increase.

5.3 Process-intensifying methods

Process-intensifying methods can be classified into four main areas: hybrid separations, multifunctional reactors, techniques using alternative energy sources, and other methods (Figure 5.3). In the case of hybrid separations and multifunctional reactors, the emphasis is on integrating operations such as reaction and separation or heat exchange. Multifunctional reactors fall under the PI method category because they synergistically combine processing units, such as combining reaction and separation into a single device.

Multifunctional reactors are intensified by adding different functions of conventional unit operations [29]. These reactors include a wide variety of integration combinations such as heat exchange for thermal intensification or reactive comminution. Reactive separation processes entail the simultaneous integration of separation and reaction in one device, e.g., reactive distillation, reactive stripping, reactive absorption, reactive extraction, reactive crystallization, and membrane reactors. Some other methods, such as reactive extrusion and fuel cells, also fall into the multifunctional reactors category.

Hybrid separations require the integration of at least two separation units, which results in a better separation performance than the application of the individual units. Some hybrid separation combination examples are the dividing wall column, which combines two distillation columns into one unit [30], and membrane distillation (MD), which is a membrane separation process enabled by phase change. Membrane adsorption and adsorptive distillation are also considered hybrid separations.

Alternative energy sources can improve numerous conventional processing techniques. Solar energy, microwave, and ultrasound are just some of the possibilities for more sustainable chemical processing. Other methods include supercritical fluids, whose unique behavior above a critical point, i.e., where there is no difference between the liquid and gas phase, can be exploited for PI. Owing to their specific physical and transport properties, supercritical fluids can be a favorable environment for mass transfer processes and chemical reactions [3]. Refer to Section 3.3 for a more detailed explanation of supercritical fluids. Dynamic reactor operations result from pulsing flows or concentrations, which can improve the outcome of reactions.

5.3.1 Membrane reactors

Membrane reactors integrate reaction and separation in a single unit. They represent a highly effective system in overcoming equilibrium limitations in reactions because the products are continuously and selectively removed, favoring a forward reaction in line with Le Chatelier's principle [31]. Membrane separation involves applying a selective barrier (membrane) to regulate the transport of substances such as gases, vapors, and liquids, at different mass transfer rates [32]. The mass transfer rates of different substances depend on the permeability of the barrier toward the feed components. Figure 5.12 shows a membrane reactor for a water-gas shift reaction.

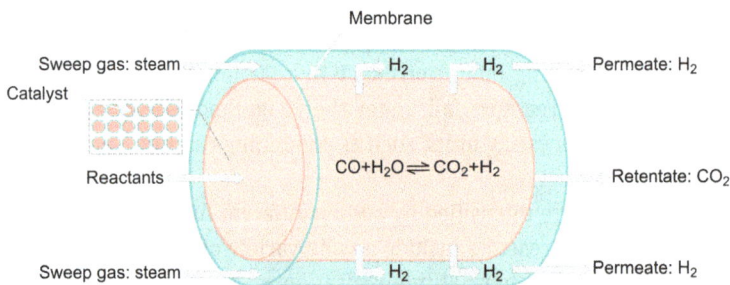

Figure 5.12: Schematic illustration of a membrane reactor for a water-gas shift reaction comprising an outer layer (the shell) and an inner layer, which is a tube packed with a catalyst. The reaction and permeation simultaneously occur in the inner layer. Adapted from [33].

The membrane can play various roles in membrane reactor systems [3], which are listed below:

- selective *in situ* separation of the reaction product, resulting in an advantageous equilibrium shift;
- controlled distribution of the feed (reactants) to increase the overall yield or selectivity of a process or to enhance mass transfer;
- shifting the equilibrium of the reaction by the continuous removal of the product or by-product;
- *in situ* recycling of catalysts, reagents, or solvents;
- catalyst can be incorporated into the membrane, evolving into a highly selective reaction-separation system (Figure 5.13).

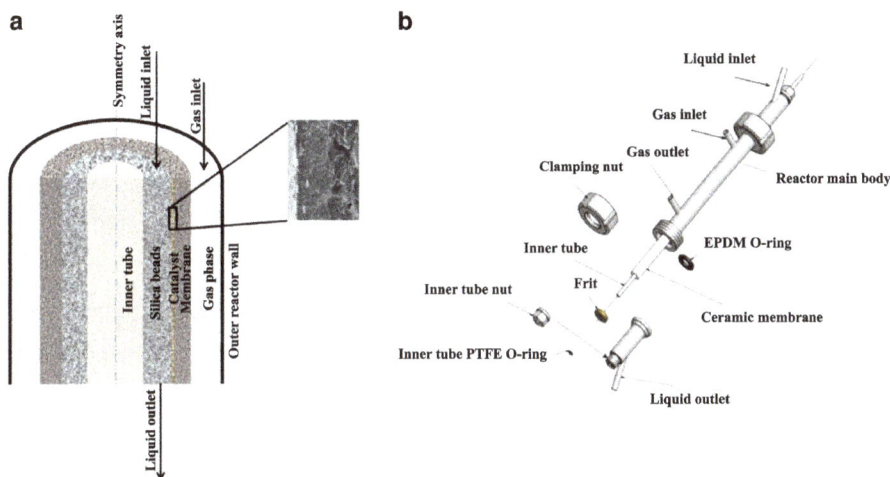

Figure 5.13: A hybrid reaction-separation system featuring a catalytic membrane. (a) SEM image of the membrane. (b) Components of the reactor. Reproduced from [34] under the CC BY 4.0 license.

These systems have a smaller footprint due to the combination of different unit operations. The green aspects of membrane technology, such as energy efficiency, continuous operation, and modularity, are also beneficial for developing an overall sustainable process. Membrane reactors can be homogeneous and heterogeneous systems often used in bioreactors in the form of an aqueous system with ultrafiltration.

Some of the drawbacks of the use of membrane technology are the short membrane lifetime under harsh conditions. Another disadvantageous property of these systems is fouling attributed to the accumulation and deposition of solutes or particles in the feed onto the membrane surface and into the membrane pores [32].

5.3.2 Hybrid separations

5.3.2.1 Membrane distillation

MD is a widely known hybrid separation technique considered an alternative to reverse osmosis and evaporation [3]. This method consists of bringing a volatile component of a liquid feed stream through a porous membrane as a vapor and condensing it on the other side into a permeated liquid. The driving force of the separation is the temperature difference. The main benefits of MD are the following:

- 100% rejection of ions, macromolecules, colloids, cells, and other non-volatiles;
- lower operating pressure than in the pressure-driven processes such as reverse osmosis;
- lower operating temperature than in temperature-driven processes such as distillation;
- the low temperatures in the range of 30–60 °C permit the reuse of residual heat flow and the use of alternative energy sources such as sun, wind, and geothermic;
- less membrane fouling due to the lack of contact with a liquid stream and larger pore size than reverse osmosis membranes (10 nm to 1 µm).

Compared with traditional distillation, MD has the characteristic benefits of membrane separation, such as simple scale-up and operation, a high membrane surface-to-volume ratio, and the potential of treating streams with heat-sensitive components [35]. Several MD configurations exist, as summarized in Figure 5.14.

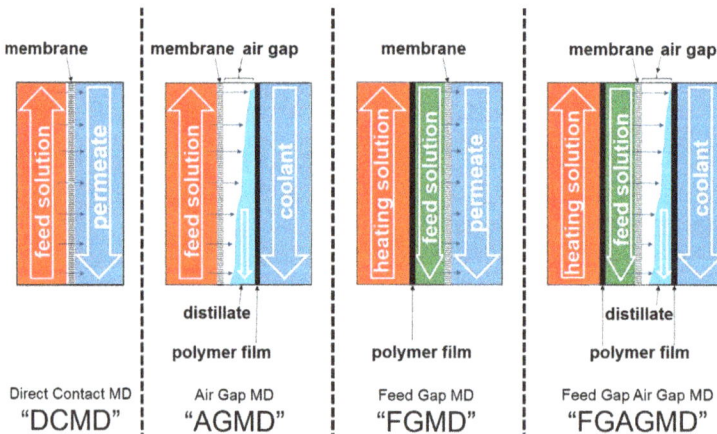

Figure 5.14: Different membrane distillation (MD) configurations, such as direct contact membrane MD, air gap MD, feed gap MD, and feed gap air gap MD. Reproduced from [36] under the CC BY 4.0 license.

Direct contact MD represents a process where the hot feed solution flows on one side, and the cooled permeate flows on the other side of the membrane in a counter-current [36]. In air gap MD, a thin polymer film is responsible for separating the distillate channel from the coolant. Feed gap MD is a system where the feed solution is circulated through the former air gap. The heating solution is introduced into the previous cooling stream, which is isolated from the feed solution using a polymer film. Feed gap air gap MD represents a combination where the feed solution is separated from the heating solution, and the coolant channel is also separated from the permeate channel.

5.3.3 Use of alternative energy: ultrasound and microwave

Sonochemistry represents a field where ultrasound is used as an alternative source of energy. The formation of cavities occurs in the liquid reaction medium via the action of ultrasound waves. These cavities behave as high-energy microreactors. Their collapse creates micro-implosions with high local energy release. These can be used for *in situ* catalyst cleaning and rejuvenation, for instance. The maximum economically and technically feasible size of the reactor vessel is still the limiting factor for industrial applications. This form of energy can be used in reactions, dissolution, purification, water treatment, and cleaning. The main advantages of ultrasound technology are good yields, short reaction times, and mild reaction conditions [37]. Sonochemistry was successfully used in oxidative desulfurization of liquid hydrocarbons [38] as well as for biomass valorization [39] (Figure 5.15).

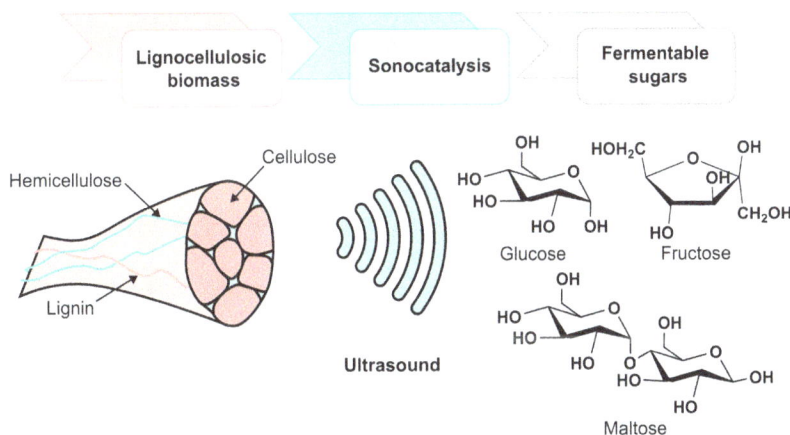

Figure 5.15: Utilizing sonocatalytic technology for biomass valorization. Lignin is removed from the lignocellulosic biomass and subsequently converted to bio-based building blocks using various pretreatment methods. Sonocatalysis is used to convert and hydrolyze lignocellulosic biomass into fermentable sugars, which can be subsequently utilized as useful products. Adapted from [39].

Oxidative desulfurization is a method to reduce the sulfur content of diesel. Organic sulfur compounds in fuels can cause pollution and acid rain, and thus their emission levels need to be controlled. Hydrodesulfurization is the conventional technology for this purpose, which requires high temperatures and high-pressure hydrogen. Not every sulfur compound can be removed by this method [40]. On the contrary, ultra-sound-assisted oxidative desulfurization has proved to be a promising alternative solution because it does not require high-pressure hydrogen or high temperatures, even though it possesses high desulfurization efficiency [38]. The chemical nature, such as the type of materials used, operating conditions (e.g., time, temperature), and geometrical parameters (e.g., reactor volume, probe size, probe immersion depth) are equally important to achieve a high level of desulfurization. Desulfurization efficiencies of 98% can be achieved with a proper probe diameter. The reactor material also has a significant role. A glass wall reactor provided approx. 16% higher desulfurization efficiency compared to a polypropylene reactor [38].

Sonocatalytic biomass valorization represents a new approach in sustainable energy production. Biomass pretreatment is one of the most critical factors to obtain bio-based building blocks for further applications; i.e., pretreatment has a crucial role in breaking the lignin down and disrupting the crystalline structure of cellulose, which makes cellulose easily accessible to enzymes for further conversion [41]. Biomass pretreatment can be performed utilizing biological (enzymatic and biodegradation), chemical (basic or acid treatment, solvent extraction, oxidation), and physical processes (irradiation, milling) [42]. Physical treatment (mechanical comminution) involves mechanical size reduction of lignocellulosic biomass through grinding, milling, or chipping to reduce the cellulose's crystallinity [43]. However, in most cases, this method uses more energy than the biomass's theoretical energy content. Irradiation techniques involve sonication, ionization, cold plasma, microwaves, shock waves, and high-energy radiation methods. They can be used due to improved digestibility of lignocellulosic biomass, although they may be expensive, and some are time-consuming and energy-intensive. Thus, the irradiation processes are often used in combination with other pretreatment methods. Chemical lignocellulosic treatment comprises catalytic reactions such as acid and alkaline hydrolysis, oxidative delignification, or organosolv processes. Acid pretreatment is utilized for the hydrolysis of the lignocellulosic material and the removal of hemicellulose, whereas the alkaline method results in more digestible cellulose due to the removal of lignin [42]. Oxidative delignification utilizes oxidizing agents such as hydrogen peroxide, ozone, oxygen, or air. In contrast, organosolv processes use organic solvents or their mixtures with water to remove lignin before enzymatic hydrolysis [44]. Biological treatment is a less energy-intensive process than chemical treatment and includes using fungi or other microorganisms to break down the lignin in lignocellulosic materials [45]. However, biological treatments are time-consuming and can take up to several days.

Ultrasound irradiation can be used to increase the yields of enzymatic hydrolysis reactions and decrease the required pretreatment time. Figure 5.15 illustrates the process of sonocatalytic biomass valorization. Exposing lignocellulosic material to ultrasound irradiation breaks it down into different products such as fermentable sugars. The ultrasound enhances the hydrolysis of lignocellulosic biomass into sugars and their subsequent fermentation. The PI results from intensified mass transfer and the activation of the catalysts in the reacting system [46]. Ultrasonic pretreatment has been successfully used for the acidic hydrolysis of oil palm empty fruit bunch at low temperature (100 °C) and atmospheric pressure [47]. The xylose yield increased more than twofold from 22% to 52%. High conversion (95%) of treated cellulose was achieved through ultrasonic heating, which is almost twice that of the untreated cellulose (43%) [48].

Microwave radiation is a highly effective heat source in chemical reactions. A microwave is a low-energy electromagnetic wave of the electromagnetic spectrum that lies between infrared and radio waves (Figure 5.16).

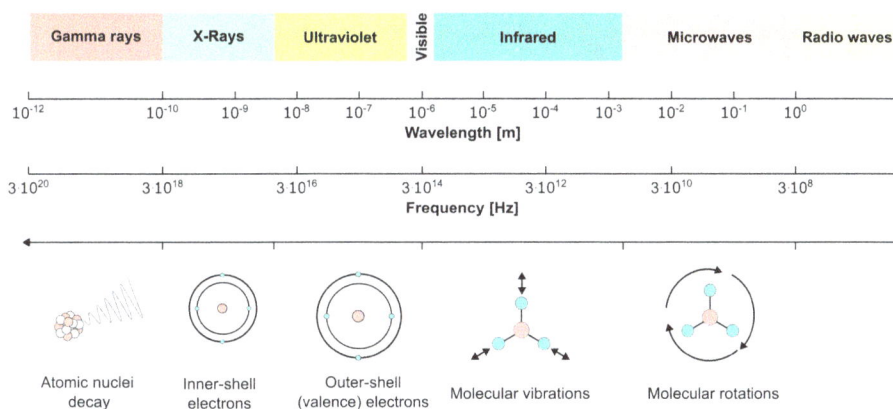

Figure 5.16: The electromagnetic spectrum represents a scale of frequencies and wavelengths of different radiation types. Microwave radiation is a region with electromagnetic wavelengths longer than infrared light but shorter than radio waves.

Microwave reactors work at a particular frequency of 2.45 GHz [49]. Microwaves can generate heat through the electric field by interacting with molecules in two different mechanisms: dipolar rotation and ionic conduction (Figure 5.17). Dipolar rotation is a phenomenon where molecules rotate back and forth to align their dipoles with the constantly oscillating electric field [50]. These rotations cause friction between molecules, resulting in heat energy generation. Ionic conduction is a translational move of free ions or ionic substances through space to synchronize with the electric field. Heat generation is also a consequence of friction between these moving species.

Microwaves interact directly with the reaction mixture's contents without intermediate processes, and therefore the heat transfer from microwaves is more efficient than conductive heating (Figure 5.18). Conventional conductive heating occurs through the reactor's wall, which is less effective and results in an undesired temperature gradient.

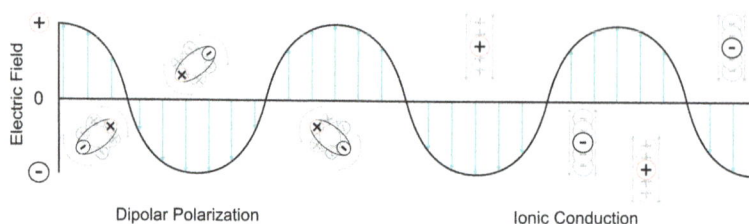

Figure 5.17: Dipolar rotation and ionic conduction in microwave heating. The dipole rotation results from the molecular rotation aligning the molecule's dipole with the electric field. Ionic conduction originates from the ion movement, and thus its alignment with the electric field. Adapted from [50].

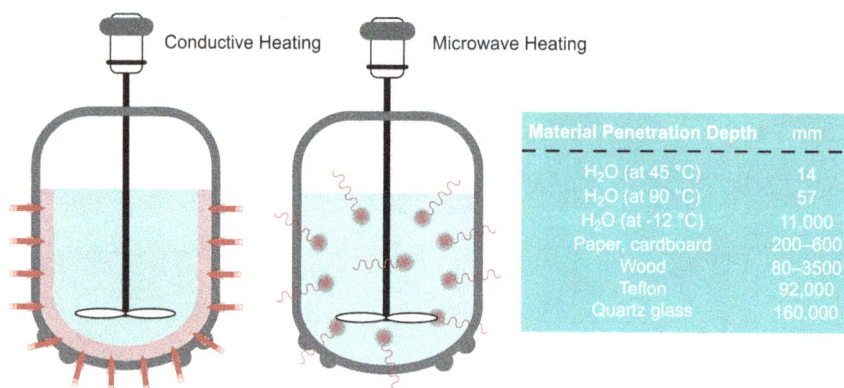

Material Penetration Depth	mm
H_2O (at 45 °C)	14
H_2O (at 90 "C)	57
H_2O (at -12 °C)	11,000
Paper, cardboard	200–600
Wood	80–3500
Teflon	92,000
Quartz glass	160,000

Figure 5.18: Difference between conductive and microwave heating. Conventional heating is illustrated in the left figure, which is based on thermal conductivity. The right figure shows microwave heating by dipolar rotation and ionic conduction, representing a more efficient heat transfer method than the previous one. Adapted from [50].

Waves can act differently in the electromagnetic field; i.e., when they enter a medium, they can be absorbed, transmitted, or reflected. Due to these interactions, some losses, such as dielectric loss, cause heat generation. This dielectric loss of a material can be high. The heating efficiency for a large-size sample is sometimes low [51], possibly due to the low penetration depth of the microwaves into the sample. Therefore, the penetration depth of the material by microwaves is crucial because it can cause scale-up issues. The dielectric constant of materials inversely changes with temperature.

Microwave technology is used for various chemical manufacturing tasks such as reactions, biodiesel production, polymerization, separations, and waste processing (Figure 5.19). The application of microwaves is common in synthetic organic chemistry for various reactions involving solvent-free and water-mediated reactions [52].

Microwaves possess many beneficial properties such as selective heating (which depends on the material's dielectric properties), an increased rate of reactions and

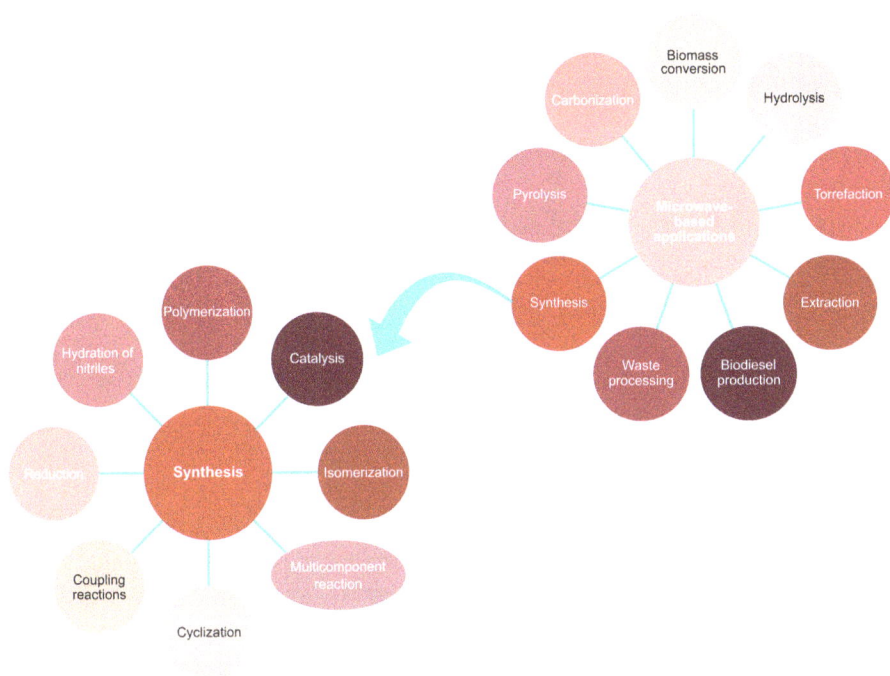

Figure 5.19: The application areas of microwave-based technologies.

percentage yield, and efficient and uniform heating. This is an eco-friendly technique for neat reactions. The batch variance with microwave-assisted processing is low.

Numerous studies demonstrate the benefits of microwave technology, for instance, a microwave-enhanced biodiesel production process [53] or Suzuki cross-coupling reactions in microwave reactors [54]. In the latter case, the microwave application increased the yield from 66% to 83% while reducing the reaction time from 5 h to 20 min. Ultrafast pyrolysis of lignocellulose is another excellent example of a sustainable process through the implementation of microwaves [55]. Figure 5.20 illustrates the actual amount of gasified materials during conventional heating and microwave heating. All generated products (except H_2) had almost comparable volumes for microwave and conventional heating, regardless of the reaction temperature. The high amount of H_2 during microwave heating at 400 °C resulted from *in situ* steam reforming from water and biochar, which originated from the cellulose bed. The large volume of generated H_2 makes the microwave more advantageous than conventional heating. The heating time of the pyrolysis reaction was reduced to 12.5% as a result of microwaves. Moreover, it reduced the process's energy consumption by 60%, and there was no need for time-consuming shredding as pretreatment. However, the technology's limitations became apparent during the pyrolysis process because of the low degree of microwave absorption of lignocellulose as a consequence of its low dielectric loss.

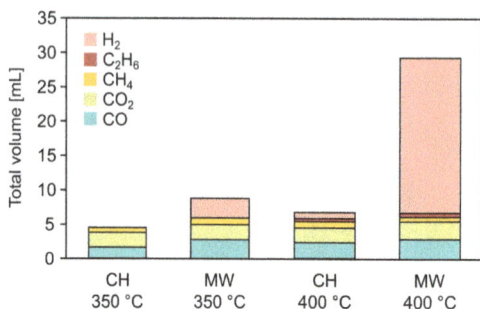

Figure 5.20: The distribution of gasified materials as the result of the ultrafast pyrolysis of lignocellulose. The high amount of produced H_2 at 400 °C makes microwave heating a beneficial technique for lignocellulose's fast pyrolysis. MW denotes microwave heating, whereas CH denotes conventional heating. Adapted from [55].

On the other hand, the bottleneck of microwave technology is that it can be applied only for materials that absorb microwaves. Sulfur, for instance, is transparent to radiation. High-pressure systems combined with a microwave can result in runaway reactions leading to a possible explosion. High-frequency microwaves can penetrate human cells and destroy them, which requires a safety risk assessment. Overheating and hotspots due to heat build-up in solid particles, such as in catalysts, can lead to side reactions or degradation of heat-sensitive compounds.

5.3.4 Other methods

Numerous other techniques fall outside the categories above. For example, supercritical fluids (see Section 3.3 for more details) play a significant role in numerous industries owing to their advantageous properties such as high diffusivity and tunable density. Supercritical fluids are widely applied in extraction [56], impregnation [57], cleaning [58], separation [59], particle formation [60], or synthesis of nanomaterials [61]. The use of supercritical fluids in the extraction of nutrients and bioactive compounds from natural raw materials and waste has demonstrated ease of operation, the advantage of low extraction temperatures, hence handling heat-sensitive molecules, and environmental friendliness [62]. Supercritical fluids are extensively applied in food and fragrance industry owing to their mild operating conditions for extraction of essential oils [63], fragrances [64], spices [65], or caffeine [66]. Contrary to traditional methods that employ solvents for dissolving and separating compounds, followed by heat-induced solvent removal to obtain the final product, supercritical fluid extraction facilitates solvent removal *via* depressurization. This approach reduces the process temperature and can be conducted without the use of organic solvents [67]. Supercritical carbon dioxide is among the most frequently used extraction media owing to its abundance in pure form, availability at low cost, non-toxicity and being inflammable [68]. Carbon dioxide is usually used for extraction of non-polar compounds, but with the addition of a co-solvent such as ethanol or water can enhance its use for polar compound extraction [56]. Moreover, supercritical fluids can be used for synthesizing new materials, supports for new catalysts, a reaction medium for different reactions, such as supercritical carbon dioxide for Friedel-Crafts alkylation, hydroformylation, and transesterification [69].

Dynamic or unsteady operation modes of chemical reactors involve intentional pulsing of flows, flow rates, concentration, temperature, and pressure. These effects can be achieved by modulation of inlet parameters, reversal of flow, or movement of the reactor. Using dynamic operation modes can enhance the selectivity and yield of reactions, improve reactant conversion and energy efficiency, and extend the lifetime of catalysts [70]. Stolte et al. [71] described a type of periodic operation involving rapid temperature pulsing to induce chemical reaction, oxidation of CO over a Pt catalyst, directly and locally at will. This implementation led to a higher reaction rate with increasing pulse frequency, and it was found that the reaction rate can be influenced almost instantaneously. Dynamic process intensification was applied in reactive distillation of methanol dehydration to dimethyl ether, a study conducted by Liu et al. [72]. They treated the target product as a mixture of two auxiliary products, and it was found that the nonsteady-state operation increased the reaction conversion and reduced energy consumption.

Bibliography

[1] Ramshaw, C. Higee' Distillation-an Example of Process Intensification. *Chem. Eng.* **1983**, *389*, 13–14.
[2] Cross, W. T.; Ramshaw, C. Process Intensification: Laminar Flow Heat Transfer. *Chem. Eng. Res. Des.* **1986**, *64* (4), 293–301.
[3] Stankiewicz, A. I.; Moulijn, J. A.; et al. Process Intensification: Transforming Chemical Engineering. *Chem. Eng. Prog.* **2000**, *96* (1), 22–34.
[4] Tsouris, C.; Porcelli, J. V. Process Intensification-Has Its Time Finally Come. *Chem. Eng. Prog.* **2003**, *99* (10), 50–55.
[5] Moulijn, J. A.; Stankiewicz, A.; Grievink, J.; Górak, A. Process Intensification and Process Systems Engineering: A Friendly Symbiosis. *Comput. Chem. Eng.* **2008**, *32* (1–2), 3–11. 10.1016/j.compchemeng.2007.05.014.
[6] Becht, S.; Franke, R.; Geißelmann, A.; Hahn, H. An Industrial View of Process Intensification. *Chem. Eng. Process. Process Intensif.* **2009**, *48* (1), 329–332. 10.1016/j.cep.2008.04.012.
[7] Lutze, P.; Gani, R.; Woodley, J. M. Process Intensification: A Perspective on Process Synthesis. *Chem. Eng. Process. Process Intensif.* **2010**, *49* (6), 547–558. 10.1016/j.cep.2010.05.002.
[8] Portha, J. F.; Falk, L.; Commenge, J. M. Local and Global Process Intensification. *Chem. Eng. Process. Process Intensif.* **2014**, *84*, 1–13. 10.1016/j.cep.2014.05.002.
[9] Reay, D.; Ramshaw, C.; Harvey, A. Process Intensification – An Overview. *Process Intensif.* **2013**, 27–55. 10.1016/b978-0-08-098304-2.00002-x.

[10] Baldea, M. From Process Integration to Process Intensification. *Comput. Chem. Eng.* **2015**, *81*, 104–114. 10.1016/j.compchemeng.2015.03.011.

[11] Van Gerven, T.; Stankiewicz, A. Structure, Energy, Synergy, Time-the Fundamentals of Process Intensification. *Ind. Eng. Chem. Res.* **2009**, *48* (5), 2465–2474. 10.1021/ie801501y.

[12] Boodhoo, K.; Harvey, A. *Process Intensification Technologies for Green Chemistry: Engineering Solutions for Sustainable Chemical Processing;* John Wiley & Sons, 2013.

[13] Grossmann, I. E.; Westerberg, A. W. Research Challenges in Process Systems Engineering. *AIChE J.* **2000**, *46* (9), 1700–1703. 10.1002/aic.690460902.

[14] Kiss, A. A. *Process Intensification Technologies for Biodiesel Production; SpringerBriefs in Applied Sciences and Technology;* Springer International Publishing: Cham, 2014. 10.1007/978–3-319-03554-3.

[15] About Flow Chemistry – ThalesNano https://thalesnano.com/applications/about-flow-chemistry/ (accessed Sep 27, 2020).

[16] Kataoka, S.; Endo, A.; Harada, A.; Inagi, Y.; Ohmori, T. Characterization of Mesoporous Catalyst Supports on Microreactor Walls. *Appl. Catal. A. Gen.* **2008**, *342* (1–2), 107–112. 10.1016/j. apcata.2008.03.011.

[17] Microfluidics – Your Clear Partner In Microfluidics Manufacturing https://microfluidics.technicolor. com/ (accessed Sep 28, 2020).

[18] Regatte, V. R.; Kaisare, N. S. Propane Combustion in Non-Adiabatic Microreactors: 1. Comparison of Channel and Posted Catalytic Inserts. *Chem. Eng. Sci.* **2011**, *66* (6), 1123–1131. 10.1016/j. ces.2010.12.017.

[19] Mason, B. P.; Price, K. E.; Steinbacher, J. L.; Bogdan, A. R.; McQuade, D. T. Greener Approaches to Organic Synthesis Using Microreactor Technology. *Chem. Rev.* **2007**, *107* (6), 2300–2318. 10.1021/ cr050944c.

[20] Cukalovic, A.; Monbaliu, J.-C. M. R.; Stevens, C. V. Microreactor Technology as an Efficient Tool for Multicomponent Reactions. In *Jurnal Online Mahasiswa (JOM) Bidang Pertanian;* **2010**; Vol. 03, pp 161–198. 10.1007/7081_2009_22.

[21] Kockmann, N.; Gottsponer, M.; Roberge, D. M. Scale-up Concept of Single-Channel Microreactors from Process Development to Industrial Production. *Chem. Eng. J.* **2011**, *167* (2–3), 718–726. 10.1016/j. cej.2010.08.089.

[22] Polymerization Plug Flow Reactors for Viscous Processing – StaMixCo Static Mixer Products & Technology http://www.stamixco-usa.com/plug-flow-reactors (accessed Sep 28, 2020).

[23] Taghavi-Moghadam, S.; Kleemann, A.; Golbig, K. G. Microreaction Technology as a Novel Approach to Drug Design, Process Development and Reliability. *Org. Process Res. Dev.* **2001**, *5* (6), 652–658. 10.1021/op010066u.

[24] Fukuyama, T.; Shinmen, M.; Nishitani, S.; Sato, M.; Ryu, I. A Copper-Free Sonogashira Coupling Reaction in Ionic Liquids and Its Application to A Microflow System for Efficient Catalyst Recycling. *Org. Lett.* **2002**, *4* (10), 1691–1694. 10.1021/ol0257732.

[25] Industrial Batch Mixers | Batch Mixer https://www.silverson.com/us/products/batch-mixers/ (accessed May 8, 2020).

[26] Qammar, H.; Gładyszewski, K.; Górak, A.; Skiborowski, M. Towards the Development of Advanced Packing Design for Distillation in Rotating Packed Beds. *Chem. Ing. Tech.* **2019**, *91* (11), 1663–1673. 10.1002/cite.201900053.

[27] Zhao, H.; Shao, L.; Chen, J. F. High-Gravity Process Intensification Technology and Application. *Chem. Eng. J.* **2010**, *156* (3), 588–593. 10.1016/j.cej.2009.04.053.

[28] Pillejera, M. K. V.; Chen, W. H.; De Luna, M. D. G. Bamboo Torrefaction in a High Gravity (Higee) Environment Using a Rotating Packed Bed. *ACS Sustain. Chem. Eng.* **2017**, *5* (8), 7052–7062. 10.1021/ acssuschemeng.7b01264.

[29] Kim, Y.; Park, L. K.; Yiacoumi, S.; Tsouris, C. Modular Chemical Process Intensification: A Review. *Annu. Rev. Chem. Biomol. Eng.* **2017**, *8* (1), 359–380. 10.1146/annurev-chembioeng-060816-101354.

[30] Reay, D.; Ramshaw, C.; Harvey, A. Intensification of Separation Processes. In *Process Intensification;* Elsevier, 2013; pp 205–249. 10.1016/B978-0-08-098304-2.00006-7.

[31] *Sustainable Development in Chemical Engineering Innovative Technologies;* Piemonte, V., De Falco, M., Basile, A., Eds.; John Wiley & Sons, Ltd: Chichester, UK; 2013. https://doi.org/10.1002/9781118629703.

[32] Shuit, S. H.; Ong, Y. T.; Lee, K. T.; Subhash, B.; Tan, S. H. Membrane Technology as A Promising Alternative in Biodiesel Production: A Review. *Biotechnol. Adv.* **2012**, *30* (6), 1364–1380. 10.1016/j. biotechadv.2012.02.009.

[33] Radcliffe, A. J.; Singh, R. P.; Berchtold, K. A.; Lima, F. V. Modeling and Optimization of High-Performance Polymer Membrane Reactor Systems for Water-Gas Shift Reaction Applications. *Processes.* **2016**, *4* (2), 1–19. 10.3390/pr4020008.

[34] Constantinou, A.; Wu, G.; Venezia, B.; Ellis, P.; Kuhn, S.; Gavriilidis, A. Aerobic Oxidation of Benzyl Alcohol in a Continuous Catalytic Membrane Reactor. *Top. Catal.* **2019**, *62* (17–20), 1126–1131. 10.1007/s11244-018-1060-9.

[35] Membrane distillation | EMIS https://emis.vito.be/en/bat/tools-overview/sheets/membrane-distillation (accessed Oct 1, 2020).

[36] Schwantes, R.; Seger, J.; Bauer, L.; Winter, D.; Hogen, T.; Koschikowski, J.; Geißen, S. U. Characterization and Assessment of a Novel Plate and Frame Md Module for Single Pass Wastewater Concentration–Feed Gap Air Gap Membrane Distillation. *Membranes (Basel).* **2019**, *9* (9). 10.3390/membranes9090118.

[37] Bonrath, W. Ultrasound Supported Catalysis. *Ultrason. Sonochem.* **2005**, *12* (1–2), 103–106. 10.1016/j. ultsonch.2004.03.008.

[38] Ebrahimi, S. L.; Khosravi-Nikou, M. R.; Hashemabadi, S. H. Sonoreactor Optimization for Ultrasound Assisted Oxidative Desulfurization of Liquid Hydrocarbon. *Pet. Sci. Technol.* **2018**, *36* (13), 959–965. 10.1080/10916466.2018.1458112.

[39] Kuna, E.; Behling, R.; Valange, S.; Chatel, G.; Colmenares, J. C. Sonocatalysis: A Potential Sustainable Pathway for the Valorization of Lignocellulosic Biomass and Derivatives. *Top. Curr. Chem.* **2017**, *375* (2), 41. 10.1007/s41061-017-0122-y.

[40] Boniek, D.; Figueiredo, D.; Dos Santos, A. F. B.; De Resende Stoianoff, M. A. Biodesulfurization: A Mini Review about the Immediate Search for the Future Technology. *Clean Technol. Environ. Policy* **2015**, *17* (1), 29–37. 10.1007/s10098-014-0812-x.

[41] Mosier, N.; Wyman, C.; Dale, B.; Elander, R.; Lee, Y. Y.; Holtzapple, M.; Ladisch, M. Features of Promising Technologies for Pretreatment of Lignocellulosic Biomass. *Bioresour. Technol.* **2005**, *96* (6), 673–686. 10.1016/j.biortech.2004.06.025.

[42] Silveira, M. H. L.; Morais, A. R. C.; Da Costa Lopes A. M.; Olekszyszen, D. N.; Bogel-Łukasik, R.; Andreaus, J.; Pereira Ramos, L. Current Pretreatment Technologies for the Development of Cellulosic Ethanol and Biorefineries. *ChemSusChem* **2015**, *8* (20), 3366–3390. 10.1002/cssc.201500282.

[43] Kumar, P.; Barrett, D. M.; Delwiche, M. J.; Stroeve, P. Methods for Pretreatment of Lignocellulosic Biomass for Efficient Hydrolysis and Biofuel Production. *Ind. Eng. Chem. Res.* **2009**, *48* (8), 3713–3729. 10.1021/ie801542g.

[44] Harmsen, P. F. H.; Huijgen, W.; Bermudez, L.; Bakker, R. *Literature Review of Physical and Chemical Pretreatment Processes for Lignocellulosic Biomass;* Report/Wageningen UR, Food & Biobased Research: 1184; Wageningen UR – Food & Biobased Research: 712, FBR BP Biorefinery & Natural Fibre Technology, 2010.

[45] Da Costa Sousa, L.; Chundawat, S. P.; Balan, V.; Dale, B. E. 'Cradle-to-grave' Assessment of Existing Lignocellulose Pretreatment Technologies. *Curr. Opin. Biotechnol.* **2009**, *20* (3), 339–347. 10.1016/j. copbio.2009.05.003.

[46] Luo, J.; Fang, Z.; Smith, R. L. Ultrasound-Enhanced Conversion of Biomass to Biofuels. *Prog. Energy Combust. Sci.* **2014**, *41* (1), 56–93. 10.1016/j.pecs.2013.11.001.

[47]   Yunus, R.; Salleh, S. F.; Abdullah, N.; Biak, D. R. A. Effect of Ultrasonic Pre-Treatment on Low
       Temperature Acid Hydrolysis of Oil Palm Empty Fruit Bunch. *Bioresour. Technol.* **2010**, *101* (24),
       9792–9796. 10.1016/j.biortech.2010.07.074.

[48]   Yang, F.; Li, L.; Li, Q.; Tan, W.; Liu, W.; Xian, M. Enhancement of Enzymatic in Situ Saccharification of
       Cellulose in Aqueous-Ionic Liquid Media by Ultrasonic Intensification. *Carbohydr. Polym.* **2010**, *81* (2),
       311–316. 10.1016/j.carbpol.2010.02.031.

[49]   Horikoshi, S.; Serpone, N. General Introduction to Microwave Chemistry. In *Microwaves in Catalysis*;
       Wiley-VCH Verlag GmbH & Co. KGaA: Weinheim, Germany, 2015; pp 1–28. 10.1002/
       9783527688111.ch1.

[50]   Microwave Heating – Mechanism and Theory https://cem.com/en/microwave-heating-mechanism-
       and-theory (accessed May 9, 2020).

[51]   Sun, J.; Wang, W.; Yue, Q. Review on Microwave-Matter Interaction Fundamentals and Efficient
       Microwave-Associated Heating Strategies. *Materials (Basel)* **2016**, *9* (4). 10.3390/ma9040231.

[52]   Gawande, M. B.; Shelke, S. N.; Zboril, R.; Varma, R. S. Microwave-Assisted Chemistry: Synthetic
       Applications for Rapid Assembly of Nanomaterials and Organics. *Acc. Chem. Res.* **2014**, *47* (4),
       1338–1348. 10.1021/ar400309b.

[53]   Gude, V. G.; Patil, P.; Deng, S. Microwave Energy Potential for Large Scale Biodiesel Production.
       *World Renew. Energy Forum, WREF 2012 Incl. World Renew. Energy Congr. XII Color. Renew. Energy Soc.
       Annu. Conf.* **2012**, *1* (c), 751–759.

[54]   Sharma, A. K.; Gowdahalli, K.; Krzeminski, J.; Amin, S. Microwave-Assisted Suzuki Cross-Coupling
       Reaction, a Key Step in the Synthesis of Polycyclic Aromatic Hydrocarbons and Their Metabolites.
       J. Org. Chem. **2007**, *72* (23), 8987–8989. 10.1021/jo701665j.

[55]   Tsubaki, S.; Nakasako, Y.; Ohara, N.; Nishioka, M.; Fujii, S.; Wada, Y. Ultra-Fast Pyrolysis of
       Lignocellulose Using Highly Tuned Microwaves: Synergistic Effect of a Cylindrical Cavity Resonator
       and a Frequency-Auto-Tracking Solid-State Microwave Generator. *Green Chem.* **2020**, *22* (2), 342–351.
       10.1039/c9gc02745a.

[56]   Vinitha, U. G.; Sathasivam, R.; Muthuraman, M. S.; Park, S. U. Intensification of Supercritical Fluid in
       the Extraction of Flavonoids: A Comprehensive Review. *Physiol. Mol. Plant Pathol.* **2022**, *118*, 101815.
       10.1016/j.pmpp.2022.101815.

[57]   Weidner, E. Impregnation via Supercritical CO2–What We Know and What We Need to Know.
       J. Supercrit. Fluids. **2018**, *134*, 220–227. 10.1016/j.supflu.2017.12.024.

[58]   Krzysztoforski, J.; Jenny, P.; Henczka, M. Mass Transfer Intensification in the Process of Membrane
       Cleaning Using Supercritical Fluids. *Theor. Found. Chem. Eng.* **2016**, *50* (6), 907–913. 10.1134/
       S0040579516060099.

[59]   Campos Domínguez, C.; Gamse, T. Process Intensification by the Use of Micro Devices for Liquid
       Fractionation with Supercritical Carbon Dioxide. *Chem. Eng. Res. Des.* **2016**, *108*, 139–145. 10.1016/j.
       cherd.2016.01.011.

[60]   Zabot, G. L.; Meireles, M. A. A. On-Line Process for Pressurized Ethanol Extraction of Onion Peels
       Extract and Particle Formation Using Supercritical Antisolvent. *J. Supercrit. Fluids.* **2016**, *110*, 230–239.
       10.1016/j.supflu.2015.11.024.

[61]   Yousefzadeh, H.; Akgün, I. S.; Barim, S. B.; Sari, T. B.; Eris, G.; Uzunlar, E.; Bozbag, S. E.; Erkey,
       C. Supercritical Fluid Reactive Deposition: A Process Intensification Technique for Synthesis of
       Nanostructured Materials. *Chem. Eng. Process. – Process Intensif.* **2022**, *176*, 108934. 10.1016/j.
       cep.2022.108934.

[62]   Baldino, L.; Della Porta, G.; Reverchon, E. Supercritical CO 2 Processing Strategies for Pyrethrins
       Selective Extraction. *J. CO2 Util.* **2017**, *20*, 14–19. 10.1016/j.jcou.2017.04.012.

[63]   Yousefi, M.; Rahimi-Nasrabadi, M.; Pourmortazavi, S. M.; Wysokowski, M.; Jesionowski, T.; Ehrlich, H.;
       Mirsadeghi, S. Supercritical Fluid Extraction of Essential Oils. *TrAC Trends Anal. Chem.* **2019**, *118*,
       182–193. 10.1016/j.trac.2019.05.038.

[64] Burger, P.; Plainfossé, H.; Brochet, X.; Chemat, F.; Fernandez, X. Extraction of Natural Fragrance Ingredients: History Overview and Future Trends. *Chem. Biodivers.* **2019**, *16* (10). 10.1002/cbdv.201900424.

[65] Luca, S. V.; Kittl, T.; Minceva, M. Supercritical CO2 Extraction of Spices: A Systematic Study with Focus on Terpenes and Piperamides from Black Pepper (Piper Nigrum L.). *Food Chem.* **2023**, *406*, 135090. 10.1016/j.foodchem.2022.135090.

[66] Dean, J. R.; Liu, B.; Ludkin, E. Supercritical Fluid Extraction of Caffeine from Instant Coffee. In *Supercritical Fluid Methods and Protocols*; Humana Press: New Jersey, 2000; pp 17–22. 10.1385/1-59259-030-6:17.

[67] Knez, Ž.; Markočič, E.; Leitgeb, M.; Primožič, M.; Knez Hrnčič, M.; Škerget, M. Industrial Applications of Supercritical Fluids: A Review. *Energy.* **2014**, *77*, 235–243. 10.1016/j.energy.2014.07.044.

[68] Chai, Y. H.; Yusup, S.; Kadir, W. N. A.; Wong, C. Y.; Rosli, S. S.; Ruslan, M. S. H.; Chin, B. L. F.; Yiin, C. L. Valorization of Tropical Biomass Waste by Supercritical Fluid Extraction Technology. *Sustainability.* **2020**, *13* (1), 233. 10.3390/su13010233.

[69] Brunner, G. Applications of Supercritical Fluids. *Annu. Rev. Chem. Biomol. Eng.* **2010**, *1* (1), 321–342. 10.1146/annurev-chembioeng-073009-101311.

[70] Haase, S.; Tolvanen, P.; Russo, V. Process Intensification in Chemical Reaction Engineering. *Processes.* **2022**, *10* (1), 99. 10.3390/pr10010099.

[71] Stolte, J.; Özkan, L.; Thüne, P. C.; Niemantsverdriet, J. W.; Backx, A. C. P. M. Pulsed Activation in Heterogeneous Catalysis. *Appl. Therm. Eng.* **2013**, *57* (1–2), 180–187. 10.1016/j.applthermaleng.2012.06.035.

[72] Liu, J.; Gao, L.; Ren, J.; Liu, W.; Liu, X.; Sun, L. Dynamic Process Intensification of Dimethyl Ether Reactive Distillation Based on Output Multiplicity. *Ind. Eng. Chem. Res.* **2020**, *59* (45), 20155–20167. 10.1021/acs.iecr.0c03973.

6 Continuous microflow processes

Gergo Ignacz, Gyorgy Szekely

6.1 Introduction

Reactor configurations have not changed significantly since scientists first started performing chemical reactions. As long ago as 50 BC, Getafix, the famous druid from Asterix and Obelix, cooked his magical strength potion in a large cauldron. Such cauldrons had the advantage of flexibility, allowing them to cook different magical potions on demand, such as potions for strength and knowledge. When the Roman army attacked their village, Getafix realized the villagers needed a large volume of strength potion to resist the Roman army. However, they soon found they needed another cauldron to meet demand. If only they had an innovative technology that could continuously produce the potion, they could have averted disaster (Figure 6.1) [1]. Unfortunately for Getafix and the villagers, they would have to wait for another 2000 years for the invention of the *continuous flow reactor*. The first mention of a continuous microfluidic device (a chip-based gas chromatograph) dates back to 1937 [2]. Manz and Wolley developed these reactors significantly in the 1990s [3, 4]. These early studies were mostly focused on capillary technologies for intended use in analytics. The widespread realization of small-scale reactors in chemistry started to emerge in the early 2000s. Follow the QR code on this page to learn about the realization of the potential benefits of the microflow technology presented by Frank Gupton, one of the pioneers of microflow chemistry.

A microfluidic device usually has a diameter of less than 0.5 mm. More precisely, for a system to be considered microfluidic, the Reynolds and Péclet numbers must be below 250 and 1000, respectively.[1] As a rule of thumb, the smaller the tubing diameter (or capillaries), the easier it is to maintain the *microfluidic regime*, and thus a higher flow rate can be used. The large surface-to-volume ratio allows better heat transfer or higher electromagnetic and acoustic wave flux, resulting in a more uniform reaction composition.

The transition from batch to continuous flow reactors is still ongoing today, requiring careful research and development. The upcycle of biomass-derived chemicals to commodity or high-added value compounds using new continuous flow process concepts could change biorefineries' future perspectives [5]. Table 6.1 shows the differences between a typical batch reactor and a continuous flow reactor.

This chapter focuses on the sustainable aspects of continuous microflow reactors and processes. The examples presented are from the pharmaceutical sector because

---

1 The Reynolds number is a dimensionless number between the inertial and viscous forces. The Péclet number is a dimensionless number for heat transport due to convective and diffusive transport.

https://doi.org/10.1515/9783111028163-006

Figure 6.1: Getafix preparing a magic potion in continuous flow (left) and a batch cauldron (right). The continuous flow reactor can supply sufficient quantities when the demand is high, while the batch reactor can only periodically supply the potion with considerable downtime between batches. Artwork based on Alberto A. Uderzo's *Astérix le Gaulois*.

Table 6.1: Typical features of batch and continuous flow processes.

Batch reactors	Features	Continuous flow reactors
Formulations, pharmaceuticals, agrochemicals, specialty chemicals	Industry sectors	Fuels, bulk chemicals, intermediates
Multipurpose plants, multiproduct	Flexibility of use	Tailored to specific products
From cryogenic to 200 °C	Operating temperature	From cryogenic to 1000 °C
Up to 10 bar for over 10 m^3 vessels	Operating pressure	Up to 1000 bar
From minutes to hours	Residence time	From milliseconds to minutes
Poor due to the complex fluid dynamics and significant inhomogeneity in the flow	Heat transfer scalability	Good due to high surface-to-volume ratio and possible different mechanisms (staged injection, intercoolers)
Very poor due to complex fluid dynamics and significant inhomogeneity	Mixing	Can be well controlled through multiple available mechanisms, available mixing times in the milliseconds range

Table 6.1 (continued)

Batch reactors	Features	Continuous flow reactors
Molecules experience all concentrations and reaction times	Residence time distribution	May approach plug flow approximation of residence time distribution
Dynamic control over the reaction progress; scheduling	Control	Maintaining steady state, response to disturbance, and long-term dynamics (e.g., catalyst deactivation)
Typically not integrated	Process integration	Frequently heat and/or mass integrated in complex facilities
Liquid phase, slurries, gas-liquid, liquid-liquid, multiphase systems	Multiphase processes	Particle flow is frequently used in complex continuous flow systems; gas and liquid-liquid systems also known; particle flow is problematic at small scale

this sector generates the most waste per kg product and is also rapidly implementing continuous flow reactors. Refer to Chapter 2 for the E-factor comparison between various industries.

6.2 The advantages and disadvantages of continuous microfluidic systems

Small size. Size matters, especially in chemical reactors. A smaller size manifests in fewer precious reagents and catalysts used. Catalyst testing, DNA and protein sequencing, or even active pharmaceutical ingredient (API) building block synthesis testing greatly benefits from a microfluidic reactor's small size. Microfluidic reactors consume far fewer reagents than their batch counterparts. Due to the continuous operation mode on a large scale, the reactor and auxiliaries take up smaller space and are easier to relocate.

Higher selectivity. The improved temperature control of microflow reactors compared to batch reactors results in improved kinetic or thermodynamic control of the reaction, manifesting in higher selectivity but not higher conversion. Since the reaction rate should not depend on the reactor configuration, the overall conversion does not change across batch and flow unless a wall-mediated reaction mechanism is involved. Generally, in flow reactors, only higher selectivity can be achieved but not higher conversion [6].

Green attributes. Owing to the smaller size and better heat transfer, the amount of energy consumed in a microflow reactor is relatively small, leading to environmental and cost benefits in the long term [6]. Similarly, the higher selectivity results in higher purity, which requires less purification and thus generates less waste. Refer to

Section 6.3 for a more detailed explanation and examples of the green and sustainable aspects of microflow systems.

Rapid reactions. Microflow reactors enable faster reactions than conventional batch reactors [6]. However, this claim is somewhat misleading because the reaction rate does not depend on the reactor's geometry unless a wall-mediated mechanism is involved. This rule is true because the kinetics does not depend on bulk diffusion. The only exception to this rule is the mass transport-limited reactions, where the diffusive effects are significant for the overall rate [7]. However, directly comparing batch and flow processes is not straightforward because batch systems are rarely optimized to achieve the best space-time yield. Moreover, they are usually left to run above the equilibrium point to ensure that the reaction proceeds well. In contrast, microflow reactors must be optimized, resulting in a better space-time yield. One must be careful to consider all possible bias in the system when comparing batch versus flow – especially in the literature.

Owing to their inherently low residence time, microfluidic devices can achieve a higher conversion rate for reactions than batch reactors. For example, organometallic lithiation reactions, which are cumbersome to perform in batch, can achieve an 80% yield with a 0.002 s residence time. This short residence time is practically impossible to achieve with conventional batch technologies [8].

Safer reactions. The small footprint and internal volume of microfluidic reactors make them inherently safe. The effective volume needed for reactions involving toxic or dangerous intermediates is minimized, and thus they can be carried out safely (refer to Section 1.3). The high surface area-to-volume ratio allows for quick cooling, enabling the extremely exothermic reaction. Reaction quenching is also enhanced in microflow reactors compared to batch systems by introducing an inlet stream of a quenching agent after the reactor.

Facile scale-up and integration. Scale-up experiments in the batch operation mode are non-trivial due to the small flask's high surface area-to-volume ratio relative to the large reactor. This burden is significant, and therefore pharma manufacturers usually operate a dedicated department for scale-up procedures. In theory, a carefully optimized microflow reactor could be applied in production the same way as in the lab. Moreover, the parallelization of microflow reactors is simple and straightforward. Generally, batch processing requires the handling and transportation of intermediates, which is labor-intensive and increases exposure risk.[2] In theory, the integration of different reactions under flow conditions is more straightforward compared to batch reactions. Solvent exchange, particle formation, and gas generation in the batch mode are not problematic, yet it remains a challenge for flow processes.

---

2 The mentioned problem could be solved to a certain extent by telescoping reactions. The procedure is called one-pot synthesis. During one-pot synthesis, a few or even several reactions are performed in the same reactor without work-up.

6.3 The green attributes of continuous flow processes

In continuous flow processes, the reactor consists of a small diameter tube or a packed bed reactor. The small reactor volume has better heat transfer, and the residence time can be controlled more precisely than in conventional batch reactors. The advantages of this technology include increased safety, quality, and productivity and decreased space-time yield. The disadvantages include the high capital cost of investment and resources [9]. The inherent safety of continuous flow technology is one of its most attractive aspects for pharmaceutical companies. For example, Eli Lilly developed a continuous hydrogenation process where the evacetrapib API was produced on a two metric ton scale. Using continuous flow technology, the amount of hydrogen in the system was reduced by 98% at a time, resulting in a safer, low-risk process [10]. Note that hydrogenation is considered a high-risk process due to the broad explosive limit of hydrogen gas (4–75%).

Another benefit of continuous flow processes is the *increased quality* resulting from the more precise heat transfer, temperature control, and mixing. The lower residence times result in fewer side reactions, which also increase the yield and purity. Solvent consumption and the E-factor are relatively low due to the decrease in downstream processing requirements needed to reach a specific quality product. Flucytosine (an antifungal medicine) is a good example to demonstrate the effect of less side-product formation. In the original process, the API was formed in four steps, including a step with $POCl_3$ and NH_3, which are both highly toxic substances. The implementation of continuous flow technology resulted in higher overall yield and significantly less waste [11]. The precise control over the reaction increased the selectivity for the monofluorinated product. In the example of atropine production, the E-factor was reduced by 99% from 2245 to 24 kg kg^{-1} by minimizing by-product formation and introducing an inline liquid-liquid extraction module [12].

Telescoping several steps into one process, i.e., by omitting purification and isolation-focused unit operations between reactions, leads to an increase in the processing speed, decreased resource intensity, and an increase in the overall sustainability of the process. The rapid reaction interrogation and optimization through process analytical technology (PAT) (refer to Chapter 9) speeds up a chemical product's research and development stage and improves asset utilization. The transfer of continuous flow processes from the laboratory scale to the pilot scale is straightforward and requires a relatively small laboratory footprint. Since continuous flow processes operate at a steady state, the system can eventually run with regular planned maintenance or cleaning. The steady-state operating condition brings another advantage: continuous processes are not as labor-intensive as batch systems, and they can be easily performed remotely. These core features are visualized in Figure 6.2, which shows the correlation between green chemistry and green engineering, highlighting the impact of flow chemistry on both.

Green Chemistry
Safer Chemistry for Accident Prevention
Real-time Analysis for Pollution Prevention

Design for Energy Efficiency

Atom Economy
Prevention

Green Engineering
Prevention Instead of Treatment
Design for Separation
Maximize efficiency

Meet Need, Minimize Excess

Integrate Material and Energy Flows

Figure 6.2: The impact of continuous flow processes on the 12 principles of green chemistry and the 12 principles of green engineering.

6.3.1 Principle 1: Prevention

The historical take-make-dispose model's sole focus was to increase profit margins by increasing the yield and production rate, resulting in high waste generation. Recent changes in legislation have pushed companies toward a more sustainable industry [13]. Prevention in microflow-operated systems can be divided into two main applications: (i) prevention by solvent-free reactions and (ii) prevention by higher efficiency and minimal reagent consumption.

6.3.1.1 Prevention by solvent-free reactions
The easiest way to prevent the generation of solvent waste is to avoid using any solvent, which is the most desirable approach in any process design. The removal of solvent(s) from the reaction is considered a process intensification method, which results in an improved process mass intensity (PMI) and lower complete E-factor. Solvent-free reactions in the batch configuration are challenging to achieve due to the management of the reactor's uncontrollable heat. Refer to the worked example in Chapter 15, which demonstrates that solvent consumption contributes significantly to waste generation, resulting in a high E-factor.

   In theory, all reactions where the starting materials are liquid or gas at the reaction temperature can be performed in a microflow system. Although a complete *solvent-free system* is challenging to achieve, if the starting materials are solid, heating the system to supply the melted chemicals is possible but cumbersome. An interesting approach is to perform reactions in *extruder machines*. Extruders are mainly used in the polymer industry to transport the melted polymers to the form. Extruders are designed to pump highly viscous substances at high temperatures. Special twin-type extruders are designed to help mix the reagents. One early example showed that extruders could be used as reactors [14]. For example, one study showed that sustainable starch-based graft co-polymers could be synthesized using reactive extrusion without any solvent [15]. The method can be employed on a large scale, and the resulting polymers are excellent packaging materials.

Another example is the large-scale reduction of aromatic aldehydes to alcohols using a water-soluble green reagent, such as sodium borohydride [16]. Since aromatic aldehydes and alcohols are poorly soluble in water, the reaction is performed in a suspension (Figure 6.3). To overcome this solubility problem, conventional methods use solvent mixtures, such as tetrahydrofuran-water or dioxane-water, in different compositions to achieve acceptable solubility for the aldehyde, the alcohol, the sodium borohydride, and the sodium hydroxide. Solvent mixtures are always undesired due to their cumbersome regeneration procedures. For example, tetrahydrofuran forms a minimum azeotropic mixture containing 6.4% water at 64 °C. Slurry-type reactions are a viable option in batch operations but not in continuous flows due to possible clogging issues.

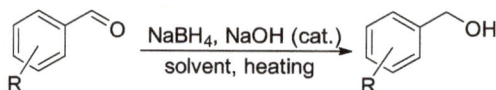

Figure 6.3: General reaction scheme of the reduction of aromatic aldehydes to alcohols. Sodium hydroxide is used as a catalyst.

A reactive extrusion can be used to perform the reactions on a slurry mixture to overcome the limitations mentioned above. Reactive extrusion is a type of extrusion where the extruder also acts as a continuous reactor. A schematic representation of a system to reduce aldehydes to alcohols is shown in Figure 6.4. Although the configuration is not a microflow system – since the extruder has larger cavities than 1 mm – the method presents an excellent example of how solids and slurries can be handled under continuous flow.

The aromatic alcohols were synthesized on a kg h^{-1} scale. The material cost decreased from 5.5 to 3.5 ton material per ton product by switching from the batch to the continuous operation mode. Similarly, energy consumption decreased from 9600 to 300 MJ ton^{-1}. The nearly 97% decrease in energy consumption was due to the more efficient heat utilization of the extruder than the batch reactor (heating and cooling).

Ionic liquid and *deep eutectic solvent* production (refer to Chapter 3 for details) may also benefit from continuous microflow systems. Conventional batch synthesis may suffer from cumbersome purification of the final solvent to achieve the desired quality. Highly controlled microflow systems can increase the yield and purity, eventually leading to lower production costs and less energy and material utilization. For instance, [EMIM][EtSO$_4$] ionic liquid was synthesized in a microflow reactor at 0.5 kg h^{-1} under completely solvent-free conditions. The overall production efficiency was three orders of magnitude higher compared to the similar-sized batch reactor.

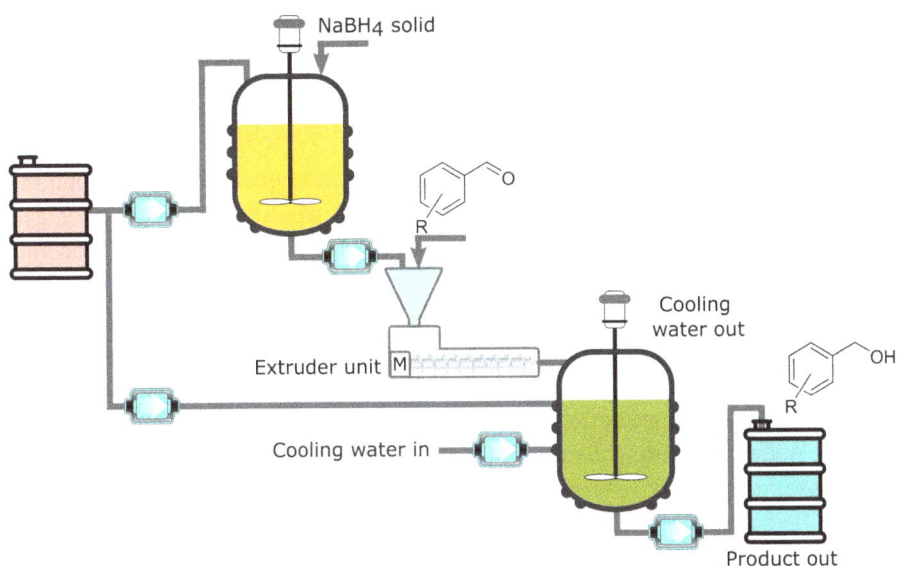

Figure 6.4: Schematic representation of the continuous flow reduction of aromatic aldehydes into alcohols.

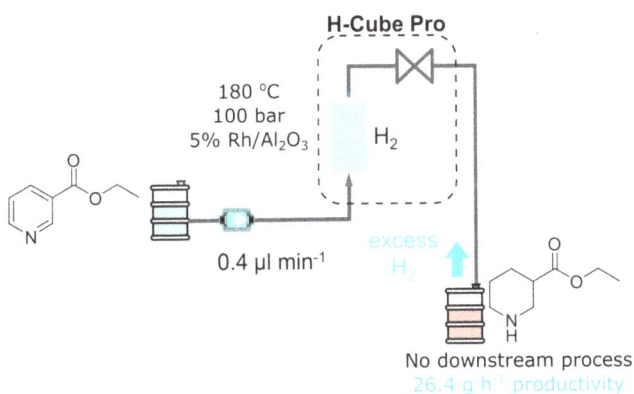

Figure 6.5: Schematic representation of the production of ethyl piperidine-3-carboxylate from ethyl nicotinate using hydrogenation under continuous flow conditions. The same reaction setup is used to obtain other hydrogenation products.

Liquid-gas reactions can also be performed under solvent-free conditions. For example, ethyl piperidine-3-carboxylate is an API building block used in the pharmaceutical industry. It is synthesized from ethyl nicotinate using activated Rh catalysts on Al_2O_3 particles under high pressure and temperature (Figure 6.5). Ethyl nicotinate is supplied from a natural source (tobacco), hydrogen is generated from water splitting, and the electricity is obtained from a renewable source, thus resulting in an overall sustainable system [17].

6.3.1.2 Minimal reagent consumption

Microflow reactors can also minimize waste in the production scale and development and laboratory scale. While batch *scouting reactions*[3] typically require tens of mg of reagents and mL of solvents, these reactions can be performed in microflow reactions using only a few micrograms of starting materials in a few μL of solvent. The ability to connect microflow reactors to inline analytical techniques (refer to Chapter 9) rapidly changes the reaction conditions, and incorporating automatic optimization algorithms allows efficient waste minimization. The integration of PAT eliminates the need for downstream processing in scouting reactions (work-up, isolation, purification, and analysis), lowering the reactions' solvent stress.

For example, a Heck reaction was optimized via feedback control in a microflow reactor. The control unit changed the parameters according to the system's actual conversion, resulting in the rapid optimization of the reaction conditions. The automated system generated 50 times (!) less waste than the conventional manual batch optimization while achieving similar or better final conversion and product purity [18].

6.3.2 Principle 2: Atom economy

Because the atom economy (AE) is inherently based on the reaction type and not on the conditions, controlling the reaction conditions (temperature, pressure, residence time, concentration, and the equivalent of reagents) does not affect the AE. How can we improve AE in microflow reactors and take advantage of the higher conversion and purity? To answer this, here we present a relevant case study on the application of diazomethane chemistry. Refer to the QR code on this page to learn *how not to work with diazomethane* by Derek Lowe.

Diazomethane is a versatile reagent, but its highly reactive nature makes it a toxic and explosive gas. Diazomethane explodes upon contact with any sharp or uneven surface, even under mild conditions such as in a solution at room temperature. It is one of the most useful reagents in the fine chemical industry because it can participate in a wide variety of reactions (Figure 6.6). The transport and storage of diazomethane are dangerous. It is usually prepared on-demand from its precursor; however, the risk of in-reaction

---

3 Small-scale scouting reactions are performed to explore the chemistry of unknown reaction paths. Several thousand reactions can be performed and worked up to determine which reaction path could lead to the best overall yield or selectivity. Scouting reactions are not optimized nor performed in high numbers, meaning their E-factor is extremely high.

explosion is still high. Unfortunately, even pilot-scale production of diazomethane is problematic, and therefore large-scale industrial alternative methods are required.

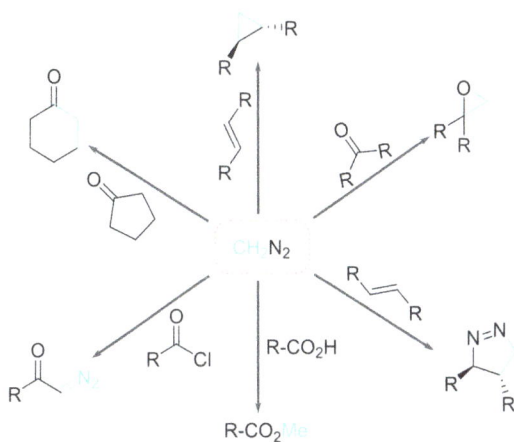

Figure 6.6: Chemical transformations using diazomethane (CH₂N₂).

Figure 6.7 illustrates the difference between the AE of a diazomethane reaction and the AE of an alternative, hypothetical route. The first reaction (Figure 6.7a) uses diazomethane as the homologation agent in the reaction with isatin (147.3 g mol^{-1}). The second reaction (Figure 6.7b) uses ethyl 2-diazoacetate (114.1 g mol^{-1}) to reach the final product (161.16 g mol^{-1}). While path (a) is only one step, path (b) requires four consecutive steps to reach the same product, where the different intermediate steps must be isolated, analyzed, and purified. The corresponding AE values for the reactions shown in Figure 6.7a and Figure 6.7b are calculated in eqs. (6.1) and (6.2), respectively:

$$AE_a = \frac{161.16}{147.13 + 42.04} \cdot 100 = 85\%, \tag{6.1}$$

$$AE_b = \frac{161.16}{147.13 + 114.10} \cdot 100 = 62\%. \tag{6.2}$$

The AE of the alternative route is more than 20% lower than that of the reaction path involving diazomethane. Note that we consider the AE only, and other factors, such as E-factor or PMI, are not considered. Refer to Chapter 15 for a worked example on this problem. Because pure diazomethane cannot be directly used in the example in Figure 6.7a, route Figure 6.7b is preferable on a large scale using a batch reactor, which ultimately results in a safer, but less efficient, multistep process.

Diazomethane can be produced from its precursors, such as from Diazald, for example, which is a relatively stable compound often used for the *in situ* generation of diazomethane in the presence of a base. Figure 6.8 shows a tube-in-tube reactor's possible implementations to generate diazomethane and immediately react it with the sub-

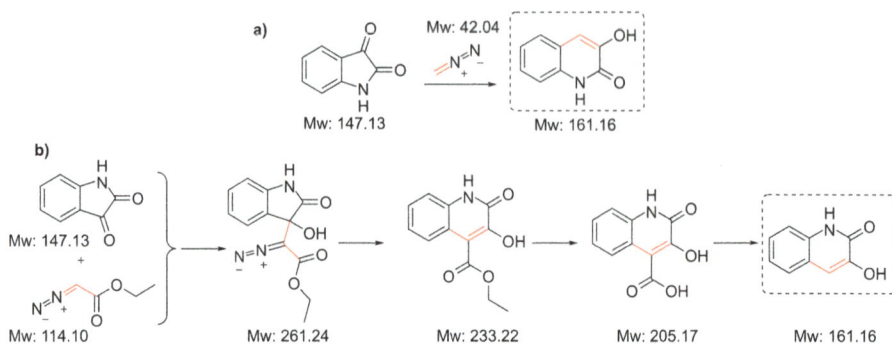

Figure 6.7: Diazomethane and atom efficiency. (a) Reaction path involving diazomethane. (b) Reaction path using an alternative route. Note that (b) is a hypothetical path.

strate. The diazomethane is generated in the inner tube (Figure 6.8b, Inner tube) and diffuses through a semi-permeable membrane into the substrate stream (Figure 6.8b, Outer tube). The membrane should only allow gases to diffuse through; therefore, a dry reaction (no water inside) can also be performed. The substrate is in the outer tube, where it reacts with the *in situ* generated diazomethane gas [19]. The flow rate in the inner and outer tubes can be fine-tuned to optimize the reaction and reduce the amount of Diazald needed. Using a 400 µL min^{-1} inlet flow rate, the final production varied between 170 and 360 mg h^{-1}, reaching 59–99% isolated yield. The basic aqueous waste was approx. 200 µL min^{-1}, containing unreacted Diazald, KOH, and potassium 4-methylbenzenesulfonate as another product of Diazald and KOH's reaction. Since the aqueous waste was still reactive, the reaction had to be quenched using acetic acid,

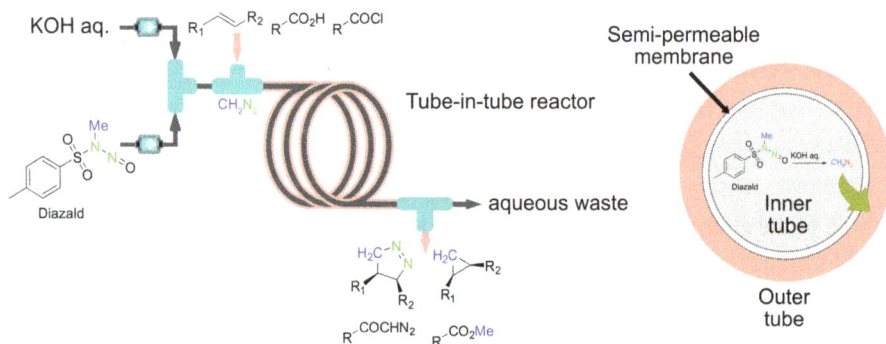

Figure 6.8: Diazomethane production using Diazald and aqueous KOH solution. The inner tube (blue) in the tube-in-tube reactor contains the diazomethane solution, and the outer tube (red) contains the reagent. The high vapor pressure of diazomethane comes in handy now! The diazomethane, generated *in situ* inside the inner tube, diffuses through a semi-permeable membrane into the outer tube, where it reacts with the reagent. Adapted from [19] .

which also reacted with the KOH and diazomethane. The organic phase (tetrahydrofuran) only contained the final product and some unreacted starting material.

To conclude, the tube-in-tube reactor technology made diazomethane chemistry safer and reduced the release of waste and toxic chemicals. This example demonstrates how to adopt green chemistry principles in terms of AE and prevention by using less hazardous chemical synthesis methods and inherently safer chemistry for accident prevention. In addition, this example is in line with green engineering efforts through the more efficient integration of energy and material flows.

6.3.3 Principle 6: Design for energy efficiency

The *energy efficiency* of microflow reactors originates from the improved heat transfer and even the reactor's temperature profile (see also Section 5.2.1). For example, an exothermic gas phase oxidation reaction can create a temperature difference of 133 °C in a 10 mm inner diameter tube reactor between the reactor's wall and center temperatures. When the reactor diameter is reduced to 1 mm, the temperature difference between the wall and the center decreases to 1.3 °C [20]. This temperature uniformity results in better heat, and thus energy, utilization.

Microflow reactors can also be transformed into potential PI systems with the ability to miniaturize complete chemical plants [21, 22]. The miniaturization of interconnected reactors, as small as a briefcase, has already been demonstrated [23, 24]. On-demand systems are specialized for a few API syntheses, which typically have a very short expiration date. The APIs' stable precursors can be stored, and the drug can be synthesized when needed, which provides enough drug supply for long expeditions both on Earth and during extraterrestrial or space exploration [25].

6.3.4 Principle 9: Catalysis

Microflow systems can also enhance the performance of certain types of catalytic reactions. For example, *photoredox catalytic reactions* are preferably performed in continuous microflow systems due to the high illuminated area-to-volume ratio. Also, photocatalysis itself is considered green because of the inherent sustainable nature of solar energy.

The main bottleneck of any type of electromagnetic wave-catalyzed reaction in batch operation is the low achievable radiated surface area [26]. Coil and microchip reactors – where the surface area-to-volume ratio is substantially higher than in batch – have many advantages in photocatalytic processes. The Lambert–Beer law defines the exponential decrease of transmittance from the light source within a medium by the following equation:

$$A = \varepsilon_i l C_i, \tag{6.3}$$

where ε_i is the absorptivity factor of species i, l is the studied length, and C_i is the concentration of species i. To illustrate this, take a standard 100 mL reactor made from a transparent glass material designed for a specific wavelength. Even if the reactor is homogeneously irradiated, the large radial distance from the wall to the reactor's center will absorb most of the light (Figure 6.9a). Because of the small internal distance in capillary microflow reactors, the light intensity further away from the surface (e.g., in the middle of the tube) has approximately the same value. Thus, the photon flux is more homogenous in a microflow reactor in the spatial distance, resulting in fewer side reactions and a higher overall conversion rate. The selectivity improves, while the reliance on the downstream separation and purification decreases. Consequently, photocatalytic processes performed in continuous flow reactors are more sustainable than the corresponding conventional batch processes. Two conventional but differently illuminated continuous flow systems are presented in Figure 6.9a.

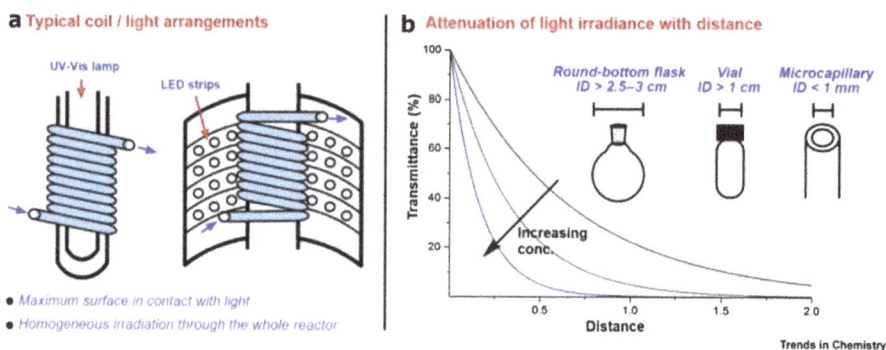

Figure 6.9: Continuous flow photoreactors. (a) Attenuation of light irradiance with distance from the light source. (b) Typical setups for flow photochemistry. Reprinted with permission from [27] .

Figure 6.9a shows a non-sensitized reaction mechanism in which the reaction is mediated using a photosensitive catalyst [28]. This field of photochemistry emerged in the last 12 years and has already reached the commercial scale. The uniqueness of photoredox catalysis is the mild redox character of the photocatalyst radical, which enhances the reaction's stereoselectivity. This paradigm shift in radical-type reactions coupled with microflow technology enables previously unattainable ("forbidden chemistry") transformations. Figure 6.10 shows microflow-based photoredox catalyst-mediated Stadler–Ziegler one-pot arylsulfide synthesis [29]. The conversion rate is high in all 14 examples, with ten of the examples above 70%, and the reaction was reported to be robust. Comparing the batch and flow productivity at similar conversions, the used batch reaction produced 0.17 mmol h^{-1}, while the flow reactor achieved a 13.2 mmol h^{-1}

conversion. Note that the direct comparison is valid here since both configurations are optimized to reach the same conversion.

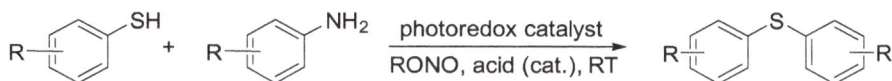

Figure 6.10: Photoredox microflow one-pot Stadler–Ziegler synthesis of arylsulfides.

Photoredox catalysis requires mild reaction conditions, desired from both economic and sustainability standpoints. The catalyst is usually an organometallic complex. The metals in these complexes are platinum, palladium, iridium, or other precious metals. Their recovery is not straightforward due to so-called photobleaching, i.e., the catalyst's degradation under light. Metal-free photoredox catalysts exist, but they are more prone to photobleaching and are not reusable. Recently, carbon nitride emerged as an efficient photocatalyst for certain types of reactions [30]. Carbon nitrides are stable against photodegradation, but they are practically insoluble in all solvents. A plug-flow microreactor system was invented to overcome this issue, which employs a heterogeneous, stable, reusable, and metal-free carbon nitride catalyst [31]. Figure 6.11 shows the structure of some photoredox catalysts.

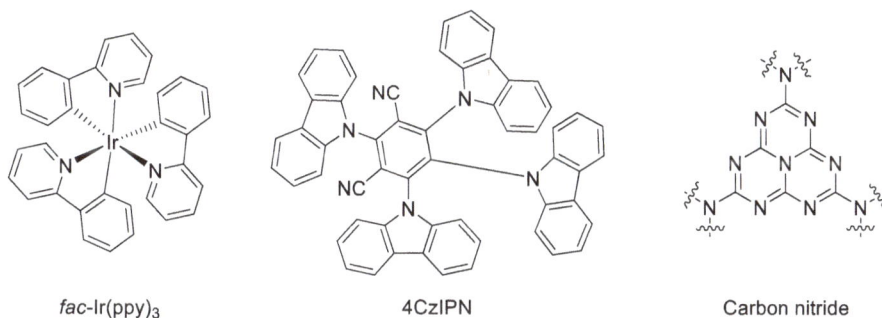

fac-Ir(ppy)₃   4CzIPN   Carbon nitride

Figure 6.11: Three different types of photoredox catalysts used in continuous microflow reactors: *fac*-Ir (ppy)₃ is an organometallic complex, 4CzIPN is a metal-free organocatalyst, and carbon nitride is a heterogeneous catalyst.

6.3.5 Principle 11: Real-time analysis for pollution prevention

PAT is discussed in detail in Chapter 9. Briefly, PATs are systems that can give real-time feedback on the reaction composition, conversion, or purity. PATs can be connected to a reactor or between reactors, and the real-time results can be analyzed or displayed using computers. The ability to monitor the reaction and solvent composition after every process step lowers the risk of accidental pollution.

6.3.6 Principle 12: Safer chemistry for accident prevention

The small effective reaction volume and the precisely controlled microflow reactors' parameters result in safer operating conditions than batch processes. The process mentioned in Section 6.3.2 is a perfect example of safely performing a generally dangerous reaction under continuous microflow conditions. Other examples include highly exothermic reactions, such as nitration, fluorination, hydrogenation, or metalation reactions.

For example, hydrogenation reactions are considered problematic due to the explosive nature of hydrogen. Storing a large volume of hydrogen in a reaction vessel containing an activated catalyst requires careful *deoxygenation* and inert gas flushing. The small volume of microfluidic reactors overcomes these issues. Several different systems have been developed for continuous flow hydrogenation. The two leading technologies are *inline gas mixing* and the tube-in-tube configuration. During inline gas mixing, the hydrogen enters the system before the reactor travels along with the reaction mixture. The tube-in-tube configuration (similar to the example above) uses a semi-permeable membrane tube to create contact between the two hydrogens and the liquid phase. The catalyst is usually a packed bed reactor filled with activated catalytic particles [32].

Many companies use hydrogen tanks to supply hydrogen to their systems; however, hydrogen storage has risks. ThalesNano developed a system to overcome this issue that does not rely on hydrogen transportation and storage (Figure 6.12) [33]. The hydrogen is generated *in situ* in the equipment via water-splitting. The hydrogen is purer than hydrogen from a commercial source due to the lack of hydrogen sulfide, a catalyst inhibitor. This method is considered environmentally friendly and sustainable because it does not rely on fossil fuel-based hydrogen (however, on a large scale, hydrogen is still produced from natural gas).

Reactors operating under extreme conditions enable researchers to develop reactions that are otherwise unattainable with conventional chemistry. Such systems are reactors that operate at extremely low or high temperature and pressure. Reactors operating under *supercritical conditions* are common examples. A pressurized and heated reactor is required to reach supercritical conditions for common solvents or carbon dioxide, which can be achieved in continuous reactors as well as in batch reactors. Several continuous methods utilizing supercritical fluids have been reported in the literature, and some have already reached the commercial scale [34]. For example, the hydrogenation of isophorone (used in the polymer industry to prepare polycarbonates) under supercritical conditions was performed by Thomas Swan & Co. as early as in 2003. Note that the reaction was not operating in the microflow regime.

The ring-opening reaction of phthalic anhydride is presented here as an example. Products derived from phthalic anhydride are used for manufacturing pharmaceuticals, plasticizers, and dyes. The reaction can be performed in a supercritical CO_2 and methanol mixture (1:1 mixture) in a continuous microflow reactor (Figure 6.13) [35].

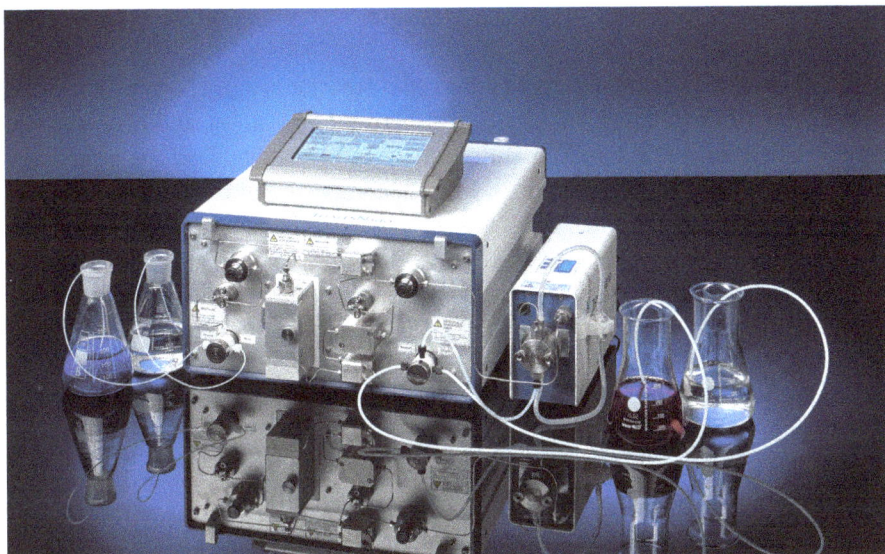

Figure 6.12: ThalesNano's H-Cube Pro. The system consists of a high-pressure pump, a hydrogen generator, a heated reactor with the catalyst inside, and a back-pressure regulator. The system can be operated at temperatures up to 150 °C and can supply hydrogen at pressures up to 100 bar.

The reactor volume is as low as 0.32 μL, where the pressure can reach 110 bar and the temperature 100 °C. For comparison, a typical water droplet has a volume of approx. 50 μL; thus, the reactor is more than 150 times smaller than a water droplet. Owing to the reactor's small size, the amount of supercritical CO_2 is also small and is generated *in situ*, which ensures safe operation.

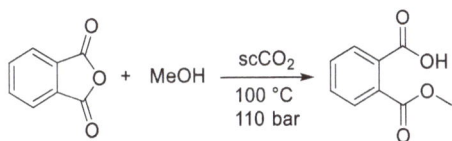

Figure 6.13: Ring-opening reaction of phthalic anhydride utilizing supercritical carbon dioxide in a microreactor.

Comparing the batch and flow operating conditions, the microflow reactor's implementation resulted in a 5400-fold increase in the reaction rate, which can be attributed to the added $scCO_2$ to the system, allowing a different reaction mechanism to occur. However, the flow rate was relatively low, in the range of 0.075 to 0.125 μL min^{-1}. Since the pressure and temperature can be controlled, and their impact on the system can be observed inline, microflow reactors have great potential to perform reactions under extreme conditions.

6.4 Microflow reactor systems

Table 6.2 shows the common parts of a microflow reactor setup. The system consists of syringe or HPLC pumps for material transport, tubing, valves, and connectors to connect the system and the reactor with heating or cooling modules. The postreactor part usually contains inline analytics purification systems and the product collector. However, some parts are interchangeable or can be omitted, and multiple reaction systems (telescopic reaction) can be repeated multiple times. This high versatility makes microflow reactors an appealing tool for organic and pharmaceutical chemists, who often encounter different reactions that can consecutively follow each other. However, there are non-trivial challenges. For example, solid particles could stop the flow and, eventually, the whole system. Product removal and inline purification are still not state-of-the-art technologies in microflow systems.

Table 6.2: General list of a simple one-step reaction configuration with inline purification and analytics.

Prereactor parts	
Injectors and pumps	Material transport and pressure
Tubing and connections	Connecting the inlet streams
Static mixers	Mixing inlet streams
Filter	Removing solid particles
Heating/cooling module	Providing heat
Reactor	Coil, packed bad, chip, rotating disc, etc.
Postreactor parts	
Backpressure regulator	Providing a specific pressure for the reactor
Inline purification system	
PAT	Chemical composition feedback
Product collector	

Figure 6.14 shows a schematic representation of a typical flow reactor.

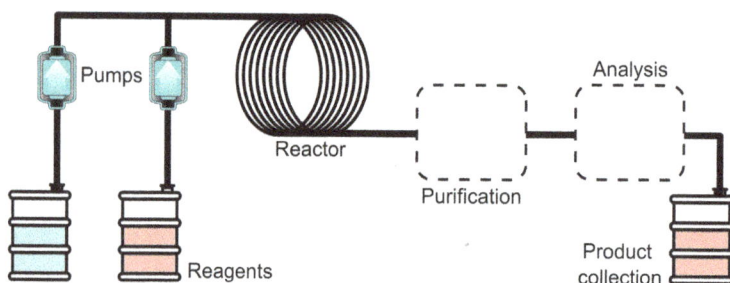

Figure 6.14: A typical flow reactor system. The reagent, pumps, heating blocks, and product collector are connected with tubing.

Several different manufacturers have developed flow reaction systems. The most common types of reactors are the tubular, packed bed, and microchip reactors. Despite the increasing number of commercial microflow reactors, the simple do-it-yourself reactor (DIYR) is still common. DIYR systems allow more configurations, easier maintenance, and lower cost. As a disadvantage, they are difficult to reproduce, and implementing a DIYR into a GMP-based pharmaceutical plant is complicated. To resolve this issue, Syrris, Vapourtec, and Labtrix offer different flow reactor configurations, which are well documented in the literature (Figure 6.15a, b, c). Another advantage is that they tend to be configurable and modular-based, therefore meeting any customer's needs. For example, Vapourtec has different specialty reactors, such as a photoreactor, an electrochemical reactor, and a microchip reactor and accessories, and they are all compatible with the central system. In addition, these systems allow telescoping reactions, which comes in handy for performing multistep reactions.

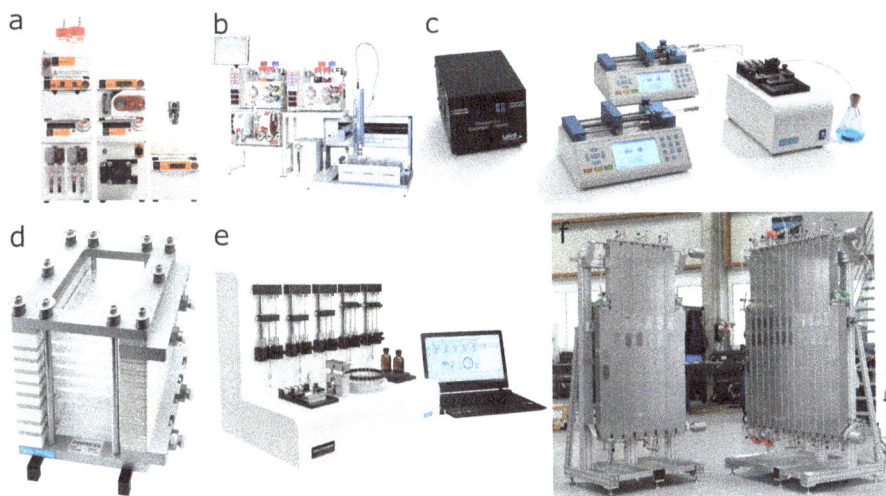

Figure 6.15: Laboratory-scale flow reactors (a, b, c) for research and development and industrial-scale flow reactors for production (d, e, f). (a) Syrris Asia modular flow electrochemistry system [36]. (b) Vapourte R-Series flow reactor [37]. (c) Labtrix Start flow reactor [38]. (d) Labtrix KiloFlow reactor [38]. (e) Labtrix S1 automated flow reactor kit [38]. (f) Fluitec ContiPlant Pilot reactor [39]. All pictures are used with direct permission from the manufacturer.

Since microflow processing is a relatively new field, it lacks industry standards, especially in the chemical industry. The variance between the reactors of different brands could result in performance discrepancies [40]. Material and performance deviations of pumps, valves, and reactors make it difficult to scale up or scale out microflow reactors in practice [41]. Although one can find premade pilot-scale reactors, the majority are custom-made for a specific purpose. This customization contrasts with the highly variable batch reactors, where one reactor could be used for several different types of reactions. Figure 6.15d, e, f shows some general-purpose kilogram-scale or pilot-scale reactors on the market.

Continuous transfer of corrosive chemicals at high temperature and pressure is a considerable challenge for pumps. Typical pumps for microflow reactors are high-performance *liquid chromatography pumps*, *syringe pumps*, and *peristaltic pumps*. HPLC pumps can be extremely precise and resistant, and they can provide high pressures up to 600 bar. The major differences between the pumps are displayed in Figure 6.16. HPLC pumps are generally more expensive than the other two, and another disadvantage is the pressure drop during the piston change. However, this can be minimized by using dual-head pumps. The weakest point of syringe pumps is the low-pressure capability. Moreover, at least two syringe pumps must be used to reach an accurate continuous reagent flow: one is in pumping mode, while the other one is in withdrawal mode.

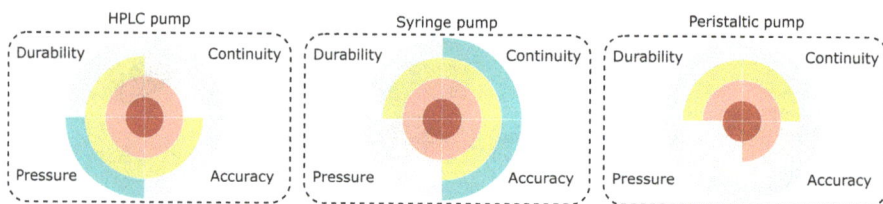

Figure 6.16: Differences between the most frequently used pumps in microflow systems. The more panel colors, the better the specific performance.

6.5 Lab-of-the-future and automated robotic platforms

Pressing environmental, governmental, and societal challenges are gradually shifting the pharmaceutical and material development industries from their usual monolithic and linear development strategies to more flexible ones. Microfluidic systems allow for the telescoping of reactions via the continuous stream of reaction mixture from one reactor into the other. By integrating separation units between two reactors, a continuous reaction–separation–reaction–separation platform can be developed. For example, an on-demand continuous-flow production of pharmaceuticals platform was built in a compact, reconfigurable system [42]. The platform was able to synthesize four different pharmaceuticals in liquid form. These systems exclusively use microfluidic reactors, pumps, and separation modules, and they are relatively small. These platforms could address pressing concerns about drug shortages, especially in high-demand zones.

The high modularity of microflow systems allows for a so-called plug-and-play approach in the development of new reactor designs. The combination of microflow systems with high-throughput screening and artificial intelligence has researchers envisioning clockwork laboratories where design, synthesis, data curation, and analysis happen in a closed loop. These lab-of-the-future approaches are introducing a paradigm shift to the pharmaceutical and material development industries, which have traditionally progressed at a slower pace. For example, it is possible to design, synthesize, and analyze thin-film materials using an automated robotic platform [43]. In another application, researchers used a robotic platform to discover novel photocatalytic materials [44]. The robot worked for eight days without stopping, performed 688 experiments, and identified photocatalytic mixtures with six times higher activity than the initial formulation. These applications are closely similar to high-throughput screenings (HTS) in the pharmaceutical industries. In an HTS, many chemical (or biological) compounds are tested against a biological target using binding assays. HTS is mostly used in drug discovery but can also be employed to test toxicity, bioavailability, or adsorption. The HTS process is automated, meaning that humans are only involved in the design and analysis phases. The lab-of-the-future projects go a step beyond the HTS world because both the human design and analysis phases are replaced by autonomous systems. Such approaches are always based on previous data, meaning that robots are using generations of human knowledge to make synthetic decisions and design choices.

In the near future, automated robotic platforms will design, synthesize, and analyze new materials with less toxicity, greater recyclability, and better alignment with both green chemistry and engineering principles.

Bibliography

[1]   Pashkova, A.; Greiner, L. Towards Small-Scale Continuous Chemical Production: Technology Gaps and Challenges. *Chem. Ing. Tech.* **2011**, *83* (9), 1337–1342. 10.1002/cite.201100037.

[2]   Philpot, J. S. L. The Use of Thin Layers in Electrophoretic Separation. *Trans. Faraday Soc.* **1937**, *33*, 524–570.

[3]   Woolley, A. T.; Mathies, R. A. Ultra-High-Speed DNA Fragment Separations Using Microfabricated Capillary Array Electrophoresis Chips. *Proc. Natl. Acad. Sci.* **1994**, *91* (24), 11348–11352. 10.1073/pnas.91.24.11348.

[4]   Manz, A.; Graber, N.; Widmer, H. M. Miniaturized Total Chemical Analysis Systems: A Novel Concept for Chemical Sensing. *Sens. Actuators B. Chem.* **1990**, *1* (1), 244–248. 10.1016/0925-4005(90)80209-I.

[5]   Gérardy, R.; Debecker, D. P.; Estager, J.; Luis, P.; Monbaliu, J.-C. M. Continuous Flow Upgrading of Selected C 2–C 6 Platform Chemicals Derived from Biomass. *Chem. Rev.* **2020**, *120* (15), 7219–7347. 10.1021/acs.chemrev.9b00846.

[6]   Valera, F. E.; Quaranta, M.; Moran, A.; Blacker, J.; Armstrong, A.; Cabral, J. T.; Blackmond, D. G. The Flow's the Thing . . . or Is It? Assessing the Merits of Homogeneous Reactions in Flask and Flow. *Angew. Chem. Int. Ed. Engl.* **2010**, *49* (14), 2478–2485. 10.1002/anie.200906095.

[7]   Hartman, R. L.; McMullen, J. P.; Jensen, K. F. Deciding whether to Go with the Flow: Evaluating the Merits of Flow Reactors for Synthesis. *Angew. Chem. Int. Ed.* **2011**, *50* (33), 7502–7519. 10.1002/anie.201004637.

[8]   Kim, H.; Nagaki, A.; Yoshida, J. A Flow-Microreactor Approach to Protecting-Group-Free Synthesis Using Organolithium Compounds. *Nat. Commun.* **2011**, *2* (1), 264. 10.1038/ncomms1264.

[9]   Dallinger, D.; Kappe, C. O. Why Flow Means Green – Evaluating the Merits of Continuous Processing in the Context of Sustainability. *Curr. Opin. Green Sustain. Chem.* **2017**, *7*, 6–12. 10.1016/j. cogsc.2017.06.003.

[10] May, S. A.; Johnson, M. D.; Buser, J. Y.; Campbell, A. N.; Frank, S. A.; Haeberle, B. D.; Hoffman, P. C.; Lambertus, G. R.; McFarland, A. D.; Moher, E. D.; White, T. D.; Hurley, D. D.; Corrigan, A. P.; Gowran, O.; Kerrigan, N. G.; Kissane, M. G.; Lynch, R. R.; Sheehan, P.; Spencer, R. D.; Pulley, S. R.; Stout, J. R. Development and Manufacturing GMP Scale-Up of a Continuous Ir-Catalyzed Homogeneous Reductive Amination Reaction. *Org. Process Res. Dev.* **2016**, *20* (11), 1870–1898. 10.1021/acs. oprd.6b00148.

[11] Harsanyi, A.; Conte, A.; Pichon, L.; Rabion, A.; Grenier, S.; Sandford, G. One-Step Continuous Flow Synthesis of Antifungal WHO Essential Medicine Flucytosine Using Fluorine. *Org. Process Res. Dev.* **2017**, *21* (2), 273–276. 10.1021/acs.oprd.6b00420.

[12] Bédard, A. C.; Longstreet, A. R.; Britton, J.; Wang, Y.; Moriguchi, H.; Hicklin, R. W.; Green, W. H.; Jamison, T. F. Minimizing E-Factor in the Continuous-Flow Synthesis of Diazepam and Atropine. *Bioorg. Med. Chem.* **2017**, *25* (23), 6233–6241. 10.1016/j.bmc.2017.02.002.

[13] Wiles, C.; Watts, P. Continuous Flow Reactors: A Perspective. *Green Chem.* **2012**, *14* (1), 38–54. 10.1039/c1gc16022b.

[14] Watts, P.; Wiles, C. Micro Reactors, Flow Reactors and Continuous Flow Synthesis. *J. Chem. Res.* **2012**, *36* (4), 181–193. 10.3184/174751912X13311365798808.

[15] Siyamak, S.; Laycock, B.; Luckman, P. Synthesis of Starch Graft-Copolymers via Reactive Extrusion: Process Development and Structural Analysis. *Carbohydr. Polym.* **2020**, *227*, 115066. 10.1016/j. carbpol.2019.115066.

[16] Isoni, V.; Mendoza, K.; Lim, E.; Teoh, S. K. Screwing $NaBH_4$ through A Barrel without A Bang: A Kneaded Alternative to Fed-Batch Carbonyl Reductions. *Org. Process Res. Dev.* **2017**, *21* (7), 992–1002. 10.1021/acs.oprd.7b00107.

[17] Ouchi, T.; Mutton, R. J.; Rojas, V.; Fitzpatrick, D. E.; Cork, D. G.; Battilocchio, C.; Ley, S. V. Solvent-Free Continuous Operations Using Small Footprint Reactors: A Key Approach for Process Intensification. *ACS Sustain. Chem. Eng.* **2016**, *4* (4), 1912–1916. 10.1021/acssuschemeng.6b00287.

[18] Nieuwland, P. J.; Koch, K.; Van Harskamp, N.; Wehrens, R.; Van Hest, J. C. M.; Rutjes, F. P. J. T. Flash Chemistry Extensively Optimized: High-Temperature Swern–Moffatt Oxidation in an Automated Microreactor Platform. *Asian J. Chem.* **2010**, *5* (4), 799–805. 10.1002/asia.200900705.

[19] Mastronardi, F.; Gutmann, B.; Kappe, C. O. Continuous Flow Generation and Reactions of Anhydrous Diazomethane Using a Teflon AF-2400 Tube-in-Tube Reactor. *Org. Lett.* **2013**, *15* (21), 5590–5593. 10.1021/ol4027914.

[20] Becht, S.; Franke, R.; Geißelmann, A.; Hahn, H. Micro Process Technology as a Means of Process Intensification. *Chem. Eng. Technol.* **2007**, *30* (3), 295–299. 10.1002/ceat.200600386.

[21] Santana, H. S.; Lopes, M. G. M.; Silva, J. L.; Taranto, O. P. Application of Microfluidics in Process Intensification. *Int. J. Chem. React. Eng.* 16 (12), 20180038. 10.1515/ijcre-2018-0038.

[22] Adamo, A.; Beingessner, R. L.; Behnam, M.; Chen, J.; Jamison, T. F.; Jensen, K. F.; Monbaliu, J. C. M.; Myerson, A. S.; Revalor, E. M.; Snead, D. R.; Stelzer, T.; Weeranoppanant, N.; Wong, S. Y.; Zhang, P. On-Demand Continuous-Flow Production of Pharmaceuticals in a Compact, Reconfigurable System. *Science (80)* **2016**, *352* (6281), 61–67. 10.1126/science.aaf1337.

[23] Crowell, L. E.; Lu, A. E.; Love, K. R.; Stockdale, A.; Timmick, S. M.; Wu, D.; Wang, Y. (Annie); Doherty, W.; Bonnyman, A.; Vecchiarello, N.; Goodwine, C.; Bradbury, L.; Brady, J. R.; Clark, J. J.; Colant, N. A.;

Cvetkovic, A.; Dalvie, N. C.; Liu, D.; Liu, Y.; Mascarenhas, C. A.; Matthews, C. B.; Mozdzierz, N. J.; Shah, K. A.; Wu, S.-L.; Hancock, W. S.; Braatz, R. D.; Cramer, S. M.; Love, J. C. On-Demand Manufacturing of Clinical-Quality Biopharmaceuticals. *Nat. Biotechnol.* **2018**, *36* (10), 988–995. 10.1038/nbt.4262.

[24] Adiga, R.; Al-adhami, M.; Andar, A.; Borhani, S.; Brown, S.; Burgenson, D.; Cooper, M. A.; Deldari, S.; Frey, D. D.; Ge, X.; Guo, H.; Gurramkonda, C.; Jensen, P.; Kostov, Y.; LaCourse, W.; Liu, Y.; Moreira, A.; Mupparapu, K.; Peñalber-Johnstone, C.; Pilli, M.; Punshon-Smith, B.; Rao, A.; Rao, G.; Rauniyar, P.; Snovida, S.; Taurani, K.; Tilahun, D.; Tolosa, L.; Tolosa, M.; Tran, K.; Vattem, K.; Veeraraghavan, S.; Wagner, B.; Wilhide, J.; Wood, D. W.; Zuber, A. Point-of-Care Production of Therapeutic Proteins of Good-Manufacturing-Practice Quality. *Nat. Biomed. Eng.* **2018**, *2* (9), 675–686. 10.1038/s41551-018-0259-1.

[25] Sipos, G.; Bihari, T.; Milánkovich, D.; Darvas, F. Flow Chemistry in Space-A Unique Opportunity to Perform Extraterrestrial Research. *J. Flow Chem.* **2017**, *7* (3–4), 151–156. 10.1556/1846.2017.00033.

[26] Plutschack, M. B.; Pieber, B.; Gilmore, K.; Seeberger, P. H. The Hitchhiker's Guide to Flow Chemistry. *Chem. Rev.* **2017**, *117* (18), 11796–11893. 10.1021/acs.chemrev.7b00183.

[27] Sambiagio, C.; Noël, T. Flow Photochemistry: Shine Some Light on Those Tubes! *Trends Chem.* **2020**, *2* (2), 92–106. 10.1016/j.trechm.2019.09.003.

[28] Romero, N. A.; Nicewicz, D. A. Organic Photoredox Catalysis. *Chem. Rev.* **2016**, *116* (17), 10075–10166. 10.1021/acs.chemrev.6b00057.

[29] Wang, X.; Cuny, G. D.; Noël, T. A. Mild, One-Pot Stadler–Ziegler Synthesis of Arylsulfides Facilitated by Photoredox Catalysis in Batch and Continuous-Flow. *Angew. Chem. Int. Ed. Engl.* **2013**, *52* (30), 7860–7864. 10.1002/anie.201303483.

[30] Mazzanti, S.; Kurpil, B.; Pieber, B.; Antonietti, M.; Savateev, A. Dichloromethylation of Enones by Carbon Nitride Photocatalysis. *Nat. Commun.* **2020**, *11* (1), 1–8. 10.1038/s41467-020-15131-0.

[31] Rosso, C.; Gisbertz, S.; Williams, J. D.; Gemoets, H. P. L.; Debrouwer, W.; Pieber, B.; Kappe, C. O. An Oscillatory Plug Flow Photoreactor Facilitates Semi-Heterogeneous Dual Nickel/Carbon Nitride Photocatalytic C–N Couplings. *React. Chem. Eng.* **2020**, *5* (3), 597–604. 10.1039/D0RE00036A.

[32] Cossar, P. J.; Hizartzidis, L.; Simone, M. I.; McCluskey, A.; Gordon, C. P. The Expanding Utility of Continuous Flow Hydrogenation. *Org. Biomol. Chem.* **2015**, *13* (26), 7119–7130. 10.1039/C5OB01067E.

[33] H-Cube® Pro – ThalesNano https://thalesnano.com/products-and-services/h-cube-pro/ (accessed Sep 19, 2020).

[34] Licence, P.; Ke, J.; Sokolova, M.; Ross, S. K.; Poliakoff, M. Chemical Reactions in Supercritical Carbon Dioxide: From Laboratory to Commercial Plant. *Green Chem.* **2003**, *5* (2), 99–104. 10.1039/B212220K.

[35] Benito-Lopez, F.; Tiggelaar, R. M.; Salbut, K.; Huskens, J.; Egberink, R. J. M.; Reinhoudt, D. N.; Gardeniers, H. J. G. E.; Verboom, W. Substantial Rate Enhancements of the Esterification Reaction of Phthalic Anhydride with Methanol at High Pressure and Using Supercritical CO2 as a Co-Solvent in a Glass Microreactor. *Lab Chip.* **2007**, *7* (10), 1345–1351. 10.1039/B703394J.

[36] Syrris – Automated Flow & Batch Chemistry Systems https://www.syrris.com/ (accessed Nov 1, 2020).

[37] International Flow Chemistry Equipment | Vapourtec Ltd https://www.vapourtec.com/ (accessed Nov 1, 2020).

[38] Chemtrix https://www.chemtrix.com/?fbclid=IwAR1UFcoyRzsIRVgelVGyCorQzx9MDbXI5AM_g0B8NHQHasjku4_RhxkGxFY (accessed Nov 1, 2020).

[39] Welcome – Fluitec International https://www.fluitec.com/ (accessed Nov 1, 2020).

[40] Rogers, L.; Jensen, K. F. Continuous Manufacturing-the Green Chemistry Promise? *Green Chem.* **2019**, *21* (13), 3481–3498. 10.1039/c9gc00773c.

[41] Al-Rawashdeh, M.; Yue, F.; Patil, N. G.; Nijhuis, T. A.; Hessel, V.; Schouten, J. C.; Rebrov, E. V. Designing Flow and Temperature Uniformities in Parallel Microchannels Reactor. *AIChE J.* **2014**, *60* (5), 1941–1952. 10.1002/aic.14443.

[42] Adamo, A., et al., On-demand Continuous-flow Production of Pharmaceuticals in a Compact, Reconfigurable System. *Science*. **2016**, *352* (6281), 61–67.

[43] MacLeod, B.P., et al., Self-driving Laboratory for Accelerated Discovery of Thin-film Materials. *Sci. Adv*. **2020**, *6* (20), eaaz8867.

[44] Burger, B., et al., A Mobile Robotic Chemist. *Nature*. **2020**, *583* (7815), 237–241.

7 Continuous separation processes

Gergo Ignacz, Gyorgy Szekely

7.1 Downstream processing in organic synthesis

In most cases, chemical transformations do not solely yield the desired product to 100%. Undesired by-products can be structurally similar to or different from the desired product. The relative difference defines the complexity of the separation and the choice of the downstream process. Generally, the larger the difference in physical and chemical features, the more straightforward the separation of products from impurities. For example, removing supercritical CO_2 ($scCO_2$) from an extraction mixture is significantly less cumbersome than the separation of two enantiomers. The other substances that require downstream processing include solvents, catalysts, additives, and remaining starting materials.

Downstream processing is used to remove the undesired compounds from the product and to recover or purify the products. Distillation is a separation process of liquid mixtures based on their different volatilities. Its main limitation is that only volatile components can be separated, while its main drawback is the high energy consumption due to the required phase change during the separation. Extraction is another common separation technique based on the different partitioning ability of a substance between two phases. A significant difference in polarity is needed for efficient separation. Crystallization is a thermodynamically driven process that occurs due to the oversaturation of a solute in a liquid. A pressure gradient drives the most common membrane separation processes, and the compounds are separated based on their molecular size, i.e., molecules larger than the membrane pores are retained. Their main advantages include modularity, mild conditions for labile compounds, and energy-efficient operation because there is no need for phase changes. In chromatography separation, a mixture is dissolved in a mobile phase, which carries the compounds through a stationary phase. The different compounds travel at different speeds; i.e., they have different affinities with the stationary phase, which is the basis of the separation.

All the above-mentioned separation processes often require the excessive use of solvents, which end up as waste for disposal. Therefore, the careful selection and potential recovery of solvents are crucial for achieving a sustainable process. Refer to Chapter 3 for solvents in general and Chapter 8 for solvent recovery.

7.2 Batch versus continuous separations

The separation and isolation of compounds can require a significant amount of energy. Therefore, the choice of the correct separation process has a profound effect on the overall manufacturing process. The energy intensity of separation unit operations,

https://doi.org/10.1515/9783111028163-007

ranked in order from high to low, is distillation, drying, evaporation, extraction, adsorption, absorption, membranes, and crystallization (Figure 7.1).

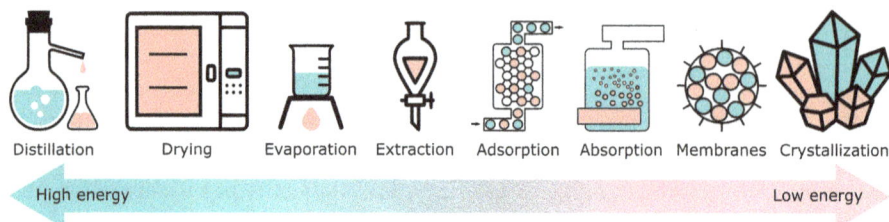

Distillation   Drying   Evaporation   Extraction   Adsorption   Absorption   Membranes   Crystallization

High energy   Low energy

Figure 7.1: The spectrum of energy requirements for separation processes. Recreated from the National Academies of Sciences, Engineering, and Medicine [1].

Similar to chemical reactor setups, separation processes can be split into batch and continuous unit operations. In the batch operation mode, the separation does not depend on space. For example, crystallization in a batch is governed by the solution's oversaturation due to solubility changes. This change occurs uniformly at every point in the reactor (ideal cases). On the other hand, when using a tubular continuous crystallization module, the solubility of the solute changes across the reactor's length due to the temperature change, which results in an uneven solution composition along the x-axis. Operating the module at steady state results in a stable and reliable system to meet industry standards.

One of the main advantages of continuous separation modules is their ability to be coupled to other continuous processes. This connectivity allows the design and modeling of systems in which every sub-process is directly connected without the need for labor-intensive manual transfer of materials. The resulting closed system is less vulnerable to contamination from the outside, needs less workforce, and runs without disruption. However, one sub-unit's failure can stop the whole system, and the restart can be time-consuming and knowledge-intensive. These issues can be rectified with automation and remote control. One of the main goals of modern chemical and pharmaceutical engineering is the full integration of separation systems in continuous flow processes [2].

Not all separation processes can be implemented in the continuous flow mode. Usually, the batch mode is preferred for time-consuming and high-viscosity systems. Thorough knowledge of every intrinsic feature of the separation process prior to continuous flow implementation is necessary. In the following chapters, we will explore the different types of existing separation techniques used in continuous flow systems.

7.3 Continuous processing with supercritical fluids

Carbon dioxide (CO_2) is a naturally occurring gas in the air at low concentration (approx. 410 ppm) but is widely produced during different chemical processes, mostly combustion of hydrocarbons. Compared to oxygen, nitrogen, or hydrogen gases, CO_2 has a relatively high supercritical temperature (31 °C) and moderate critical pressure (71 bar). These characteristics make CO_2 easy to liquefy through reaching supercritical conditions. Under supercritical conditions (above the critical temperature and pressure), CO_2 liquid and its vapor share the same density and other intrinsic features, and therefore, they are indistinguishable. The solubility properties of *supercritical CO_2* (scCO_2) are superior to normal liquid solvents in some aspects [3]. For example, changing the temperature and the pressure of scCO_2 results in different solubility properties. Therefore, scCO_2 can be used to extract a wide range of materials, from polar caffeine to non-polar hydrocarbons. Since CO_2 is a gas under standard temperature and pressure, it does not leave trace contamination in the reaction mixture or the equipment, although it can be trapped in crystals. CO_2 is non-toxic (but it can lead to suffocation), non-flammable (but certain metals can burn in CO_2), non-corrosive (but it can damage certain materials), and chemically stable (but it is reactive), and, most importantly, its use in chemical manufacturing is considered environmentally friendly. Supercritical fluid extraction (SFE) is not limited to carbon dioxide. Although it is possible to use any stable material that can reach a supercritical state, the energy used in these processes is high, and the supercritical liquid properties are not sufficient. Using and reusing carbon dioxide for SFE right from combustion gases offers a cheap, alternative solution to managing global greenhouse gas emissions. Supercritical solvents are used in many applications such as extraction, deposition, decomposition, chromatography, reaction media, reagents, particle formation, refrigeration, crystallization, and even for drying clothes.

Figure 7.2 compares materials deposition using conventional liquid solvents and supercritical fluids. The solvent molecules cannot effectively penetrate the material; therefore, the precursor deposition is less efficient and requires several deposition-solvent removal steps. Additionally, the vaporization energy of the liquid solvent is approximately an order of magnitude higher than that of carbon dioxide, making the conventional process based on the liquid solvent energy-intensive. In comparison, scCO_2 has a significantly lower dynamic viscosity than a typical liquid solvent. For example, ethyl acetate, a low-viscosity solvent widely used for extraction processes, has a dynamic viscosity of approx. 0.44 mPa-s at 27 °C, whereas the dynamic viscosity of scCO_2 ranges between 0.02 mPa-s and 0.15 mPa-s, depending on the applied pressure, at 27 °C [4]. This order of magnitude difference and the low surface tension of scCO_2 results in the extensive penetration of scCO_2 into porous materials.

Based on the advantages outlined above, scCO_2 is now widely used in the chemical and food industry. For example, extraction of caffeine from coffee beans is usually done by SFE using scCO_2. The process has substantial benefits over using liquid solvent technology. The extracted caffeine and the residual beans can also be used (for decaf-

a

Precursor dissolution   Precursor depostion   Washing steps: Waste solvents

b

Precursor dissolution   Precursor depostion

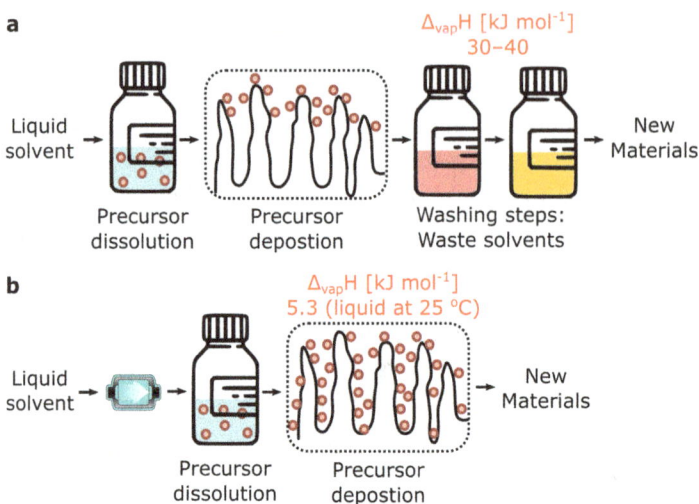

Figure 7.2: Schematic representation of a precursor deposition in a porous material using conventional liquid solvent (a) and scCO$_2$ (b). Adapted from [5].

feinated coffee, for instance) because there is no leftover solvent, which deteriorates the taste and smell of the product and poses health concerns. Another example where continuous scCO$_2$ extraction is used is vanilla extraction from the dry pulp [6]. The extraction of natural products is usually more sustainable than their petrochemical-based synthesis. Figure 7.3 shows a schematic diagram of the continuous SFE module using scCO$_2$. The system uses a specific emulsification unit that contains a micromixer to distribute the components evenly. The residence time is as low as 10 s, and the extraction efficiency is 97% (compared to the thermodynamic limit) at 37 °C and 200 bar. The process can be adapted for use with other liquids that contain hydrophilic solutes.

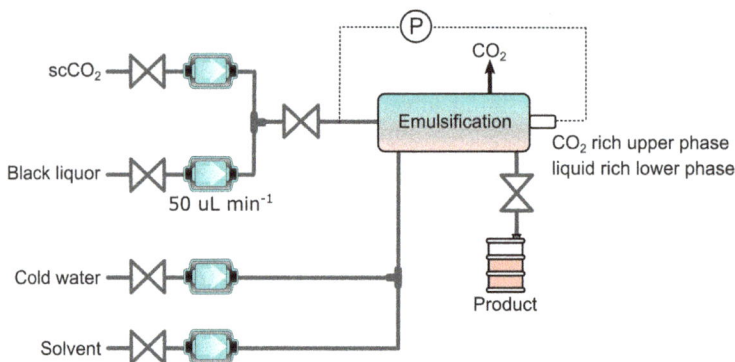

Figure 7.3: Schematic representation of the continuous extraction of vanillin from dry pulp using scCO$_2$. Adapted from [6].

Supercritical carbon dioxide is an *alternative solvent* with tunable solubility properties that vary depending on the applied pressure and temperature. Extractions and
other processes using scCO$_2$ will not solve the global warming problem but may potentially be useful for the exploitation of CO$_2$ gas.

7.4 Continuous membrane separations

Membrane technologies have been used for separations for decades, but continuous
processes have been rapidly emerging over the past decade. This section provides an
overview of continuous membrane separations, introduces the basic concepts, and
highlights the main ideas through case studies. *Dead-end filtration* is the simplest
way to perform membrane separations, but due to concentration polarization and
scale-up difficulties, *cross-flow filtration* is mainly utilized on an industrial scale. In
cross-flow filtration, a continuous stream of feed enters the system and is split into
permeate and retentate streams. This operation mode allows the continuous streams
to be diverted into any other separation unit's reaction, ensuring simple integration
into any continuous automated series of unit operations. For example, reverse osmosis operates in the continuous filtration mode without exception and is implemented
into water treatment plants to continuously supply fresh water [7]. Most membrane
separations do not require a phase change, which allows the processing of labile compounds and also saves energy. Moreover, the applied pressure is relatively low (approx.
0–40 bar), contributing to this technology's energy-efficient operation. Consequently,
membrane separations are considered green alternatives to distillation or recrystallization [8]. Replacing conventional distillation methods with membrane technology could
decrease industrial energy consumption by as much as 90% [9].

The scale-up of membrane processes is straightforward. The throughput depends
on the membrane area. Membranes are manufactured in spiral wound membrane
modules (SWMMs) with a given surface area in the range of 3–60 m^2 in a multileaf
arrangement. Figure 7.4 shows the layer-by-layer structure as well as the inlet and
outlet streams of an SWMM. The feed solution enters the axial side of the module and
flows along the module's length. The retentate side is always enriched with the less
permeable component, while the permeate side purges the smaller components. Industries widely implement membrane processes for gas separation and water treatment. However, industrial implementation of membrane-based liquid separations in
harsh pH or organic media is scarce due to the shorter membrane lifetime, lack of
industrial know-how, and time-consuming process development [10].

The area normalized flow rate (N) is called the flux (Q), which is generally a property
of the membrane and process (eq. (7.1)). Normalizing the flux with the applied pressure
difference across the membrane gives the permeance (J). Permeance is a membrane
property with the strict criteria that the pressure gradient across the membrane will not
change the membrane uniformity. The normalized thickness permeance is denoted as

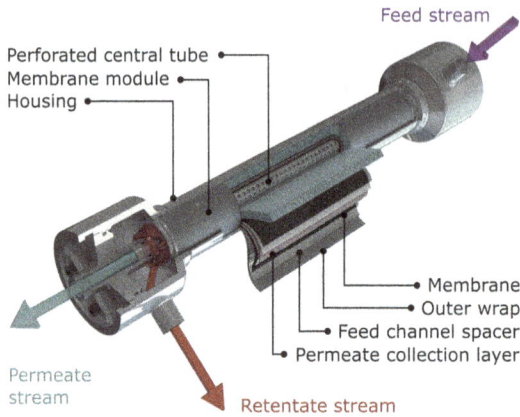

Feed stream

Perforated central tube
Membrane module
Housing

Membrane
Outer wrap
Feed channel spacer
Permeate collection layer

Permeate
stream

Retentate stream

Figure 7.4: Schematic representation of a spiral wound membrane module (SWMM). Between the outward-facing membrane, the layers include the inert spacer and permeate collection material. The flat-sheet membranes are wrapped around a perforated central tube, glued together at the ends, and placed into a plastic or metal casket.

permeability (P), which is a material (membrane, solvent, and solute) property. The complexity of the measurement of these descriptors follows the same order. Equation (7.1) shows the relationship between the formulas. If the membrane is asymmetric in structure or built from different layers, one should follow the resistance in a series model, such as that developed for gas separation but also valid for liquid-liquid separation [11]. The area normalized flow rate can be calculated as follows:

$$N = \frac{V}{t} \rightarrow Q = \frac{V}{A \cdot t} \rightarrow J = \frac{V}{A \cdot t \cdot \Delta p} \rightarrow P = \frac{V \cdot l}{A \cdot t \cdot \Delta p} \left[\frac{m^3 \cdot m}{m^2 \cdot s \cdot Pa} \right], \tag{7.1}$$

where V is the volume (m³), t is time (s), A is the area (m²), Δp is the pressure difference across the membrane cross-section (Pa), and l is the membrane thickness (for a monolayered, dense isotropic membrane).

Since the main idea of separation technology is to fractionate different components, the rejection of a substance must be defined. Rejection (R) is an arbitrary value showing the relative concentration ratio before and after separation, defined in eq. (7.2), where j is the substance and C is its concentration (mol m⁻³). In liquid separations, flux and permeance are usually calculated for a solvent, while rejection is calculated for the solutes.

$$R_j = 1 - \frac{C_{j,\text{permeate}}}{C_{j,\text{retentate}}} [\%]. \tag{7.2}$$

The pharmaceutical and food industries often use harsh conditions and organic media, which generally requires special reactors and tubing. Resistant and stable polymeric- or ceramic-type materials are required to implement membrane separations in these industrial sectors. This technology is called organic solvent nanofiltration (OSN), and it uses solvent-resistant membranes that can separate solutes with a molecular weight between 50 and 2000 g mol^{-1} [12]. Hence, OSN requires robust membranes that retain their separation characteristics in harsh organic media such as acetone, chloroform, N,N-dimethylformamide, or N-methyl-2-pyrrolidinone. Figure 7.5 illustrates the operating principle behind OSN, showing that small-size impurities pass through the membrane while large solutes are retained. Separating on the molecular level is an attractive field for purification [13], catalyst recovery, [14] or solvent recovery [15] .

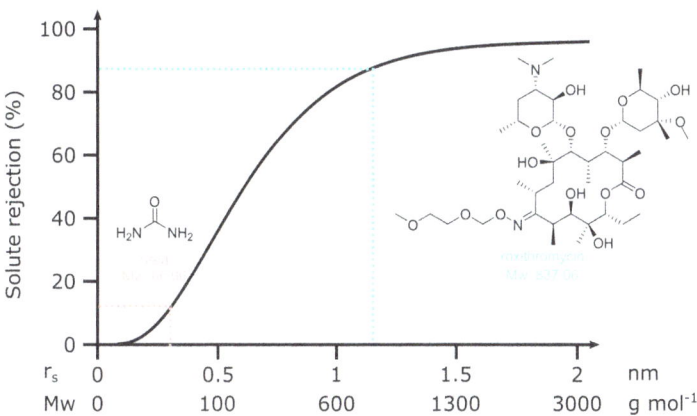

Figure 7.5: Separation principles of organic solvent nanofiltration (r_s is molecular size, MW is molecular weight).

The driving force in OSN membrane separation is the *pressure difference* across the membrane. The direction of the flow is always from the high-pressure side to the low-pressure side. A plethora of physicochemical properties of the membranes, solvents, and solutes alter the separation performance during membrane filtration. Two major separation models are accepted nowadays. One is the Knudsen diffusion-based pore flow model, [16] and the other is the solution diffusion model [17]. Reverse osmosis separation with less than 2 nm pore size is widely accepted to be based on the solution diffusion model, while ultrafiltration with a pore size of more than 50 nm is based on the pore flow model. However, nanofiltration membranes generally lie between these two pore sizes, most probably compromising solution diffusion for smaller pore sizes and pore flow separation for larger pore sizes. During Knudsen diffusion, the particles' diffusivity coefficient can be approximated using the Stokes–Einstein relationship, which assumes spherical-shaped particles:

$$D_i = \mu_i k_B T, \tag{7.3}$$

where D_i is the diffusivity coefficient of species i, μ_i is the mobility of species i, k_B is the Boltzmann constant, and T is the temperature. The mobility of a molecule depends on the size; therefore, the diffusivity is a kinetic parameter.

In dense membranes, as the solution diffusion model states, the solute has to dissolve in the membrane material first,[1] diffuse through the membrane material, and then desorb on the permeate side. This process is driven by the interaction between the polymer material and the solute itself. These parameters are well defined for several different polymer-solvent systems (see Hansen solubility parameters). Therefore, we can conclude that the permeability and the selectivity for a specific substance are defined as follows:

$$P_i = D_i + S_i, \tag{7.4}$$

where S_i is the solubility of species i.

Figure 7.6 shows a continuous membrane-based liquid separation assembly as a rendered image (a) and a process flow schematic (b). The feed stream is transported by a pump, which provides the feed flow rate and the pressure as the driving force for the separation. The feed stream then enters the circular membrane cells where the separation occurs. A certain amount of solvent and solute permeates through the membrane, forming the permeate stream (downstream of the membrane material). The corresponding retentate stream passes through the membrane cell without permeating through the membrane (upstream of the membrane material). A given fraction of the retentate stream is recirculated in a so-called retentate loop, as shown by the red arrows. A gear pump provides a high flow rate to ensure that the mixing in the retentate loop, and therefore the membrane cell, is thorough, resulting in a homogeneous concentration upstream of the membrane. The cross-flow operation reduces the effect of concentration polarization [19]. The process flow diagram shows the coupling of a cross-flow nanofiltration unit to a continuous flow reactor [18]. Refer to Chapters 5 and 6 for details on flow reactors. The reactor is a packed bed reactor filled with an organocatalyst attached to the surface of silica gel. The catalyzed model reaction is a Michael addition of nitromethane to a conjugated ketone in different solvents (Figure 7.7). The crude reaction mixture is diverted into a nanofiltration unit to recycle *in situ* the excess nitromethane (note: 1.4 equivalents of nitromethane reagent was used) and solvent (Figure 7.6b). Solvent and reagent recycling reached up to 90% over six weeks. The integration of the continuous membrane separation unit resulted in reducing the E-factor and the carbon footprint by 91% and 19%, respectively. At the end of most manufacturing processes, the products need to be isolated or concentrated, which involves removing the solvent. Removal of the solvent is conventionally

---

1 This is very similar to the Flory–Huggins model where the interaction parameter is based on the addition of one molecule (usually a solvent) to the infinite number of polymer chains.

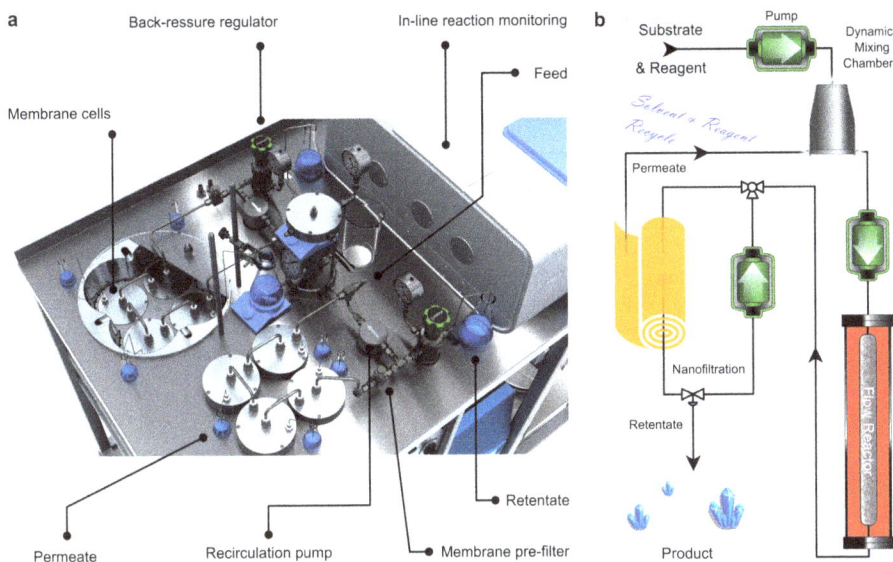

Figure 7.6: Rendered image of a sub-pilot-scale liquid membrane separation unit (a) and a schematic process flow diagram of a continuous nanofiltration unit with solvent and reagent recycle (b). Adapted from [18].

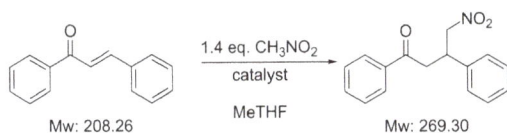

Figure 7.7: Michael addition of nitromethane to a conjugated ketone, the reaction performed in the continuous flow membrane reactor depicted in Figure 7.6b.

done by distillation, evaporation, or crystallization. However, in this case, the membrane process was designed to concentrate the product by an order of magnitude.

Table 7.1 summarizes the sustainability of two different processes: one without the membrane module and one with the membrane-assisted continuous solvent and reagent recovery. The incorporation of the nanofiltration unit reduced the E-factor from 317 to 29 kg kg^{-1}. Note that the overall contribution of the solvent to the E-factor remained above 99%. This seemingly minor change (99.7% → 99.3%) reduced the solvent consumption from 316 to 29 kg kg^{-1}, corresponding to a 91% improvement in solvent consumption. Simultaneously, the carbon footprint was reduced from 2885 to 2315 kg kg^{-1}. This modest 20% reduction in carbon footprint can be attributed to the reduced use of solvent. The main contributor to the carbon footprint is the heating through the application of a thermostat. This example demonstrates that the solvent contributes mostly to the generated waste in organic reactions and that the E-factor can be significantly reduced by recycling the solvent. Moreover, the case study high-

Table 7.1: Comparison of energy consumption, the E-factor, and the carbon footprint for the reaction performed in the continuous flow reactor depicted in Figure 7.6b and Figure 7.7. Two reaction setups are compared: one with *in situ* solvent and reagent recycling and one without solvent. Notation: n. a. denotes "not applicable."

	Without *in situ* solvent and reagent recovery			With *in situ* solvent and reagent recovery		
	Energy consumption (kWh kg^{-1})	E-factor (kg kg^{-1})	Carbon footprint (kg kg^{-1})	Energy consumption (kWh kg^{-1})	E-factor (kg kg^{-1})	Carbon footprint (kg kg^{-1})
High-pressure pump	0.0117	n.a.	0.0080	0.4678	n.a.	0.3200
Recirculation pump	n. a.*	n.a.	n.a.	12.30	n.a.	0.4400
Thermostat	3300	n.a.	2257	3300	n. a.	2257
	(99.9%)		(78.2%)	(99.6%)		(97.2%)
Solvent	n.a.	316	626	n.a.	29	57
		(99.7%)	(21.7%)		(99.3%)	(2.5%)
Reagent	n.a.	0.9100	1.8000	n.a.	0.0900	0.1840
Catalyst	n.a.	0.1175	0.1175	n.a.	0.1175	0.1175
Membrane module	n.a.	n.a.	n.a.	n.a.	0.0002	0.0065
Total	3300	317	2885	3312	29	2315

lights the need for reducing energy consumption as it is an equally important contributor to the total carbon footprint as waste generation.

This conclusion is in close agreement with the presented calculation in Section 14.2, where we demonstrate that the solvent usually primarily contributes to the generated waste. This observation is generally true for the pharmaceutical industry, where an excess amount of solvent is used during the reaction and the purification of the product. Most solvent waste ends up incinerated, thereby increasing greenhouse gas emissions and potentially leading to toxic substance release. However, solvent recovery by various technologies is currently on the rise.

Another notable example of continuous membrane separation is the *recovery of homogenous catalysts* from the postreaction mixture of hydroformylation reactions (Figure 7.8a). This reaction class is important in producing surfactants, detergents, and plasticizers. It uses C6–C10 olefins to generate high-C number olefins through metathesis reactions that could be functionalized to form various valuable products [20].

The hydroformylation reaction comprises the addition of a formyl group to the double bond of an olefin using carbon dioxide, hydrogen, and a catalyst. This aldehyde could be subjected to further transformation into various functionalities such as hydroxyl or carboxyl. The catalyst usually has a transition metal (e.g., Rh, Ru, Pd, Pt) complexed with phosphate ligands (e.g., Xantphos, XPhos, triphenylphosphine, BiPhePhos) and carbon monoxide. The recovery of these metal complexes is of utmost importance due to their scarce nature, high price, and toxicity. OSN can recycle the catalyst back to the reactor, thus lowering the amount of catalyst needed and increasing the turnover number (in the case of a stable catalyst) while keeping the reaction rate at a quasi-constant level.

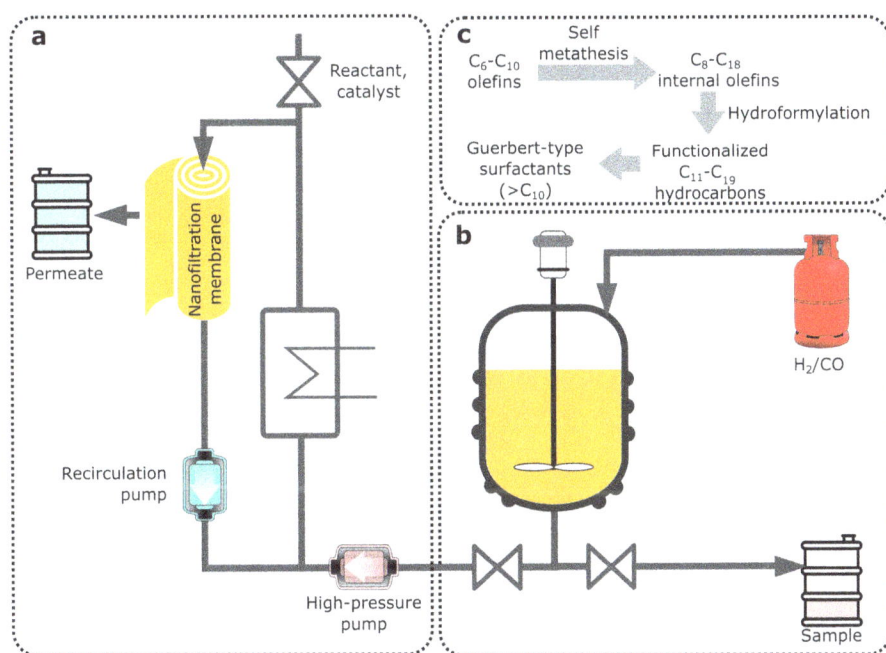

Figure 7.8: Membrane-assisted hydroformylation. (a) Schematic representation of surfactant production through metathesis and hydroformylation reactions. (b) Schematic process flow diagram of a hydroformylation reactor coupled with organic solvent nanofiltration. The high-pressure pump provides the flow rate and the pressure for the filtration system. The recirculation pump acts as a homogenizer set at a high flow rate, which reduces the effect of concentration polarization. Adapted from [21].

Figure 7.8b shows the schematic diagram of a hydroformylation reactor coupled with an OSN system. The permeate stream contains the solvent and the product, while the retentate stream is enriched in the catalyst. The unreacted CO and H_2 can be easily recovered from both the permeate and the retentate by lowering the pressure. Compared to a conventional distillation procedure, the OSN separation technique consumes less energy, resulting in significant cost savings of about 85%. Moreover, OSN ensures that metal contamination is kept below 5 ppm. Figure 7.9 shows the energy usage and total operating cost for different solvents processed by OSN and distillation [20]. The efficiency of the process depends on the solvent's physical and chemical properties, which could significantly alter the operating costs and could even triple the operating expenditure (1-decene versus 1-undecanol). Membrane processes require less energy and depend considerably less on the type of solvent used.

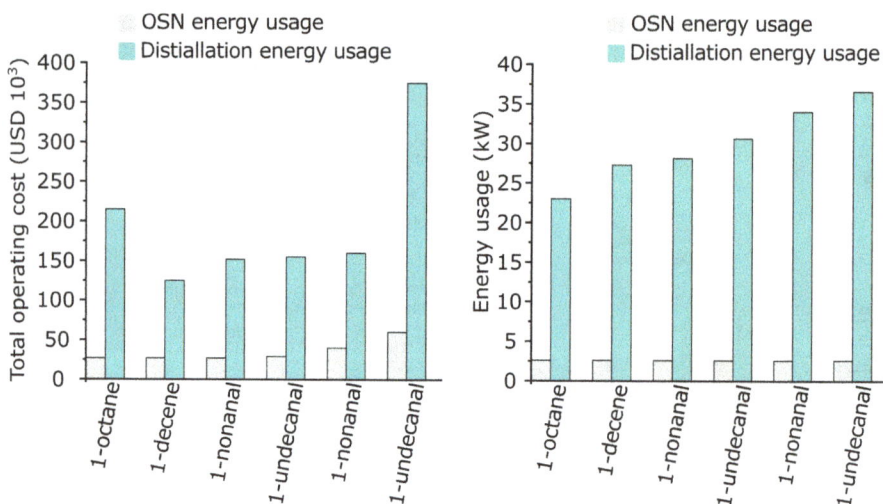

Figure 7.9: Comparison of the energy usage (a) and total operating cost (b) for processing different solvents with OSN and distillation.

If the reaction rate is relatively low, the conventional batch reactor can be substituted with a CSTR, CSTR in series, or tubular reactor. The permeate enriched in the valuable product can be used immediately in another reaction, and the pressurized retentate can be recycled back into the reactor. Overall, continuous membrane separation is a powerful tool to reduce solvent consumption and to separate chemicals at the molecular level. These hybrid processes offer an *in situ* solution for the recovery and recycling of solvents, catalysts, and reagents, thereby increasing the system's overall effectiveness.

7.5 Continuous crystallization processes

Continuous crystallization is a solid-liquid separation technology to purify a desired product from one or more undesired components, such as side-products, unreacted compounds, additives, catalysts, or solvents. Crystallization is a thermodynamically driven process that occurs due to a solute's oversaturation in a liquid. Crystallization is frequently employed in the fine chemical industry, and it has a special place in the pharmaceutical industry [22]. Since most pharmaceutical products are formulated in solid form, active pharmaceutical ingredients (APIs) must be crystallized during the manufacturing process. If we consider that the pharmaceutical industry is moving toward continuous technology, implementing continuous crystallization is imperative. Thus, process development engineers make tremendous effort to design and implement continuous crystallization technology into continuous production lines [23]. Crystallization processes aim to deliver a product with the right particle size, morphology, purity, and density. These parameters are affected by the crystallization technology used. For example, a faster cooling procedure might induce a different crystal polymorph, resulting in different physical properties [24, 25]. Since APIs fall under strict regulations by various authorities, changing any final product parameters is prohibited. Thus, all criteria agreed with the authorities should be met at GMP standards, irrespective of the crystallization technology. Follow the QR code on this page to browse the European Medicines Agency (EMA) website, which regulates APIs and medicines in the European Union.

Figure 7.10 shows the comparison of different continuous crystallizers. The *mixed-suspension, mixed-product removal* (MSMPR) crystallizers are considered semi-continuous processes, similar to continuous stirred-tank reactors. MSMPR is compatible with most batch systems, and therefore their implementation in batch-to-batch technology is straightforward. In theory, if we increase the number of units in MSMPR to infinity, we eventually reach a *continuous plug flow crystallizer* (CPFC). CPFCs are simple tubular reactors where the crystallization occurs along the length of the reactor. Since the formed solid particles (crystals) tend to subside, CPFCs can easily get clogged, result-

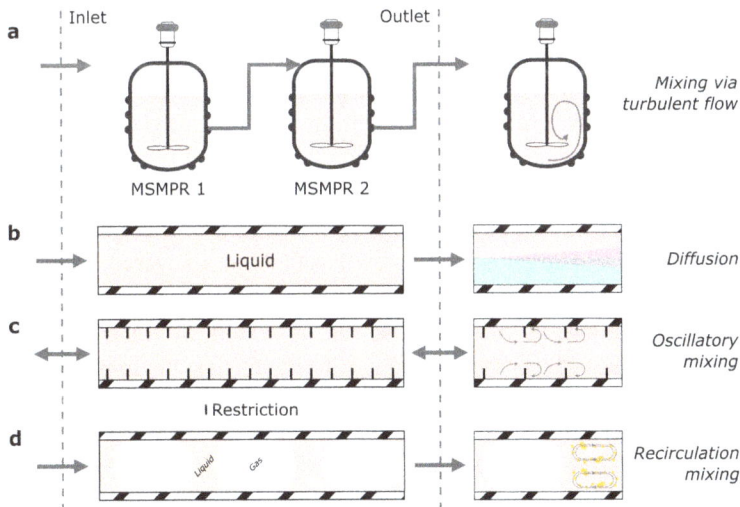

Figure 7.10: Schematic diagrams of different types of continuous crystallizers and the corresponding mixing patterns. (a) Mixed-suspension, mixed-product removal (MSMPR) crystallizer cascade. (b) Continuous plug flow crystallizer (CPFC). (c) Continuous oscillatory baffled crystallizer (COBC). (d) Continuous segmented flow crystallizer (CSFC).

ing in the production line's halt. Oscillatory-type flow can be introduced to prevent the equipment from clogging, known as *continuous oscillatory baffled crystallizer* (COBC). To further increase the mixing, a so-called slug-flow method can be used, called the *continuous segmented flow crystallizer* (CSFC). The fluid comprises two phases before crystallization and forms separated segments, which will significantly increase each segment's mixing. The phases are usually immiscible solvents or liquid-gas systems [26].

Fouling is one of the main issues in CPFC because nucleation can occur on the tubing wall due to a gradual cooling effect along the tubing (Figure 7.11). Fouling can alter the heat transfer and the flow dynamics of the system, thus increasing the energy consumption. Fouling can also decrease the residence time, which will result in lower yield and product quality. Eventually, extensive fouling can block the whole system, leading to emergency shutdown or equipment damage. Thus, it is crucial to develop methods that are fouling-proof.

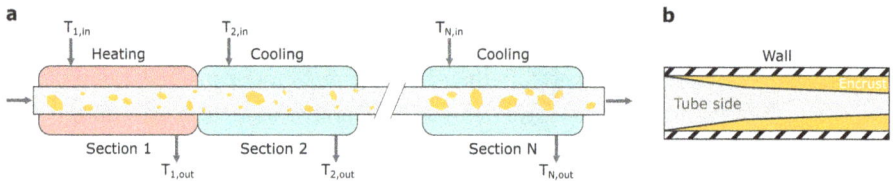

Figure 7.11: The problem of fouling in continuous crystallizers. (a) Antifouling and crystal-size distribution control via a sequential cooling-heating system. (b) Fouling along the CPFC tubing.

Figure 7.11 shows a sequential cooling-heating CPFC module. Nucleation, i.e., the initiation of crystal formation, starts in Section 1. Larger particles are thermodynamically more stable, and thus the growth rate increases with increasing particle size, which can lead to broad particle size distribution. In Section 2, the partial reheating of the mixture dissolves the smaller particles because they are thermodynamically less favorable. Another cooling section follows the heating section. Since there are fewer small crystals in the system and larger crystals' growth is favored, an overall improvement in the particle size distribution occurs. The procedure is named Ostwald ripening after the Baltic German chemist Wilhelm Ostwald, who won the Nobel Prize in Chemistry in 1909. The highest energy is required to start the nucleation. Implementation of Ostwald ripening in a series enables control of the particle size distribution and prevents fouling.

Slow nucleation kinetics in a tubular-type crystallizer can cause difficulties because the residence times are limited. Ultrasound-assisted nucleation induction was invented to overcome this issue, which is often called sonocrystallization [27, 28]. Ultrasound induces the possibility of heterogeneous nucleation in the system and can significantly reduce the induction time (Figure 7.12). The beneficial reduction in induction time is the result of accelerated diffusion in the presence of ultrasound. So-

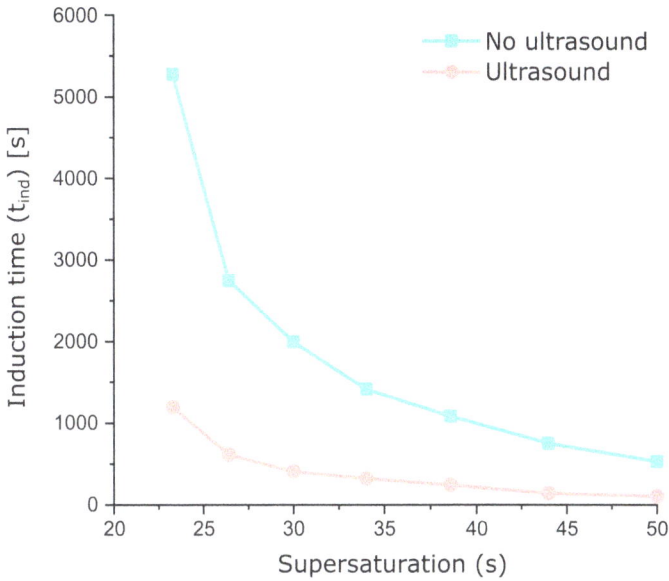

Figure 7.12: The effect of ultrasound on crystallization processes. The induction time (t_{ind}) of roxithromycin antibiotic crystallization as a function of the degree of supersaturation (S) in the absence (squares) or presence (triangles) of ultrasound (20 kHz). Reprinted with permission from [29].

nocrystallization can reduce induction time and increase the nucleation rate, which is crucial for obtaining small crystals in the sub-micrometer and nanometer range. High nucleation rates usually result in a large number of smaller crystals. Follow the QR code on this page for more details on sonocrystallization and the related equipment designed for continuous flow operation.

7.6 Centrifugal partition chromatography

Centrifugal partition chromatography (CPC) is a *counter-current-type liquid-liquid adsorption process*. CPC stands out among the different continuous separations because it does not require any stationary phase (silica gel), unlike other chromatography methods. If the solvent waste is regenerated, CPC does not produce any waste at all. This zero-waste emission, coupled with the inherent continuous aspect of CPC, appeals to the pharmaceutical sector [30].

During CPC, the stationary phase is a liquid withheld (i.e., immobilized) by centrifugal force, and therefore it must have a higher density than the mobile phase (Figure 7.13). Imagine a spinning set of small, connected chambers, filled with a solvent acting as the stationary phase. The mobile phase is pumped through these chambers from the inner (descending mode) or the outer (ascending mode) end with respect to the spinning. In the

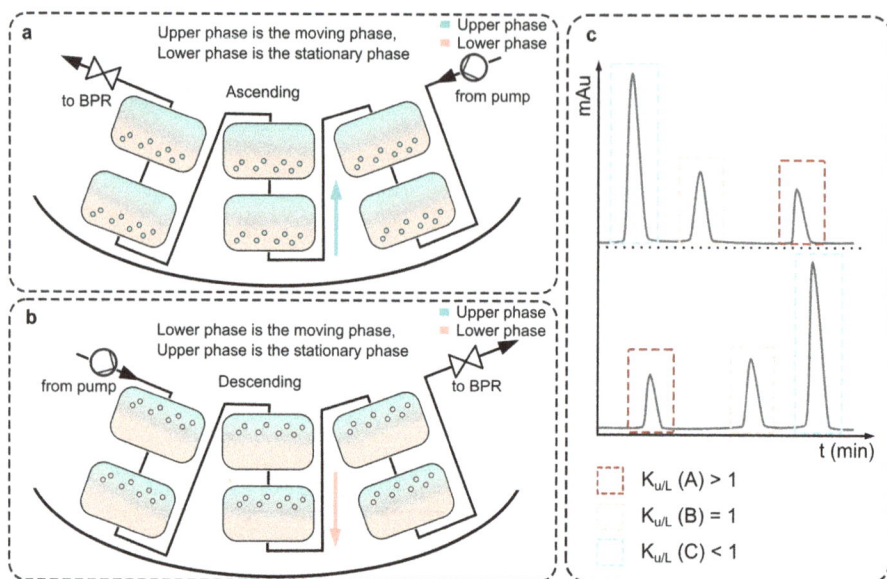

Figure 7.13: Schematic representation of the two modes in centrifugal partition chromatography. Reprinted with permission from [31].

ascending mode, the stationary phase is forced to the chambers' outer end, while the mobile phase is forced to travel to the next chamber due to the continuous incoming stream. The descending mode is operated in the same manner, but the densities and the direction of the flow are reversed [31]. Follow the QR code on this page to watch a video explaining how industrial-scale CPC works.

The typical rotational speed for a general CPC machine varies between 500 and 2000 rpm, depending on the density difference between the two solvents. Every other aspect of the CPC is similar to other chromatography techniques. In particular, the separation's efficiency is based on the partition coefficient of the substance in the used solvents. The partition coefficient (K_i) is defined as the concentration ratio of species i for the two solvents as follows:

$$K_i = \frac{c^i_{\text{Solvent}\,A}}{c^i_{\text{Solvent}\,B}} .$$

(7.5)

A typical *reaction-separation module* is presented in Figure 7.14. The first reaction occurs at 100 °C in a tube reactor with 10 min residence time, followed by catalytic hydrogenation under high pressure using an H-Cube Pro (Section 6.3.6). The crude reaction stream is diverted into the separation unit, presented as a combination of one ascending mode (AM) module and one descending mode (DM) module. Before entering the CPC module, the crude reaction stream is continuously charged with solvent (water) as the extraction medium. The two outlets from AM and DM consist of the

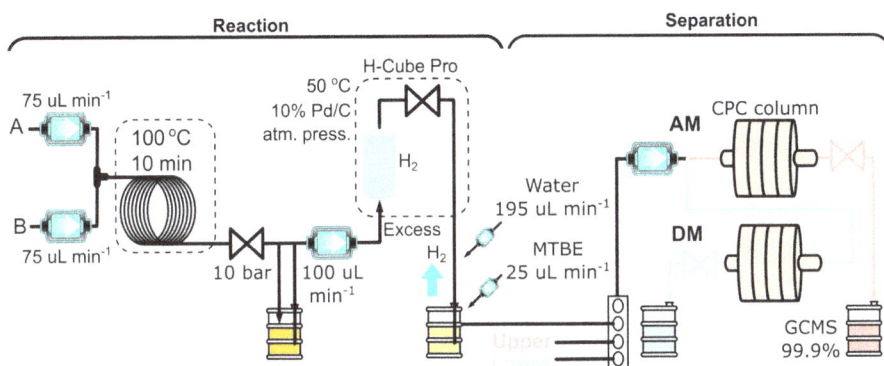

Figure 7.14: Continuous synthesis and purification by coupling a multistep flow reaction with centrifugal partition chromatography. Adapted from [31]. Refer to Chapter 6 for the details on continuous reactors.

final product and the impurities, respectively. Note that the used solvents cannot be miscible with each other.

7.7 Pressure and temperature swing adsorption

The field of adsorption science is vast; in this section, we focus primarily on continuous and semi-continuous temperature and pressure swing adsorption processes. Adsorption can be used in downstream processing to remove gases from postreaction mixtures, remove impurities, or isolate valuable fine chemicals [32].

Temperature swing adsorption (TSA) and pressure swing adsorption (PSA) are based on the difference in the thermodynamic limit of the specimen uptake at different temperatures or pressures, respectively. Different molecules have a different affinity toward the surface of adsorbent materials at different pressures or temperatures, which serves as the basis for TSA and PSA in separation processes. Figure 7.15 shows the working principle for TSA and PSA. A PSA cycle starts with pressurization and adsorption at high pressure, followed by depressurization and desorption at low pressure. PSA is considered a robust technology, which means it can be operated under changing conditions [33].

Applying pressure on the system at a specific temperature (moving along on the T_1 isochore surface), the uptake will reach the thermodynamic limit. By increasing the temperature while staying under isobaric conditions, the adsorption uptake will be less. Thus, the previously adsorbed molecules on the surface are released. Similarly, by decreasing the pressure and the temperature, the system reaches the original starting point. Repeating the adsorption-desorption cycles could lead to an efficient way of removing undesired molecules from a mixture. During TSA, the heating and cooling cycles generally take longer than the pressurizing and blowdown times during a PSA process. This time difference significantly affects the bed size: a small bed size is usu-

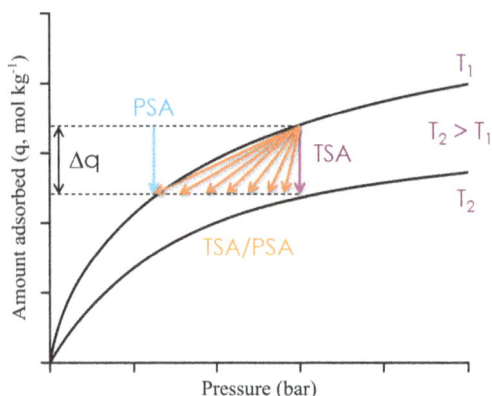

Figure 7.15: Definition of the temperature swing adsorption (TSA) and pressure swing adsorption (PSA) processes.

ally sufficient for TSA, while PSA requires a large bed size. Therefore, it is challenging to combine the two processes. If the pressure is below ambient pressure, the process is called vacuum pressure adsorption (VSA). The pressurizing step during PSA can require a considerable amount of energy compared to TSA, and therefore TSA is preferred from a sustainability point of view. This energy difference is the manifestation of the temperature sensitivity of the adsorption capacity of different materials. In other words, reaching the same Δq with TSA requires less energy than with PSA, and low-grade heat utilization can also be considered.

Owing to the cyclical nature of PSA and TSA, they can be classified under periodically operated dynamic processes, such as simulated moving bed reactors. PSA is spatially distributed and non-linear due to the non-isothermal adsorption cycles. Note that PSA and TSA do not reach steady state at normal operating conditions. Highly porous materials such as activated carbon, zeolites, silica gel, alumina, or different polymeric adsorbents and resins are often used for PSA and TSA. PSA has been successfully used for CO_2 capture [34, 35] as well as methane [36] and hydrogen [37] purification. PSA has a high adsorbed material recovery rate (above 90%), flexible operating conditions, and better compatibility with thermally labile compounds than TSA.

The second half of this section gives a detailed example of TSA used in agricultural waste upcycling. The conventional linear mentality of chemical technology (take, make, dispose) often neglects the power of downstream processing to make a difference [38]. For example, agricultural waste utilization via the extraction of fine chemicals or pharmaceuticals is a growing field. Olive leaves and other agricultural waste from olive harvests contain a significant amount of biophenols such as oleuropein, which are natural bioactive compounds and valuable building blocks for the pharmaceutical industry. Low-value agriculture waste can be upcycled into a high-value product through the extraction of these compounds. However, isolating a single compound from a complex mixture found in waste is challenging.

Olive tree leaves can be digested at 40 °C in ethyl acetate using sustainable ultrasound technology, followed by the adsorption of oleuropein using a molecularly imprinted polymer (MIP, Figure 7.16). Read about the use of ultrasound in process intensification in Section 5.3.3. MIPs are highly selective polymer-based materials featuring a fingerprint-like structure of a particular molecule embedded into a microstructure during polymerization. In principle, an imprinted polymer has a complementary cavity in terms of topography and chemical functionality to the template molecule used during its preparation. To learn more about MIPs, the reader is referred to the review by Chen and co-workers [39] and the MIP database [40]. Follow the QR code on this page to watch an animation on the preparation and application of MIPs.

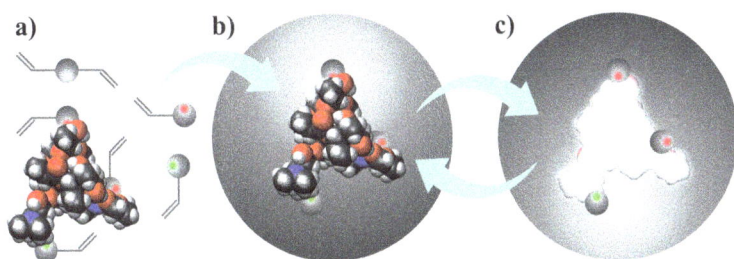

Figure 7.16: Schematic overview of the preparation of a molecularly imprinted polymer (MIP). (a) The mixture of the monomers (red and green), crosslinkers (gray), and the template molecule are mixed and allowed to self-assemble through secondary interactions. (b) Polymerization in the presence of a template, which creates the fingerprint of the molecule. (c) Removing the template molecule from the MIP creates a template-shaped cavity for the selective recognition of the target molecule. Adapted from [38].

The oleuropein-imprinted MIPs are filled into two stainless steel columns (Figure 7.17). The inlets of the columns are connected to two reservoirs through a four-way valve. One of the reservoirs contains the olive leaf extract, while the other supplies the ethyl acetate solvent for the column regeneration. Switching the four-way valve to olive leaf extract or eluent determines whether the columns are in adsorption or desorption mode, respectively. The two columns' outlets are also connected to a four-way valve, which diverts the waste and the desorbed product streams to separate membrane units. One of the membrane units concentrates the waste stream, while the other membrane unit concentrates the product. Both membrane units' permeate streams contain quasi-pure ethyl acetate solvent, which can be recycled *in situ*. Two thermostats ensure that the columns are always at the correct temperature. The column is set at 25 °C in the adsorption mode, while the other column is in the desorption mode at 45 °C.

First, the four-way valve is set to Position 1, allowing the adsorption mode on Column 1 and the desorption mode on Column 2. The purpose of the desorption is to regenerate the column and simultaneously collect the valuable product. The olive leaf extract is pumped through Column 1, filled with MIPs. The adsorption takes place at

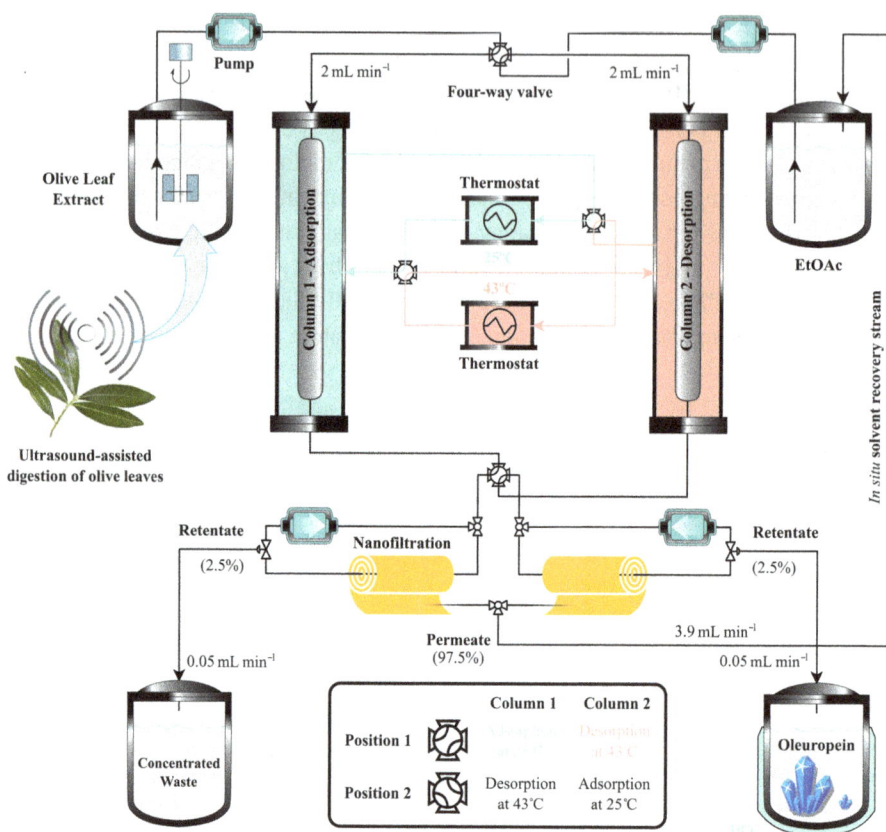

Figure 7.17: Schematic representation of a continuous temperature swing adsorption process coupled with a membrane-based *in situ* solvent recovery system. The used flow rate resulted in 1.75 g product per kg adsorbent in 1 h. The overall size of the system did not exceed 2 m². Adapted from reference [38].

low temperatures. When the column reaches an acceptable load (i.e., bed utilization), the four-way valves are set to Position 2, which alternates the adsorption-desorption cycle between the two columns. Ethyl acetate is used for the desorption. The 25 °C thermostat is always connected to the column in the adsorption mode, while the 43 °C thermostat is connected to the column in the desorption mode. Continuous membrane separations are explained in detail in this chapter. The pure oleuropein solution is recrystallized from EtOAc by cooling down the solution.

The adsorption units, featuring columns filled with MIPs that are highly selective toward oleuropein, ensure that only the oleuropein is adsorbed. Processes based on adsorption are solvent-intensive, and therefore green metric calculations often perform poorly in terms of sustainability. Comparisons of the E-factor and carbon footprint with and without *in situ* solvent recycling reveal the significant impact of solvents on the overall process sustainability (Figure 7.18). Coupling the membrane

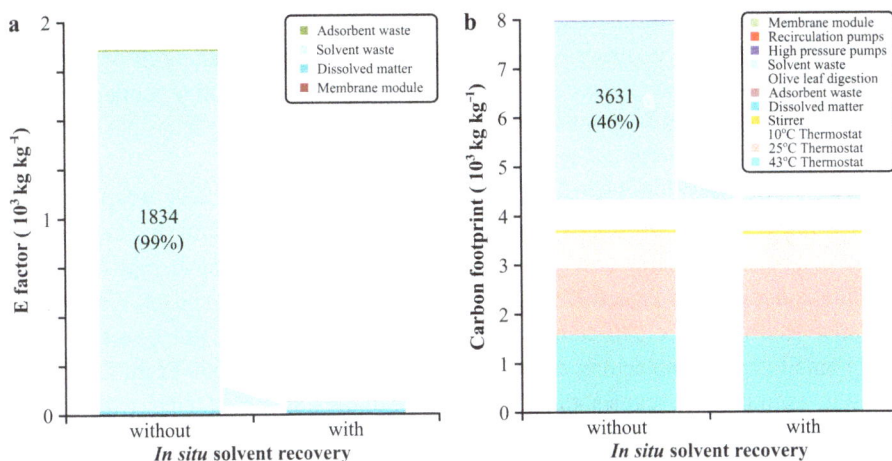

Figure 7.18: The effect of continuous solvent recovery on the E-factor and the carbon footprint of the temperature swing adsorption process presented in Figure 7.17.

filtration system to the TSA process ensures reduced solvent consumption by recycling the ethyl acetate into the system. The membrane unit allowed the reduction in E-factor and carbon footprint by 99% and 46%, respectively. The same principles can also be applied for the simultaneous extraction and isolation of different antioxidants [41]. Refer to Chapter 3 to learn more about the importance of solvent selection and green solvents. For further details on adsorption and its role in process intensification, refer to references [42] and [43].

7.8 Artificial intelligence in chemical and separation technologies

With the recent surge of artificial intelligence (AI) in various sectors, a new horizon has appeared for the chemical industries. As we have seen in Chapter 6.5, robotic platforms and automated laboratories are expected to emerge in the near future, offering excellent tools to complement human researchers by handling labor-intensive tasks. These advancements are underpinned by the ever-evolving computational power and the emergence of large datasets. ML algorithms excel at solving hyperdimensional problems previously considered unsolvable [44]. ML is an algorithm that learns from experience in relation to a class of tasks and a performance measure. During the learning process, the performance measure on the task improves [45]. These learning types can be either supervised or unsupervised, depending on the nature of the data used during the learning process. For example, classifying chemical substances based on their toxicity is a classic example of supervised learning. Unsupervised learning includes, for instance, clustering different molecules based on chemical structural similarities. ML techniques can be predictive or generative in nature. Predictive ML tasks typically involve estimating properties such as toxicity or solubility. In contrast, generative models focus on creating (potentially) novel materials with predefined properties, such as low toxicity and high solubility. While predictive models can be integrated into high-throughput screening (HTS), generative approaches have the potential to design new materials with improved properties [46].

ML in the chemical industry is dedicated to both predictive and generative methods. For instance, it is now feasible to design new reaction routes adhering to the principles of green chemistry and green engineering [47,48]. A typical high-level ML pipeline is depicted in Figure 7.19. A specific chemical domain is chosen with representative examples and sufficiently large datasets. For example, this domain might target small organic molecules, with billions of examples available. From these vast datasets, examples are translated into a machine-understandable format with the aid of cheminformatics tools. Such translations can be based on chemical descriptors [49], or graph representations [50]. The ML model learns to represent this data in a way that can be applied to various downstream applications. Popular examples include ML-aided virtual screening, process optimization, and molecular design [51–53].

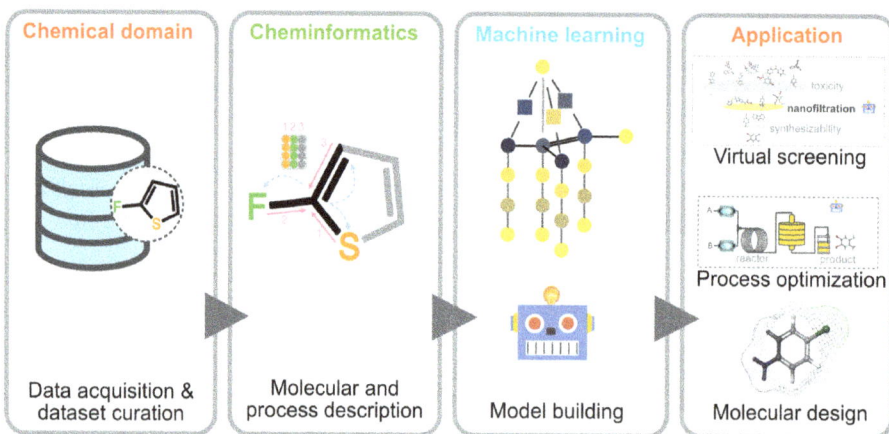

Figure 7.19: Schematics of ML applications for the chemical industry. The chemical domain can represent either specific molecules (catalysts) or a more general data (inorganic substances). The application examples can range from virtual screening through process optimization to molecules design.

ML is increasingly influential in separation applications. For example, Metal-Organic Frameworks (MOFs) are at the forefront of ML-driven separation research due to their vast variability. This variability renders them combinatorial in nature, with a high-dimensional space emerging from the vast differences in chemical structures. The combinations of many molecules with differing physicochemical properties result in a hyperdimensional design space, which is practically impossible for humans to comprehend and navigate. However, these are scenarios where ML excels; there are several exemplary cases of ML-based design for new MOF materials [54], aligning with the previously discussed generative approaches. Other hyperdimensional challenges relate to the process property prediction in separation processes. For instance, the selectivity of reverse osmosis and nanofiltration membrane processes is well-established [55]. Such predictive capabilities were once unimaginable using purely analytical methods due to the hyperdimensionality of the domain.

One of the primary criticisms of ML and AI has been the black-box nature of these methods. That is, although ML models can yield excellent results, the underlying decision-making processes are often too complex to decipher. However, with advancements in ML and the rise of explainable AI, it is now possible to distill knowledge and elucidate the decision-making in ML predictions [56]. Another major concern is the data-driven nature of ML, implying that increasingly larger and more diverse datasets are required to enhance predictive accuracy, which extends overall training time. Consequently, the high power consumption of data clusters and computers emits significant CO_2 into the atmosphere [57].

Despite potential drawbacks, ML is securing a significant role in the chemical industry, from designing molecules to predicting process parameters for molecular separation. The environmental impacts can be mitigated through the development of more efficient algorithms and the utilization of renewable energy sources (refer to Section 12).

Bibliography

[1]   *A Research Agenda for Transforming Separation Science*; National Academies Press, 2019. https://doi.org/10.17226/25421.

[2]   Jiménez-González, C.; Poechlauer, P.; Broxterman, Q. B.; Yang, B. S.; Am Ende, D.; Baird, J.; Bertsch, C.; Hannah, R. E.; Dell'Orco, P.; Noorman, H.; Yee, S.; Reintjens, R.; Wells, A.; Massonneau, V.; Manley, J. Key Green Engineering Research Areas for Sustainable Manufacturing: A Perspective from Pharmaceutical and Fine Chemicals Manufacturers. *Org. Process Res. Dev.* **2011**, *15* (4), 900–911. 10.1021/op100327d.

[3]   Bartle, K. D.; Clifford, A. A.; Jafar, S. A.; Shilstone, G. F. Solubilities of Solids and Liquids of Low Volatility in Supercritical Carbon Dioxide. *J. Phys. Chem. Ref. Data.* **1991**, *20* (4), 713–756. 10.1063/1.555893.

[4]   Heidaryan, E.; Hatami, T.; Rahimi, M.; Moghadasi, J. Viscosity of Pure Carbon Dioxide at Supercritical Region: Measurement and Correlation Approach. *J. Supercrit. Fluids.* **2011**, *56* (2), 144–151. 10.1016/j.supflu.2010.12.006.

[5]   Ushiki, I.; Matsuyama, K.; Smith, R. L. *Sustainable Approaches for Materials Engineering with Supercritical Carbon Dioxide*; Elsevier Inc., 2019. 10.1016/B978-0-12-814681-1.00015-1.

[6]   Fujii, T.; Matsuo, Y.; Kawasaki, S. I. Rapid Continuous Supercritical CO2 Extraction and Separation of Organic Compounds from Liquid Solutions. *Ind Eng Chem Res.* **2018**, *57* (16), 5717–5721. 10.1021/acs.iecr.8b00812.

[7]   Elimelech, M.; Phillip, W. A. The Future of Seawater Desalination: Energy, Technology, and the Environment. *Science 80.* **2011**, *333* (6043), 712–717. 10.1126/science.1200488.

[8]   Szekely, G.; Jimenez-Solomon, M. F.; Marchetti, P.; Kim, J. F.; Livingston, A. G. Sustainability
       Assessment of Organic Solvent Nanofiltration: From Fabrication to Application. *Green Chem.* **2014**, *16*
       (10), 4440–4473. 10.1039/c4gc00701h.
[9]   Sholl, D. S.; Lively, R. P. Seven Chemical Separations to Change the World. *Nature.* **2016**, *532* (7600),
       435–437. 10.1038/532435a.
[10]   Le Phuong, H. A.; Blanford, C. F.; Szekely, G. Reporting the Unreported: The Reliability and
       Comparability of the Literature on Organic Solvent Nanofiltration. *Green Chem.* **2020**, *22* (11),
       3397–3409. 10.1039/D0GC00775G.
[11]   Henis, J. M. S.; Tripodi, M. K. Composite Hollow Fiber Membranes for Gas Separation: The Resistance
       Model Approach. *J. Membr. Sci.* **1981**, *8* (3), 233–246. 10.1016/S0376-7388(00)82312-1.
[12]   Marchetti, P.; Jimenez Solomon, M. F.; Szekely, G.; Livingston A. G. Molecular Separation with
       Organic Solvent Nanofiltration: A Critical Review. *Cheme Rev.* **2014**, *114* (21), 10735–10806. 10.1021/
       cr500006j.
[13]   Székely, G.; Bandarra, J.; Heggie, W.; Sellergren, B.; Ferreira, F. C. Organic Solvent Nanofiltration: A
       Platform for Removal of Genotoxins from Active Pharmaceutical Ingredients. *J. Membr. Sci.* **2011**, *381*
       (1–2), 21–33. 10.1016/j.memsci.2011.07.007.
[14]   Tsoukala, A.; Peeva, L.; Livingston, A. G.; Bjørsvik, H. -R. Separation of Reaction Product and
       Palladium Catalyst after a Heck Coupling Reaction by Means of Organic Solvent Nanofiltration.
       ChemSusChem. **2012**, *5* (1), 188–193. 10.1002/cssc.201100355.
[15]   Kim, J. F.; Szekely, G.; Schaepertoens, M.; Valtcheva, I. B.; Jimenez-Solomon, M. F.; Livingston, A. G. In
       Situ Solvent Recovery by Organic Solvent Nanofiltration. *ACS Sustain. Chem. Eng.* **2014**, *2* (10),
       2371–2379. 10.1021/sc5004083.
[16]   Sarbolouki, M. N. Pore Flow Models and Their Applicability. *Ion Exch. Membr.* **1975**, *2* (2), 117–122.
[17]   Wijmans, J. G.; Baker, R. W. The Solution-Diffusion Model: A Review. *J. Membr. Sci.* **1995**, *107* (1), 1–21.
       10.1016/0376-7388(95)00102-I.
[18]   Fodi, T.; Didaskalou, C.; Kupai, J.; Balogh, G. T.; Huszthy, P.; Szekely, G. Nanofiltration-Enabled in Situ
       Solvent and Reagent Recycle for Sustainable Continuous-Flow Synthesis. *ChemSusChem.* **2017**, *10*
       (17), 3435–3444. 10.1002/cssc.201701120.
[19]   Song, L.; Elimelech, M. Theory of Concentration Polarization in Crossflow Filtration. *J. Chem. Soc.
       Faraday Trans.* **1995**, *91* (19), 3389–3398. 10.1039/FT9959103389.
[20]   Peddie, W. L.; Van Rensburg, J. N.; Vosloo, H. C. M.; Van der Gryp, P. Technological Evaluation of
       Organic Solvent Nanofiltration for the Recovery of Homogeneous Hydroformylation Catalysts. *Chem.
       Eng. Res. Des.* **2017**, *121*, 219–232. 10.1016/j.cherd.2017.03.015.
[21]   Priske, M.; Wiese, K. D.; Drews, A.; Kraume, M.; Baumgarten, G. Reaction Integrated Separation of
       Homogenous Catalysts in the Hydroformylation of Higher Olefins by Means of Organophilic
       Nanofiltration. *J. Membr. Sci.* **2010**, *360* (1–2), 77–83. 10.1016/j.memsci.2010.05.002.
[22]   Jiang, M.; Braatz, R. D. Designs of Continuous-Flow Pharmaceutical Crystallizers: Developments and
       Practice. *CrystEngComm.* **2019**, *21* (23), 3534–3551. 10.1039/C8CE00042E.
[23]   Ma, Y.; Wu, S.; Macaringue, E. G. J.; Zhang, T.; Gong, J.; Wang, J. Recent Progress in Continuous
       Crystallization of Pharmaceutical Products: Precise Preparation and Control. *Org. Process Res. Dev.*
       2020. 10.1021/acs.oprd.9b00362.
[24]   Köllges, T.; Vetter, T. Polymorph Selection and Process Intensification in A Continuous
       Crystallization–Milling Process: A Case Study on l-Glutamic Acid Crystallized from Water. *Org. Process
       Res. Dev.* **2019**, *23* (3), 361–374. 10.1021/acs.oprd.8b00420.
[25]   O'Mahony, M. A.; Croker, D. M.; Rasmuson, Å. C.; Veesler, S.; Hodnett, B. K. Measuring the Solubility
       of a Quickly Transforming Metastable Polymorph of Carbamazepine. *Org. Process Res. Dev.* **2013**, *17*
       (3), 512–518. 10.1021/op300228g.
[26]   Kashid, M. N.; Platte, F.; Agar, D. W.; Turek, S. Computational Modelling of Slug Flow in a Capillary
       Microreactor. *J. Comput. Appl. Math.* **2007**, *203* (2), 487–497. 10.1016/j.cam.2006.04.010.

[27]   Sander, J. R. G.; Zeiger, B. W.; Suslick, K. S.; Sonocrystallization and Sonofragmentation. In *Ultrasonics Sonochemistry*; Elsevier B.V., 2014; Vol. 21, pp 1908–1915. 10.1016/j.ultsonch.2014.02.005.

[28]   Belca, L. M.; Ručigaj, A.; Teslič, D.; Krajnc, M. The Use of Ultrasound in the Crystallization Process of an Active Pharmaceutical Ingredient. *Ultrason. Sonochem.* **2019**, *58* (June). 10.1016/j.ultsonch.2019.104642.

[29]   Guo, Z.; Zhang, M.; Li, H.; Wang, J.; Kougoulos, E. Effect of Ultrasound on Anti-Solvent Crystallization Process. *J. Cryst. Growth* **2005**, *273* (3–4), 555–563. 10.1016/j.jcrysgro.2004.09.049.

[30]   Ito, Y. Golden Rules and Pitfalls in Selecting Optimum Conditions for High-Speed Counter-Current Chromatography. *J. Chromatogr. A* **2005**, *1065* (2), 145–168. 10.1016/j.chroma.2004.12.044.

[31]   Örkényi, R.; Éles, J.; Faigl, F.; Vincze, P.; Prechl, A.; Szakács, Z.; Kóti, J.; Greiner, I. Continuous Synthesis and Purification by Coupling a Multistep Flow Reaction with Centrifugal Partition Chromatography. *Angew Chem. Int. Ed. Engl.* **2017**, *56* (30), 8742–8745. 10.1002/anie.201703852.

[32]   Jiang, L.; Wang, R. Q.; Gonzalez-Diaz, A.; Smallbone, A.; Lamidi, R. O.; Roskilly, A. P. Comparative Analysis on Temperature Swing Adsorption Cycle for Carbon Capture by Using Internal Heat/Mass Recovery. *Appl. Therm. Eng.* **2020**, *169*, 114973. 10.1016/j.applthermaleng.2020.114973.

[33]   Peng, H.; Couenne, F.; Le Gorrec, Y. Robust Control of a Pressure Swing Adsorption Process. *IFAC Proc. Vol.* **2011**, *44* (1), 7310–7315. 10.3182/20110828-6-IT-1002.02965.

[34]   Ko, D.; Siriwardane, R.; Biegler, L. T. Optimization of a Pressure-Swing Adsorption Process Using Zeolite 13X for CO2 Sequestration. *Ind. Eng. Chem. Res.* **2003**, 42 (2), 339–348. 10.1021/ie0204540.

[35]   Ko, D.; Siriwardane, R.; Biegler, L. T. Optimization of Pressure Swing Adsorption and Fractionated Vacuum Pressure Swing Adsorption Processes for CO2 Capture. *Ind. Eng. Chem. Res.* **2005**, *44* (21), 8084–8094. 10.1021/ie050012z.

[36]   Olajossy, A.; Gawdzik, A.; Budner, Z.; Dula, J. Methane Separation from Coal Mine Methane Gas by Vacuum Pressure Swing Adsorption. *Chem. Eng. Res. Des.* **2003**, *81* (4), 474–482. 10.1205/026387603765173736.

[37]   Khajuria, H.; Pistikopoulos, E. N. Optimization and Control of Pressure Swing Adsorption Processes under Uncertainty. *AIChE J.* **2013**, *59* (1), 120–131. 10.1002/aic.13783.

[38]   Didaskalou, C.; Buyuktiryaki, S.; Kecili, R.; Fonte, C. P.; Szekely, G. Valorisation of Agricultural Waste with an Adsorption/Nanofiltration Hybrid Process: From Materials to Sustainable Process Design. *Green Chem.* **2017**, *19* (13), 3116–3125. 10.1039/c7gc00912g.

[39]   Chen, L.; Wang, X.; Lu, W.; Wu, X.; Li, J. Molecular Imprinting: Perspectives and Applications. *Chem. Soc. Rev.* **2016**, *45*, 2137–2211. 10.1039/C6CS00061D.

[40]   MIPdatabase home http://www.mipdatabase.com/ (accessed Aug 25, 2020).

[41]   Voros, V.; Drioli, E.; Fonte, C.; Szekely, G. Process Intensification via Continuous and Simultaneous Isolation of Antioxidants: An Upcycling Approach for Olive Leaf Waste. *ACS Sustain. Chem. Eng.* **2019**, *7* (22), 18444–18452. 10.1021/acssuschemeng.9b04245.

[42]   Toth, J. *Adsorption*; 1st ed., CRC Press. 10.1201/b12439.

[43]   Tian, Y.; Demirel, S. E.; Hasan, M. M. F.; Pistikopoulos, E. N. An Overview of Process Systems Engineering Approaches for Process Intensification: State of the Art. *Chem. Eng. Process.* **2018**, *133*, 160–210. 10.1016/j.cep.2018.07.014.

[44]   Jumper, J. et al. Highly Accurate Protein Structure Prediction with AlphaFold. *Nature.* **2021**, *596* (7873), 583–589.

[45]   Mitchell, T. M.; Mitchell, T. M. *Machine Learning*; McGraw-hill New York, Vol. 1. 1997.

[46]   Silva, T. C.; M. Eppink; Ottens, M. Automation and Miniaturization: Enabling Tools for Fast, High-throughput Process Development in Integrated Continuous Biomanufacturing. *J. Chem. Technol. Biotechnol.* **2022**, *97* (9), 2365–2375.

[47]   Li, S. -W. et al. Reaction Performance Prediction with an Extrapolative and Interpretable Graph Model Based on Chemical Knowledge. *Nat. Commun.* **2023**, *14* (1), 3569.

[48]   Braconi, E.; Godineau E. Bayesian Optimization as A Sustainable Strategy for Early-Stage Process Development? A Case Study of Cu-Catalyzed C–N Coupling of Sterically Hindered Pyrazines. *ACS Sustain. Chem. Eng.* **2023**, *11* (28), 10545–10554.

[49]   Todeschini, R.; Consonni, V. *Handbook of Molecular Descriptors*; John Wiley & Sons, Vol. *11*. 2008.

[50]   Coley, C. W. et al. A Graph-convolutional Neural Network Model for the Prediction of Chemical Reactivity. *Chem. Sci.* **2019**, *10* (2), 370–377.

[51]   Schmidt, J. et al. Recent Advances and Applications of Machine Learning in Solid-state Materials Science. *Npj Comput. Mater.* **2019**, *5* (1), 83.

[52]   Hardian, R. et al. Artificial Intelligence: The Silver Bullet for Sustainable Materials Development. *Green Chem.* **2020**, *22* (21), 7521–7528.

[53]   Ignacz, G.; Szekely, G. Deep Learning Meets Quantitative Structure–activity Relationship (QSAR) for Leveraging Structure-based Prediction of Solute Rejection in Organic Solvent Nanofiltration. *J. Membr Sci.* **2022**, *646*, 120268.

[54]   Jablonka, K. M. et al. Big-Data Science in Porous Materials: Materials Genomics and Machine Learning. *Chem. Rev.* **2020**, *120* (16), 8066–8129.

[55]   Jeong, N.; T. -H. Chung; Tong, T. Predicting Micropollutant Removal by Reverse Osmosis and Nanofiltration Membranes: Is Machine Learning Viable? *Environ. Sci. Technol.* **2021**, *55* (16), 11348–11359.

[56]   Ignacz, G.; N. Alqadhi; Szekely, G. Explainable Machine Learning for Unraveling Solvent Effects in Polyimide Organic Solvent Nanofiltration Membranes. *Adv. Membr.* **2023**, *3*, 100061.

[57]   Available from: https://cacm.acm.org/magazines/2023/8/274925-the-carbon-footprint-of-artificial-intelligence/fulltext.

8 Solvent recovery and recycling

Diana Gulyas Oldal, Gyorgy Szekely

Solvents play an essential role in every stage of chemical production. Thus, there is an increasing amount of organic solvent waste to handle each year. The vast majority of the world's solvent production eventually ends up being disposed of through incineration or dispersal into the biosphere [1]. There is a negligible accumulation of solvents in long-term artifacts, and thus the annual discharge of solvents closely matches their production rate. The conventional step-by-step approach for solvent recovery is shown in Figure 8.1, which includes four stages: solid removal, recovery, purification, and refinement. The role of solvents in sustainable chemical production is explained in Chapter 3.

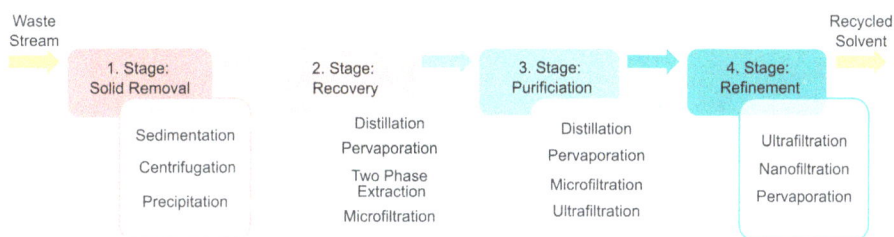

Figure 8.1: Conventional step-by-step approach for solvent recovery and the common techniques for each stage. Adapted from [2].

Lonza Engineering introduced a saving-potential ranking for various solvent handling approaches, ranging from off-site incineration to *in situ* recycling (Figure 8.2) [3]. The disposal of solvent waste via on-site or off-site incineration is expensive and ultimately significantly impacts a company's carbon footprint [4]. Consequently, the solvents should be recycled on-site, specific to the process. When the solvent is purified and recirculated within the same process, it is often referred to as *in situ* solvent recycling, which has the highest saving potential and the smallest environmental footprint. Figure 8.3 compares the different solvent recovery and recycling routes from a process engineering point of view. Keep in mind that each solvent storage, including spent and recovered solvents, needs logistics, risk assessment, safety precautions, and quality control, all of which have a significant physical footprint in a manufacturing plant. These measures are considerably less for *in situ* solvent recycling. Therefore, it should be the first option to consider by process engineers. The most prevalent technologies for solvent recovery and recycling are distillation (such as pressure swing, azeotropic and extractive distillations), adsorption, and membrane processes (including organic solvent nanofiltration and organophilic pervaporation) [4]. These are further explained in the following sections. Follow the QR code on this page to watch a

https://doi.org/10.1515/9783111028163-008

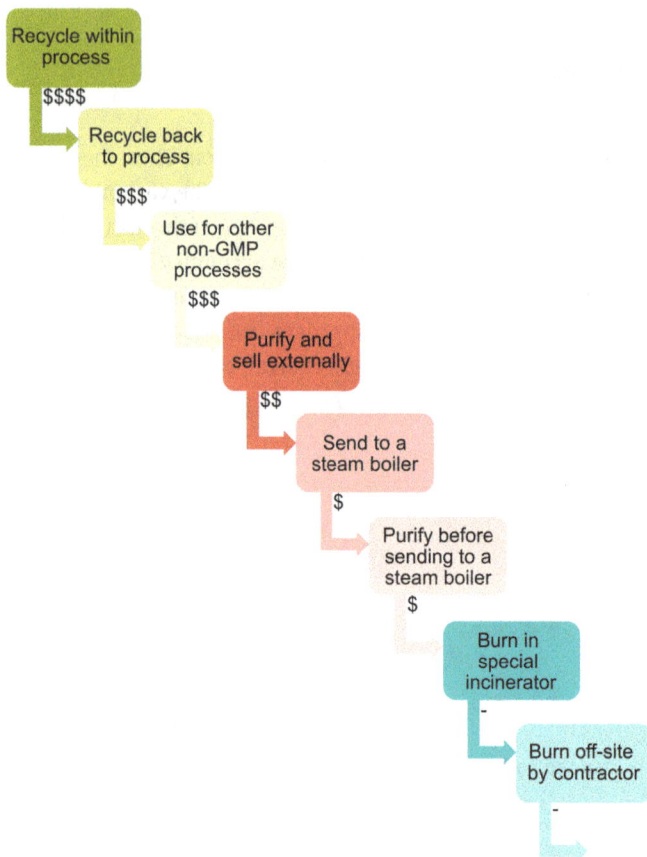

Figure 8.2: Ranking of organic solvent waste handling approaches based on their money-saving potential. Adapted from [3].

video on the industrial-scale environmentally benign treatment of contaminated solvents by Veolia. For further reading on solvent recovery, refer to the following literature [1, 5].

8.1 Distillation processes

Distillation is a separation process based on differences in the volatilities and boiling points of chemicals in liquid mixtures. The relative volatility[1] of the components in a liquid mixture is measured by comparing their vapor pressure values. It indicates the

---

1 Relative volatility for a mixture of two components (i, more volatile component; j, less volatile component; y, vapor phase; x, liquid phase) can be calculated as follows: $a = \dfrac{(y_i/x_i)}{(y_j/x_j)}$.

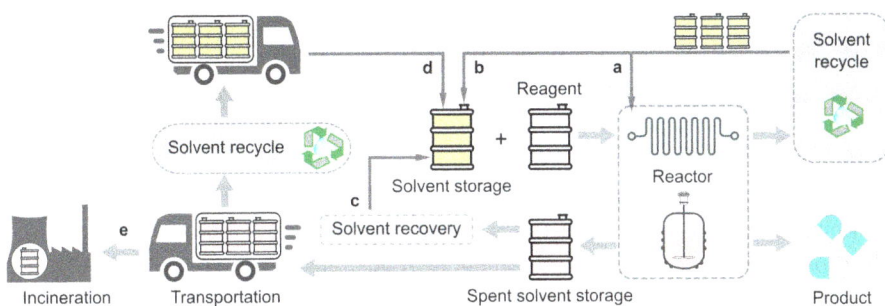

Figure 8.3: Comparison of the different solvent recovery and recycling routes from a process engineering perspective. (a) Continuous *in situ* solvent recycling is the best option from both an economic and a sustainability perspective. (b) Continuous solvent recovery for on-site storage and reuse. (c) Batch solvent recovery with on-site storage of both spent solvent waste and recovered solvent for reuse. (d) Off-site solvent recycling, which has a high cost and a large carbon footprint. (e) Incineration of waste solvent, which is the worst option for handling solvent waste.

effectiveness of the separation between the more volatile (lower boiling point) and the less volatile (higher boiling point) components. Experimental data show that if the relative volatility is higher than 1.2, distillation should be suitable for separation. If this value falls between 1.2 and 1.05, distillation is still a good option, although some other separation units should be considered. If the relative volatility falls below 1.05, then distillation processes should be avoided [6]. The primary distillation techniques include simple distillation, fractional distillation, vacuum distillation, steam distillation, and extractive distillation.

Simple distillation is a process where the volatile compound is evaporated, and the vapor is channeled into a condenser. Thus, the composition of the distillate is the same as the composition of the vapors. This technique can be applied to separate mixtures containing non-volatile compounds (e.g., impurities, solids) and components that have a difference in boiling points of more than 70 °C.

Fractional distillation is a technique used to separate components with similar boiling points, i.e., with a difference of less than 25 °C. This separation is performed by repeated vaporizations and condensations in a fractionating column. These columns consist of an array of plates, and the more volatile component will tend to move toward the top of the column while the less volatile component stays at the bottom. Therefore, better separation of liquids can be achieved with this method than with simple distillation.

Vacuum distillation is used for the separation of components with a high boiling point. These types of compounds can be boiled by lowering the pressure instead of increasing the temperature. The vapor pressure of the chemical at a given temperature must be considered. This technique is applied for distilling dimethyl sulfoxide (boiling point [atm.] = 189 °C) and glycerol (boiling point [atm.] = 290 °C), for instance.

Steam distillation is applied for the separation of heat-sensitive compounds. Steam provides good heat transfer rates without the need for high temperatures. Thus, the use of steam in the mixture enables vaporization at lower temperatures, and subsequently, the heat-sensitive compounds do not decompose. Steam distillation is commonly applied in the manufacturing of essential oils, perfumes, and skincare products.

Extractive distillation is a separation process that uses a third component that is miscible with the feed steam, has a high boiling point, and is relatively non-volatile. These solvent additives form an azeotrope[2] with the other components in the mixture. This solvent is often called *entrainer*, which facilitates the separation by altering the relative volatility of the close-boiling or azeotrope-forming components. Therefore, extractive distillation is a suitable method for the separation of mixtures with low relative volatility.

Every listed distillation technique can be performed either in batch or continuous mode, except for extractive distillation, which can only be carried out continuously [7]. Vacuum and steam distillation can also be used as an energy and cost-reducing method since less input energy is required than for the simple distillation technique. Regarding waste generation, extractive distillation and the selection of an entrainer requires careful consideration because it can result in increased waste generation if the solvent additive is difficult to recycle. The substitution of conventional solvents with bio-based solvents for entrainers is a sustainable solution. Cyrene and sulfolane can be used as entrainers for extractive distillation in aromatic/aliphatic and olefin/paraffin separation [8]. The application of cyrene reduces the minimum reflux ratio by 43% compared to sulfolane, demonstrating the improvement in the process's energy efficiency. The recovery of acetone and dimethyl ether organic solvents by extractive and *pressure swing distillation* has been demonstrated to be economically feasible with considerable profit. The optimization of that process was assisted by a genetic algorithm, which shows the practical potential of artificial intelligence in sustainable process development [9]. There is a trend toward the exploitation of artificial intelligence in both separation processes and reaction engineering.

Distillation processes are among the most evolved techniques among separation unit operations, although they require a large amount of energy. Nonetheless, several industrial sectors still widely use distillation for solvent recycling because high recovery and purity can be achieved [10]. Moreover, 95% of all the chemical industry's separation processes involve distillation [11], and therefore it is among the most commonly used processes for solvent recovery in the pharmaceutical industry [7]. Pfizer uses distillation solvent recovery systems to reuse acetonitrile, isopropanol, and toluene from small-volume waste streams [12]. These solvents were applied during the manufactur-

---

2 An azeotrope is a mixture of two or more liquids which has a constant boiling point, the vapor phase has the same composition as the liquid phase, and thus, the components cannot be separated by simple distillation.

ing of three active pharmaceutical ingredients (APIs), and as a result of the solvent recovery, the company's carbon emission was reduced by approx. 677,000 kg per year. Pfizer reported case studies for three solvent waste recovery options using distillation: a mixture of acetone and MeCN from the production of selamectin, IPA and THF originating from the manufacture of nelfinavir, and toluene and acetone from the production of hydrocortisone [13]. Distillation proved to be an adequate separation choice because these binary systems do not form azeotropes. The evaluations were carried out using a software toolbox of Aspen Plus and SimaPro with the proper databases, and the recovered solvent purity was set to at least 98 wt% (Table 8.1). Both the cost savings and the carbon footprint reduction made the distillation processes superior to waste incineration. A high recovery rate above 98% was obtained by distillation, which significantly reduced the amount of both fresh solvent to be purchased and used solvent to be disposed of by incineration. According to the study, for example, the manufacture of 1 kg of THF and IPA was responsible for 5.65 and 2.20 kg total emissions, respectively. Most emissions were released to the air. In the first case of THF, 5.52 kg of total air emission was calculated, where 5.46 kg was associated with CO_2 and the remaining 0.06 kg with other GHGs and pollutants. In the case of IPA, 1.66 kg made up the total air emission, where 1.63 kg was attributed to CO_2. Owing to the mass production of these APIs on the ton scale, the overall environmental implications of recovering or disposing these solvents were tremendous.

Table 8.1: Pfizer's case studies on solvent recovery by distillation [13].

Case Study	Selamectin	Nelfinavir	Hydrocortisone
Waste composition (wt%)	Acetonitrile (72) + Acetone (28)	IPA (86) + THF (14)	Toluene (91) + Acetone (9)
Waste mass (kg year^{-1})	84,500	78,700	257,600
Desired recovery (wt%)	Acetonitrile, acetone (98, 98)	IPA (98)	Toluene (99.8), Acetone (98)
Carbon footprint reduction (kg eq. CO_2 year^{-1})	253,300	253,259	1,145,000
Cost savings (k$ year^{-1})	236.8	98.5	293.3

Some companies offer different solvent recovery distillation systems [14–16]. Solutex designed two types of systems with versatile capacities for non-flammable and flammable solvents [14]. GEA's *multiple-effect distillation* technology is designed to recover solvents up to 98% in the medical membrane manufacturing process [17]. Their technology functions by having multiple stages. Every stage reuses the energy from the previous stage; i.e., the introduced heat is used several times at different temperatures and pressure levels. Such optimization of the process heat system results in lower overall energy consumption and eventually lower environmental burden. Progressive Recovery Inc. provides an example of integrating a solvent recycling unit into a closed-loop solvent-recycling plant [18]. Figure 8.4 presents the process integration where the waste solvent is collected into a spent solvent tank and then transferred to the distillation unit. The solvent is then separated from the impurities, and this solvent is directed to the clean solvent tank. In this process, only the contaminants are disposed of and the newly obtained solvent is reused. Note that the system requires several vessels to collect and store spent and recovered solvents, which has a significant footprint on-site.

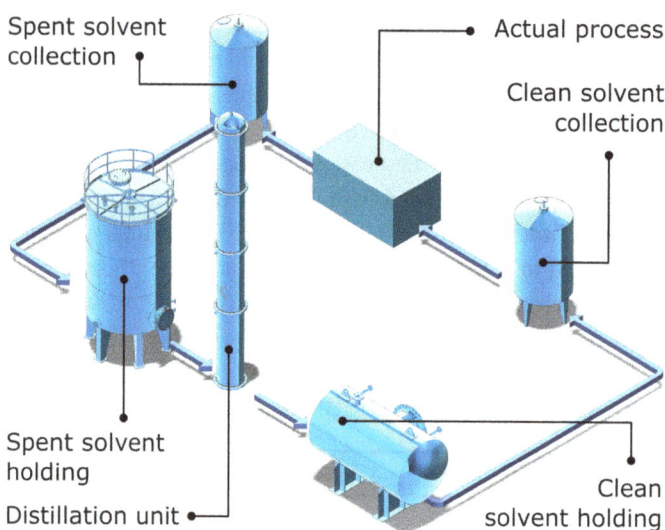

Figure 8.4: Closed-loop solvent recycling introduced by PRI. Adapted from [18].

8.2 Adsorption processes

Adsorption processes are often used for solvent recovery due to their efficiency and significantly lower energy requirement than distillation. Adsorption refers to the enrichment of gaseous or dissolved substances on the boundary surface of a solid, known as the adsorbent [5]. These solids have so-called *active sites* on their surface where the atoms' binding forces are not fully saturated, thus allowing them to bind to other atoms or molecules. Desorption represents the opposite process of releasing the bound molecules.

A high surface area and porosity are needed to achieve high efficiency. Adsorption-based solvent recovery is most often performed with non-selective adsorbents; i.e., the applied solid does not differentiate the impurities and scavenges almost everything from the bulk liquid. However, in some instances, there are solutes of high value in the liquid that need to be recovered simultaneously. High-selectivity adsorbents, such as *molecularly imprinted polymers*, are needed to capture and recover these solutes in high purity. We can differentiate many adsorbents based on their chemical nature, application area, selectivity, and surface area. Activated carbon is the most common non-selective adsorbent for solvent recovery because it can efficiently adsorb organic and other non-polar compounds to its hydrophobic surface [5]. The morphology and the chemical nature of activated carbon vary from those of crude graphite; activated carbon has a random amorphous structure and highly porous materials with pore sizes from large visible cavities to those at the molecular level. *Activated carbon* can be derived from renewable resources such as waste biomass, lignin, fruit, and nutshells [19]. Silica gel and activated alumina adsorb polar substances such as water. *Molecular sieve* zeolites function by releasing their water content to allow the adsorption of other molecules. However, dry molecular sieves can be used to capture water molecules in organic solvents. Solvent recovery includes dehydration of organic solvents, which can be done via molecular sieves.

Adsorbents can be used to treat bulk liquid streams or vapors in batch or continuous processes. The latter process requires the adsorbents to be packed into columns. The gas stream is passed through an adsorption unit containing activated carbon to remove and recycle solvent vapors from the air. Following the adsorption of solvents, solvent recovery is achieved by steam or hot gas desorption, which means that the solvents are stripped from the activated carbon. The released material is separated into an aqueous and an organic phase [5]. Furthermore, water removal from solvent waste represents one of the most important application fields of adsorption. Water content can have some undesired consequences during the solvent's reuse, which may manifest in solubility issues [4].

In the following example, an adsorption-based solvent recovery was implemented during the continuous purification process of roxithromycin API using a continuous membrane cascade [20]. A packed bed adsorption column was utilized for the solvent recovery, and an inexpensive adsorbent (charcoal) was used in the adsorption column. Charcoal can be obtained from renewable sources such as coconut shells. The hybrid, continuous process generated significantly less waste than the diafiltration process without solvent recovery. In particular, the E-factor with the solvent recovery stage was approx. 20 kg kg^{-1}, whereas the E-factor without the solvent recovery unit was found to be approximately ten times higher. The solvent waste was approx. 99.9% of the E-factor of the continuous cascade without the solvent recovery unit. This example demonstrates the importance of solvent recovery during separation processes. Incorporating the adsorption column allowed the regeneration of the solvent waste, and only 70% of the E-factor originated from the solvent waste. The charcoal waste accounted for 30% of the E-factor. The process could have been further improved by regenerating and reusing the activated charcoal. Overall, the amount of solvent waste was reduced by 93% by applying an adsorption-based solvent recovery unit.

8.3 Membrane-based solvent recovery processes and their comparison with distillation and adsorption

Membrane processes are among the most critical technologies to drive process intensification in solvent recovery, among other applications [21]. They are simple, easy to scale up, and energy-efficient. Various membrane technologies have been employed for solvent recovery: organic solvent nanofiltration (OSN) [22], organic solvent pressure-assisted osmosis (OSPAO) [23], organic solvent forward osmosis (OSFO) [24], and vapor permeation [25], as depicted in Figure 8.5. Refer to Section 7.4 for the fundamentals of membrane processing, and follow the QR codes on this page to watch animations on how membrane modules work. Most membrane-based separations are driven by a pressure gradient, which is an advantage over conventional temperature gradient-driven distillation. There is no need for phase change, and therefore the formation of impurities from thermal degradation does not occur. Various process configurations can be constructed from multiple membrane units. For instance, membrane units arranged in a series form of membrane cascades achieve better separation efficiency and mitigate solvent consumption [21, 26].

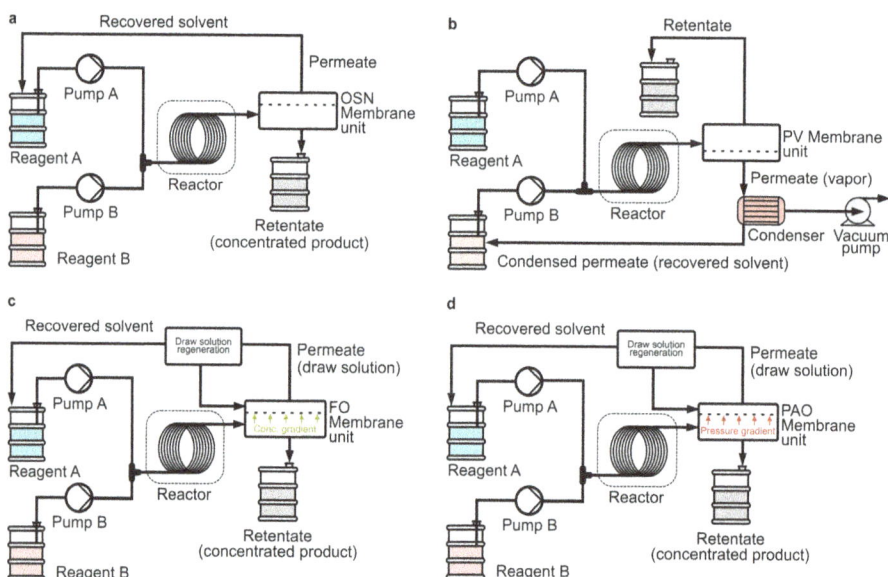

Figure 8.5: Schematic process diagrams comparing membrane-based solvent recovery processes. (a) OSN operates at pressures up to 40 bar. (b) Pervaporation (PV) is a separation process where only part of the feed solution is vaporized and then condensed, resulting in the formation of the permeate. (c) Organic solvent forward-osmosis (FO) utilizes the natural energy of osmotic pressure for the solvent transport. (d) Organic solvent pressure-assisted osmosis (PAO) applies external hydraulic pressure to enhance the solvent flux. The diluted draw solution is regenerated by various means such as distillation, direct filtration, or evaporation.

Membrane processes can be easily coupled with reactors, distillation, adsorption, or chromatography to perform hybrid processes [21]. Nanofiltration coupled with continuous flow reactors can assist with *in situ* solvent, catalyst, and reagent recycling [35, 42], significantly reducing the overall process's E-factor and carbon footprint. For more details on hybrid processes with membranes, see Section 7.4. Besides reactors, nanofiltration can be coupled with continuous adsorption units. Adsorptive agricultural waste utilization using olive leaves was performed with the assistance of membranes [27]. *In situ* solvent recovery was realized with nanofiltration, which recovered 97.5% of the ethyl acetate solvent used during the previous extraction and adsorption processes. The carbon footprint was reduced by 44.5%.

Membrane technologies are sustainable process solutions for product concentration, impurity removal, and solvent exchange, among others. However, *constant-volume diafiltration* (CVD) has a high solvent intensity [21] because it requires a continuous inlet stream of solvent during the separation process. Figure 8.6 shows the correlation between the solvent consumption of CVD and the rate of impurity removal during a CVD for the case of impurity with a 50% rejection. The solvent consumption exponentially increases with the impurity removal. Consequently, processes to achieve high purity, which is a standard requirement for fine chemicals, require large amounts of solvents, in particular APIs. Initially, the coupling of adsorptive solvent recovery utilizing charcoal was proposed to overcome CVD's high solvent consumption [21]. Process optimization revealed that a 70% solvent recovery is the optimum, considering the cost of charcoal, fresh solvent purchase, and waste disposal. This example demonstrates that although 100% solvent recovery is possible to achieve, it is economically and environmentally less sustainable. Besides the adsorption-based solvent recovery, membrane cascades are also suitable for *in situ* solvent recovery during CVD [26].

Concerning other solvent recovery technologies, the effectiveness of OSN is demonstrated through the comparison of carbon footprint (Figure 8.7). Distillation processes have a larger carbon footprint than adsorption processes up to 70% solvent recovery because of the required phase change, resulting in high energy consumption [21]. After that point, the CO_2 intensity of adsorptive solvent recovery sharply rises because the recovery of the remaining 30% of solvent waste necessitates the use of a large adsorbent amount due to the lower packed bed utilization. Lower bed utilization results in higher purity and helps to achieve a higher solvent recovery rate. Still, there is a *trade-off* as it requires more adsorbent per unit volume of recovered solvent. As a solvent recovery technique, adsorption produces a substantial amount of solid waste, and at over 70% solvent recovery, the solid waste generation outweighs the solvent savings [21]. The dependence of the CO_2 intensity of nanofiltration on the solvent recovery rate is negligible. Note that the line is not flat but has a slope, which results from both the waste generated by disposing the membranes at the end of their lifetime and the energy consumption of the high-pressure and recirculation pumps. The comparative analysis of the three solvent recovery processes demonstrates CO_2 intensity's usefulness as a green metric. Distillation, adsorption, and membrane pro-

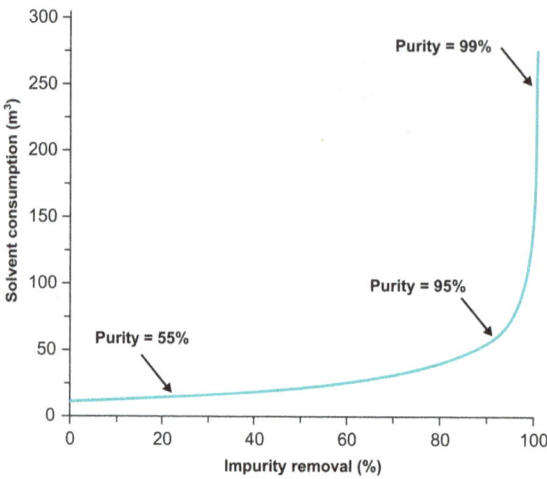

Figure 8.6: Correlation between solvent consumption and impurity removal in a diafiltration process (impurity rejection = 50%, product and impurity concentration = 1 gL^{-1}, system volume = 10 m^3). Adapted from [21]. The cost of high product purity comes at the price of high solvent consumption during diafiltration.

cesses have different pros and cons and pose different environmental burdens, from waste generation to energy consumption. These burdens cannot be directly compared, and therefore the use of equivalent CO_2 is a useful metric that allows the relative comparison of apples and oranges.

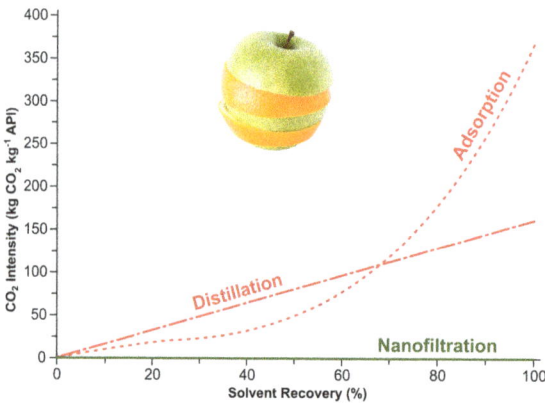

Figure 8.7: Comparison of carbon footprints of solvent recovery technologies calculated for methanol. Adapted from [21].

The amount of top solvent waste generated in the United States is reported in Table 8.2. The energy required and the associated carbon footprint to recover the solvent waste

Table 8.2: Comparison of energy requirements and equivalent carbon footprints for solvent recovery of the top solvent waste generated in the United States. The comparison considers distillation and OSN processes and does not take into account heat integration. [22] Q_{dist} and Q_{OSN} are energy requirements for solvent recovery by distillation and OSN.

Solvent	Waste solvent generated in 2006 [10^6 kg y^{-1}]	Q_{dist} [kWh]	Q_{OSN} [kWh]	Q_{dist}/Q_{OSN}	CO_2 footprint [10^6 kg y^{-1}]
Methanol	44.8	150	0.023	6453	18
Dichloromethane	22.3	111	0.014	8010	3
Toluene	12.1	197	0.021	9278	12
Acetonitrile	7.9	141	0.023	6029	3
Chloroform	3.71	131	0.012	10543	0.4
n-Hexane	2.99	149	0.028	5300	3
n-Butyl alcohol	2.86	223	0.023	9788	2
N,N-Dimethylformamide	2.79	244	0.019	12569	2
N-Methyl-2-pyrrolidone	2.02	303	0.018	16930	1
Xylene	1.47	208	0.021	9748	1
1,1,2-Trichloroethane	1.23	194	0.013	15090	0.2
Methyl tert-butyl ether	1.2	126	0.025	5062	1
Ethylene glycol	0.82	337	0.017	20285	0.3

using distillation and OSN are compared. In distillation, the energy demand increases with increasing heat of the solvent's vaporization because separating less volatile compounds requires more energy investment. Note that this basic calculation and comparison do not consider heat integration, which can significantly reduce the energy consumption during distillation processes. Most OSN processes use alcohols, followed by polar aprotic solvents (Figure 8.8) [28]. Although most alcohols fall into the preferred solvent category, only 5.6% of polar aprotic solvents, such as 2-methyltetrahydrofuran, propylene, and dimethyl carbonate, are preferred. The figure shows that OSN membranes can process a wide variety of solvents covering different chemical classes, polarities, and viscosities.

The experimentally validated nanofiltration-based solvent recovery systems are presented in Table 8.3. The table demonstrates that the field of nanofiltration has come a long way, and substantial improvements are accomplished in versatile application areas by removing small-size impurities, concentrating products, and exchanging solvents between unit operations. Note that all solvent recovery applications by membranes require *quasi-complete rejection* (eq. (7.2)) of the solutes (>99%). Rejections lower than 100% compromise the purity and result in contamination of the recovered solvent with solutes. The lower the rejection from the ideal 100%, the lower the purity of the recovered solvent. Larger solutes with high molecular weights have higher rejections, which makes the solvent recovery easier.

Even though organic solvent nanofiltration is a promising and widely used technique for organic solvent recovery, it requires high pressure (up to 40 bars). Thus, its

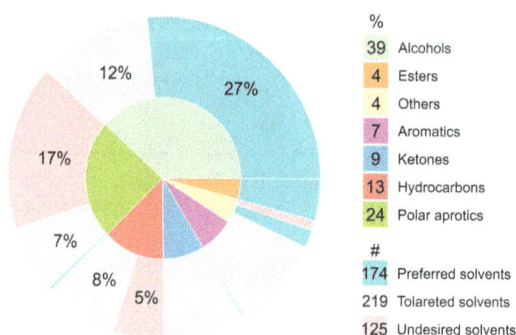

Figure 8.8: The frequency of use of the different classes of organic solvents in membrane processes. The breakdown focuses on organic solvent nanofiltration and therefore excludes water. Green, gray, and pink indicate preferred, tolerated, and undesired solvents based on industrial solvent selection guides [29]. Data obtained from [28].

Table 8.3: Solvent recovery examples using organic solvent nanofiltration (MW, molecular weight; n. d., not determined).

Year	Application	MW [g mol^{-1}]	Rejection [%]	Membrane	Solvent	Solvent saving [%]
2000	Lube oil dewaxing [30]	500	95	Polyimide-based	Methyl ethyl ketone, toluene	n. d.
2010	Pharmaceutical [31]	>1000, 675	>99	DuraMem 300, 1000	Tetrahyrofuran	n. d.
2013	Catalyst recovery [32]	1044, 300	>99	DuraMem 300, 500	Tetrahydrofuran	96
2012	Crystallization [33]	600	>99	PuraMem 280	Isopropyl alcohol	80
2014	Pharmaceutical [21]	260	>99	DuraMem 150	Methanol	
2015	Cyclic peptide formation [34]	1069	>96	Inopor, TiO$_2$ ceramic membrane	Ethanol, acetic acid, water	74
2016	Fine chemical [26]	110	>99	Polyimide-poly (ether-ether-ketone)	Acetone	n. d.
2019	Catalyst recovery [35]	1600	>99	DuraMem 900	2-Methyltetrahydrofuran	50
2019	Natural product extraction [36]	286–540	97	Polybenzimidazole	Ethyl acetate	99

operation can result in additional costs. This issue can be solved by applying forward osmosis (FO), which utilizes the potential chemical difference across a semi-permeable membrane to transport a solvent from a lower to a higher concentration. In this example, OSFO was used for the simultaneous concentration of the API and solvent recovery [24]. The organic solvent was transported from the feed side to the draw side, whereas the API was rejected by the membrane and concentrated at the feed side. Thus, the solvent molecules were spontaneously transported against the osmotic pressure gradient, and there was no need for external hydraulic pressure. Furthermore, OSFO had a low fouling tendency and minimal irreversible fouling, and thus, it could be used for treating high-concentration feed solutions.

Although OSFO was found to possess advantages over other membrane-based separation processes, it suffered from some limitations, such as low flux. This issue could be overcome by applying an external hydraulic pressure on the feed side to help transport solvent molecules through the membrane, i.e., in a process called OSPAO. This technique can be utilized for the simultaneous concentration of APIs and solvent recovery. Compared to OSFO, OSPAO employs a low external pressure (1 bar) in addition to the osmotic pressure difference [23]. The results showed that the solvent flux significantly increased with increasing external pressure and with increasing concentrations of the draw solution.

Pervaporation is a membrane-based separation technique with transport through porous or dense membranes [37]. Pervaporation represents a combination of permeation and vaporization. The feed stream is a liquid mixture, while the permeate stream is recovered in a vapor form by applying a vacuum or sweeping gas; i.e., a phase change from liquid to vapor occurs. Thus, in this process, only a part of the feed mixture is vaporized. Processes with phase changes, such as in distillation, require a large amount of energy. However, pervaporation overcomes this drawback by processing only minor components (less than 10 wt%) of the liquid mixtures and using high-selectivity membranes [38]. The former aspect is responsible for reducing energy consumption because only the feed stream's minor components permeate through the membrane. In other words, only the minor components are subject to a phase change and consume the latent heat, which results in lower energy consumption than distillation. This technique enables the separation of mixtures, which is difficult to achieve using distillation, extraction, or adsorption. Hence, it is suitable for the separation of azeotrope and close-boiling mixtures, thermally sensitive compounds, molecules with a similar shapes or sizes, and removing low-concentration species [39]. Pervaporation found its main application in the dehydration of organic solvents (e.g., alcohols, ethers, acids, esters). However, it is also used to remove organic compounds from diluted aqueous streams (e.g., volatile organic compounds) and separate organic-organic mixtures. Pervaporation is used as a sustainable drying process for the recovery of tetrahydrofuran [40]. It is integrated with constant-volume distillation resulting in a hybrid process, which allows annual process waste reduction by 122,300 kg, energy reduction by 684,500 kWh, and operating cost reduction by 509,150 USD, which correspond to a de-

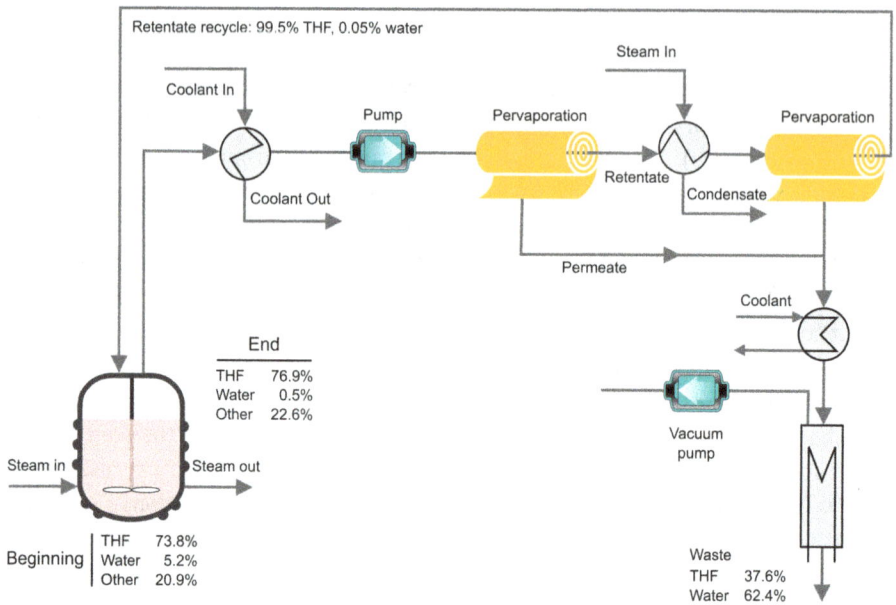

Figure 8.9: Schematic process diagram for a pilot-scale constant-volume distillation coupled with pervaporation used in the sustainable drying and recovery of tetrahydrofuran solvent. Adapted from [40].

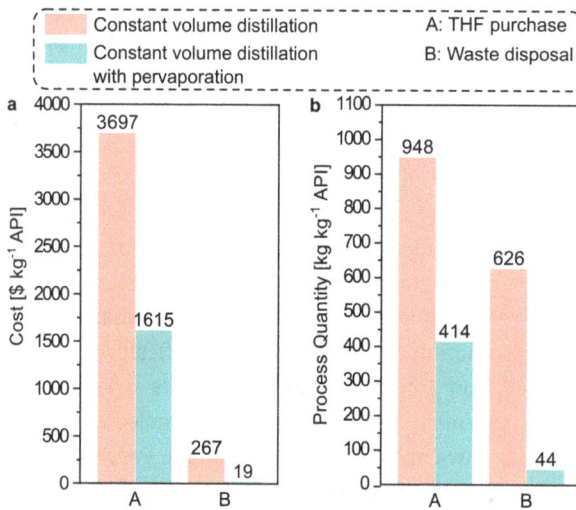

Figure 8.10: Comparison of cost (a) and process quantity (b) improvements at pilot-scale constant-volume distillation due to coupling with pervaporation. The sustainability of the drying and recovery of tetrahydrofuran solvent considerably improved. Adapted from [40]. See Figure 8.9 for the schematic process diagram.

crease in 93%, 80%, and 99%, respectively (Figures. 8.9 and 8.10). Moreover, hybrid silica membranes can be used to purify and dehydrate solvents like acetone and methyl ethyl ketone from binary and multicomponent mixtures utilizing the pervaporation technique [41].

8.4 Tools for solvent recovery process design

Solvent waste comprises a significant portion of chemical waste. Finding the most viable solution for solvent recovery can be challenging because of diverse application areas. Therefore, solvent recovery tools can assist in identifying the solvent waste mitigation strategies to reduce the solvent waste released into the environment. Computational tools can offer a useful solution to design the solvent recovery process. Stengel et al. [43] developed a software tool to quickly model potential solvent recovery systems by inputing the data of a given waste stream, the purity and the desired recovery.

Chea et al. [44] developed a framework for comparing solvent recovery options for any process by collecting data such as solvent use, properties, disposal methods, and prices. They employed General Algebraic Modeling Systems (GAMS) to evaluate feasible recovery pathways by including mathematical models that focus on material and energy balance, utility requirements, equipment design, and costs involved in common separation technologies such as distillation, ultrafiltration, membranes, evaporation, and extraction. In the conducted case study, it was revealed that for recovery of isopropanol from a celecoxib waste stream, the optimal pathway was to use aqueous two-phase extraction, followed by membrane separation and decantation to achieve the targeted isopropanol purity. To comprehensively evaluate environmental impact of different solvent recovery technologies, life cycle inventory (LCI) analysis models can help to calculate waste-solvent specific inventory flows, such as emission flows and ancillary uses, as a function of few input parameters (such as waste-solvent composition and treatment technology) [45]. Ecosolvent is a LCI tool developed at ETH Zurich, which combines LCI models for distillation and thermal treatment. Various waste solvent treatment options can be carried out for specific waste-solvent compositions to identify the most environmentally preferable one [46]. Furthermore, machine learning (ML) has emerged as an important tool for solvent and technology selection. ML can be used to design and develop OSN membranes for solvent recovery [47,48]. Refer to Section 7.8 for more information on machine learning in membrane separations.

Bibliography

[1]   Smallwood, I. M.; *Solvent Recovery Handbook*, 2nd Edition, Blackwell Science: Florida, USA, 2002.

[2]   Chea, J. D.; Lehr, A. L.; Stengel, J. P.; Savelski, M. J.; Slater, C. S.; Yenkie, K. M. Evaluation of Solvent Recovery Options for Economic Feasibility through a Superstructure-Based Optimization Framework. *Ind. Eng. Chem. Res.* **2020**, *59* (13), 5931–5944. 10.1021/acs.iecr.9b06725.

[3]   Lonza Engineering Ltd. Solvent Recovery https://lonza.picturepark.com/Website/?Action=downloa dAsset&AssetId=31328.

[4]   Cseri, L.; Razali, M.; Pogany, P.; Szekely, G. *Organic Solvents in Sustainable Synthesis and Engineering*; Elsevier Inc, 2018; 10.1016/B978-0-12-809270-5.00020-0.

[5]   Henning, K. D. Solvent Recycling, Removal, And Degradation. In Handbook of Solvents; Elsevier, 2014; Vol. 2, pp 787–861. 10.1016/B978-1-895198-65-2.50011-9.

[6]   Johnson, J. A.; Ablin, D. W.; Ernst, G. A. et al. *Efficient Petrochemical Processes: Technology, Design and Operation*; John Wiley & Sons, 2019.

[7]   Dunn, P. J.; Wells, A. S.; Williams, M. T. *Green Chemistry in the Pharmaceutical Industry*; Dunn, P. J.; Wells, A. S.; Williams, M. T., Eds.; Wiley, 2010; 10.1002/9783527629688.

[8]   Brouwer, T.; Schuur, B. Bio-Based Solvents as Entrainers for Extractive Distillation in Aromatic/ Aliphatic and Olefin/Paraffin Separation. *Green Chem.* **2020**, *22* (16), 5369–5375. 10.1039/d0gc01769h.

[9]   Modla, G.; Lang, P. Removal and Recovery of Organic Solvents from Aqueous Waste Mixtures by Extractive and Pressure Swing Distillation. *Ind. Eng. Chem. Res.* **2012**, *51* (35), 11473–11481. 10.1021/ ie300331d.

[10]   Savelski, M. J.; Slater, C. S.; Tozzi, P. V.; Wisniewski, C. M. On the Simulation, Economic Analysis, and Life Cycle Assessment of Batch-Mode Organic Solvent Recovery Alternatives for the Pharmaceutical Industry. *Clean Technol. Environ. Policy.* **2017**, *19* (10), 2467–2477. 10.1007/s10098-017-1444-8.

[11]   Ramzan, N.; Degenkolbe, S.; Witt, W. Evaluating and Improving Environmental Performance of HC's Recovery System: A Case Study of Distillation Unit. *Chem. Eng. J.* **2008**, *140* (1–3), 201–213. 10.1016/j. cej.2007.09.042.

[12]   Continuous Processing | Solvent Recovery at Pfizer: A Continuous Solution for Small Waste Streams | Pharmaceutical Manufacturing https://www.pharmamanufacturing.com/articles/2011/150/?start=1 (accessed May 23, 2020).

[13]   Cavanagh, E. *A New Software Tool to Environmentally and Economically Evaluate Solvent Recovery in the Pharmaceutical Industry*; Rowan University, 2014.

[14]   Solvent Recycling Systems – Solvent Recovery – Solutex Ltd https://www.solutex.co.uk/products/sol vent-recycling-systems/ (accessed May 23, 2020).

[15]   Solvent Recovery at Pharmaceutical and Chemical Production Plants – Thermal Kinetics https://thermalkinetics.net/solvent-recovery-at-pharmaceutical-and-chemical-production-plants/ (accessed May 23, 2020).

[16]   GEA. Solvent Recovery by means of distillation https://www.gea.com/en/binaries/solvent-recovery-distillation-isopropanol-ethanol_tcm11-49825.pdf.

[17]   As good as new: solvent recovery https://www.gea.com/en/news/insights/2017/as-good-as-new-solvent-revovery.jsp (accessed May 24, 2020).

[18]   Solvent Recovery Systems | Efficient Fast ROI | Progressive Recovery, Inc. https://prisystems.com/ products/industrial-process-systems/solvent-recycling/ (accessed May 18, 2020).

[19]   Ukanwa, K. S.; Patchigolla, K.; Sakrabani, R.; Anthony, E.; Mandavgane, S. A Review of Chemicals to Produce Activated Carbon from Agricultural Waste Biomass. *Sustain.* **2019**, *11* (22), 1–35. 10.3390/ su11226204.

[20]   Peeva, L.; da Silva Burgal, J.; Valtcheva, I.; Livingston, A. G. Continuous Purification of Active Pharmaceutical Ingredients Using Multistage Organic Solvent Nanofiltration Membrane Cascade. *Chem. Eng. Sci.* **2014**, 116, 183–194. 10.1016/j.ces.2014.04.022.

[21] Kim, J. F.; Szekely, G.; Schaepertoens, M.; Valtcheva, I. B.; Jimenez-Solomon, M. F.; Livingston, A. G. In Situ Solvent Recovery by Organic Solvent Nanofiltration. *ACS Sustain. Chem. Eng.* **2014**, *2* (10), 2371–2379. 10.1021/sc5004083.

[22] Szekely, G.; Jimenez-Solomon, M. F.; Marchetti, P.; Kim, J. F.; Livingston, A. G. Sustainability Assessment of Organic Solvent Nanofiltration: From Fabrication to Application. *Green Chem.* **2014**, *16* (10), 4440–4473. 10.1039/c4gc00701h.

[23] Cui, Y.; Chung, T. S. Solvent Recovery via Organic Solvent Pressure Assisted Osmosis. *Ind. Eng. Chem. Res.* **2019**, *58* (12), 4970–4978. 10.1021/acs.iecr.8b06115.

[24] Cui, Y.; Chung, T. S. Pharmaceutical Concentration Using Organic Solvent Forward Osmosis for Solvent Recovery. *Nat. Commun.* **2018**, *9* (1), 1–9. 10.1038/s41467-018-03612-2.

[25] Leemann, M.; Eigenberger, G.; Strathmann, H. Vapour Permeation for the Recovery of Organic Solvents from Waste Air Streams: Separation Capacities and Process Optimization. *J. Membr. Sci.* **1996**, *113* (2), 313–322. 10.1016/0376-7388(95)00130-1.

[26] Schaepertoens, M.; Didaskalou, C.; Kim, J. F.; Livingston, A. G.; Szekely, G. Solvent Recycle with Imperfect Membranes: A Semi-Continuous Workaround for Diafiltration. *J. Membr. Sci.* **2016**, *514*, 646–658. 10.1016/j.memsci.2016.04.056.

[27] Didaskalou, C.; Buyuktiryaki, S.; Kecili, R.; Fonte, C. P.; Szekely, G. Valorisation of Agricultural Waste with an Adsorption/Nanofiltration Hybrid Process: From Materials to Sustainable Process Design. *Green Chem.* **2017**, *19* (13), 3116–3125. 10.1039/c7gc00912g.

[28] Le Phuong, H.; Blanford, C.; Szekely, G. Reporting the Unreported: The Reliability and Comparability of the Literature on Organic Solvent Nanofiltration. *Green Chem.* **2020**, *22* (11), 3397–3409. 10.1039/D0GC00775G.

[29] Alder, C. M.; Hayler, J. D.; Henderson, R. K.; Redman, A. M.; Shukla, L.; Shuster, L. E.; Sneddon, H. F. Updating and Further Expanding GSK's Solvent Sustainability Guide. *Green Chem.* **2016**, *18* (13), 3879–3890. 10.1039/c6gc00611f.

[30] White, L. S.; Nitsch, A. R. Solvent Recovery from Lube Oil Filtrates with a Polyimide Membrane. *J. Membr. Sci.* **2000**, *179* (1–2), 267–274. 10.1016/S0376-7388(00)00517-2.

[31] Sereewatthanawut, I.; Lim, F. W.; Bhole, Y. S.; Ormerod, D.; Horvath, A.; Boam, A. T.; Livingston, A. T. Demonstration of Molecular Purification in Polar Aprotic Solvents by Organic Solvent Nanofiltration. *Org. Process Res. Dev.* **2010**, *14* (3), 600–611. 10.1021/op100028p.

[32] Siew, W. E.; Ates, C.; Merschaert, A.; Livingston, A. G. Efficient and Productive Asymmetric Michael Addition: Development of a Highly Enantioselective Quinidine-Based Organocatalyst for Homogeneous Recycling via Nanofiltration. *Green Chem.* **2013**, *15* (3), 663–674. 10.1039/c2gc36407g.

[33] Rundquist, E. M.; Pink, C. J.; Livingston, A. G. Organic Solvent Nanofiltration: A Potential Alternative to Distillation for Solvent Recovery from Crystallisation Mother Liquors. *Green Chem.* **2012**, *14* (8), 2197–2205. 10.1039/c2gc35216h.

[34] Ormerod, D.; Noten, B.; Dorbec, M.; Andersson, L.; Buekenhoudt, A.; Goetelen, L. Cyclic Peptide Formation in Reduced Solvent Volumes via In-Line Solvent Recycling by Organic Solvent Nanofiltration. *Org. Process Res. Dev.* **2015**, *19* (7), 841–848. 10.1021/acs.oprd.5b00103.

[35] Kisszekelyi, P.; Alammar, A.; Kupai, J.; Huszthy, P.; Barabas, J.; Holtzl, T.; Szente, L.; Bawn, C.; Adams, R.; Szekely, G. Asymmetric Synthesis with Cinchona-Decorated Cyclodextrin in a Continuous-Flow Membrane Reactor. *J. Catal.* **2019**, *371*, 255–261. 10.1016/j.jcat.2019.01.041.

[36] Voros, V.; Drioli, E.; Fonte, C.; Szekely, G. Process Intensification via Continuous and Simultaneous Isolation of Antioxidants: An Upcycling Approach for Olive Leaf Waste. *ACS Sustain. Chem. Eng.* **2019**, *7* (22), 18444–18452. 10.1021/acssuschemeng.9b04245.

[37] Crespo, J. G.; Brazinha, C. Fundamentals of Pervaporation. In *Pervaporation, Vapour Permeation and Membrane Distillation*; Elsevier, 2015; pp 3–17. 10.1016/B978-1-78242-246-4.00001-5.

[38] Shao, P.; Huang, R. Y. M. Polymeric Membrane Pervaporation. *J. Membr. Sci.* **2007**, *287* (2), 162–179. 10.1016/j.memsci.2006.10.043.

[39] Nagy, E. Pervaporation. In *Basic Equations of Mass Transport Through a Membrane Layer*; Elsevier, 2019; pp 429–445. 10.1016/B978-0-12-813722-2.00016-9.

[40] Stewart Slater, C.; Savelski, M. J.; Moroz, T. M.; Raymond, M. J. Pervaporation as a Green Drying Process for Tetrahydrofuran Recovery in Pharmaceutical Synthesis. *Green Chem. Lett. Rev.* **2012**, *5* (1), 55–64. 10.1080/17518253.2011.578590.

[41] La Rocca, T.; Carretier, E.; Dhaler, D.; Louradour, E.; Truong, T.; Moulin, P. Purification of Pharmaceutical Solvents by Pervaporation through Hybrid Silica Membranes. *Membranes.* **2019**, *9* (7), 10.3390/membranes9070076.

[42] Fodi, T.; Didaskalou, C.; Kupai, J.; Balogh, G.; Huszthy, P.; Szekely, G. Nanofiltration-Enabled in Situ Solvent and Reagent Recycle for Sustainable Continuous-Flow Synthesis. *ChemSusChem.* **2017**, *10* (17), 3435–3444. 10.1002/cssc.201701120.

[43] Stengel, J. P.; Lehr, A. L.; Aboagye, E. A.; Chea, J. D.; Yenkie, K. M. Development of an Interactive Software Tool for Designing Solvent Recovery Processes. *Ind. Eng. Chem. Res.* **2023**, *62* (5), 2090–2103. 10.1021/acs.iecr.2c02920.

[44] Chea, J. D.; Christon, A.; Pierce, V.; Reilly, J. H.; Russ, M.; Savelski, M.; Slater, C. S.; Yenkie, K. M. *Framework for Solvent Recovery, Reuse, and Recycling In Industries*; 2019; pp 199–204. 10.1016/B978-0-12-818597-1.50032-1.

[45] Capello, C.; Hellweg, S.; Badertscher, B.; Betschart, H.; Hungerbühler, K. Environmental Assessment of Waste-Solvent Treatment Options. *J. Ind. Ecol.* **2007**, *11* (4), 26–38. 10.1162/jiec.2007.1231.

[46] Ecosolvent – Safety and Environmental Technology Group | ETH Zurich https://emeritus.setg.ethz.ch/research/downloads/software---tools/ecosolvent.html (accessed Jan 31, 2024).

[47] Wang, M.; Shi, G. M.; Zhao, D.; Liu, X.; Jiang, J. Machine Learning-Assisted Design of Thin-Film Composite Membranes for Solvent Recovery. *Environ. Sci. Technol.* **2023**, *57* (42), 15914–15924. 10.1021/acs.est.3c04773.

[48] Xu, Q.; Gao, J.; Feng, F.; Chung, T. -S.; Jiang, J. Synergizing Machine Learning, Molecular Simulation and Experiment to Develop Polymer Membranes for Solvent Recovery. *J. Memb. Sci.* **2023**, *678*, 121678. 10.1016/j.memsci.2023.121678.

9 Process analytical technology

Martin Gede, Gyorgy Szekely

9.1 Introduction

Chemists and chemical engineers aim to minimize the variability and maximize the efficiency of the industrial processes they design and operate. A thorough understanding of these manufacturing and chemical processes is essential to effectively and safely control their related systems. Conventionally, chromatographic methods are the primary tool for monitoring reactions, identifying impurities, and designing optimal process parameters. The major drawback of these techniques lies in their offline nature. Samples must be collected and prepared prior to analysis, which introduces an inevitable time-lag between the sampling and the analysis, and the analysis and the control actions. *Real-time analytical tools* provide an attractive alternative to traditional offline methods by providing a continuous stream of real-time data. Such tools are referred to as process analytical technology (PAT) analyzers. PAT includes simple measuring devices, such as temperature and pressure sensors, but more commonly involves inline analyzers such as mid-infrared (IR), near-IR (NIR), nuclear magnetic resonance (NMR), and Raman spectrometers [1]. These analytical tools provide information on the levels of single ions and molecules. However, other methods are suitable for characterizing structural features ranging from 0.1 to 100 nm in scale (Figure 9.1) [2]. Scattering methods can track morphological changes and aggregation in real-time, while imaging microscopy techniques provide information about the structural changes in lengths in the nanoscopic to the microscopic range.

In 2004, the Food and Drug Administration (FDA) launched active pharmaceutical ingredient (API) manufacturing guidelines that define PAT "as a system for designing, analyzing, and controlling manufacturing processes through timely measurements of critical quality and performance attributes" [3]. This PAT system can be challenging to grasp. Therefore, a simple example is provided and illustrated in Figure 9.2. In this example, a chemical compound with a simple reduction of the starting material needs to be manufactured. First, the substrate is dissolved, and the catalyst is added, followed by the application of gaseous hydrogen until the reaction is complete. Samples from the mixture are taken to determine the end-point of the reaction, and after appropriate sample preparation, the conversion is monitored using a chromatography technique. The sampling process in this example takes approx. 10 min. However, once the product is formed, a less reactive functional group (–Cl) also starts to be slowly reduced, resulting in undesired by-product formation. Even if we were lucky enough to catch the reaction end-point precisely, the reaction would have proceeded further by the time the analytical results had become available, generating the unwanted by-product. Predicting the end-point of the reaction is possible based on experience, but with some uncertainty; however, heterogeneous reactions are known to be difficult to predict. A PAT

https://doi.org/10.1515/9783111028163-009

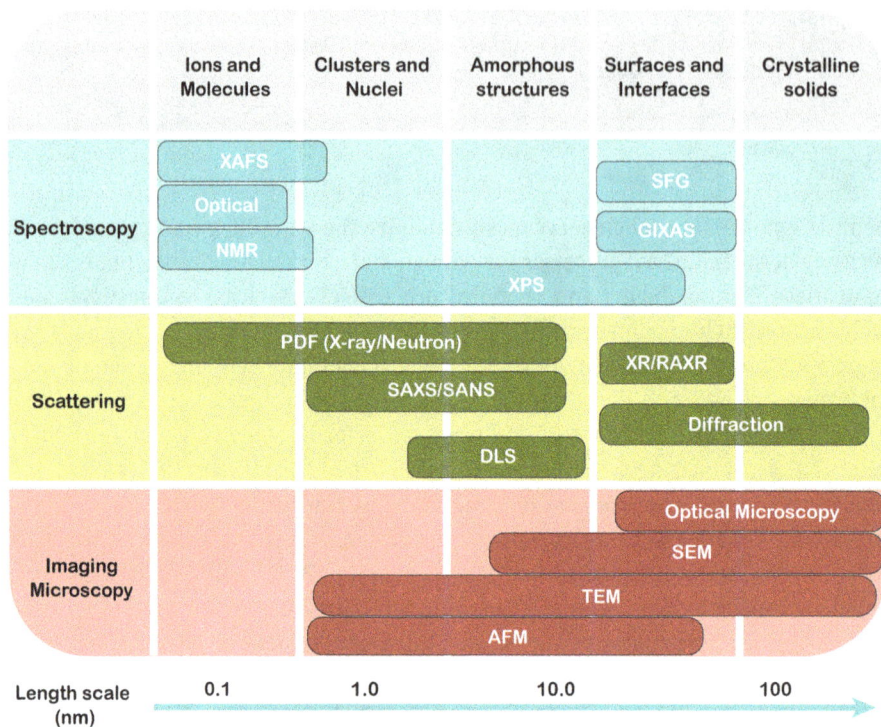

Figure 9.1: Schematic of commonly used analytical tools, including spectroscopy, scattering, and imaging microscopy techniques. XAFS, X-ray absorption fine structure; SFG, sum frequency generation; GIXAS, glancing incident X-ray absorption spectroscopy; XPS, X-ray photoelectron spectroscopy; PDF, pair distribution function; SAXS/SANS, small-angle X-ray scattering/small-angle neutron scattering; XR/RAXR, X-ray reflectivity/resonant anomalous X-ray reflectivity; DLS, dynamic light scattering; SEM, scanning electron microscopy; TEM, transmission electron microscopy; AFM, atomic-force microscopy.

method can overcome this issue by monitoring the conversion in real-time and enabling the operator to stop the reaction precisely when it is complete. Since chemical reactions involve converting functional groups, spectroscopy-based PAT methods provide an adequate solution in this example. PAT can help achieve a higher yield and a product with no or minimum impurity formation and thus remove the need for time-consuming purification processes. Avoiding both the tedious manual sampling for off-line analysis and the sometimes hazardous reactions is also beneficial.

The sustainability aspects of PAT, as well as the practical case studies, are discussed in the following sections. For further information and in-depth understanding, the reader is referred to the following literature: Assessment of recent process analytical technology (PAT) trends: A multiauthor review [1]; Industry Perspectives on Process Analytical Technology: Tools and Applications in API Development [4]; and Use of Process Analytical Technology (PAT) in Small Molecule Drug Substance Reaction Development

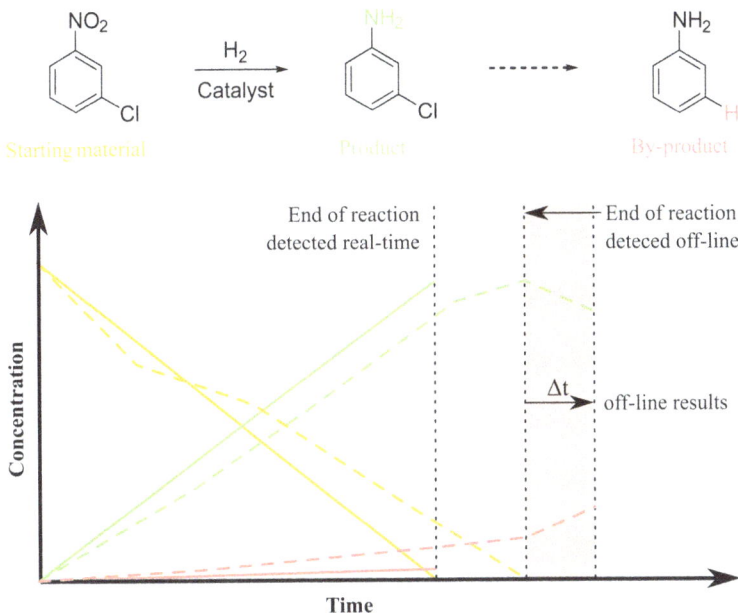

Figure 9.2: A hydrogenation reaction and its concentration profile with (solid line) and without (dashed line) the implementation of process analytical technology (PAT).

[5]. Follow the QR code on this page to learn the latest industrial trends on PAT through the European Pharmaceutical Manufacturer magazine.

9.2 PAT for green chemistry and engineering

PAT can provide invaluable insight and deeper understanding of a product's research and development process, which is crucial for industrially feasible production [1]. In the manufacturing phase, the primary purpose of PAT is to provide safe and effective process control. Between these two phases lies the scale-up campaign, where the use of PAT may mitigate any arising complications, which ultimately allows for a more seamless transition from lab to production [5]. Having a thorough process understanding and *robust control strategy* combined with real-time analytical methodologies creates opportunities to design sustainable processes. In this context, PAT adheres to multiple green chemistry and green engineering principles (Figure 9.3). Preventing waste formation, designing safer chemistry, and maximizing efficiency are paramount. Inline monitoring is capable of detecting deviations in the process stream and thus is able to prevent by-product formation, which in turn can minimize the occurrence of chemical accidents and environmental pollution. PAT also provides a *continuous stream of data*, which is an invaluable source of information for optimizing process parameters and maximizing efficiency. Figure 9.4 gives an overview of the sustainability benefits of PAT for chemical processes. More detailed examples can be found in the case studies in Section 9.6.

Green Chemistry
Safer Chemistry for Accident Prevention
Real-time Analysis for Pollution Prevention
Design for Degradation
Catalysis
Reduce Derivatives
Use of Post-waste Feedstocks
Design for Energy Efficiency
Safer Solvents and Auxiliaries
Design Safer Chemicals
Less Hazardous Chemical Synthesis
Atom Economy
Prevention

Green Engineering
Inherent Rather Than Circumstantial
Prevention Instead of Treatment
Maximize efficiency
Output-pulled vs. Input-pushed
Conserve Complexity
Durability Rather Than Immortality
Meet Need, Minimize Excess
Minimize Material Diversity
Integrate Material and Energy Flows
Design for Commercial Afterlife
Renewable Rather Than Depleting

Figure 9.3: The application of process analytical technology (PAT) addresses multiple principles of green chemistry (prevention, less hazardous chemical synthesis, real-time analysis for pollution prevention, inherently safer chemistry for accident prevention) and green engineering (maximize efficiency, prevention instead of treatment).

Green analytical chemistry can be achieved by implementing most of the principles of green chemistry (Table 1.1). However, only the 11th principle addresses the analytical methodologies directly. This principle can be broken down into further benefits of sustainability, such as reducing labor and energy consumption, eliminating hazardous reagents and solvents, limiting the consumption of the organic solvents, and reducing waste emissions from analytical laboratories (Figure 9.5) [6]. PAT tools do not require sampling and sample treatments, and thus, there is no consumption of organic solvents. The inline method also eliminates operator exposure. Moreover, there is no need for analyte derivatization, which consumes additional time and solvent, and often needs highly reactive and toxic chemical reagents. Analytical sample storage often requires refrigeration at 4 °C or − 20 °C, which consumes significant energy. The continuous stream of digital data allows for facile automation; with the implementation of a control system, artificial intelligence can be employed to optimize or intervene, without an operator's involvement [7, 8].

Process understanding
- Inline measurements enable the detection of reactive or short-lived intermediates that cannot be precisely identified by chromatographic methods. Optimization and controlling the reaction begins with understanding the underlying chemistry.

Optimization
- PAT provides an immense amount of valuable data during the process optimization campaign, which greatly reduces the time needed for these experiments

Maximize efficiency
- The principle of Green Engineering states that products, processes, and systems should be designed to maximize mass, energy, space, and time efficiency. Real-time control strategy provides long-term economic benefits, without sacrificing quality.

Better quality product and less impurity
- Better impurity profile means less energy consumption by removing by-products, which eventually prevents additional waste generated by purification.

Preventing waste
- Incidental errors can be corrected while the process is still in progress, so that material losses can be avoided

Non-hazardous sampling
- Analytical probe installed directly into the process stream provides data without exposing operators to potentially dangerous chemicals

Encouraging renewable materials
- Slight to major variations in the quality of renewable materials, depending on their specific origin, can be mitigated by flexible process control strategy

Process safety
- Inline monitoring can detect the harmful and potentially dangerous accumulation of reactive substances

Figure 9.4: Overview of the sustainability benefits of PAT tools in chemical processes.

Figure 9.5: The 11th principle of green chemistry (Table 1.1) highlights the need for real-time, in-process analytical methodologies that significantly impact various aspects of process sustainability.

9.3 Development of PAT systems

The first step in implementing a PAT system is identifying the *critical quality attributes* (CQAs) and the *critical process parameters* (CPPs). The guideline published by the International Council for Harmonization of Technical Requirements for Pharmaceuticals for Human Use, often abbreviated as ICH, defines CQA as a physical, chemical, biological, or microbiological property or characteristic that should be within an appropriate limit, range, or distribution to ensure the desired product quality [9]. These attributes cover a wide range of parameters, such as a product's physical attributes, appearance, impurity profile, and water content – everything essentially – considered essential qualities of the final product (Figure 9.6a). The CPPs are the parameters whose variability impacts CQAs, and therefore should be monitored and controlled to ensure the process produces the desired quality. These parameters include, but are not limited to, temperature, pressure, flow rate, reaction time, and agitation (Figure 9.6b). If an unexpected deviation occurs during the process, such as a sudden rise in temperature, the end-product quality will be affected. Follow the QR code on this page to browse the ICH website, which brings together the regulatory authorities and manufacturers discussing API production's technical aspects.

The second step is to design a measurement system for the process parameters, including the analytical instrument, software and hardware, and suitable operator training. Because PAT provides a continuous stream of data, it enables an effective in-process control (IPC), such as control of the rate of reaction temperature heat-up and flow rate of the reagent feed stream. The development of mathematical relationships between product quality and process parameters can be assessed by chemometric

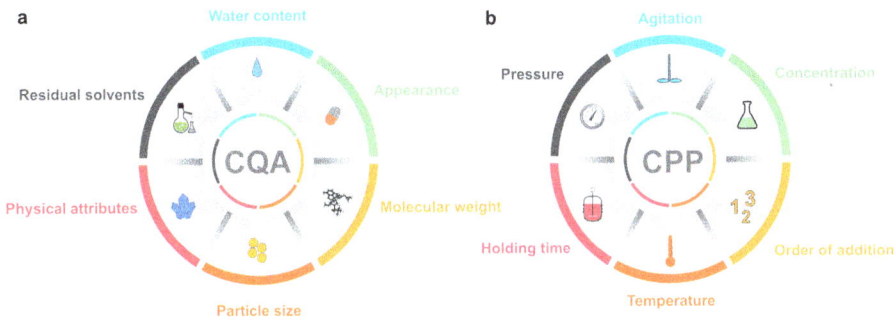

Figure 9.6: (a) Examples of critical quality attributes (CQAs) and (b) examples of critical process parameters (CPPs). CQAs are the crucial quality attributes of a product and should be within an appropriate range; CPPs are the parameters whose variability impacts CQAs.

techniques [10], including partial least squares (PLS) [11], artificial neural networks [12], and principal component analysis (PCA) [13].

It is important to clarify and understand the differences between sampling methods, namely, offline, at-line, online, and inline techniques (Figure 9.7). The *offline* sampling method has been the most commonly used and is characterized by manual sampling, followed by sample preparation and transportation to the analytical department. This sampling method introduces inconsistency within batch-to-batch production due to the varying time-lag between sampling and results. The *at-line* technique is more advanced because the analytical instruments are close, preferably next to the reaction vessel; therefore, the waiting time can be significantly reduced. In *online* meas-

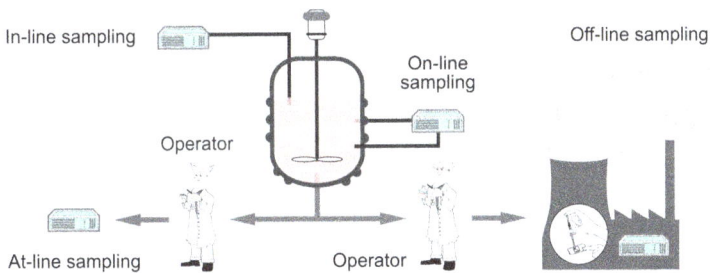

Method	Location	Duration
Offline	Analysis performed in a laboratory	Days
At-line	Analysis performed on a factory floor	Minutes
Online	Automated sampling and analysis	Seconds
Inline	No sampling – the probe is in the process	< 1 second

Figure 9.7: Schematic presentation of inline, online, at-line, and offline sampling methods in a chemical process. For real-time analysis, only inline and online methods are adequate.

urements, a small portion of the reaction mixture is diverted (and sometimes returned) from the process, and the results can be acquired in a matter of seconds. In the *inline* method, a probe is inserted directly into the reactor or the process stream [14]. For IPC, both online and inline methods are adequate. Most PAT tools belong to the category of inline analyzers.

9.4 Industry outlook

Over recent centuries, we have witnessed several industrial revolutions, and pharmaceutical manufacturers have currently begun adopting the ethos of *Industry 4.0*. Industry 4.0 focuses on utilizing new technologies, linking manufacturing components into a cyber-physical system, acquiring data in real-time, big data computing, and automation. Follow the QR code on this page to watch a video explaining what Industry 4.0 is. PAT tools are expected to play an essential role in the fourth industrial revolution because they can be integrated into production lines as smart sensors [15]. A smart sensor consists of a physical and digital twin, where the physical twin gathers process information, which can be a probe inserted into the process stream. Meanwhile, the digital twin can process this information and generate knowledge from the data. This opportunity allows the smart sensor to be self-calibrating and self-optimizing and to maintain operations autonomously.

Industry 4.0 innovations are expected to create positive sustainability impacts throughout the whole manufacturing process [16]. The application of PAT also strongly aligns with the concepts of quality by design (QbD), which promotes a scientific, systematic, and risk-oriented approach to production, moving away from outdated empirical techniques [17]. This methodology advises a more in-depth process understanding by identifying the sources of variations and holistically implementing quality control strategies, including the application of PAT [18]. Moreover, there has been a paradigm shift in the pharmaceutical industry from traditional batch reactions to continuous flow manufacturing [19]. The advantages of flow technology include fast heat and mass transfer, instantaneous mixing, prevention of hot-spots and heat accumulation, and improved safety and selectivity of reactions. Refer to Chapter 6 to learn more about such continuous processes. The availability and implementation of PAT can also facilitate the transition from batch to continuous manufacturing [20]. Follow the QR code on this page to learn more about the implementation of PAT in pharmaceutical manufacturing.

9.5 PAT methods

9.5.1 Infrared spectroscopy

IR spectroscopy is one of the most convenient methods for non-destructive real-time inline reaction monitoring. IR spectroscopy is based on the absorption of light in the 400–4000 cm^{-1} region, which corresponds to molecular vibrations. The absorption bands are unique to the functional groups' vibrational modes, which results in a high degree of structural specificity. The amount of absorption is proportional to the concentration. Therefore, the appearance or disappearance of various functional group bands can provide valuable information about the reaction. Follow the QR code on this page to watch a video explaining the basics of IR.

IR spectroscopy offers a powerful tool for deeper process understanding by determining the reaction kinetics and identifying the structures of intermediates. Furthermore, IR spectroscopy allows deducing whether a reaction has reached completion or not (Figure 9.8). For real-time analysis, the sampling method is mainly based on attenuated total reflectance (ATR) measurements, as depicted in Figure 9.9a [21]. In this case, the light is transmitted through a solid material, which is in direct contact with the sample. The light reflects off the surface and interacts with the sample, creating an evanescent wave, which carries the information. Mid-IR spectroscopy is a suitable choice for monitoring liquid phase reactions and is particularly useful for continuous flow processes [22], detecting unstable intermediates [1], and even providing feedback control of supersaturation [23]. An example of a mid-IR instrument is presented in Figure 9.9b.

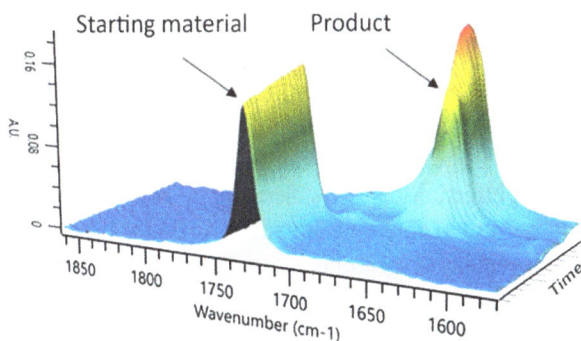

Figure 9.8: Real-time mid-infrared spectra of the progress of a reaction. Reprinted with permission from [24] under the CC-BY 3.0 license.

a

IR Energy to Detector   IR Energy In

ZnSe Support Element

Hastelloy housing

Infrared Beam

Diamond Sensing
Element / ATR Crystal

Depth of Penetration
(DP) = 2μm

Sample in contact with
evanescent wave

6 Reflections

b

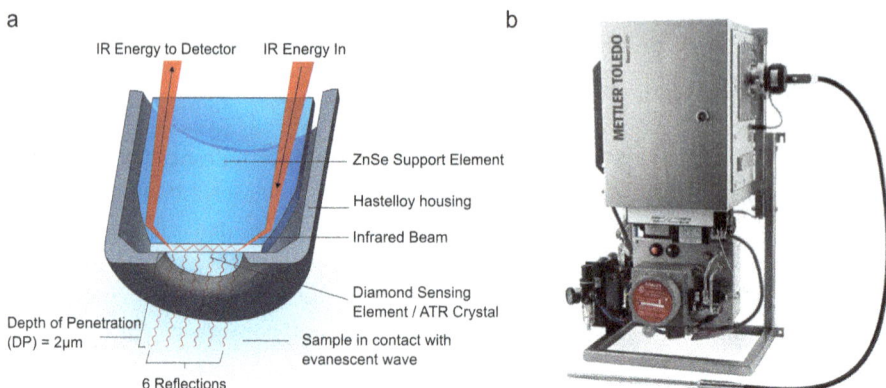

Figure 9.9: (a) Schematic illustration of attenuated total reflectance (ATR) measurement. (b) Mettler Toledo 45P mid-IR instrument. The illustrations were reused with permission from the manufacturer [27].

The presence of water in samples presents the main challenge when using IR spectroscopy, which strongly absorbs in the IR region, making the analysis difficult. PLS regression, PCA regression, multiple linear regression (MLR), and solvent subtraction techniques may be used to deconvolute the IR spectra of a reaction mixture [25]. Further challenges include the overlapping of the signals of different intermediates, which may also cause problems [24]. Moreover, low concentrations, typically less than 0.05 M, are difficult to detect accurately [22]. In heterogeneous reaction mixtures, air bubbles also adversely affect the spectra [26].

NIR spectroscopy is a useful PAT tool with wide application in process development, quality control, and raw material testing [4]. The NIR region of the IR spectrum covers the 4000–13000 cm^{-1} spectrum of wavelengths, where the absorption bands represent overtones or combinations of fundamental vibrational bands corresponding to the 1700–3000 cm^{-1} range. Due to the overtone nature of vibrations, NIR spectroscopy is not suitable for structural elucidation or identification of compounds. Nonetheless, this technique is well suited to perform quantitative analysis aided by chemometric techniques [5]. The application of NIR spectroscopy covers a broad array of fields, including API content uniformity determination during solid dosage formulation [28], quantitative measurement of solvent composition [4], water content elucidation [14], and the controlling and monitoring of milk renneting [29]. Factors such as temperature, humidity, sample thickness, and solid-state properties are significant sources of variability in the spectra, and measurements are usually valid only for a defined calibration model [10]. Similar to mid-IR, the absorbance of water in the NIR region is a disadvantage.

9.5.2 Raman spectroscopy

Raman spectroscopy measures the wavelength shifts associated with Raman scattering. When a sample is excited with monochromatic light (laser light), a small portion of the scattered radiation undergoes changes in frequency, resulting in vibrational and rotational transitions. The Raman effect accounts for the new frequencies in the spectrum and is reported as Raman shift. The scattered photons are returned to the spectrometer, and the Rayleigh scattered[1] photons are filtered out. Raman scattering relies on a change in a molecule's dipole polarizability. Follow the QR code on this page to watch a video explaining the basics of Raman spectroscopy. The strength of the Raman effect depends on the excitation wavelength and is dramatically increased at high excitation frequencies. Ideally, a lower-wavelength light source provides a stronger Raman spectrum. However, in many cases, samples are fluorescence active, which can overwhelm the Raman scattered photons, thus limiting the applied laser's wavelength. Raman instruments use a laser line of 532, 785, or 1000 nm, with the most common being 785 nm, which gives a good balance between Raman intensity and low fluorescence. Since glass is Raman inactive, samples can be analyzed through the glass wall of a vial. Water is also Raman inactive, and therefore this is an excellent technique for analyzing biotransformation streams such as fermentation [5]. Raman scattering also yields information on the crystal structure and, as a result, can be used to identify and quantify relative amounts of different polymorphs in solid products such as most APIs [30].

9.5.3 Nuclear magnetic resonance spectroscopy

NMR spectroscopy is based on the absorption of radiofrequency radiation to produce absorption on the nuclear spin level when nuclei are exposed to a strong magnetic field. Follow the QR code on this page to watch a video explaining the basics of NMR spectroscopy. NMR is among the most powerful methods for structural determination, allowing qualitative and quantitative analysis. However, the sensitivity is lower compared with other spectroscopic techniques [31]. Recently, benchtop low-field NMR instruments have been developed to offer an alternative to their expensive, large footprint, and offline, but high-resolution, counterparts. These benchtop instruments are manufactured by Magritek [32], Bruker [33], ThermoFisher Scientific [34], Nanalysis [35], and Oxford Instruments [36]. Their advantages include their lower cost compared to high-resolution instruments and their better integration into continuous production systems [37]. However, due to the lower magnetic field and probe sensitivity, their application is limited to relatively simple structures or concentration measurements. Applications of inline NMR spectroscopy in-

---

1 Scattered light without the change in wavelength, in this particular case it is the wavelength of the applied laser light.

clude the optimization of reactions in continuous flow [11] for monitoring product quality and maximizing plant profit in continuous production of chemicals [37] and providing a complementary inline analytical tool in multistep reactions [24].

An example of online NMR is presented in Figure 9.10 [38]. In this case, an NMR flow tube was used, and 245 spectra were recorded in 47 h, which allowed monitoring the conversion of a diamine compound. The online NMR measurement gave valuable insight into the mechanistic and kinetic aspects of the reaction. Consumption of the starting material (**1**) can be seen by the disappearing peaks at 6.5 ppm, followed by the immediate formation of the monoimine (**3**) at 6.6 ppm. The reaction eventually resulted in an equilibrium between the monoamine and diamine (**4**, approx. 7.0 ppm).

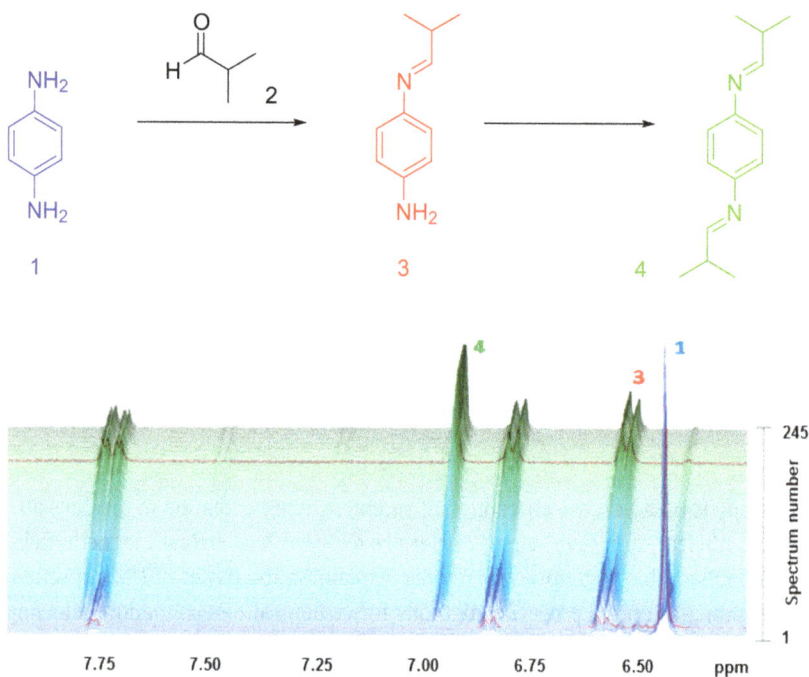

Figure 9.10: Reaction monitoring of an imine formation using inline NMR spectroscopy. The disappearing and appearing peaks of the compounds are numbered according to the ^{1}H NMR spectra's reaction scheme [38]. Reprinted with permission under the CC-BY 3.0 license.

9.5.4 Ultraviolet-visible spectroscopy

Ultraviolet-visible (UV/Vis) spectroscopy measures the absorption of a sample in the UV/Vis spectrum covering the 200–700 nm region. Chemical compounds containing bonding, non-bonding, and conjugated electrons can absorb light, exciting electrons into higher antibonding molecular orbitals. The absorbed wavelength is affected by the electronic properties of the compounds. Colorful, highly conjugated molecules absorb light at higher wavelengths (e.g., β-carotene absorbs light between 400 and 500 nm) [39], while less conjugated systems absorb light at lower wavelengths (e.g., toluene absorbs light between 230 and 270 nm[40]). Although UV/Vis spectroscopy lacks chemical specificity, it is commonly used to monitor liquid phase reactions due to its simplicity, sensitivity, and high time resolution [5]. Despite some compounds' spectral overlap, absorbance changes can be assigned to specific components, making UV/Vis spectroscopy suitable for online reaction monitoring [41] and fast detection of conversion deviations in continuous hydrogenations [42]. Most chemical reactions are performed in organic solvents, and some of these solvents absorb UV light. For quantitative analysis, care must be taken to focus the wavelength above the UV cut-off limit[2] or reduce the path length.

9.6 Case studies

The following case studies highlight the use of PAT tools in complex reaction systems. They include PAT application examples to establish IPCs, covering the different process stages from the description of the method development to method implementation in a manufacturing setting.

9.6.1 Control of ammonia content and reaction monitoring with FTIR

Consider the reaction in Figure 9.11, which shows the synthesis of a pharmaceutical ingredient [5]. The reductive amination reaction is performed in two consecutive steps. First, the ketimine intermediate is formed with the use of gaseous ammonia. Second, the reduction of the imine group takes place with heterogeneous catalytic hydrogenation.

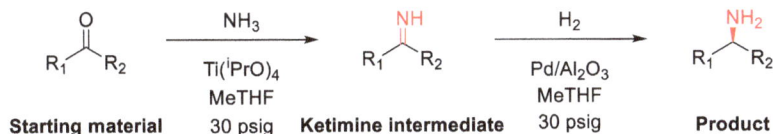

Figure 9.11: Reductive amination process in two consecutive steps.

---

2 Above the UV cut-off wavelength, compounds no longer absorb light. For example, the UV cut-off limits for common solvents are the following: water (191 nm), methanol (203 nm), ethyl acetate (256 nm), acetone (330 nm).[5].

This reaction requires an effective and robust purging control due to the significant amount of dissolved ammonia – concentrations higher than 1,000 mol m^{-3} – in the reactor. Residual ammonia is undesirable in the downstream process to ensure appropriate emission controls and safe handling. Moreover, a dehalogenated impurity is formed during the filtration of the Pd/Al$_2$O$_3$ catalyst, which is accelerated by the residual ammonia. Interestingly, this impurity is not observed in laboratory-scale experiments because of the difference in filtration times. At the laboratory scale, the filtration is performed in a few minutes, whereas it takes hours at the manufacturing scale.

During manufacturing, the ammonia is removed by a sequence of nitrogen purges, i.e., by venting the headspace to atmospheric pressure with the agitation turned off, flushing the headspace with nitrogen, and then turning on the agitation until the pressure stabilizes. Owing to ammonia's high solubility in the MeTHF green solvent, a large number of purge cycles are needed. Consequently, the manufacturer monitors the ammonia levels in the postammonolysis stream and implements a control strategy to establish an acceptable ammonia threshold concentration. Inline testing of the reaction mixture with spectroscopic methods allows the manufacturer to assess and identify the key compounds quickly.

Inspection of the Fourier transform IR (FTIR) spectra in the 1500–1800 cm^{-1} region indicates that the carbonyl band at 1700 cm^{-1} is well separated from the other peaks, which confirms FTIR in an adequate monitoring strategy for the conversion of the ammonolysis (Figure 9.12a). Inline FTIR monitoring enables the rapid collection of kinetic data from the reaction mixture, which provides invaluable information for the reaction development. Ammonia pressures greater than 50 psig and more than 4.5 equivalents of Ti(^{i}PrO)$_4$ do not further reduce the reaction time (Figure 9.12c, d). In contrast, offline sampling of the pressurized reaction would have posed a safety risk due to the high dissolved ammonia levels. Moreover, the moisture sensitivity of the manufacturing process and the frequency of sampling could have jeopardized the efficiency. Nonetheless, the overlap in the FTIR spectra of the ammonia (1550–1650 cm^{-1}) and imine (1600–1650 cm^{-1}) compounds during the reaction monitoring hindered the quantitative determination of the ammonia content.

Examination of the NIR spectra reveals that the specificity with respect to ammonia content is excellent, which appears as a strong peak at approx. 5000 cm^{-1} (Figure 9.12b). The next step in the method development is to construct a calibration curve. Offline measurements are impractical due to the gas-liquid equilibrium nature of the dissolution process and the dependence of ammonia concentrations on headspace pressure. Moreover, solution handling can also introduce significant errors during calibration. However, the use of inline spectroscopic measurements eliminates these uncertainties. Using a pressure vessel equipped with a supply line of gaseous ammonia, a pressure gauge, and an NIR probe is an efficient method. A known amount of ammonia is introduced into the system, followed by agitation until the equilibrium pressure is reached. Application of the ideal gas law allows for determining the ammonia concentrations for each operation, which is used to produce the calibration curve.

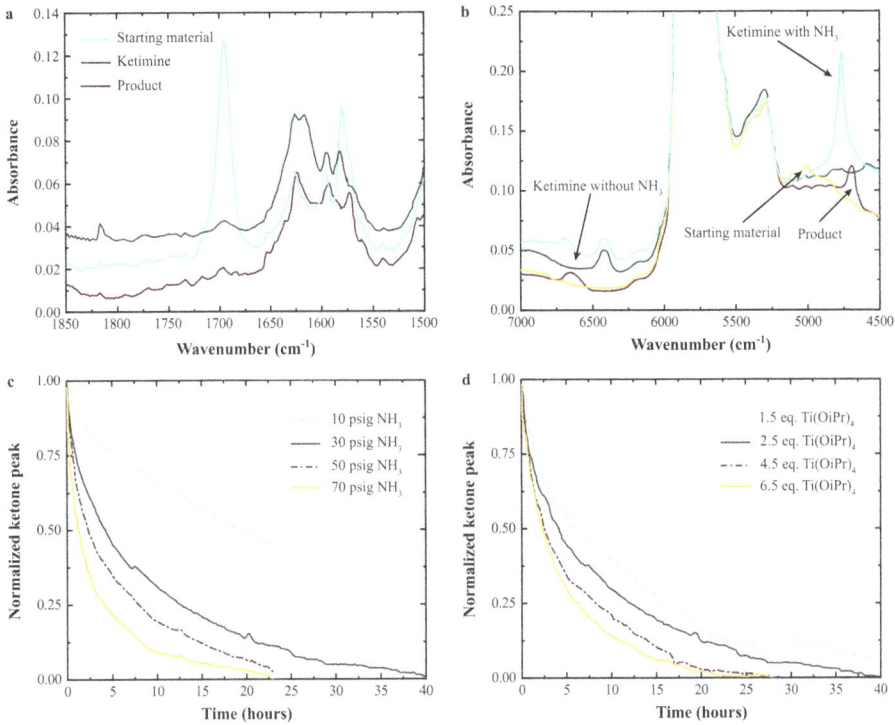

Figure 9.12: FTIR spectra of the reaction components during ammonolysis (a). The carbonyl band of the starting material (marked with C = O) separates from the other peaks well, making it suitable for conversion monitoring. NIR spectra of reaction components after the ammonolysis reaction (b). Ammonolysis reaction progress versus ammonia pressure (c) and Ti(^{i}PrO)$_4$ equivalents (d).

Having developed the PAT for inline ammonia detection, stability studies for posthydrogenation streams with variable ammonia content are performed. This case study concludes that ammonia concentrations lower than 0.1 M achieve good product stability. Implementation of NIR spectroscopy-assisted IPC achieved 3.5 times less impurity in the isolated product at the production scale. The inline FTIR spectroscopy results in a shorter reaction development phase without the hazardous sampling methods by assessing Ti(^{i}PrO)$_4$ amount and ammonia pressure. The PAT also provides the carbonyl compound's conversion data, which enables the operators to act promptly in response to any deviation during the ammonolysis reaction.

9.6.2 FTIR spectroscopy-enabled control strategy for brivanib alaninate manufacturing

The final synthesis step of the anticancer drug brivanib alaninate involves converting the parent drug into its alaninate prodrug[3] form (Figure 9.13) [5, 43]. The chemical development team at Bristol-Myers Squibb Company telescoped two reactions in a single step without isolation of the intermediate species. First, they performed the esterification with carboxybenzyl group (Cbz) protected alaninate, followed by removing the protective group with hydrogenolysis. The insufficient hydrogen supply at the early stages of the reaction resulted in catalyst poisoning by CO_2, which significantly inhibited the reaction rate. Prolonged aging of the hydrogenation reaction mixture, i.e., an extended reaction time, could mitigate this problem. However, longer reaction times could also result in higher levels of impurities.

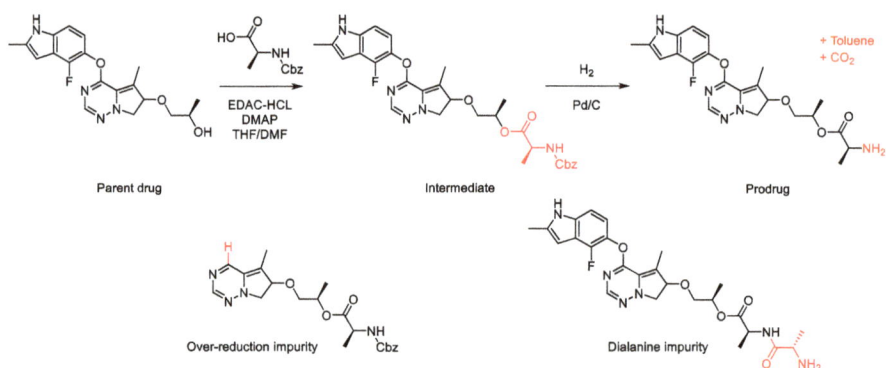

Figure 9.13: The final synthesis step of brivanib involves the conversion of the parent drug into its prodrug form. The parent drug reacts with Cbz protected alanine, followed by the removal of the protective group with catalytic hydrogenation. The possible impurities are the result of overreduction and dialanine formation.

To address this risk and ensure a robust manufacturing process, they implemented inline FTIR spectroscopy into the system. The reaction progress and kinetics were monitored to eliminate stalling caused by catalyst poisoning and estimate the reaction end-point to prevent overaging. Additionally, FTIR spectroscopy proved to be useful in determining residual levels of CO_2 at the end of the hydrogenation reaction, which accelerated the product's undesirable hydrolysis, deeming the parent drug as an impurity.

---

3 A prodrug is a pharmacologically inert form of a drug (precursor) that must undergo a metabolic process to release the active molecule.

To develop a suitable PAT that employs FTIR spectroscopy for controlling both the reaction kinetics and CO_2 concentration, the reaction components were monitored through inline measurements. Figure 9.14 shows the spectra of the THF solvent, the intermediate ester, and the end-of-reaction mixture. The product peak appears in the 1100–1900 cm^{-1} region. The three peaks (marked with asterisks), which correspond to the intermediate, can be monitored during the reaction's progress. Construction of the calibration curve was carried out using the PLS regression model, which included a range of different reaction parameters such as temperature (20–35 °C) and catalyst loading (5–17 wt%). Nonetheless, the FTIR detection limit for the starting material was approx. 2.5 wt%, which is unacceptably low for pharmaceutical production.

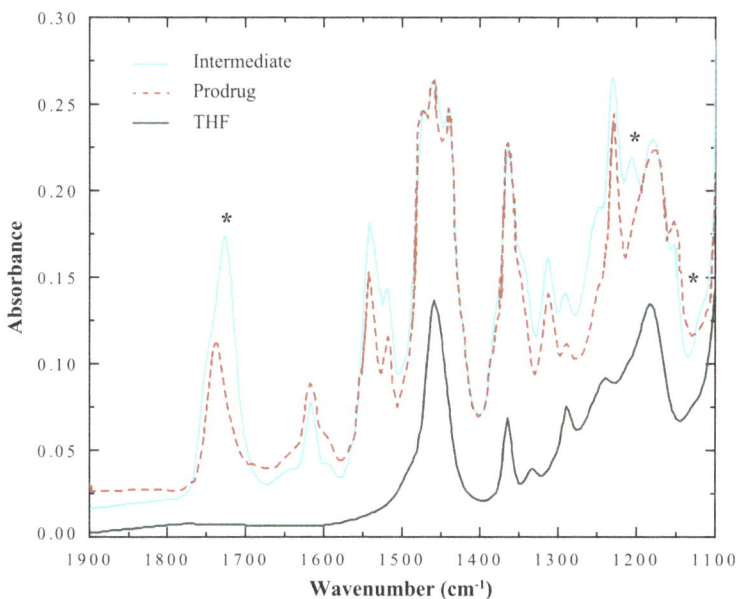

Figure 9.14: FTIR spectra of the hydrogenolysis components during brivanib synthesis. The marked peaks correspond to the intermediate and can be monitored during the conversion.

To overcome this limitation, the manufacturers developed a prediction-based end-of-reaction IPC. Their goal was to evaluate the reaction rate and conversion levels close to the sensitivity limit and predict the reaction completion time. The CO_2 peak at 2350 cm^{-1} was well resolved from the rest of the IR bands, allowing for an accurate calibration curve at low concentrations. Inline FTIR was highly sensitive to CO_2 detection, down to ppm levels, which made it possible to determine whether the CO_2 concentration level was low enough (less than 250 ppm) for extended storage of the posthydrogenation stream.

During the scale-up campaign, all the IPCs performed well and proved to be suitably effective and robust elements to ensure product quality, with conversion rates greater than 99%. Moreover, a remarkable variation in reaction time between 40 and 125 min was found. This inherent fluctuation between different hydrogenations can be explained by catalyst degradation and the process's heterogeneous nature. Real-time reaction monitoring addressed this risk and prevented by-product formation. PAT implementation resulted in significant process improvements by establishing a robust control strategy, preventing waste formation, minimizing downstream processing of the product, and ensuring product quality.

9.6.3 Implementation of Raman spectroscopy in reaction monitoring

Specialty chemical manufacturers and pharmaceutical companies sometimes outsource various synthetic steps to contract manufacturing organizations (CMOs). The technology transfer between companies is not always as smooth and straightforward as one might expect. Thorough process understanding and reaction monitoring are of the utmost importance to accomplish a seamless transfer and ensure drug product quality.

In this case study, the manufacturing of an API involved a complex synthetic reaction. The starting material could easily have undergone degradation, which would have resulted in the reaction stalling and the production of unreacted components and by-products that are difficult to purge. If the reactant became overcharged, thus driving the reaction to a quicker completion, removing the remaining reactant would also become burdensome, resulting in discoloration of the product and high impurity levels. If the reaction had proceeded too long, it could have produced by-products, which could have interfered with subsequent stages of the synthesis.

For reaction monitoring, a Raman spectrometer was employed by the CMO. By monitoring the starting material's consumption rate with Raman, the reaction's endpoint was detected, as seen in Figure 9.15a [1]. A problematic production batch is

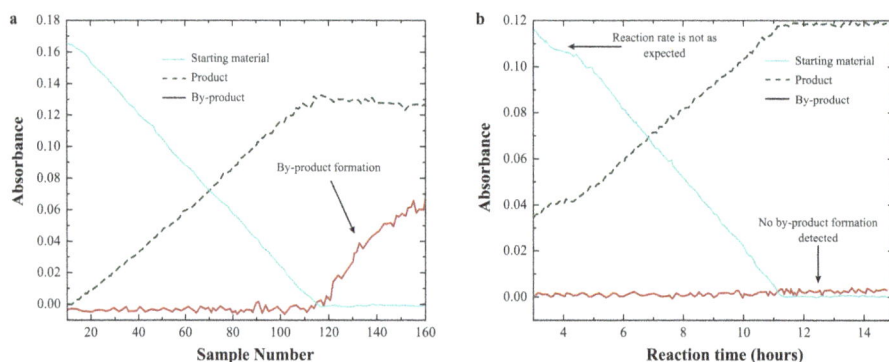

Figure 9.15: Raman spectral trend monitoring the reaction progress (a) and a problematic production batch (b).

shown in Figure 9.15b, revealing an unexpected deviation in the starting material's consumption rate after 5 min. This observation was communicated to the plant operators, and they realized that the agitation speed was set too low. Once this parameter was corrected, the rate of consumption proceeded to match the expected rate.

The in-process correction eventually saved the reaction batch, which passed the quality specifications. Without the inline reaction monitoring, this operation mistake would have gone unnoticed, resulting in the accumulation of unwanted impurities and batch discharge. This seemingly small deviation, and the subsequent correction, ultimately compensated for the effort to implement the Raman PAT system. The spectrometer's application allowed for the successful control of the reaction while eliminating hazardous sampling and facilitating technology transfer without the need for time- and energy-consuming experiments.

9.6.4 Process control of continuous synthesis and solid drug formulation by IR and Raman spectroscopy

There are only a few attempts in developing end-to-end production of solid drug formulation, where the synthesis of an API and its solid formulation is coupled in a single process (Figure 9.16). The continuous manufacturing of acetylsalicylic acid (aspirin) starts with the acetylation of salicylic acid in a flow reactor, followed by the quenching of the reaction and the introduction of a polymer excipient [44]. The solution, which contains the polymer material and the acetylsalicylic acid, was electrospun into solid nanofibers and deposited into a thin pullulan[4] sheet. Follow the QR code on this page to watch a video explaining the basics of electrospinning. These layered films were continuously cut into orally dissolving, small pieces. The conversion and the impurity profile were continuously monitored with IR spectroscopy to prevent the formulation of low-quality fibers. The application of Raman spectroscopy enabled the monitoring of the fiber deposition, which allows the instant disposal of the layered films in the case of uneven fiber distribution.

In greater detail, the acetylation step requires fine-tuned parameters in order to maximize the conversion and minimize the process time and impurity content. The acetylation was achieved with excess acetic anhydride (Ac_2O) and H_3PO_4 as acid catalyst. The solvent has to readily dissolve the reagents, and must also be suitable for electrospinning. Ethyl acetate fulfills these requirements. After the reactants were mixed, the reaction itself takes place in a capillary tube reactor (depicted as *Reaction* in Figure 9.16), which enables higher pressures and a wider temperature range. The quenching step involves the removal of excess acetic anhydride and the conversion of possible

---

4 Pullulan itself is an edible polysaccharide, and its best known commercial application is the production of thin film minty breath refreshers.

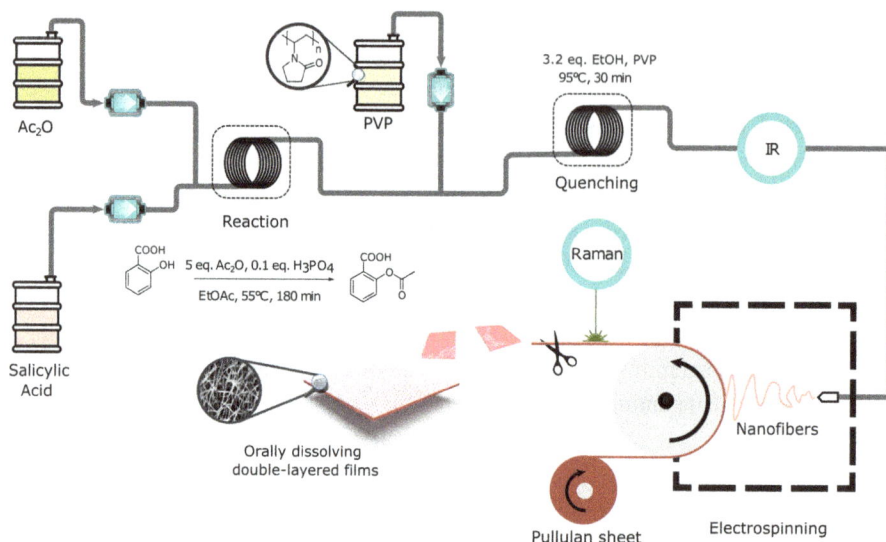

Figure 9.16: The continuous production of orally dissolving double layered acetylsalicylic film starts with the acetylation of the salicylic acid, with excess Ac₂O. The reaction takes place in a heated capillary tube, followed by quenching, which removes the excess Ac₂O, and introduces the polymer excipient (PVP) into the reaction stream. FTIR spectroscopy determines the quantity of the product, and in problematic cases, stops the formulation of the nanofibers. The nanofibers were prepared with electrospinning, and deposited on a pullulan sheet. Raman spectroscopy enabled the investigation of the fiber distribution in the film, which was eventually cut into single pieces with 5–50 mg doses of acetylsalicylic acid.

impurities, and the addition of the polymeric material, namely, polyvinylpyrrolidone (PVP). PVP is widely used in the pharmaceutical industry, it dissolves in water, and it is suitable for electrospinning. The quenching agent was ethanol, which reacts with acetic anhydride by generating ethyl acetate and acetic acid. The majority of ethanol, ethyl acetate, and acetic acid were removed during the electrospinning, and their quantities in the final product remained below the regulatory limits. In order to continuously monitor the purity of the reaction, an FTIR flow cell was introduced into the reaction stream, as a PAT tool, which enabled quantitative determination of the product. In the event of an unacceptable purity level, the voltage generator turned off automatically, thus preventing the formation of low-quality nanofibers.

The formulation and collection of the nanofibers were achieved with electrospinning, which enables effective, yet gentle drying of the material. The deposition of the fibers takes place on a rotating wheel, covered with a pullulan thin film, as a carrier. The formulated double layered film was continuously cut into single pieces (30 × 30 mm) with 5–50 mg of acetylsalicylic acid doses. The dose can be easily adjusted by altering the size of the pieces or the convection of the film (the rotating speed of the wheel), allowing the production of personalized medicine. Before the cutting, the fiber (and API) distribution of the double layered film was continuously investigated with a motorized Raman probe, which enabled transversal movement through the surface of the fibrous film. If the content uniformity is unacceptable, the system automatically detects it, and redirects the problematic sheets to the bin, instead of the product container.

The synthesis, the electrospinning of the fibers, and all of the instruments (including the PAT tools) can be placed on a 150×50 cm bench, greatly reducing the space requirements of drug manufacturing. The PAT tools not only enabled effective process control, but also ensured the overall product quality.

Bibliography

[1] Simon, L. L.; Pataki, H.; Marosi, G.; Meemken, F.; Hungerbühler, K.; Baiker, A.; Tummala, S.; Glennon, B.; Kuentz, M.; Steele, G.; Kramer, H. J. M.; Rydzak, J. W.; Chen, Z.; Morris, J.; Kjell, F.; Singh, R.; Gani, R.; Gernaey, K. V.; Louhi-Kultanen, M.; O'Reilly, J.; Sandler, N.; Antikainen, O.; Yliruusi, J.; Frohberg, P.; Ulrich, J.; Braatz, R. D.; Leyssens, T.; Von Stosch, M.; Oliveira, R.; Tan, R. B. H.; Wu, H.; Khan, M.; O'Grady, D.; Pandey, A.; Westra, R.; Delle-Case, E.; Pape, D.; Angelosante, D.; Maret, Y.; Steiger, O.; Lenner, M.; Abbou-Oucherif, K.; Nagy, Z. K.; Litster, J. D.; Kamaraju, V. K.; Chiu, M. -S. Assessment of Recent Process Analytical Technology (PAT) Trends: A Multiauthor Review. *Org. Process Res. Dev.* **2015**, *19* (1), 3–62. 10.1021/op500261y.

[2] *A Research Agenda for Transforming Separation Science*; National Academies Press, 2019. https://doi. org/10.17226/25421.

[3] Administration, F. and D. Guidance for Industry, PAT-A Framework for Innovative Pharmaceutical Development, Manufacturing and Quality Assurance. **2004**, No. September.

[4] Chanda, A.; Daly, A. M.; Foley, D. A.; LaPack, M. A.; Mukherjee, S.; Orr, J. D.; Reid, G. L.; Thompson, D. R.; Ward, H. W. Industry Perspectives on Process Analytical Technology: Tools and Applications in API Development. *Org. Process Res. Dev.* **2015**, *19* (1), 63–83. 10.1021/op400358b.

[5] Skliar, D.; Nye, J.; Ramirez, A. Use Of Process Analytical Technology (PAT) In Small Molecule Drug Substance Reaction Development. In David, J., Mari, T., Eds.; *Chemical Engineering in the Pharmaceutical Industry: Active Pharmaceutical Ingredients*; John Wiley & Sons, Inc.:Hoboken, NJ, USA, 2019; 937–955. 10.1002/9781119600800.ch42.

[6] He, Y.; Tang, L.; Wu, X.; Hou, X.; Lee, Y. I. Spectroscopy: The Best Way toward Green Analytical Chemistry?. *Appl. Spectrosc. Rev.* **2007**, *42* (2), 119–138. 10.1080/05704920601184259.

[7] O'Mahony, N.; Murphy, T.; Panduru, K.; Riordan, D.; Walsh, J. Machine Learning Algorithms for Process Analytical Technology. In *2016 World Congress on Industrial Control Systems Security (WCICSS)*; IEEE, 2016; pp 1–7. 10.1109/WCICSS.2016.7882607.

[8] Lynch, L.; McGuinness, F.; Clifford, J.; Rao, M.; Walsh, J.; Toal, D.; Newe, T. Integration of Autonomous Intelligent Vehicles into Manufacturing Environments: Challenges. *Procedia. Manuf.* **2019**, 38, 1683–1690. 10.1016/j.promfg.2020.01.115.

[9] Administration, F. and D. Guidance for Industry, PAT-A Framework for Innovative Pharmaceutical Development, Manufacturing and Quality Assurance. **2004**, No. September.

[10] Henriques, J.; Sousa, J.; Veiga, F.; Cardoso, C.; Vitorino, C. Process Analytical Technologies and Injectable Drug Products: Is There a Future?. *Int. J. Pharm.* **2019**, 554, 21–35. 10.1016/j.ijpharm.2018.10.070.

[11] Sagmeister, P.; Poms, J.; Williams, J. D .; Kappe, C. O. Multivariate Analysis of Inline Benchtop NMR Data Enables Rapid Optimization of a Complex Nitration in Flow. *React. Chem. Eng.* **2020**, 5 (4), 677–684. 10.1039/D0RE00048E.

[12] Nagy, B.; Petra, D.; Galata, D. L.; Démuth, B.; Borbás, E.; Marosi, G.; Nagy, Z. K.; Farkas, A. Application of Artificial Neural Networks for Process Analytical Technology-Based Dissolution Testing. *Int. J. Pharm.* **2019**, 567(March), 10.1016/j.ijpharm.2019.118464.

[13] Wang, X.; Esquerre, C.; Downey, G.; Henihan, L.; O'Callaghan, D.; O'Donnell, C. Assessment of Infant Formula Quality and Composition Using Vis-NIR, MIR and Raman Process Analytical Technologies. *Talanta*, **2018**, 183(February), 320–328. 10.1016/j.talanta.2018.02.080.

[14] Skibsted, E.; Engelsen, S. B. Spectroscopy for Process Analytical Technology (PAT). In *Encyclopedia of Spectroscopy and Spectrometry*; Elsevier, 2017; pp 188–197. 10.1016/B978-0-12-803224-4.00026-1.

[15] Eisen, K.; Eifert, T.; Herwig, C.; Maiwald, M. Current and Future Requirements to Industrial Analytical Infrastructure – Part 2: Smart Sensors. *Anal. Bioanal. Chem.* **2020**, *412* (9), 2037–2045. 10.1007/s00216-020-02421-1.

[16] Tsai, W. -H. Modelling and Analysis of Sustainability Related Issues in New Era. *Sustainability*. **2019**, *11* (7), 2134. 10.3390/su11072134.

[17] Calhan, S. D.; Eker, E. D.; Sahin, N. O. Quality by Design (Qbd) and Process Analytical Technology (PAT) Applications in Pharmaceutical Industry. *Eur. J. Chem.* **2017**, *8* (4), 430–433. 10.5155/eurjchem.8.4.430-433.1667.

[18] Matas, M. De; Beer, T. De; Folestad, S.; Ketolainen, J.; Lindén, H.; Almeida, J.; Oostra, W.; Weimer, M.; Öhrngren, P.; Rantanen, J. Strategic Framework for Education and Training in Quality by Design (QbD) and Process Analytical Technology (PAT). *Eur. J. Pharm. Sci.* **2016**, 90, 2–7. 10.1016/j.ejps.2016.04.024.

[19] Gutmann, B.; Cantillo, D.; Kappe, C. O. Continuous-Flow Technology-A Tool for the Safe Manufacturing of Active Pharmaceutical Ingredients. *Angew. Chem., Int. Ed. Engl.* **2015**, *54* (23), 6688–6728. 10.1002/anie.201409318.

[20] Gouveia, F. F.; Rahbek, J. P.; Mortensen, A. R.; Pedersen, M. T.; Felizardo, P. M.; Bro, R.; Mealy, M. J. Using PAT to Accelerate the Transition to Continuous API Manufacturing. *Anal. Bioanal. Chem.* **2017**, *409* (3), 821–832. 10.1007/s00216-016-9834-z.

[21] Tromp, S. A.; Mul, G.; Zhang-Steenwinkel, Y.; Kreutzer, M. T.; Moulijn, J. A. Bottom-Mounted ATR Probes: Pitfalls that Arise from Gravitational Effects. *Catal. Today.* **2007**, *126* (1–2), 184–190. 10.1016/j.cattod.2007.01.072.

[22] Carter, C. F.; Lange, H.; Ley, S. V.; Baxendale, I. R.; Wittkamp, B.; Goode, J. G.; Gaunt, N. L. ReactIR Flow Cell: A New Analytical Tool for Continuous Flow Chemical Processing. *Org. Process Res. Dev.* **2010**, *14* (2), 393–404. 10.1021/op900305v.

[23] Duffy, D.; Barrett, M.; Glennon, B. Novel, Calibration-Free Strategies for Supersaturation Control in Antisolvent Crystallization Processes. *Cryst. Growth Des.* **2013**, *13* (8), 3321–3332. 10.1021/cg301673g.

[24] Sagmeister, P.; Williams, J. D.; Hone, C. A.; Kappe, C. O. Laboratory of the Future: A Modular Flow Platform with Multiple Integrated PAT Tools for Multistep Reactions. *React. Chem. Eng.* 2019, 10.1039/c9re00087a.

[25] Lindon, J. C.; Tranter, G. E.; Koppenaal, D. W. *Encyclopedia of Spectroscopy and Spectrometry*.

[26] Baker, M. J.; Trevisan, J.; Bassan, P.; Bhargava, R.; Butler, H. J.; Dorling, K. M.; Fielden, P. R.; Fogarty, S. W.; Fullwood, N. J.; Heys, K. A.; Hughes, C.; Lasch, P.; Martin-Hirsch, P. L.; Obinaju, B.; Sockalingum, G. D.; Sulé-Suso, J.; Strong, R. J.; Walsh, M. J.; Wood, B. R.; Gardner, P.; Martin, F. L. Using Fourier Transform IR Spectroscopy to Analyze Biological Materials. *Nat. Protoc.* **2014**, *9* (8), 1771–1791. 10.1038/nprot.2014.110.

[27] METTLER TOLEDO Balances & Scales for Industry, Lab, Retail – METTLER TOLEDO https://www.mt.com/us/en/home.html (accessed Nov 3, 2020).

[28] Roggo, Y.; Pauli, V.; Jelsch, M.; Pellegatti, L.; Elbaz, F.; Ensslin, S.; Kleinebudde, P.; Krumme, M. Continuous Manufacturing Process Monitoring of Pharmaceutical Solid Dosage Form: A Case Study. *J. Pharm. Biomed. Anal.* **2020**, 179, 112971. 10.1016/j.jpba.2019.112971.

[29] Grassi, S.; Strani, L.; Casiraghi, E.; Alamprese, C. Control and Monitoring of Milk Renneting Using FT-NIR Spectroscopy as a Process Analytical Technology Tool. *Foods.* **2019**, *8* (9), 405. 10.3390/foods8090405.

[30] Saerens, L.; Dierickx, L.; Lenain, B.; Vervaet, C.; Remon, J. P.; De Beer, T. Raman Spectroscopy for the In-Line Polymer–Drug Quantification and Solid State Characterization during a Pharmaceutical Hot-Melt Extrusion Process. *Eur. J. Pharm. Biopharm.* **2011**, *77* (1), 158–163. 10.1016/j.ejpb.2010.09.015.

[31] Gomez, M. V.; De la Hoz, A. NMR Reaction Monitoring in Flow Synthesis. *Beilstein J. Org. Chem.* **2017**, 13, 285–300. 10.3762/bjoc.13.31.

[32] 80 MHz Benchtop NMR – Introducing Spinsolve 80 | Magritek http://www.magritek.com/2017/05/30/80-mhz-benchtop-nmr-introducing-spinsolve-80/ (accessed Jun 11, 2020).

[33] Benchtop NMR | Spectrometer | Nuclear Magnetic Resonance | Bruker https://www.bruker.com/products/mr/nmr/benchtop-nmr.html (accessed Jun 11, 2020).

[34] picoSpin™ 45 Series II NMR Spectrometer https://www.thermofisher.com/order/catalog/product/912A0911?SID=srch-hj-912A0911#/912A0911?SID=srch-hj-912A0911 (accessed Jun 11, 2020).

[35] NMReady-60 Products – Benchtop NMR Spectrometer | Nanalysis https://www.nanalysis.com/products-overview/ (accessed Jun 11, 2020).

[36] X-Pulse is the world's first benchtop NMR system to offer true multinuclear capability – Magnetic Resonance – Oxford Instruments https://nmr.oxinst.com/x-pulse (accessed Jun 11, 2020).

[37] Kern, S.; Wander, L.; Meyer, K.; Guhl, S.; Mukkula, A. R. G.; Holtkamp, M.; Salge, M.; Fleischer, C.; Weber, N.; King, R.; Engell, S.; Paul, A.; Remelhe, M. P.; Maiwald, M. Flexible Automation with Compact NMR Spectroscopy for Continuous Production of Pharmaceuticals. *Anal. Bioanal. Chem.* **2019**, *411* (14), 3037–3046. 10.1007/s00216-019-01752-y.

[38] Foley, D. A.; Bez, E.; Codina, A.; Colson, K. L.; Fey, M.; Krull, R.; Piroli, D.; Zell, M. T.; Marquez, B. L. NMR Flow Tube for Online NMR Reaction Monitoring. *Anal. Chem.* **2014**, *86* (24), 12008–12013. 10.1021/ac502300q.

[39] Wang, X. -F.; Matsuda, A.; Koyama, Y.; Nagae, H.; Sasaki, S.; Tamiaki, H.; Wada, Y. Effects of Plant Carotenoid Spacers on the Performance of a Dye-Sensitized Solar Cell Using a Chlorophyll Derivative: Enhancement of Photocurrent Determined by One Electron-Oxidation Potential of Each Carotenoid. *Chem. Phys. Lett.* **2006**, *423* (4–6), 470–475. 10.1016/j.cplett.2006.04.008.

[40] Frezzato, D. Photoexcitation Free Energies of Solvated Molecules from Raw Absorption Spectra: Can a Jarzynski-like Equality Be Employed?. *Chem. Phys. Lett.* **2012**, 533, 106–113. 10.1016/j.cplett.2012.02.071.

[41] Benito-Lopez, F.; Verboom, W.; Kakuta, M.; Gardeniers, J.; Han, G. E.; Egberink, R. J. M.; Oosterbroek, E. R.; van den Berg, A.; Reinhoudt, D. N. Optical Fiber-Based on-Line UV/Vis Spectroscopic Monitoring of Chemical Reaction Kinetics under High Pressure in a Capillary Microreactor. *Chem. Commun.* **2005**, 22, 2857. 10.1039/b500429b.

[42] Said, M. B.; Baramov, T.; Herrmann, T.; Gottfried, M.; Hassfeld, J.; Roggan, S. Continuous Selective Hydrogenation of Refametinib Iodo-Nitroaniline Key Intermediate DIM-NA over Raney Cobalt Catalyst at Kg/Day Scale with Online UV-Visible Conversion Control. *Org. Process Res. Dev.* **2017**, *21* (5), 705–714. 10.1021/acs.oprd.7b00039.

[43] Lobben, P. C.; Barlow, E.; Bergum, J. S.; Braem, A.; Chang, S.-Y.; Gibson, F.; Kopp, N.; Lai, C.; LaPorte, T. L.; Leahy, D. K.; Müslehiddinoğlu, J.; Quiroz, F.; Skliar, D.; Spangler, L.; Srivastava, S.; Wasser, D.; Wasylyk, J.; Wethman, R.; Xu, Z. Control Strategy for the Manufacture of Brivanib Alaninate, a Novel Pyrrolotriazine VEGFR/FGFR Inhibitor. *Org. Process Res. Dev.* **2015**, *19* (8), 900–907. 10.1021/op500126u.

[44] Balogh, A.; Domokos, A.; Farkas, B.; Farkas, A.; Rapi, Z.; Kiss, D.; Nyiri, Z.; Eke, Z.; Szarka, G.; Örkényi, R.; Mátravölgyi, B.; Faigl, F.; Marosi, G.; Nagy, Z. K. Continuous End-to-End Production of Solid Drug Dosage Forms: Coupling Flow Synthesis and Formulation by Electrospinning. *Chem. Eng. J.* **2018**, *350* (March), 290–299. 10.1016/j.cej.2018.05.188.

10 Sustainable nuclear fuels

Diana Gulyas Oldal, Abdulhadi M. Alhaji, Nawaf M. AlGhamdi, Gyorgy Szekely

The progress of humanity and the resulting activities have profoundly impacted the Earth's climate, as manifested in, for instance, changes to the planet's *radiative forcing*,[1] among many other impacts. Figure 10.1 shows the different contributors to radiative forcing. Carbon dioxide and other greenhouse gases directly or indirectly released by human activities make a significant positive contribution to radiative forcing and consequently accelerate global warming. This phenomenon has triggered innovations in sustainable technologies with fewer emissions and less environmental impact.

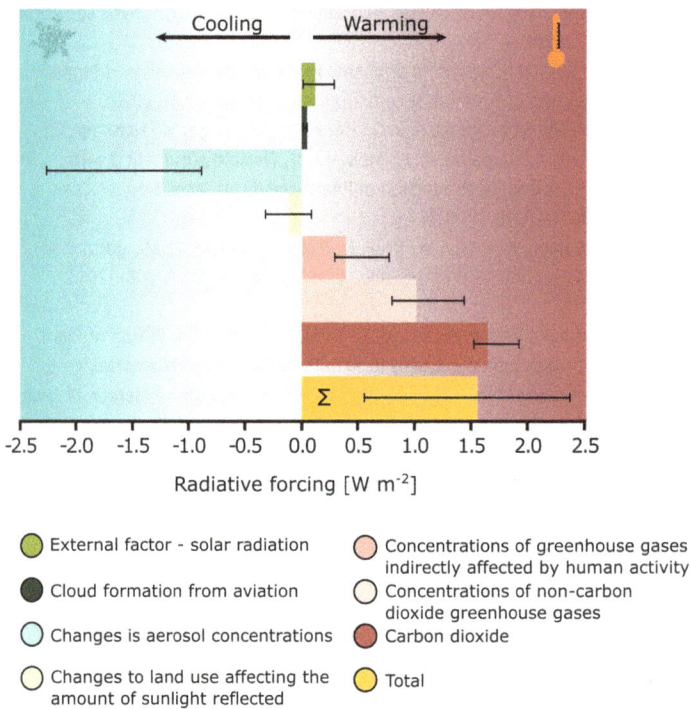

Figure 10.1: Contribution of anthropogenic and natural factors to global warming. Adapted from [3].

Nuclear energy technology has developed rapidly since it was first implemented in the 1950s. Nuclear energy was initially designed as a weapon in World War II. However, shortly after the war ended, its enormous potential as an electricity source was acknowledged. Nowadays, it is considered a cleaner, low-carbon electricity source be-

---

1 Radiative forcing is the measure of the change in the energy balance of the atmosphere.

https://doi.org/10.1515/9783111028163-010

cause it does not generate greenhouse gases or other atmospheric pollutants. The fission[2] of heavy elements (such as uranium) typically releases around 200 MeV of energy more than the energy released by chemical oxidation (such as hydrocarbon combustion), which only releases 20 eV per reaction [1]. In other words, nuclear fission energy releases 10 million times more energy per event. Another advantage of nuclear fission is the significant energy content of heavy metals. According to the Nuclear Energy Institute, one uranium pellet (which is about the size of a fingertip) generates the same amount of energy as one ton of coal, 481 cubic meters of natural gas, or 564 liters of oil [2].

Before discussing nuclear energy technologies, we should first take a look at how energy is generally harnessed from different sources (Figure 10.2). Emitted energy comes from the transition between two states, where one energy state is higher than the other; the maximum extractable energy is the difference between State 2 and State 1. The extraction is contingent upon overcoming the *activation energy* necessary for the transition. In the case of combustion (Figure 10.2b), the latent chemical bonding energy between the elements in a combustible material (such as octane) is released by a chemical reaction with oxygen to form more stable combustion products, such as CO_2. The required activation energy can be provided by a spark, and the energy difference between the reactants (such as octane and oxygen) and the product (CO_2) is released in the form of heat, which can be further utilized for different purposes.

In the case of nuclear power, the extractable energy originates from the nuclear binding energy between the nucleons (protons or neutrons) that constitute the nucleus of the atom. Figure 10.3 illustrates the average binding energy per nucleon plotted against the number of nucleons in each nucleus. Points on the graph are showing elements and their isotopes. Many of these isotopes' lifetime is considerably longer than the age of the solar system [4]. There is an initial rise in binding energy with the peak in the area at ^{56}Fe and ^{62}Ni; i.e., the binding energy significantly increases from the lighter to the heavier elements.[3] Following that region, there is a slight decrease in binding energy. Consequently, the formation of elements lighter than ^{56}Fe and ^{62}Ni releases energy, while their destruction will consume energy. For elements heavier than ^{62}Ni, the energy release originates from the splitting of their nucleus. Accordingly, nuclear energy can be classified as fission and fusion energy. *Fission* produces energy by dividing heavy elements into more tightly bound elements, whereas *fusion*

---

2 Nuclear fission is a reaction where the nucleus of an atom breaks into two or more smaller fragments accompanied by the release of a large amount of energy.
3 ^{62}Ni is the most tightly bound nucleus owing to its highest value of binding energy, and it is followed by ^{58}Fe and ^{56}Fe.[47].

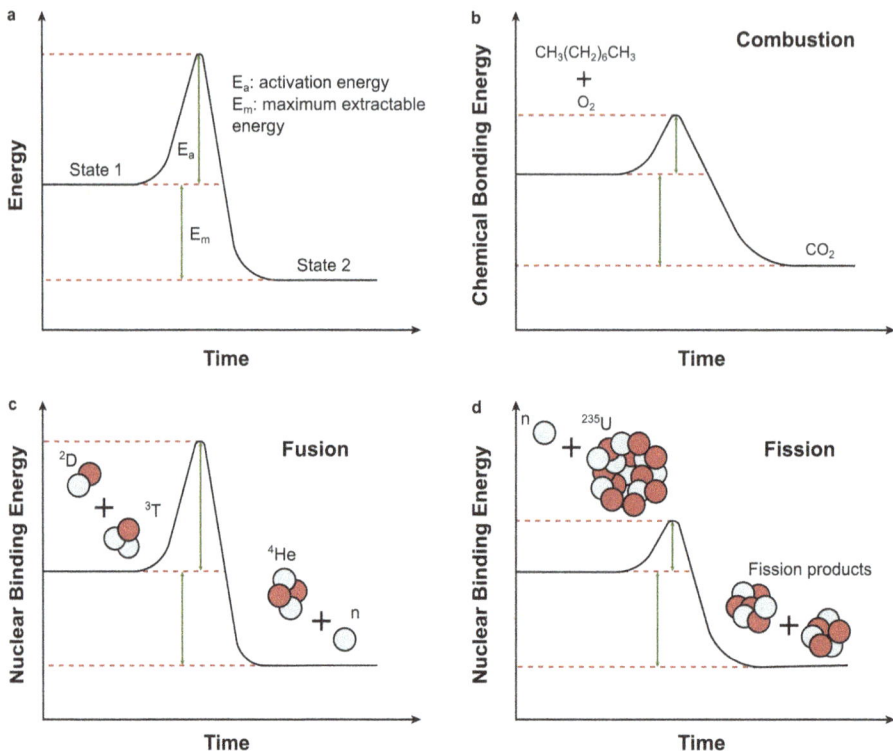

Figure 10.2: Free energy diagrams illustrating how energy is obtained. (a) General concept. (b) Combustion. (c) Nuclear fusion between deuterium and tritium resulting in the generation of helium and a neutron accompanied by the release of energy. (d) Nuclear fission of ^{235}U induced with a neutron.

releases energy by joining lighter elements. Generally speaking, the primary goal is to produce more elements that are more tightly bound than the initial elements in order to achieve the maximum binding energy because the higher the binding energy of a nucleus, the more energy is required to divide it into nucleons.

One of the most common examples of fusion is the fusion of the hydrogen isotopes: deuterium and tritium (eq. (10.1), Figure 10.2c). When deuterium and tritium undergo fusion, a helium atom and a neutron are formed along with the release of 17.59 MeV of energy, by the following reaction [5]:

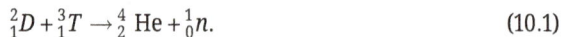

$$^2_1D + ^3_1T \rightarrow ^4_2He + ^1_0n. \tag{10.1}$$

The major obstacle of this process is overcoming the energy barrier originating from Coulombic forces. Nuclei are positively charged because of their protons, and thus they repel each other, forming an energy barrier. To overcome the Coulomb energy

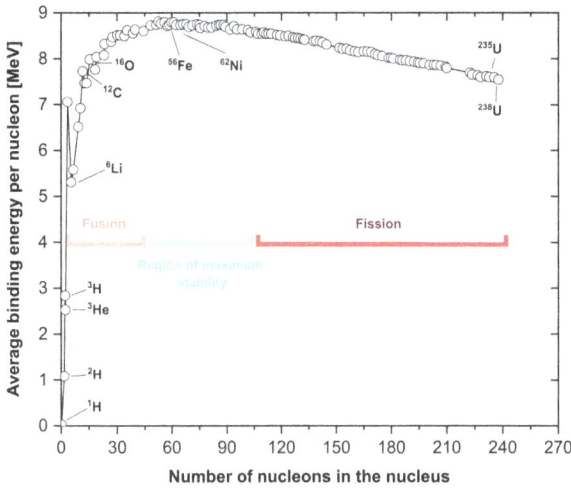

Figure 10.3: The average binding energy per nucleon as a function of the mass number. ^{56}Fe and ^{62}Ni possess the highest binding energy values per nucleon. Based on [4].

barrier, atoms should have enough incident kinetic energy for the fusion reaction to occur, which can be provided by a particle accelerator or high temperatures [5].

Nuclear fission refers to the process when the nucleus of a heavy element splits into two or more smaller and lighter nuclei (Figure 10.2d). The most common example for nuclear fission is the splitting of ^{235}U into lighter elements that are statistically distributed with neutrons, accompanied by the release of large amounts of energy. Unlike fusion, the activation barrier is considerably smaller because the nucleus is bombarded with a neutron. Fission energy is released through a spontaneous or induced fission reaction.

Figure 10.2d illustrates the free energy diagram of a fission reaction that is followed by the release of energy originating from the difference between the initial and final states. Comparing with Figure 10.2c, we can see that the fusion reaction requires a higher activation energy than that required for fission. Thus, the higher activation energy required for nuclear fusion compared to nuclear fission is the main reason why nuclear fusion technology is not applied in practice, and why all current nuclear power generation relies entirely on nuclear fission. Follow the QR code on this page to watch a video by National Geographic explaining how nuclear energy works.

10.1 Benefits of nuclear energy

There are several advantages to the utilization of nuclear energy for power generation. First, nuclear energy is produced through nuclear fission (the splitting of uranium atoms) rather than chemical burning, and thus it can be considered as a *clean energy source* that does not release greenhouse gases. The energy released from splitting can be used to provide reliable baseload electricity with relatively few emissions of pollutants (sulfur, NO_x, heavy metals, etc.), thus minimizing the impact on air quality. According to the Nuclear Energy Institute, the United States has avoided 476.2 million metric tons of CO_2 emission, 0.218 million metric tons of NO_x, and 0.199 million metric tons of SO_2 by using nuclear energy [6].

Second, a nuclear power plant is considered a *secure energy source*. The capacity factor (eq. (10.2)) is used to compare the reliability of different energy-producing technologies and is defined as the ratio between the actual electricity production and the maximum possible electricity output over a given period. This factor provides information on the actual energy production without any interruption (a plant operating at maximum power). In eq. (10.2), capacity denotes the maximum electricity amount that the given power plant can produce.

$$\text{Capacity Factor} = \frac{\text{Actual electricity production [MWh]}}{\text{Capacity [MW]} \cdot \text{Period [h]}} \cdot 100\,\%. \qquad (10.2)$$

Figure 10.4 shows the capacity factors of common power generation technologies for an average annual period in 2019. Nuclear energy has by far the highest capacity factor value at 93.5%, which means that nuclear energy is the most reliable energy source, producing maximum energy during 93.5% of a given year. Nuclear energy is almost three times as reliable as wind and solar energy sources, and twice as reliable as biomass, natural gas, and coal plants. One of the main reasons behind this high capacity factor is that nuclear energy requires less maintenance and refueling. In contrast, fossil fuel–based plants require substantially more offline periods and downtime. Likewise, intermittent and renewable energy sources are generally less reliable because they depend on various factors beyond control. For example, the "fuel" itself (water, wind, sun) depends on the weather conditions, e.g., the sun does not shine continuously, and thus power cannot be generated at night. Similarly, if a day is overcast or dusty, the efficiency of solar power can decrease. Follow the QR code on this page for a video that compares nuclear energy with solar energy. Refer to Chapter 13 to learn more about the sustainability prospects of solar powered engineering and its limitations. Nuclear power is dispatchable, which means that the power generated can be controlled. In contrast, the power generated from wind energy depends on wind conditions and cannot be controlled.

Furthermore, power generation by nuclear energy has a *small land footprint*. Wind farms require up to 360 times as much land area to generate the same amount of power as nuclear reactors, whereas solar photovoltaic plants require an area 75

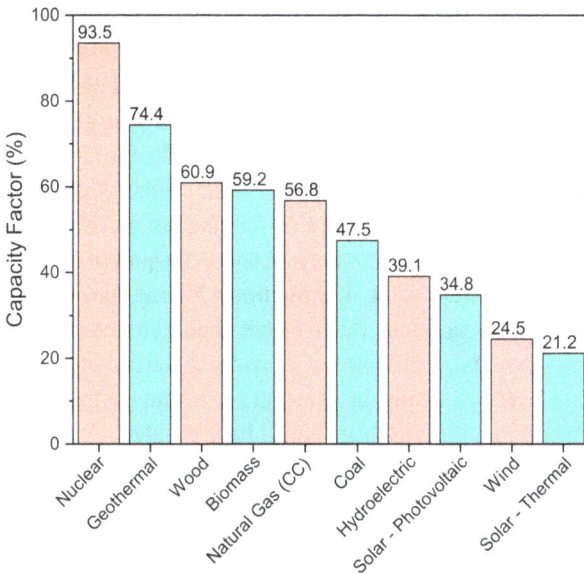

Figure 10.4: Annual capacity factors in 2019 by energy sources where nuclear energy takes first place (CC denotes combined cycle). The original data were published in references [7, 8].

times larger [9]. Another comparison study concluded that 3.125 million solar photovoltaic panels or 431 wind turbines are required to be capable of generating the equivalent amount of power as a single nuclear reactor [10]. Also, nuclear energy generates minimal waste due to the enormous density of nuclear fuels. Because one uranium pellet has the same energy amount as one ton of coal, the generated waste amount is proportionally much smaller. The entire amount of used nuclear fuel that has been generated since the late 1950s could fit within a football field to a depth of approx. 9 meters, which is much less than in the case of a coal plant generating the same amount of waste on an hourly basis [11]. Refer to Section 10.4 for more details on nuclear waste management.

10.2 Disadvantages of nuclear energy

Besides the numerous benefits of nuclear energy, there are several drawbacks, which mainly manifest in their operation. First, there are *safety and security concerns* associated with this technology. There have been few major accidents since the emergence of nuclear energy, including the Three Mile Island incident in the United States, the Chernobyl disaster in the former Soviet Union, and Fukushima in Japan. These incidences have cast doubt in the minds of the public on the safety of nuclear reactors and have tarnished their public image. Moreover, there are international security concerns related to the proliferation of nuclear weapons. Another drawback of nuclear power plants is the economic costs related to their construction. These plants generally require high initial capital investment (approx. 2 billion USD) and have long commissioning and decommissioning times (which can reach up to a 20 year gap).

Even though nuclear power plants have relatively few *emissions*, the limited emissions that they do emit are comparable to those of renewable technologies. A comprehensive assessment of greenhouse gas-equivalent emissions for nuclear power plants has shown that nuclear power plants have mean emissions of 66 g CO_2 eq. kWh^{-1}, ranging from 1.4 to 288 g CO_2 eq. kWh^{-1} [12]. This value is three times higher than carbon equivalent emissions from wind turbines and four times higher than hydroelectric resources [13]. In addition to these emissions, *radioactive fuel* waste materials produced as a result of nuclear power plant operations must be safeguarded for an extended period, and their release into the surrounding environment must be prevented.

Emissions from nuclear power plants are mostly associated with *construction* at the frontend and *decommissioning* at the backend of the plant life cycle (Figure 10.5a). These include processes such as *fuel mining* and processing, sourcing materials for their construction, the building process, and waste disposal. Figure 10.5b illustrates the life cycle estimations for various types of electricity generation. Nuclear power is a better electricity generator than conventional fossil fuel–based energy resources such as coal and natural gas. However, it still ranks lower than renewable electricity generators such as wind or solar power.

A typical nuclear power plant is *material-intensive* in terms of building materials, requiring 170,000 tons of concrete, 32,000 tons of steel, 1,363 tons of copper, and a total of 205,464 tons of other materials, such as aluminum, lithium, and silver [14].

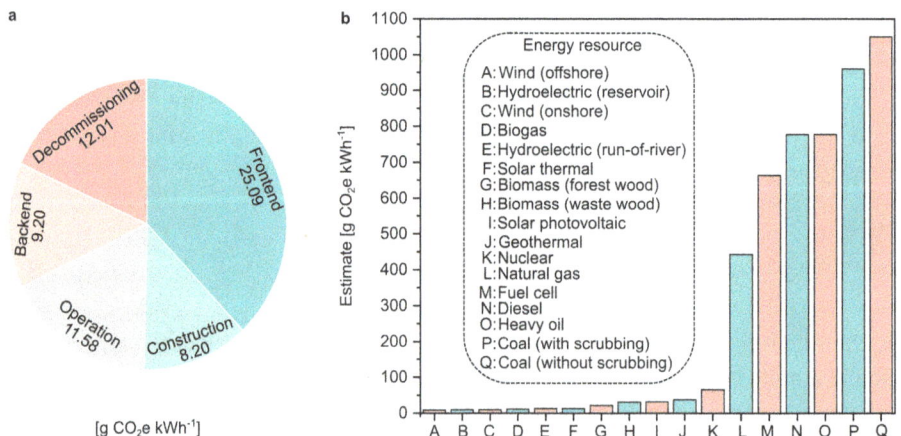

Figure 10.5: (a) The sources of mean emissions reported from different studies of nuclear fuel cycles. (b) Life cycle estimates for electricity generators. Nuclear energy is ranked between renewable and fossil-based electricity generators. Data obtained from [12].

10.3 Uranium as a nuclear fuel

Uranium is a gray metal in the actinide series of the periodic table and is a common radioactive material that can be found in the Earth's crust. Natural uranium mainly consists of the following isotopes: 0.72% of ^{235}U, 99.2745% of ^{238}U, and approx. 0.0055% of ^{234}U. The majority of nuclear fuels used in nuclear reactors mainly comprise the fissile isotopes ^{235}U and ^{239}Pu (the transmutation outcome of ^{238}U). Since only a small concentration of fissile uranium can be found in nature, it needs to be concentrated prior to use in a nuclear reactor. This process of concentrating is called *enrichment*. Most commonly, the concentration of ^{235}U is increased from 0.7% to 3–5%, which can be utilized by most reactors. Only particular types of reactors can utilize naturally extracted uranium. The sourcing of uranium and the enrichment process have a considerable impact on the sustainability of nuclear power plants. The following sections discuss the various approaches for sourcing uranium.

10.3.1 Availability of uranium

Uranium is mainly extracted from geological deposits by *mining*. This type of uranium is the primary fuel source in nuclear reactors. Figure 10.6 illustrates the distribution of the identified uranium resources below 130 USDkg^{-1} U cost category. Australia ranks top for uranium geological resources, with approx. 30% of the total identified resources globally. Approx. 74% of Australia's uranium can be found in the Fe-oxide breccia complex in the Olympic Dam deposit [15]. Australia is followed by Kazakhstan in the global ranking with approx. 14%, followed by Canada with 11% of the uranium share.

10.3.2 Current methods for uranium sourcing

Before the extracted uranium can be used as fuel, it undergoes numerous processing steps, such as recovery, milling, conversion, and enrichment. Several methods are known to extract uranium from geological deposits, such as underground mining, open-pit mining, *in situ* leaching, heap leaching, or obtaining it as a co- or by-product through the extraction process of other metals (e.g., silver or copper). In the case of conventional *mining*, the uranium ore is crushed in a mill and then ground and leached, usually with sulfuric acid to dissolve the uranium oxides [16]. After, at the mine/leaching plant, uranium is separated by ion exchange, followed by drying and packing, typically as U_3O_8. In the case of co- or by-product extraction of other materials, the process is more complicated.

Mining is an important source for uranium extraction. Over two-thirds of the world's uranium production comes from mines located in only three countries: 43%

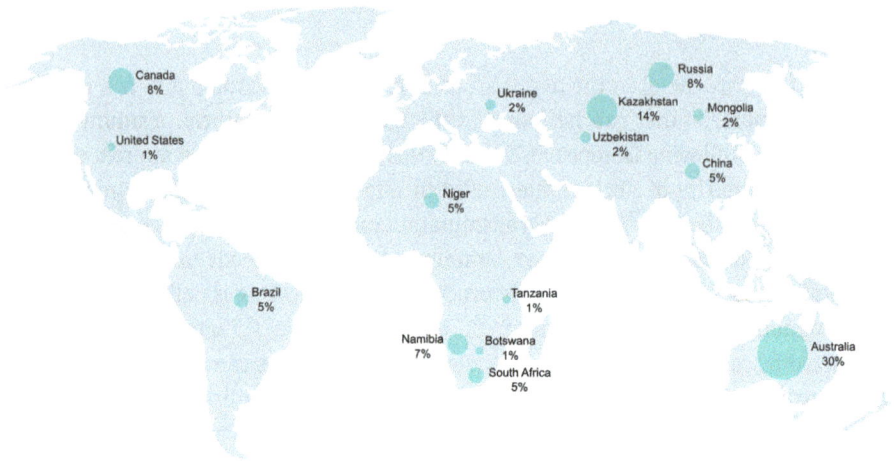

Figure 10.6: Map of the global distribution of identified uranium resources illustrating the main uranium-producing countries or countries with significant plans to develop and increase nuclear capacity as of January 2017. Adapted from [15].

from Kazakhstan (with approx. 23 kt), 13% from Canada (with approx. 7 kt), and 12% from Australia (with approx. 6 kt) [16]. Nowadays, *in situ leaching* is becoming more and more popular; over 50% of uranium is produced via this method [16]. In contrast, 42% of uranium originates from mines and 7% through the extraction of other minerals [17].

In situ leaching is used when the uranium ores are present in groundwater or in porous unconsolidated material such as gravel or sand. These ores can be accessed through the dissolution of the uranium contained within [18]. The general process for *in situ* leaching consists of the dissolution of the underground ore with an oxidant (e.g., H_2O_2) and fixation of the uranium minerals with a complexing agent, which can be acidic or alkaline depending on the nature of the ore. The solution containing uranium is transferred to the surface after delivering the leachate to the underground location. Follow the QR code on this page for a video about *in situ* leaching of uranium.

Besides conventional uranium deposits, other unconventional uranium sources exist that have been less depleted, such as *phosphate deposits*, which are traditionally used to produce phosphoric acid for fertilizers. Phosphate deposits alone are estimated to contain 22 million tonnes of uranium [19]. Another source of uranium is *coal ash*, which, in fact, has a higher maximum potential energy present than the energy released by burning coal. Moreover, uranium can be extracted from *seawater*. Extraction from seawater is a promising technique that has the potential to expand the current supply of uranium by more than 250 times. It is also considered a sustainable source. We will discuss this further in the following section.

10.3.3 Sustainable extraction of uranium from seawater

Supplies of conventional fuel are expected to be depleted in the near future (e.g., fossil-based energy sources), and thus, their use will diminish over time. Nuclear energy represents a low-emission type of reliable energy source; however, it requires further improvements to be a truly sustainable energy source. Uranium extraction from seawater is considered a sustainable source because it does not jeopardize the energy source or drinking water production. Seawater contains a wide variety of minerals, as summarized in Table 10.1, including an overall large amount of uranium at over 4 billion tonnes. However, it is severely diluted and thus exists only in *low concentrations* (3–4 ppb), which poses a challenge for its effective extraction [15]. The extraction of uranium from seawater requires an advanced adsorbent with high selectivity and capacity, which requires costly research and development, as well as implementation. Follow the QR code on this page for a video about uranium extraction from the oceans.

Table 10.1: Average concentration of the elements present in seawater. Data from [20].

Element	Concentration [ppb]	Element	Concentration [ppb]
Cl	$1.91 \cdot 10^7$	Fe	1–2
Na	$1.08 \cdot 10^7$	Ni	0.5–1.7
Mg	$1.33 \cdot 10^6$	V	1.5
Ca	$4.22 \cdot 10^5$	Ti	1
K	$3.80 \cdot 10^5$	Cu	0.6
Li	170	Mn	0.25
Zn	4	Co	0.05
U	3–3.3	Pb	0.03
Al	2		

The uranium content of seawater is continuously renewed by deposits from rivers that collect sediment through the erosion of the Earth's mantle; i.e., some of the uranium in the Earth's crust is dissolved by the water in rivers, which eventually feed the uranium into the ocean. The recovery of uranium from seawater is sustainable because the process of leaching, i.e., the process of complexation and transferring the uranium from underground ores to the surface of the Earth, is performed by nature itself, thus avoiding the need for toxic chemicals or human intervention.

To get a better perspective on the uranium recovery technologies, we can perform some back-of-the-envelope calculations. Assume the following data for a recovery process:

- the uranium concentration in seawater is 3.3 ppb;
- the uptake capacity is 5 kg U t^{-1};
- the adsorbent can recover 100% of the uranium in a given volume of seawater;

- all of that uranium can be recovered from the adsorbent;
- the adsorbent can be reused ten times before it loses its performance.

Table 10.2 summarizes the calculation results for the estimation of the *adsorbent* and seawater needs for the robust extraction of uranium. A typical nuclear reactor with an electric generating capacity of 1000 MWe consumes approx. 250 tonnes of natural uranium per year, and about 5,000 tonnes of adsorbent is needed to process 76 billion cubic meters of water [21]. For reference, the largest reverse osmosis plant in the world in Ashkelon, Israel, produces 108 million cubic meters of water, and the water price is 0.527 USD m^{-3} [22]. Scaling the production to a global supply of 53,000 tonnes, 1.06 million tonnes of adsorbent would be needed to process 16 trillion cubic meters of seawater. These simple calculations show that actively pumping seawater over an adsorbent to extract uranium is neither practical nor sustainable. Thus, *passive deployment* is a more promising technique. Passive deployment, in this case, means that the adsorbent is placed in the sea instead of forcibly pumping the seawater through the absorbent.

Table 10.2: Estimation of the need for uranium adsorbents along with the seawater demand.

	Mass U year^{-1} (tonnes)	Mass adsorbent year^{-1} (tonnes)	Volume of seawater (m^3)
Light-water nuclear reactor	250	5,000	$76 \cdot 10^9$
Annual worldwide production	53,000	1,060,000	$16 \cdot 10^{12}$

Large-scale marine experiments deploying an adsorption platform in seawater have been conducted in Japan [23]. Figure 10.7a illustrates the floating frame and adsorbent beds, along with the used adsorbent stacks. An amidoxime-functionalized polyethylene/poly-propylene fabric was used as the adsorbent. The platform contained a total of 350 kg of adsorbent distributed in three beds, which were lowered 20 m below the surface of the sea. The beds were left to adsorb uranium for 20–40 days, after which the uranium was extracted by recovering the adsorbent from the sea. During the testing period, a total of 1 kg of uranium with an average capacity of 0.5 kg U t^{-1} adsorbent was recovered. The platform without a floating frame and adsorbent beds enabled the reduction of the total cost by 40% [24]. Consequently, the adsorbent was redesigned in the form of kelp-like braids, which were anchored to the seabed with a chain, which could be easily disconnected remotely for recovery of the uranium (Figure 10.7b). This redesign reduced the collection cost and increased the adsorbent capacity to 1.5 kg U t^{-1} thanks to the increased contact between the adsorbent and the seawater.

There are many different types of adsorbents that can be used for uranium extraction. *Synthetic polymers* are commonly used for the extraction of uranium from seawater due to their following characteristics:

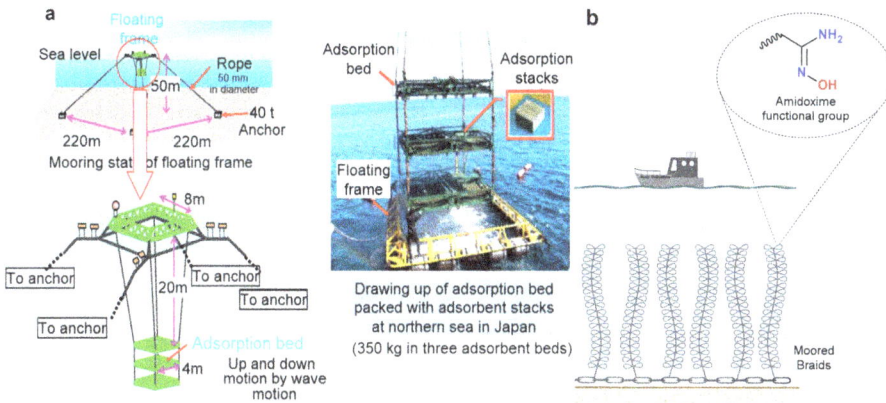

Figure 10.7: (a) Adsorbent bed technology for uranium recovery from seawater representing an adsorbent stack (on the left), and showing the floating adsorbent frame with three adsorbent beds (on the right). Reprinted with permission from [25]. (b) The rearranged deployment system for braided polymer adsorber. Reprinted with permission from [26].

- selective functional groups can be added to polymer chains to enhance uranium adsorption capacity and affinity;
- both robust and ductile polymers can be used as adsorbents;
- diverse shapes of polymer adsorbents can be fabricated in large amounts;
- the density of the polymer adsorbents is similar to the density of seawater, thereby facilitating deployment.

There are several types of polymerization for uranium adsorbents; however, radiation-induced graft polymerization (RIGP) and atom transfer radical polymerization are the most common ones. RIGP is a highly favored method due to its versatility; it can be used to functionalize the grafted chains on various shapes of polymers with different functional groups, such as amidoxime, sulfonic acid, phosphoric acid, or iminodiacetate [20].

Amidoxime-containing adsorbents are widely used for uranium recovery owing to their high affinity for chelating uranyl ions in seawater (amidoximated synthetic polymers proved to have adsorption capacity of 28.1 kg U t^{-1}) [27]. Various inorganic materials are used as adsorbents due to their beneficial properties such as tunable pore structure, high surface area, and low cost. Hydrous titanium oxides have traditionally attracted the most interest, whereas nowadays, silicates are emerging as more popular candidates. Biopolymers are also used as adsorbents because they are renewable and environmentally friendly (biocompatible and biodegradable). Refer to Chapter 12 for in-depth discussions on sustainable polymers and their manufacture.

Chitin can be used as a surface-modified fiber for the extraction of uranium from seawater [28]. *Chitosan* and agarose cryomatrix can be synthesized by cryotropic polymerization, and its surface can be functionalized with melanin, which results in 97% uranium adsorption at pH 5.5 [29]. Without melanin, only 40% uranium adsorption can be achieved. The strong sorption achieved by the melanin-modified biopolymer matrix can be attributed to the electrostatic interaction between the uranium cation and the anionic surface ligands, such as hydroxyl, and phenolic and carboxylic acid groups in the melanin molecule. In addition, the interaction of uranyl ions with the nitrogen and the oxygen of the chitosan and agarose biopolymers also enhances the adsorption of uranium. However, in real seawater, the uranium uptake is reduced to 86% due to the presence of salts. Both chitin and chitosan are natural polymers and are considered sustainable. Phenylarsonic acid–type chitosan resin has also been found to have a high affinity toward uranium (virtually 100% over a broad pH range from 4 to 8) because phenylarsonic acid and its derivatives form stable complexes with U(VI) [30].

Hydrogels, a hydrophilic network of crosslinked polymer chains, are capable of absorbing large amounts of water. A bio-based antibacterial cellulose paper-poly(amidoxime) composite hydrogel has been demonstrated to have good uranium adsorption of 12.9 kg U t^{-1} poly(amidoxime). This high adsorption may be due to the fact that the hydrogel network can disperse the amidoxime-functionalized molecules, and thus facilitate the uranyl ions to reach the inner part [31]. According to the study, the *antibacterial property* has a significant effect on the uranium uptake, whereby the antibacterial adsorbent is more than 31.5% more effective than the non-antibacterial adsorbent. This adsorption value approaches the conventional amidoxime-based adsorbents' value and therefore provides a renewable and green alternative solution for uranium extraction. However, further research is required in the field of renewable uranium adsorbents to ensure their sustainable development by addressing the high uptake value, low-cost, reusability, selectivity, and large-scale production along with *inertness* to the surrounding environment.

Nanostructured materials are often used for uranium recovery. *Metal-organic frameworks* (MOFs) are porous materials consisting of metal ion/clusters, which are coordinated with organic ligands to form infinite 1D, 2D, or 3D structures. They are characterized by their high surface area and their high and tunable porosity. *Selective* uranium adsorption is another critical property for sustainable nuclear energy development. Thanks to amidoxime's great chelating affinity for uranium, it has also been used in MOFs. An amidoxime-appended MOF, called UiO-66-AO, has been shown to exhibit an adsorption capacity of 2.68 kg U t^{-1} in real seawater and a reusability that maintains more than 80% uranium desorption after at least three cycles [32]. Amidoxime-based MOFs are a promising tool for uranium extraction owing to their high affinity for uranyl cations and their high surface areas. Another amidoxime-functionalized MOF (MIL-101-OA) has been synthesized by grafting acrylonitrile into the mesopores of MOF, which exhibited uptake of 4.6 kg U t^{-1} adsorbent in real seawater [33]. Other nanomaterials for the adsorption of uranium include bio-inspired nano-trap amidoxime-based MOFs with an adsorption capacity of 4.36 kg U t^{-1} in real seawater [34].

Several challenges have been encountered in the development of adsorbents. The chemistry of seawater changes over time and space due to marine flora and fauna, water transport, weather patterns, currents, etc. The pH value, temperature, and salinity all have an impact on adsorption performance. Selective ion collection is a challenge because *ion competition* (with vanadium, in particular) can reduce the uranium adsorbent capacity and affect regeneration. Vanadium has complex solution chemistry, which poses additional difficulties regarding selective ion recovery [35]. The relative abundance of the metals adsorbed by amidoxime-based polymer adsorbents (in molar percent) is as follows: vanadium (14.9%) $\gg$ iron (1.6%) > uranium (1.0%). Moreover, vanadium is 15 times more abundant in seawater than uranium [35]. *Biofouling* is also a potential issue, which results from microorganisms in seawater collecting on the adsorbent. Furthermore, the regeneration of the adsorbent can be problematic if its configuration or structure is not explicitly designed for regeneration. Ensuring large-scale production while also ensuring sustainability and high performance, all at a low cost, is a challenge yet to be overcome.

10.4 Waste management

Nuclear radioactive waste can be described by its specific radioactivity, the unit of measure of which is becquerel per cubic meter ($Bq\ m^{-3}$), and is classified into three groups:

– high-level radioactive waste higher than $10^{14}\ Bq\ m^{-3}$;
– medium-level radioactive waste between 10^{10} and $10^{14}\ Bq\ m^{-3}$;
– low-level radioactive waste below $10^{11}\ Bq\ m^{-3}$.

The most abundant waste produced by volume (90%) comprises low-level waste such as tools and work clothing, which represents only 1% of the total radioactivity of all the waste generated [36]. On the contrary, high-level waste comprises only 3% of the total waste volume produced and primarily includes spent nuclear fuels. Nuclear waste is solid and can potentially be recycled, and its amount is relatively small due to its high energy density. Nuclear waste can be safely disposed of by final geological disposal [37]. Geological disposal sites provide multibarrier isolation of nuclear waste, deep underneath the Earth's surface in a tectonically stable location. The International Atomic Agency estimates that currently, there is a total of 22,000 m^3 high-level solid nuclear waste held in storage [38]. Approx. 28,464,000 m^3 of medium- and low-level radioactive waste is disposed of due to its considerably lower radioactivity, whereas 6,295,000 m^3 is kept in storage.

Nuclear waste is disposed of in designated disposal sites. Low-level radioactive nuclear waste can be disposed of in near-surface repositories, taking care to avoid water and soil contamination. However, highly radioactive spent nuclear fuel after removal from a reactor will continue to generate heat for decades and, accordingly, must be cooled by storing it in water pools. This process prevents radiation exposure but requires a significant amount of time. Therefore, the disposal of depleted nuclear fuel in geological repositories could serve as a solution for handling high-level radioactive waste. Finland is close to finishing the first deep geological repository for the disposal of spent nuclear fuel. The used fuel will be encapsulated in corrosion-resistant containers, and it will be stored 500 m below ground level [39]. Any water used for cooling the spent nuclear fuel can also be isolated, and thus, the waste disposal becomes a more sustainable process.

Nuclear energy is one of the most promising and reliable energy forms currently available. However, significant challenges remain, including the proliferation of radioactive material, energy resource security, environmental contamination issues, and especially the issue of nuclear waste and risks at nuclear plants (such as accidents). The redox state of used nuclear fuel affects the fuel's mobility in the environment, making it water-soluble and, subsequently, a potential environmental contaminant [40].

There are some promising methods to mitigate the adverse effects of nuclear waste. The use of *microbes* in nuclear waste management is emerging due to their low cost and low environmental impact [41]. There are two main mechanisms for the use of microbes to mitigate nuclear waste: demobilization and mobilization. In *demobilization*, the goal is to stop nuclear waste from reaching streams that could harm humans and nature. Demobilization includes bio-reduction (Figure 10.8c) of the toxic metals used in the plant, and bio-mineralization (Figure 10.8d) and bio-sorption (Figure 10.8e), where the waste is adsorbed in the cell walls of microbes and demobilized. In contrast, *mobilization* requires moving the waste from point A to point B, which is done by the bio-oxidation (Figure 10.8a) of metals or bio-demineralization (Figure 10.8b).

Bio-reduction is a reduction process of the soluble, oxidized U(VI) to insoluble U(IV), which typically involves sulfide, molecular hydrogen, or organic molecules as the reducing agents [43]. Electroactive microbes can bio-reduce aqueous U(VI) to insoluble U(IV) using electron transport from contaminated groundwaters [44].

Bio-mineralization is another key process for waste management where the uranium is precipitated with phosphates produced by microorganisms. Thus, uranium can be precipitated via microbially generated ligands like sulfide or phosphate, or carbonates or hydroxides, in response to local alkaline conditions at the cell surface (Figure 10.8d) [42]. Diverse bacteria can precipitate uranium as uranyl phosphate or as autunite.

Bio-sorption involves the removal of metals from solution using microorganisms that adsorb uranium onto the surface of the living cell. Bio-sorption can follow either physicochemical interactions or electrostatic/binding interactions. Physicochemical interactions involve physical adsorption, ion exchange, or complexation between the metal and functional groups of the cell surface [45]. Physical adsorption consists of creating bonds between the metal and active sites of the bio-sorbent (in this case, a microorganism). Ion exchange is the primary mechanism in bio-sorption, driven by the competition between the cation and the metal ions for the binding sites [46]. The sorp-

Bio-mobilization Mechanisms

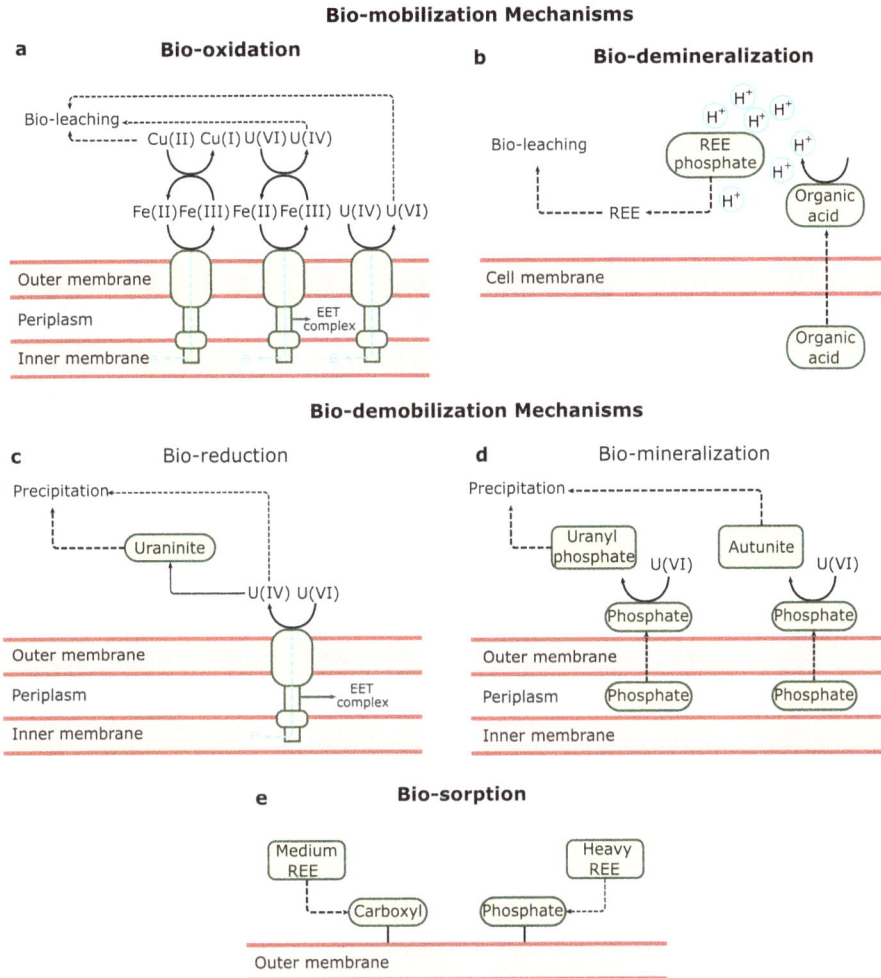

Figure 10.8: Metal mobilization and demobilization via microbes. (a) Metal-oxidizing organisms can dissolve metals (e.g., uranium) through indirect oxidation or direct oxidation. (b) Mineral-solubilizing microbes can precipitate metals from ores. (c) Bio-reduction from U(VI) to U(IV) by microorganisms, which immediately precipitates forming insoluble uranium-crystals. (d) Phosphates produced by microbes can precipitate uranyl phosphate. (e) Carboxyl and phosphate groups on the surface of organisms can bind middle rare and heavy earth elements. Adapted from [42]. REE, rare earth elements; EET, extracellular electron transfer; bioleaching is the conversion of insoluble metal into a soluble form.

tion based on electrostatic interaction is the result of the microorganisms' negative charge on their cell surface (because of the anionic structure), which allows them to bind the metal cations (Figure 10.8e). Chemical groups on the cell surface that possess a negative charge include carboxyl, amine, hydroxyl, phosphate, and sulfhydryl groups, and these can adsorb the metal cations.

Bio-oxidation is a process whereby the aqueous metals are turned insoluble through oxidation reactions (Figure 10.8a). Another mobilization process is *bio-demineralization*, which occurs when an organic acid produced by the microorganism dissolves the minerals in the ore, thus precipitating the metal (Figure 10.8b).

Although sustainable solutions, these bio-oriented approaches face many challenges for their wider application in the nuclear energy field. Bio-reduction converts the water-soluble uranium into monomeric U(IV) or uraninite (UO_2), which only possesses medium stability; i.e., they can reoxidize (in the case of uraninite, reoxidization is less likely due to its crystalline structure) [40]. Additionally, to achieve more extended stability in bio-reduction, electron donors are introduced, and thus, there can be electron donor replenishment issues if the electron donors are depleted too rapidly [42]. Nuclear waste has two main issues: radiation and metal toxicity. With the microbes tested to date, there tends to be a trade-off between the microbes' tolerance to radiation or metal toxicity. Some microbes can be tolerant to both, but there are complications with genetically deciphering them. Therefore, genetic engineering efforts have instead focused on understanding this trade-off. Thus, the handling of nuclear waste remains a challenge, and perhaps presents an innovation opportunity where emerging discoveries can make use of the spent fuel, as opposed to relying solely on waste disposal.

Bibliography

[1]   Ben-Menahem, A. *Historical Encyclopedia of Natural and Mathematical Sciences*; Springer: Berlin Heidelberg: Berlin, Heidelberg, 2009. 10.1007/978-3-540-68832-7.
[2]   Nuclear Fuel https://www.nei.org/fundamentals/nuclear-fuel (accessed Jul 4, 2020).
[3]   Garnaut, R. *The Garnaut Climate Change Review*; Cambridge University Press, 2008.
[4]   De Sanctis, E.; Monti, S.; Ripani, M. Energy from Nuclear Fission. In *Undergraduate Lecture Notes in Physics*; Springer International Publishing: Cham, 2016; 10.1007/978-3-319-30651-3.
[5]   Kenneth Shultis, J.; Faw, R. *Fundamentals of Nuclear Science and Engineering*; CRC Press, 2002. 10.1201/9780203910351.
[6]   Emissions Avoided by U.S. Nuclear Industry https://www.nei.org/resources/statistics/emissions-avoided-by-us-nuclear-industry (accessed Aug 17, 2020).
[7]   Electric Power Monthly – U.S. Energy Information Administration (EIA) https://www.eia.gov/electricity/monthly/epm_table_grapher.php?t=epmt_6_07_b (accessed Aug 17, 2020).
[8]   Electric Power Monthly – U.S. Energy Information Administration (EIA) https://www.eia.gov/electricity/monthly/epm_table_grapher.php?t=epmt_6_07_a (accessed Aug 17, 2020).
[9]   Land Needs for Wind, Solar Dwarf Nuclear Plant's Footprint https://www.nei.org/news/2015/land-needs-for-wind-solar-dwarf-nuclear-plants (accessed Aug 18, 2020).
[10] INFOGRAPHIC: How Much Power Does A Nuclear Reactor Produce? | Department of Energy https://www.energy.gov/ne/articles/infographic-how-much-power-does-nuclear-reactor-produce (accessed Aug 18, 2020).
[11] Nuclear Waste https://www.nei.org/fundamentals/nuclear-waste (accessed Aug 19, 2020).
[12] Sovacool, B. K. Valuing the Greenhouse Gas Emissions from Nuclear Power: A Critical Survey. *Energy Policy*. **2008**, *36* (8), 2950–2963. 10.1016/j.enpol.2008.04.017.

[13] [ISA]. Integrated Sustainability Analysis. *Life-Cycle Energy Balance And Greenhouse Gas Emissions of Nuclear Energy in Australia*; Sydney, 2006.

[14] Sovacool, B. K. *The National Politics of Nuclear Power*; Routledge, 2012. 10.4324/9780203115268.

[15] OECD. *Uranium 2018*; Uranium; OECD, 2019. 10.1787/uranium-2018-en.

[16] World Uranium Mining – World Nuclear Association. https://www.world-nuclear.org/information-library/nuclear-fuel-cycle/mining-of-uranium/world-uranium-mining-production.aspx (accessed Aug 25, 2020).

[17] Joyce, M. Nuclear Fuel Manufacture. In Nuclear Engineering; Elsevier, 2018; pp 297–305. 10.1016/B978-0-08-100962-8.00012-3.

[18] Uranium Mining Overview – World Nuclear Association. https://www.world-nuclear.org/information-library/nuclear-fuel-cycle/mining-of-uranium/uranium-mining-overview.aspx (accessed Aug 27, 2020).

[19] Uranium from Phosphates | Phosphorite Uranium – World Nuclear Association https://www.world-nuclear.org/information-library/nuclear-fuel-cycle/uranium-resources/uranium-from-phosphates.aspx (accessed Aug 31, 2020).

[20] Kim, J.; Tsouris, C.; Mayes, R. T.; Oyola, Y.; Saito, T.; Janke, C. J.; Dai, S.; Schneider, E.; Sachde, D. Recovery of Uranium from Seawater: A Review of Current Status and Future Research Needs. *Sep. Sci. Technol.* **2013**, *48* (3), 367–387. 10.1080/01496395.2012.712599.

[21] Fuel Consumption of Conventional Reactor – Nuclear Power https://www.nuclear-power.net/nuclear-power-plant/nuclear-fuel/fuel-consumption-of-conventional-reactor/ (accessed Aug 31, 2020).

[22] Ashkelon Seawater Reverse Osmosis (SWRO) Plant, Israel – Water Technology https://www.water-technology.net/projects/israel/ (accessed Aug 28, 2020).

[23] Seko, N.; Katakai, A.; Hasegawa, S.; Tamada, M.; Kasai, N.; Takeda, H.; Sugo, T.; Saito, K. Aquaculture of Uranium in Seawater by a Fabric-Adsorbent Submerged System. *Nucl. Technol.* **2003**, *144* (2), 274–278. 10.13182/NT03-2.

[24] Tamada, M. Current Status of Technology for Collection of Uranium from Seawater. **2010**, 243–252. https://doi.org/10.1142/9789814327503_0026.

[25] Abney, C. W.; Mayes, R. T.; Saito, T.; Dai, S. Materials for the Recovery of Uranium from Seawater. *Chem. Rev.* **2017**, *117* (23), 13935–14013. 10.1021/acs.chemrev.7b00355.

[26] Dungan, K.; Butler, G.; Livens, F. R.; Warren, L. M. Uranium from Seawater – Infinite Resource or Improbable Aspiration? *Prog. Nucl. Energy.* **2017**, *99*, 81–85. 10.1016/j.pnucene.2017.04.016.

[27] Kavakli, P. .; Seko, N.; Tamada, M.; Güven, O. A Highly Efficient Chelating Polymer for the Adsorption of Uranyl and Vanadyl Ions at Low Concentrations. *Adsorption.* **2005**, *10* (4), 309–315. 10.1007/s10450-005-4816-z.

[28] Barber, P. S.; Kelley, S. P.; Griggs, C. S.; Wallace, S.; Rogers, R. D. Surface Modification of Ionic Liquid-Spun Chitin Fibers for the Extraction of Uranium from Seawater: Seeking the Strength of Chitin and the Chemical Functionality of Chitosan. *Green Chem.* **2014**, *16* (4), 1828–1836. 10.1039/c4gc00092g.

[29] Tripathi, A.; Melo, J. S. Synthesis of a Low-Density Biopolymeric Chitosan- Agarose Cryomatrix and Its Surface Functionalization with Bio-Transformed Melanin for the Enhanced Recovery of Uranium(VI) from Aqueous Subsurfaces. *RSC Adv.* **2016**, *6* (43), 37067–37078. 10.1039/c6ra04686j.

[30] Oshita, K.; Seo, K.; Sabarudin, A.; Oshima, M.; Takayanagi, T.; Motomizu, S. Synthesis of Chitosan Resin Possessing a Phenylarsonic Acid Moiety for Collection/Concentration of Uranium and Its Determination by ICP-AES. *Anal. Bioanal. Chem.* **2008**, *390* (7), 1927–1932. 10.1007/s00216-008-1931-1.

[31] Gao, J.; Yuan, Y.; Yu, Q.; Yan, B.; Qian, Y.; Wen, J.; Ma, C.; Jiang, S.; Wang, X.; Wang, N. Bio-Inspired Antibacterial Cellulose Paper-Poly(Amidoxime) Composite Hydrogel for Highly Efficient Uranium(VI) Capture from Seawater. *Chem. Commun.* **2020**, *56* (28), 3935–3938. 10.1039/c9cc09936k.

[32] Chen, L.; Bai, Z.; Zhu, L.; Zhang, L.; Cai, Y.; Li, Y.; Liu, W.; Wang, Y.; Chen, L.; Diwu, J.; Wang, J.; Chai, Z.; Wang, S. Ultrafast and Efficient Extraction of Uranium from Seawater Using an Amidoxime

Appended Metal–Organic Framework. *ACS Appl. Mater. Interfaces*. **2017**, *9* (38), 32446–32451. 10.1021/acsami.7b12396.

[33] Wu, H.; Chi, F.; Zhang, S.; Wen, J.; Xiong, J.; Hu, S. Control of Pore Chemistry in Metal-Organic Frameworks for Selective Uranium Extraction from Seawater. *Microporous Mesoporous Mater*. **2019**, *288* (June), 109567. 10.1016/j.micromeso.2019.109567.

[34] Sun, Q.; Aguila, B.; Perman, J.; Ivanov, A. S.; Bryantsev, V. S .; Earl, L. D.; Abney, C. W.; Wojtas, L.; Ma, S. Bio-Inspired Nano-Traps for Uranium Extraction from Seawater and Recovery from Nuclear Waste. *Nat. Commun*. **2018**, *9* (1), 1–9. 10.1038/s41467-018-04032-y.

[35] Parker, B. F .; Zhang, Z.; Rao, L.; Arnold, J. An Overview and Recent Progress in the Chemistry of Uranium Extraction from Seawater. *Dalton Trans*. **2018**, *47* (3), 639–644. 10.1039/c7dt04058j.

[36] What is nuclear waste and what do we do with it? – World Nuclear Association https://www.world-nuclear.org/nuclear-essentials/what-is-nuclear-waste-and-what-do-we-do-with-it.aspx (accessed Aug 29, 2020).

[37] Radioactive Waste Management | Nuclear Waste Disposal – World Nuclear Association iu (accessed Aug 31, 2020).

[38] Status and Trends in Spent Fuel and Radioactive Waste Management; Nuclear Energy Series; INTERNATIONAL ATOMIC ENERGY AGENCY: Vienna, 2018.

[39] Developing the First Ever Facility for the Safe Disposal of Spent Fuel | IAEA https://www.iaea.org/newscenter/news/developing-the-first-ever-facility-for-the-safe-disposal-of-spent-fuel (accessed Aug 31, 2020).

[40] Newsome, L.; Morris, K.; Lloyd, J. R. The Biogeochemistry and Bioremediation of Uranium and Other Priority Radionuclides. *Chem. Geol*. **2014**, *363*, 164–184. 10.1016/j.chemgeo.2013.10.034.

[41] Mishra, S.; Panda, S.; Pradhan, N.; Biswal, S. K.; Sukla, L. B.; Mishra, B. K. Environmental Microbial Biotechnology. Soil Biology, vol 45; Springer, Cham. Online ISBN: 978-3-319-19018-1.

[42] Adesina, O.; Anzai, I. A .; Avalos, J. L .; Barstow, B. Embracing Biological Solutions to the Sustainable Energy Challenge. *Chemistry*. **2017**, *2* (1), 20–51. 10.1016/j.chempr.2016.12.009.

[43] Myllykylae, E. *Reduction of Uranium in Disposal Conditions of Spent Nuclear Fuel*; Finland, 2008.

[44] Lovley, D. R .; Phillips, E. J . P .; Gorby, Y. A.; Landa, E. R. Microbial Reduction of Uranium. *Nature*. **1991**, *350* (6317), 413–416. 10.1038/350413a0.

[45] Ansari, M. I .; Masood, F.; Malik, A. Bacterial Biosorption: A Technique for Remediation of Heavy Metals. In Ahmad, I., Ahmad, F., Pichtel, J., Eds.; *Microbes and Microbial Technology: Agricultural and Environmental Applications*; Springer New York: New York, 2011; pp 283–319. 10.1007/978-1-4419-7931-5_12.

[46] Schiewer, S.; Volesky, B. Biosorption Processes for Heavy Metal Removal. In *Environmental Microbe - Metal Interactions*; ASM Press: Washington, DC, USA, 2014; pp 329–362. 10.1128/9781555818098.ch14.

[47] Fewell, M. P. The Atomic Nuclide with the Highest Mean Binding Energy. *Am. J. Phys*. **1995**, *63* (7), 653–658. 10.1119/1.17828.

11 Toward sustainable biofuel production processes

Martin Gede, Diana Gulyas Oldal, Gyorgy Szekely

There is increasing awareness that our energy systems need to be improved or even revolutionized. The grand sustainability challenge of the twenty-first century embraces novel approaches to improve energy security and mitigate any detrimental impacts on the environment. The primary goal of these novel approaches is to mitigate global warming and climate change. Fossil fuels can easily meet current energy demands, but they have harmful effects on the environment (e.g., greenhouse gas emissions). Consequently, there is a need to discover sustainable energy resources. Renewable-based products and energy forms have great potential to reduce society's environmental burden.

Biofuels are energy sources derived from living matter through biological processes. They can be solid (wood, dried plant material), liquid (bioethanol and biodiesel), or gaseous (biogas), and are all sustainable. Biofuels are classified into two groups: primary and secondary biofuels. *Primary biofuels* are not processed and are directly utilized without any modification for heating, electricity production, or cooking. Unprocessed biomass such as firewood, wood chips and pellets, animal waste, landfill gas, and forest and crop residues represent the primary biofuel feedstocks. *Secondary biofuels* are the modified forms of the primary biofuels, and they can be processed and produced as solids (e.g., charcoal), liquids (e.g., bioethanol and biodiesel), or gases (e.g., biogas, synthesis gas, and hydrogen). These fuels are used in numerous industrial processes and transportation. Secondary biofuels can be further classified into first-, second-, third-, and fourth-generation based on the used feedstock sources and production technologies [1].

First-generation biofuels derive energy from food sources (e.g., corn, wheat, palm, corn, soybean, sugarcane, rapeseed, oil crops). Bioethanol is a commonly known first-generation biofuel obtained from starch fermentation (e.g., wheat or corn) or by fermentation of sugars (e.g., sugarcane) [2]. Biodiesel produced from the transesterification of oil crops (e.g., rapeseed, soybeans) represents another well-known first-generation biofuel [3].

Fuels derived from non-food feedstocks and waste from leftover food resources are called *second-generation biofuels*. These sources include agricultural residues, forest harvesting residues, wood processing waste (such as leaves, straw, wood chips), and the non-edible parts of corn or sugarcane [4]. Second-generation biofuels are more favorable than first-generation biofuels because they are produced from lignocellulosic biomass or the woody part of plants, which do not directly jeopardize or compete with food production. From a sustainability perspective, one of the advantages of utilizing these by-products to produce biofuels is that no additional water, soil, or fertilizers are required to grow the feedstock. On the other hand, the expensive process to produce second-

https://doi.org/10.1515/9783111028163-011

generation biofuels, such as converting and pretreating woody biomass into fermentable sugars, restricts their widespread use on a large, commercial scale. This generation of biofuels includes bioethanol, biodiesel, dimethyl ether, biosyngas Fischer–Tropsch (FT) biodiesel, and biomass to liquid (BtL) biodiesel [5].

Third-generation biofuels are primarily derived from microalgae. Microalgae are photosynthetic microorganisms that require simple primary growing conditions (light, sugars, CO_2, N, P, and K). They can produce lipids, proteins, and carbohydrates in large amounts over short periods [6]. Third-generation fuels can also be produced from some other aquatic microorganisms such as cyanobacteria, bacteria, fungi, and yeast [7]. Microalgae are a promising sustainable energy source because they can produce 15–300 times more oil (by area) for biodiesel production than traditional plants [8]. Moreover, microalgae have a short harvesting time (1–10 days), which means that they can be harvested continuously and multiple times with higher yields than traditional crops (their harvesting cycle is once or twice annually). Follow the QR code on this page to learn more about the production of biofuel from algae from the American Scientist. Third-generation biofuels include biomethane, bioethanol, biobutanol, vegetable oil gasoline, biodiesel, biomethanol, and jet fuels. The main benefits of third-generation biofuel feedstocks are that they do not compete for agricultural land use or risk food production. Moreover, algae can contribute to wastewater treatment by removing nutrients from water, such as N and P. Furthermore, they are highly biodegradable and can be grown with minimal land and water demand. Follow the QR code on this page to watch a TED talk about energy from floating algae pods by Jonathan Trent.

Fourth-generation biofuels represent a broad category of biofuels that can also capture and store carbon dioxide. These biofuel feedstock crops are genetically designed to consume more CO_2 from the atmosphere than they generate during their combustion as fuels [9]. Consequently, they contribute to the reduction of emissions by decreasing the CO_2 in the environment. The most prominent examples of fourth-generation biofuels are green biodiesel, biogasoline, and green aviation fuels produced from vegetable oils and biodiesel. This class of biofuels is still relatively new and under development. In the following sections, the conventional and more efficient biofuel production techniques are discussed. For further reading, Singh's review [10] and Gude's book [11] are recommended. To learn about biofuel economics from David Ruzic at the University of Illinois Urbana-Champaign, follow the QR code on this page.

11.1 Production of alcohols as fuels

One of the most common alcohol biofuels is bioethanol. Bioethanol has an oxygen atom and therefore has a lower oxygen demand. Thus, bioethanol facilitates the combustion process better than hydrocarbons. First-generation bioethanol is derived from different sugar and starch grains, such as corn, wheat, or sugarcane. The starch from the grains provides the raw material for grain-based bioethanol production, but in sugar-based plants, the sugar content is extracted from sugarcane. Second-generation bioethanol is extracted from lignocellulosic biomass; i.e., it is produced from the woody parts of plants, wood waste, or agricultural residues. Figure 11.1 illustrates the process of bioethanol production from diverse raw materials. Bioethanol production mainly depends on raw materials and consists of three main steps:

1. preparation of the solution containing sugars for fermentation;
2. fermentation of sugars to ethanol;
3. separation and purification of the obtained ethanol.

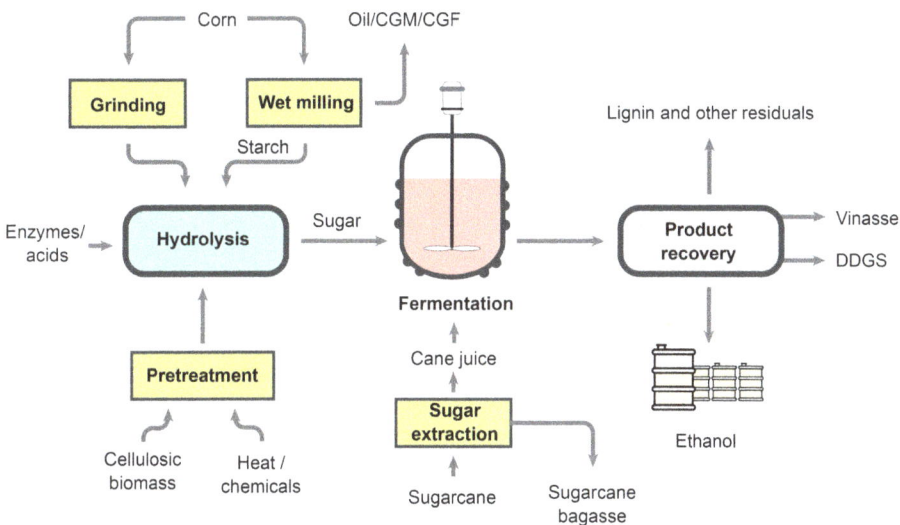

Figure 11.1: The production of bioethanol from different raw materials such as sugarcane, corn, and cellulosic biomass. The fermentation and ethanol recovery processes are similar in each case. CGM, corn gluten meal; CGF, corn gluten feed; DDGS, dried distillers grains with solubles (animal feed).

Sugar-based crops are commonly used for bioethanol production due to the high yield of sugar per acre and cheap conversion processes [12]. However, their use is limited due to their seasonal availability. In a broader sense, the following steps are the general constituents of bioethanol production:

1. *Grinding*: The raw material is ground into a fine powder.
2. *Pretreatment/liquefaction*: This fine powder is mixed with water and cooked at elevated temperature to produce a gravy, followed by partial hydrolysis by alpha-amylase enzymes that break down the long starch chains.

3. *Hydrolysis/saccharification*: The obtained liquid mixture is cooled, and then glucoamylase is introduced to the system to transform starch molecules to fermentable sugars.
4. *Fermentation*: Yeast or bacteria is added to the fermentable mixture to convert sugars to ethanol and CO_2 via alcoholic fermentation.
5. *Distillation*: The fermented solution contains 10 vol% alcohol, which is separated from other substances (e.g., non-fermentable compounds, yeast molecules) by distillation, resulting in 95 vol% ethanol.
6. *Dehydration*: Industrially applicable anhydrous bioethanol contains 99.5 vol% ethanol, with water content below 0.5 vol%; however, water forms an azeotropic mixture with ethanol (95 vol% ethanol, 5 vol% water), and therefore simple distillation is not sufficient to achieve a higher alcohol content. Hence, dehydration with molecular sieves or azeotropic distillation with an entrainer compound is used to remove water and produce anhydrous alcohol because the water content could damage metal parts of internal combustion engines.
7. *Denaturation*: Ethanol used as fuel is denatured with a small amount of gasoline (for instance, 2–5 vol%) to make it unsuitable for human consumption, which consequently incurs lower taxes [13].

11.1.1 Biochemical conversion of lignocellulosic biomass

There are two major production processes for lignocellulosic biomass sources: biochemical conversion (sugar platform) and thermochemical conversion (syngas platform) [14]. The biochemical process utilizes enzymes to convert the pretreated lignocellulosic biomass to fermentable sugars through cellulose, which ends up as bioethanol. In contrast, the thermochemical conversion gasifies the resources to produce syngas – a mixture of CO, H_2, and CO_2 – that can be converted to ethanol through chemical catalysis or biological reaction by microorganisms. The process of biochemical conversion is illustrated in Figure 11.2. Bioethanol production from lignocellulosic biomass does not compete with food production and is considered more sustainable than production from starch- and sugar-based plants. Follow the QR code on this page to watch a video illustrating the biofuel production process.

11.1.2 Grinding

The biochemical conversion process starts with the grinding of the lignocellulosic material. Grinding is an essential step to reduce the biomass recalcitrance, i.e., plant cell walls' natural resistance to microbial and enzymatic breakdown [15]. The process decreases the particle size and increases the porosity and cellulose accessibility to achieve effective hydrolysis in the following step. Common techniques to reduce par-

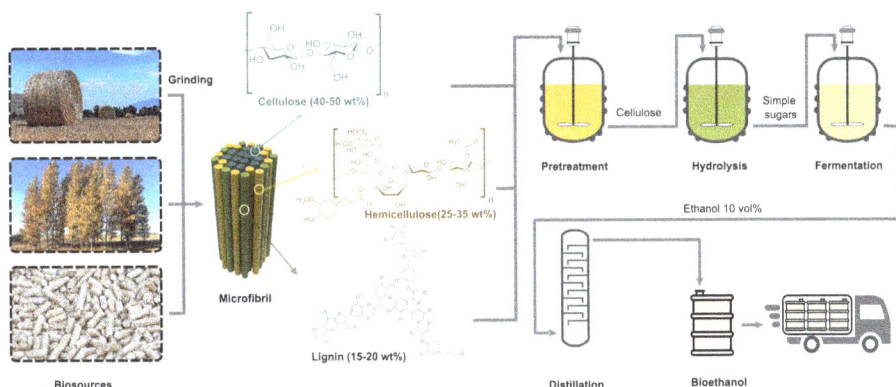

Figure 11.2: Bioethanol production from lignocellulosic material starts with the grinding of bio-derived sources such as lignin, hemicellulose, and cellulose. The next step is the pretreatment to remove lignin, followed by hydrolysis of cellulose into sugars. In the fermentation process, the sugars are converted into ethanol, which has to be concentrated and dehydrated to make it suitable for fuel applications.

ticle size from 1–3 cm to 0.2–2 mm are the combination of chipping, grinding, and milling [16]. Energy consumption depends heavily on the starting material's nature; for instance, the treatment of hardwood material and corn stover for the same particle size consumes 130 kWh ton^{-1} and 14 kWh ton^{-1}, respectively [17].

11.1.3 Pretreatment

The presence of lignin in the biomass limits cellulose accessibility. Lignin also binds to the cellulose enzymes, resulting in lower saccharification rates [18]. Therefore, pretreatment of the lignocellulosic biomass is needed to reduce the physical and chemical barriers during enzymatic conversion. Alkaline pretreatment with NaOH, Ca(OH)$_2$, KOH, and aqueous ammonia and acid pretreatment with HCl, H$_2$SO$_4$, and H$_3$PO$_4$ have been used to digest the lignin matrix [19]. Drawbacks of this approach include the insufficient separation of cellulose and lignin, high energy consumption and waste generation, and the formation of toxic inhibitors (furfural, organic acids, phenolics) that eventually decrease the enzymatic conversion. Although biological conversion with white rot and brown rot fungi is considered environmentally benign, the long pretreatment time and the large amount of space required to implement the process make it impractical for industrial purposes [19].

The development of advanced pretreatment technologies is essential to overcome these drawbacks. The application of ionic liquids and other *green solvents* has the potential to benefit this process. Ionic liquids are considered green due to their low vapor pressures, high thermal stability, and relatively low toxicity. Refer to Chapter 3 to learn about green solvents and their role in sustainable manufacturing. Ionic liquids can selectively remove lignin and hemicellulose from lignocellulosic biomass, although they are currently too expensive for large-scale processing [20]. However, a life cycle assessment revealed that triethyl ammonium hydrogen sulfate has the lowest total cost compared to acetone and glycerol [21]. Moreover, bio-based *ionic liquids* can be produced from lignin by applying controlled depolymerization and synthetic steps. Depolymerization mainly yields vanillin and *p*-anisaldehyde, which can be converted into tertiary amines via reductive amination. Treating the amines with phosphoric acid produces the ionic liquid, which can be used for biomass pretreatment. This method is ideal for creating *closed-loop* biorefineries [20].

Cyrene (dihydrolevoglucosenone), a bio-renewable solvent that is considered an alternative to dipolar organic solvents, is derived mainly from cellulose and dissolves lignin readily. Cyrene's utilization in biomass pretreatment requires milder temperatures (120 °C compared to conventional sulfuric acid treatment at 140–200 °C) [15], improves lignin removal, and significantly increases the accessibility of cellulose, resulting in virtually 100% conversion [22].

The obtained dissolved lignin is a valuable starting material in the valorization of biomass waste, which increases the cost-competitiveness and sustainable aspects of biorefineries [23]. Valorization techniques include the thermochemical conversion of lignin biomass via pyrolysis, which produces syngas, bio-oil, and biochar. *Microwave* pyrolysis of agricultural waste is a promising technology since it can favorably alter pyrolysis products' yields and properties [24]. The separation of high-purity phenol (97 wt%) and catechol is also feasible from the aqueous waste stream generated by fast catalytic pyrolysis [25]. Phenol is primarily used in polycarbonate resins, while catechol is primarily used in insecticides and pharmaceuticals and as polymerization inhibitors [26]. Refer to Chapter 12 to learn more about green polymer and building block manufacturing. The high-value product vanillin can be obtained directly from lignin by applying electrolysis without toxic and expensive oxidizers [27].

11.1.4 Hydrolysis/saccharification

Chemical hydrolysis or enzymatic hydrolysis can be applied to convert cellulose into simple fermentable sugars. Conventionally diluted acid hydrolysis is preferred, although this process struggles to achieve glucose yields of more than 70% while avoiding glucose decomposition. Concentrated acid treatment positively affects the glucose yields; however, it leads to significant acid consumption, waste generation, and equipment corrosion [28]. *Enzymatic hydrolysis* is potentially more environmentally benign since it is conducted under milder reaction conditions (40–50 °C, pH 4–5), which means less energy consumption and better corrosion properties. The drawback of enzymatic reactions is the slow reaction rate and product inhibition problem. Product inhibition occurs during enzyme catalysis when the product binds to the enzyme's active site, thus preventing it from further reaction and leading to lower yields of sugar

[29, 30]. *Membrane integrated bioreactors* can effectively contribute to tackling this problem [30]. The continuous removal of newly generated sugars during the process can maintain or even increase the conversion rate [31]. The basic configuration of such a reactor system is illustrated in Figure 11.3.

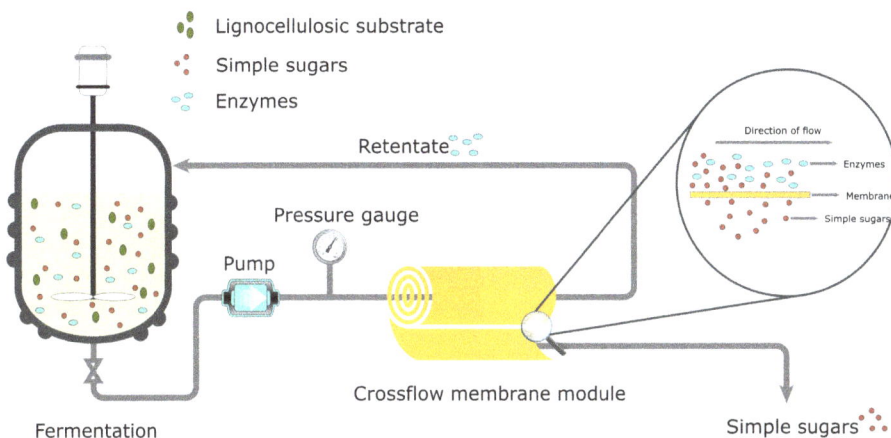

Figure 11.3: A typical membrane bioreactor system for the enzymatic hydrolysis of lignocellulosic biomass.

Suitable ultrafiltration membranes facilitate the permeation of newly formed sugars with approx. 180 g mol^{-1} molecular weight while retaining cellulose enzymes with approx. 35,000–65,000 g mol^{-1} molecular weight. A polysulfone ultrafiltration membrane with a molecular weight cut-off of 10 kDa was successfully implemented in a similar reactor design. A 53% conversion rate was achieved, much better than the 35% achieved in a traditional batch operation [31].

11.1.5 Fermentation

In the fermentation step, simple sugars are converted into ethanol by bacteria or yeast through biological processes. The microorganism type plays a prominent role in producing ethanol: ideally, the microorganism should have a broad substrate capacity, high ethanol productivity, and the capability to withstand the reaction temperature (28–52 °C) and the high concentration of ethanol [30]. *Product inhibition*, which is a major limiting factor in producing ethanol, is observed when the ethanol concentration reaches 12 vol%, eventually bringing the conversion to a halt and leaving a significant amount of unconverted sugars in the fermentation batch [32]. A multistage membrane integrated bioreactor can be implemented to overcome these problems, which simplifies and replaces several high-energy-consuming downstream processes, such as centrifugation, distillation, neutralization, and dehydration. The target molecules can be fractionated at different stages by combining microfiltration, ultrafiltration, and nanofiltration membranes, as illustrated in Figure 11.4. The yeast cells and the residual sugars are recycled, and the water–ethanol mixture is separated by *direct contact membrane distillation* (DCMD). The continuous removal of ethanol reduces the problem of product inhibition and increases sugar conversion, and the overall process consumes less energy than conventional batch reactions [33].

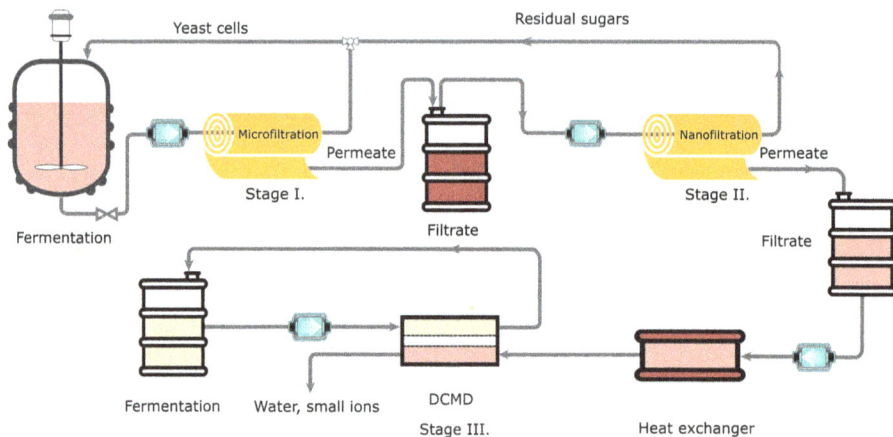

Figure 11.4: Schematic representation of a three-stage membrane bioreactor. The first stage is microfiltration, where the retention of yeast cells takes place. The second stage is nanofiltration, where residual sugars are recycled, and finally, the separation of water and ethanol through direct contact membrane distillation (DCMD) occurs.

11.1.6 Distillation/dehydration

The energy required to increase the ethanol concentration from 12 vol% to 96 vol% by conventional distillation contributes to approx. 40% of the total energy consumption in bioethanol production [34]. The water content of industrially applicable ethanol should not exceed 0.5 vol%. Advanced *extractive distillation* techniques can be used with the addition of ethylene glycol and *n*-pentane to overcome the azeotropism of a water–ethanol mixture. However, this technique also generates waste and has a high energy demand.

Membrane-based advanced processes have been extensively studied for the direct recovery and concentration of bioethanol. Based on the partial pressure difference between ethanol and water, ethanol vapor can be separated through *membrane distillation*. In contrast, the concentration difference between the feed and permeate allows separation through *pervaporation* [35]. Membrane distillation can be carried out at a much lower temperature than conventional distillation and is also considered a sustainable green alternative approach in separation processes [36]. Ethanol productivity can be improved during the fermentation process by coupling with direct membrane distillation, allowing selective removal of ethanol and thus lowering the product inhibition effect. Owing to their low surface tension values, polymeric membranes for these applications are made of polytetrafuoroethylene, polypropylene (PP), and polyvinylidenefluoride (PVDF) [30].

Membrane pervaporation performance depends on the membrane's characteristics, such as membrane thickness, selectivity, feed temperature, and ethanol concentration. Pervaporation membranes are either hydrophobic (cellulose acetate butyrate, polyimide, polydimethylsiloxane) or hydrophilic (cellulose ester) [37]. Applying both hydrophilic and hydrophobic membranes in pervaporation separation, as illustrated in Figure 11.5, can provide fuel-grade quality ethanol with lower energy demand [38].

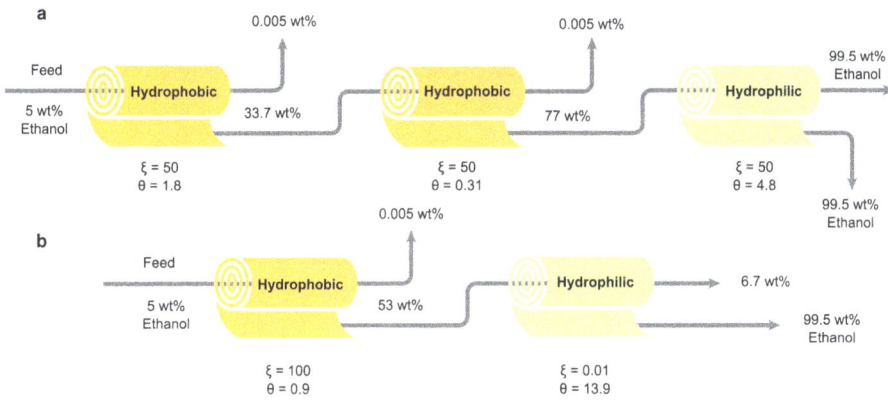

Figure 11.5: Multistage separation by pervaporation applying both hydrophobic and hydrophilic membranes, with separation factors (ξ) 50 (A route) and 100 (B route); Θ = flux of water to the flux of ethanol ratio.

11.1.7 Case study of a membrane integrated bioreactor system for the continuous production of bioethanol

The industrial batch production of bioethanol suffers from several problems such as product inhibition, slow reaction rate, and difficulties in separation and purification. The integration of a modular bioreactor with a multistage membrane system is a promising technology offering simplicity, flexibility, efficiency, and eco-friendliness.

In this case study, the production of bioethanol was designed in a continuous operation from sugarcane juice, as illustrated in Figure 11.6 [33]. The production can be separated into two main parts: the upstream process, where the preparation of the medium and the fermentation take place, and the downstream process, which consists of the isolation of ethanol from the fermentation batch.

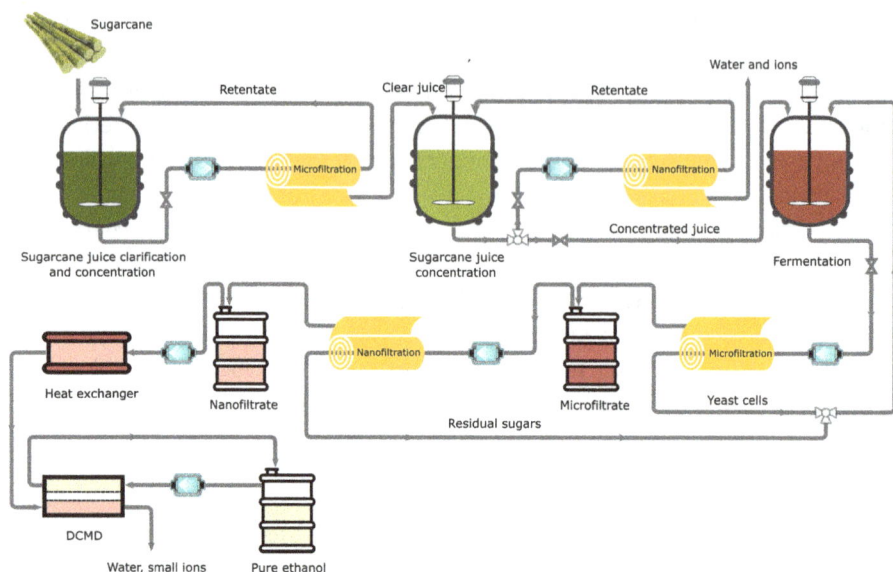

Figure 11.6: Schematic representation of bioethanol production starting from sugarcane. The sugarcane juice is clarified via microfiltration, where the permeate is further concentrated by nanofiltration. The fermentation is a continuous multistage membrane process, which directly yields pure ethanol. During fermentation, the yeast cells are recycled back into the fermentation, while the permeate stream is directed into the nanofiltration stage. Residual sugars are also recycled back into the fermentation, and the permeate, which contains ethanol, water, and small ions, is separated via direct contact membrane distillation.

The upstream process starts with the *clarification*, *sterilization*, and *concentration* of sugarcane juice. The conventional process involves multiple heat treatments (often 130 °C), chemical addition, and centrifugation. However, in this membrane-based system, the raw juice was clarified and cold-sterilized using a microfiltration membrane (PVDF, pore size: 0.45 μm, MWCO = 5×10^5 g mol^{-1}) in a cross-flow mode to remove dirt, bagasse particles, and microbes. The filtrated juice sugar concentration was increased from 12.5 wt% to 36 wt% using nanofiltration (polyamide membrane, pore size: 0.53 nm, MWCO = 150–300 g mol^{-1}, cross-flow).

In the next step, the *fermentation* and the *separation* of ethanol were coupled through a multistage membrane process in three distinct stages: microfiltration, nanofiltration, and DCMD. During microfiltration (PVDF, pore size: 0.45 μm), the yeast cells were separated and recycled back to the fermenter. In the second stage, nanofiltration was applied (polyamide, pore size: 0.57 nm, MWCO = 150–300 g mol^{-1}) to separate and recycle back the residual sugar and permeate water and ethanol. The fouled membranes were cleaned by soaking in 0.1 N NaOH, 0.01 N NaOCl, and 0.01 M HNO$_3$ aqueous solution for 1 h at 45 °C. The microfiltration membranes were cleaned every 10–12 h and nanofiltration membranes every 45–50 h to control the membrane *fouling*.

The fermentation begins with the addition of yeast cells. Once it reaches the desired amount of biomass (24–26 h), fermentation can be switched to the continuous mode by adding fresh sugar feed. The dilution rate must be equal to the microfiltrate permeate volume. The separation of ethanol from water is challenging because ethanol and water form an azeotropic mixture in a ratio of 95:5 vol% during conventional distillation. Membrane-based distillation is an alternative to energy-intensive conventional distillation. Membrane distillation should not be used immediately after fermentation due to a deterioration in permeability, selectivity, efficiency, and membrane fouling. Nonetheless, if the feed is filtered and purified in advance through microfiltration and nanofiltration, the fouling and wettability problems can be significantly reduced. The permeate after the nanofiltration contains 97 gL^{-1} ethanol, 2 gL^{-1} sugar, and some ions (Ca^{2+}, Na^+, Mg^{2+}). A hydrophobic membrane (PTFE/PET) was used during DCMD in counter-current flow mode. A solar panel provided the energy for the heating (temperature of the hot stream is 60 °C). This novel design enabled a high evaporation rate (flux was 29.5 kg EtOH m^{-2} day^{-1}), with the ethanol concentration reaching 98 vol% without any residual sugar and ions. The separated water and the solutes were recycled back to the preparation of the fresh fermentation feed. Refer to Chapter 13 for more examples of solar powered engineering.

The overall production process consumes 76% less energy than a conventional system because there is no need for unit operations with high energy demand. The atom economy of the membrane system is close to 1 (compared to 0.23 in the previous process) due to minimal waste and by-product generation. The lower number of operation units simplifies the process and greatly reduces the overall plant size to 0.2 m^2 m^{-3} of ethanol compared to 1.8 m^2 m^{-3} for the conventional system, also contributing to cost-effectiveness.

11.2 Biodiesel and its conventional production

Biodiesel has many advantages and is an adequate alternative to conventional diesel derived from fossil fuels. According to the EPA, biodiesel is less toxic than table salt and is more biodegradable than sugar [39]. Additionally, biodiesel is made from plant sources and does not contain sulfur, aromatic hydrocarbon, metal, or crude oil [40]. Biodiesel contributes to energy security and environmental preservation by decreasing net CO_2 emissions by 78% compared to petroleum diesel [41].

Biodiesel production complies with some of the green chemistry and engineering principles shown in Figure 11.7. Biodiesel's renewable nature and energy efficiency are the major sustainability benefits of biodiesel production. Biodiesel's biodegradability also contributes to its environmentally benign characteristic as it does not leave behind hazardous waste compounds.

The direct exploitation of vegetable oils as fuels without the need for further processing would be ideal from a sustainability perspective. However, raw vegetable

Green Chemistry

Safer Chemistry for Accident Prevention
Real-time Analysis for Pollution Prevention
Design for Degradation
Catalysis
Reduce Derivatives
Use of Renewable Feedstocks
Design for Energy Efficiency
Safer Solvents and Auxiliaries
Design Safer Chemicals
Less Hazardous Chemical Synthesis
Atom Economy
Prevention

Green Engineering

Inherent Rather Than Circumstantial
Prevention Instead of Treatment
Design for Separation
Maximize efficiency
Output-pulled vs Input-pushed
Conserve Complexity
Durability Rather Than Immortality
Meet Need, Minimize Excess
Minimize Material Diversity
Integrate Material and Energy Flows
Design for Commercial Afterlife
Renewable Rather Than Depleting

Figure 11.7: Green chemistry and engineering principles in biodiesel production.

oils in diesel engines cause decreased engine performance and increased carbon monoxide and hydrocarbon emissions, albeit with lower NO_x emissions [42]. Not all the fuel is combusted during the combustion process, irrespective of the engine type (petrol or diesel). This uncombusted fuel leaves behind carbon, which over time builds up and causes several issues. The *chemistry of biodiesel* is explained in Figure 11.8. Vegetable oils consist of esters of glycerol and fatty acids, which produce glycerol upon hydrolysis. This glycerol results in carbon build-up in the pistons and fuel injectors of diesel engines due to its low thermal stability and low combustion efficiency. Carbon build-up causes drivability issues, i.e., the engine does not run smoothly, engine vibrations or shaking, louder engine operation, and smoky exhaust, translating to increased emissions and a drop in fuel economy.

Vegetable oils are compounds consisting of different mixtures of triglycerides. Vegetable oil molecules are much larger than biodiesel molecules, giving rise to gelation in cold weather. The almost tenfold higher viscosity of vegetable oils has detrimental effects on an engine, such as poor fuel atomization, incomplete combustion, carbon deposition on the injector and valve seats, and engine fouling. Vegetable oil conversion to biodiesel produces smaller molecules, closer to conventional diesel sizes, which mitigates most of the drawbacks of vegetable oil as fuel.

Biodiesel can be efficiently combusted in diesel engines without modification because it has a similar structure to conventional diesel (Figure 11.8). Both molecules possess a long chain of carbon atoms, and only the functional group at the end of the chain differs. Their chemical and structural similarity is the reason why biodiesel can be used in diesel engines. The chemical conversion of vegetable oil to biodiesel is called *transesterification*, which is essentially the reaction of the oil with alcohol in the presence of a catalyst. Three monoesters and glycerol are obtained; these esters have similar viscosities as regular diesel and thus can be utilized as fuels.

Before transesterification, vegetable oils, triglycerides from algae, and waste cooking oil need to undergo pretreatment to minimize water content (less than 0.1 wt%) and remove any non-lipid substances [43]. Production of biodiesel from *microalgae* begins

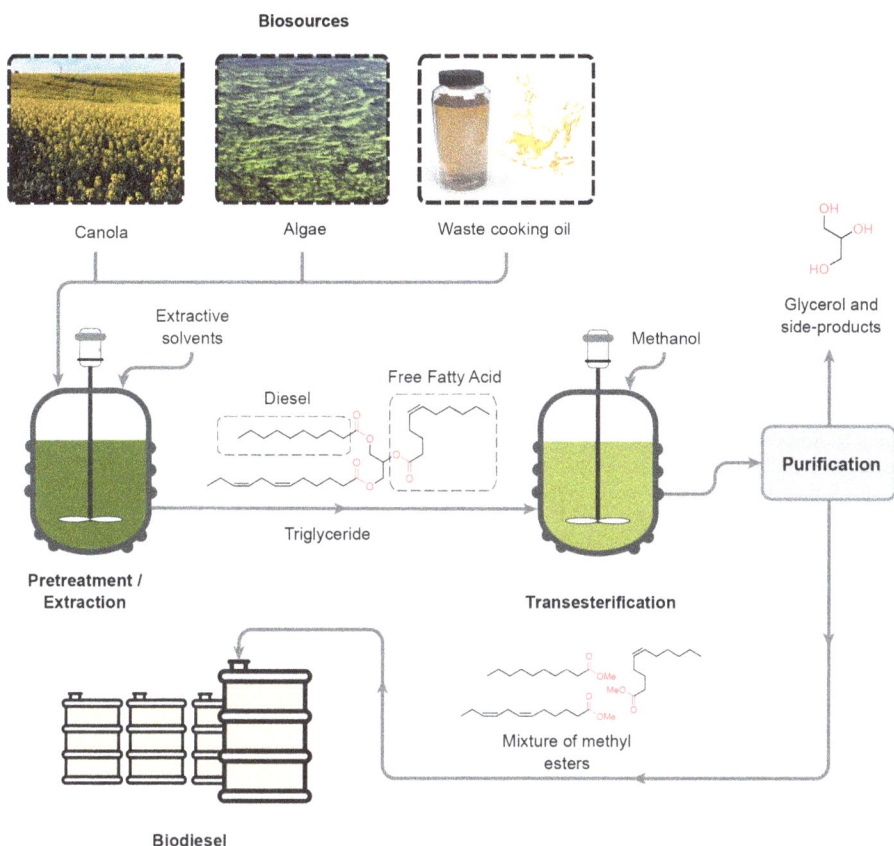

Figure 11.8: Biodiesel chemistry showing the diesel molecule, which is a hydrocarbon without an ester group; a fatty acid that is similar to a diesel molecule but has a carboxyl group; vegetable oil consisting of triacylglycerides, which are glycerol esters incorporating three fatty acid chains; and the transesterification reaction, which converts the vegetable oils to biodiesel.

with their harvest, followed by the disruption of plant cells to release the oils and extraction with an appropriate solvent [44]. Petroleum solvents such as hexane and diethyl ether are currently used in the industry [45]; however, the exploitation of ionic liquids [46], supercritical CO_2 [47], and supercritical ethanol [47] also show promising potential in this application. After the reaction, the products are separated, the glycerol is excreted, and the fatty acid alkyl esters are purified by water washing, resulting in the final biodiesel product. The purification step consumes a large amount of clean water, resulting in undesired wastewater. For each liter of biodiesel, approx. 10 L of wastewater is produced [48].

There are two conventional methods for conventional transesterification reactions. The first is the *acid-catalyzed method* using sulfuric acid, ferric sulfate, methanolic boron trifluoride, or methanolic sulfuric acid [49]. The disadvantages of acid-catalyzed transesterification include the need for a high molar ratio of alcohol to oil, a prolonged reaction time, and high reaction temperatures. The second method is the *base-catalyzed method*, which uses sodium and potassium hydroxide, alkoxides, potassium, or sodium carbonates. This method's advantages include lower operating temperatures, smaller amounts of catalyst, and shorter reaction times (4000 times faster reaction). However, possible soap formation is the main drawback of this method [43]. Soap formation occurs when the free fatty acid content is more than 3 wt%. In this case, the transesterification is a two-step process: acid esterification followed by the base-catalyzed method. These catalysts have to be removed from the final product with repeated washing using distilled water, resulting in a significant amount of wastewater. Alternatively, heterogeneous catalysis can be used in the transesterification reaction. Activated calcium oxide and magnesium oxide can be filtered out after the reaction, simplifying the purification step and reducing the water consumption.

11.2.1 Alternative routes for biodiesel production

Conventional biodiesel production requires high temperatures and pressures, excessive alcohol to increase the yield, a substantial amount of washing water, and alcohol and catalyst removal. Alternative process routes for biodiesel production have been sought to overcome these drawbacks. The use of sustainable approaches, such as supercritical fluids, microwave energy, and energy-efficient membrane separations, is discussed in the following sections.

11.2.1.1 Supercritical alcohol transesterification

Conventional technologies for biodiesel production that apply alkali catalysts cannot process raw materials with high water and free fatty acid content [43]. To overcome this bottleneck, supercritical methanol and ethanol can be utilized without the use of any catalyst. The absence of pretreatment, soap, and a catalyst significantly reduces the environmental burden. However, the high pressure (80 bar) and temperature (250 °C) result in high operating costs [50]. Nonetheless, high conversion can be achieved quickly (10–40 min) [50], which makes the overall process economically feasible and a viable alternative to base-catalyzed commercial processes [51].

Under these extreme conditions, liquid alcohol will reach a critical point where both the gas and liquid phases become indistinguishable and exhibit both liquid and gas properties. Supercritical alcohol can dissolve materials like a fluid and can penetrate solids like a gas [52, 53]. To learn more about supercritical fluids and their sustainable use in the extraction process, refer to Section 7.3. During transesterification, the hydrogen bonding between alcohol molecules decreases in the supercritical environment, and thus it can act as an acid catalyst. Optimization of the process conditions (e.g., temperature, alcohol:oil ratio, reaction time) is crucial to maximize the conversion and yield. Statistical methods [54] and artificial neural networks provide significant advantages for finding the optimal parameters. For instance, a 98% experimental yield has been achieved at 285 °C using methanol:oil ratio of 23.4 and a reaction time as low as 26.5 min [55].

11.2.1.2 Microwave energy for biodiesel production

The conventional transesterification method to obtain biodiesel uses jacketed reactors and indirect steam heating methods. Microwave irradiation, combined with an acid, base, and heterogeneous catalysis, is an alternative green method [56]. The conventional heating used in transesterification processes requires more energy input and a long preheat and reaction time of around 1 h to achieve a 96% yield to obtain biodiesel [57]. On the contrary, the microwave method is faster (a few minutes) and more energy-efficient [58]. Microwave irradiation can decrease both the reaction and separation time by 97% and 94%, respectively. Moreover, it is possible to increase the yield reaching 100% compared to conventional heating methods. The microwave method does not influence the quality of the obtained biodiesel. Read more about the green principles and sustainable use of microwave energy in Section 5.3.3.

Microwave irradiation can be used in the following two steps of biodiesel production: oil extraction and chemical transesterification. The integration of these two steps into a single-step extractive transesterification reaction would be beneficial. The process of biodiesel production can be carried out using microwave irradiation as the heating method [56]. After preparing the raw material, the oilseeds are mixed with the extraction solvent (usually hexane) and exposed to microwave radiation. Disruption of the oilseed cells occurs when the water molecules inside the cell start to boil, leading to the rupture of the cell walls. The selected compounds then migrate from the sample matrix to the extraction solvent [56]. Hexane is evaporated from the mixture, and the extracted oil extracted mixture is then placed in a microwave reactor for the transesterification process. The process ends with the separation and purification of the resulting products (biodiesel and glycerol).

Microwave technology can be applied for the thermal pretreatment or process promotion technique to extract oils and lipids from biodiesel feedstock with low energy and solvent demand [56]. The extraction of oils and lipids from plant feedstocks is carried out by disrupting the oilseed or algal cells. When the temperature reaches the boiling point of water, the cell wall ruptures, resulting in the extraction of the organic compounds from the sample matrix to the solvent. The production of biodiesel via microwave technology is a feasible technology because microwave thermal effects are adequate for the rapid disruption of cells, resulting in efficient lipid and oil extraction.

Microwave technology can also be used for the chemical process of transesterification. Microwaves can enhance the transesterification reaction by the thermal effect or evaporation of the solvent (methanol) owing to the strong microwave interaction with the material. Microwave irradiation results in a substantial decrease in the activation energy due to the higher dipolar polarization during the interaction with the reaction compounds. In other words, microwave irradiation reduces the energy demand of the process. The solvent selection can significantly influence the extent of the reduction of the activation energy. For instance, methanol absorbs microwaves better than ethanol and therefore is preferred to ethanol for transesterification reactions.

Table 11.1 summarizes the main properties, benefits, and drawbacks of the different heating methods [56]. Conventional heating is a time- and energy-consuming process with low efficiency and demands additional steam generation operations, circulating, and raising the heat transfer fluid temperature. In contrast, supercritical heating shows a slight improvement in terms of reaction time and process efficiency. Moreover, biodiesel production processes that use supercritical heating do not require catalysts because, under supercritical conditions, the strength of hydrogen bonding between the alcohol molecules decreases, thus acting as an acid catalyst by itself. However, the main drawbacks, such as higher energy demand (temperature and pressure) and high-pressure safety risks, need to be considered. The microwave heating method is superior to the other two methods due to its time- and energy-efficient operation, which results in a highly efficient process.

Table 11.1: Comparison of the three heating types used in biodiesel production.

Parameter	Conventional heating	Supercritical heating	Microwave heating
Reaction time	1–2 h (long)	Less than 1 h (short)	0.05–0.1 h (very short)
Reaction temperature	40–100 °C	250–400 °C	40–100 °C
Reaction pressure	Atmospheric	35–60 MPa (high)	Atmospheric
Catalyst required	Yes	No	Yes/No
Heat loss	High	Moderate	Low
Energy form	Electrical energy to thermal energy	Electrical energy to thermal energy	Electrical energy applied through microwave
Process efficiency	Low	Moderate	High
Catalyst removal	Yes	No	Yes
Soap removal	Yes	No	Yes

Table 11.1 (continued)

Parameter	Conventional heating	Supercritical heating	Microwave heating
Advantages	Simple operation; low energy usage	Short reaction time; easy product separation	Short reaction time; cleaner products; energy efficiency
Limitations	High energy requirement; saponified products	High capital costs; Pressure vessel safety	May not be efficient with feedstock containing solids

11.2.1.3 Membrane-based biodiesel production

Process intensification in biodiesel production can be achieved by integrating membrane systems at different stages for different purposes, from pretreatment to final product purification [59]. The unconverted residual oil can be recycled *in situ* for further transesterification, improving the overall process economy and sustainability. Green solvents acting as green catalysts, such as ionic liquids and deep eutectic solvents, can be used for transesterification reactions. It is possible to reuse green solvents through the implementation of membrane units. Refer to Section 8.3 to learn more about membrane-assisted continuous solvent recovery processes. Cross-flow membrane rigs mitigate fouling and can efficiently remove impurities, such as glycerol, catalysts, and soaps, from the biodiesel. The *direct contact membrane distillation* technique is suitable for the recovery of unreacted alcohol from the biodiesel mixture. The membrane-based system allows obtaining the desired alkyl esters with a lower environmental footprint and at a lower price.

The conventional purification step in biodiesel production can be classified into wet and dry washing categories: washing with distilled water (wet) or using a solid adsorbent (dry). Distilled warm water is the best for purification, as both glycerol and alcohol are highly soluble in water, and the warm water prevents saturated fatty esters from precipitating. The *water consumption* is large since seven consecutive washing steps are required to achieve the threshold limit of 0.02 wt% free glycerol [60]. The conventional biodiesel separation technology for free glycerol removal utilizes 10 liters of water to treat 1 liter of biodiesel. In contrast, the membrane -produced biodiesel only requires 0.002 liters of water per liter of biodiesel [61]. Poly(ether-sulfone) ultrafiltration membranes with a 10 kDa cut-off can effectively reduce the glycerol content below 0.007 wt% upon the addition of 0.2 wt% of water to the crude biodiesel feed. However, severe membrane fouling can occur with a decrease in the flux from 70 to 30 $kg\,m^{-2}\,h^{-1}$ after only 1 h of operation [62].

Membrane reactors (Figure 11.9) can increase conversion by shifting the equilibrium reaction. Transesterification is an *equilibrium reaction*, which can be shifted toward biodiesel formation by continuously removing the product from the reaction mixture. The separation process depends on the type of membrane. Microporous membranes such as mixed cellulose ester membranes can separate fatty acid methyl esters and glycerol. In contrast, dense polymer pervaporation membranes such as polyethylene terephthalate membranes that are supported with PP separate glycerol and alcohol. Membrane-based biodiesel production is performed under mild conditions at approx. 70 °C and 173 kPa, consuming less energy than supercritical or heterogeneous (180–200 °C) transesterification [63]. Catalytic membrane reactors can also improve biodiesel production sustainability by decreasing the amount of catalyst used in the transesterification. For instance, a level of 0.05 wt% of base catalyst is sufficient for a continuous membrane reactor, which is significantly lower than the 0.5–1 wt% NaOH catalyst applied in conventional biodiesel transesterification processes [63].

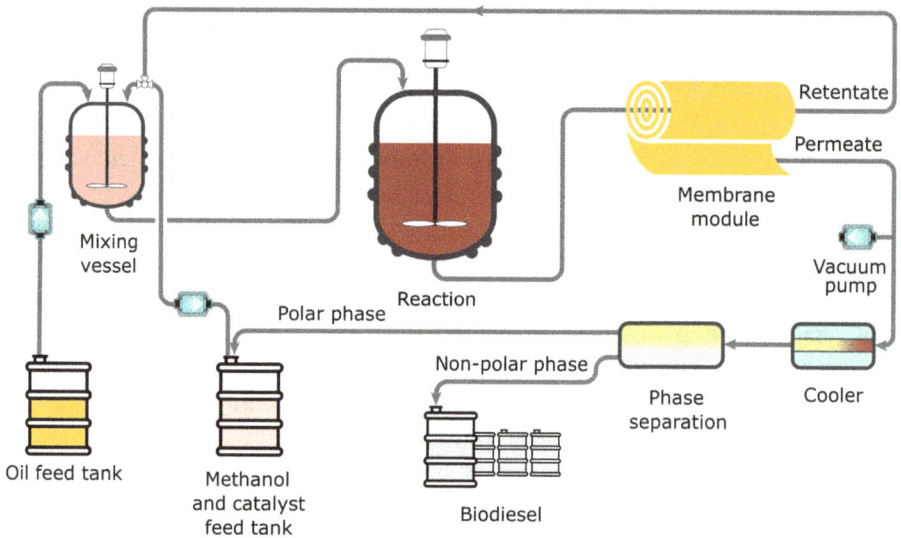

Figure 11.9: Schematic illustration of a transesterification reaction with a membrane module. The oil and a mixture of methanol/catalyst are charged into the mixing vessel and heated to the desired reaction temperature before being transferred into the membrane reactor. The permeate stream consists of biodiesel, glycerol, methanol, and the catalyst. The retentate stream contains the unreacted oil droplets, which are transferred back to the mixing vessel. Subsequently, the permeate stream is separated into polar and non-polar phases. The methanol from the polar phase is separated and recycled back to the feed tank.

Alcohols (usually methanol or ethanol) are used in the transesterification process, and their amount greatly depends on the applied technology. Methanol is preferred over ethanol because of its lower cost and lower viscosity than ethyl esters. The transesterification process is performed near the lower boiling point of methanol [43]. However, the toxic nature and the non-renewable production of methanol present some obstacles to its sustainable and safe production. As a green substituent to methanol, ethanol has higher miscibility with mixtures containing biodiesel, residual alcohol, and glycerol. However, the resulting emulsified soap and water can cause problems, which can be overcome by separating and recovering the unreacted ethanol by membrane distillation or pervaporation. The molar ratio of the used alcohol and oil is an important parameter in the transesterification process. The optimum molar ratio for the supercritical method is 40:1 of methanol to oil [64]. Membrane reactors utilize much less alcohol (as low as 6:1) than the supercritical methanol method [63].

Bibliography

[1]   Dahman, Y.; Dignan, C.; Fiayaz, A.; Chaudhry, A. An Introduction to Biofuels, Foods, Livestock, and the Environment. In Biomass, Biopolymer-Based Materials, and Bioenergy; Elsevier, 2019; pp 241–276. 10.1016/B978-0-08-102426-3.00013-8.

[2]   Schmidt, L. D.; Dauenhauer, P. J. Hybrid Routes to Biofuels. *Nature*. **2007**, 447(7147), 914–915. 10.1038/447914a.

[3]   Kralova, I.; Sjöblom, J. Biofuels–Renewable Energy Sources: A Review. *J. Dispers. Sci. Technol*. **2010**, 31(3), 409–425. 10.1080/01932690903119674.

[4]   Begum, S.; Dahman, Y. Enhanced Biobutanol Production Using Novel Clostridial Fusants in Simultaneous Saccharification and Fermentation of Green Renewable Agriculture Residues. *Biofuels, Bioprod. Biorefining*. **2015**, 9(5), 529–544. 10.1002/bbb.1564.

[5]   Ruan, R.; Zhang, Y.; Chen, P.; Liu, S.; Fan, L.; Zhou, N.; Ding, K.; Peng, P.; Addy, M.; Cheng, Y.; Anderson, E.; Wang, Y.; Liu, Y.; Lei, H.; Li, B. Biofuels: Introduction. In Biofuels: Alternative Feedstocks and Conversion Processes for the Production of Liquid and Gaseous Biofuels;; Elsevier, 2019; pp 3–43. 10.1016/B978-0-12-816856-1.00001-4.

[6]   Brennan, L.; Owende, P. Biofuels from Microalgae-A Review of Technologies for Production, Processing, and Extractions of Biofuels and Co-Products. *Renew. Sustain. Energy Rev*. **2010**, 14(2), 557–577. 10.1016/j.rser.2009.10.009.

[7]   Najafpour, G. D. Biofuel Production** This Chapter Was Written with Contributions from: Arash Mollahoseini, Biofuel Research Team (BRTeam), Karaj, Iran; Meisam Tabatabaei, Biofuel Research Team (BRTeam), Karaj, Iran and Agricultural Biotechnology Research Institute of Iran. In Biochemical Engineering and Biotechnology; Elsevier, 2015; pp 597–630. 10.1016/B978-0-444-63357-6.00020-1.

[8]   Dragone, G.; Fernandes, B. D.; Vicente, A. A.; Teixeira, J. A. Third Generation Biofuels from Microalgae. **2010**.

[9]   Demirbas, M. F. Biofuels from Algae for Sustainable Development. *Appl. Energy*. **2011**, 88(10), 3473–3480. 10.1016/j.apenergy.2011.01.059.

[10] Singh, B.; Guldhe, A.; Rawat, I.; Bux, F. Towards a Sustainable Approach for Development of Biodiesel from Plant and Microalgae. *Renew. Sustain. Energy Rev*. **2014**, 29, 216–245. 10.1016/j.rser.2013.08.067.

[11] Gude, V. G. In *Green Chemistry for Sustainable Biofuel Production*; Gude, V. G., Ed.; Apple Academic Press: Toronto/New Jersey, 2018. 10.1201/b22351.

[12] Vohra, M.; Manwar, J.; Manmode, R.; Padgilwar, S.; Patil, S. Bioethanol Production: Feedstock and Current Technologies. *J. Environ. Chem. Eng*. **2014**, 2(1), 573–584. 10.1016/j.jece.2013.10.013.

[13] Beer, T.; Grant, T.; Morgan, G.; Lapszewicz, J.; Anyon, P.; Edwards, J.; Nelson, P.; Watson, H.; Williams, D. Comparison of Transport Fuels. *Final Rep. Aust. Greenh. Off*, 2001, 485.

[14] Vohra, M.; Manwar, J.; Manmode, R.; Padgilwar, S.; Patil, S. Bioethanol Production: Feedstock and Current Technologies. *J. Environ. Chem. Eng*. **2014**, 2(1), 573–584. 10.1016/j.jece.2013.10.013.

[15] Himmel, M. E.; Ding, S.-Y.; Johnson, D. K.; Adney, W. S.; Nimlos, M. R.; Brady, J. W.; Foust, T. D. Biomass Recalcitrance: Engineering Plants and Enzymes for Biofuels Production. *Science (80).* **2007**, 315(5813), 804–807. 10.1126/science.1137016.

[16] Kumar, P.; Barrett, D. M.; Delwiche, M. J.; Stroeve, P. Methods for Pretreatment of Lignocellulosic Biomass for Efficient Hydrolysis and Biofuel Production. *Ind. Eng. Chem. Res.* **2009**, 48(8), 3713–3729. 10.1021/ie801542g.

[17] Balat, M. Production of Bioethanol from Lignocellulosic Materials via the Biochemical Pathway: A Review. *Energy Convers. Manag.* **2011**, 52(2), 858–875. 10.1016/j.enconman.2010.08.013.

[18] Yoo, C. G.; Meng, X.; Pu, Y.; Ragauskas, A. J. The Critical Role of Lignin in Lignocellulosic Biomass Conversion and Recent Pretreatment Strategies: A Comprehensive Review. *Bioresour. Technol.* **2020**, 301, 122784. 10.1016/j.biortech.2020.122784.

[19] Sun, S.; Sun, S.; Cao, X.; Sun, R. The Role of Pretreatment in Improving the Enzymatic Hydrolysis of Lignocellulosic Materials. *Bioresour. Technol.* **2016**, 199, 49–58. 10.1016/j.biortech.2015.08.061.

[20] Socha, A. M.; Parthasarathi, R.; Shi, J.; Pattathil, S.; Whyte, D.; Bergeron, M.; George, A.; Tran, K.; Stavila, V.; Venkatachalam, S.; Hahn, M. G.; Simmons, B. A.; Singh, S. Efficient Biomass Pretreatment Using Ionic Liquids Derived from Lignin and Hemicellulose. *Proc. Natl. Acad. Sci.* **2014**, 111(35), E3587–E3595. 10.1073/pnas.1405685111.

[21] Baaqel, H.; Díaz, I.; Tulus, V.; Chachuat, B.; Guillén-Gosálbez, G.; Hallett, J. P. Role of Life-Cycle Externalities in the Valuation of Protic Ionic Liquids-a Case Study in Biomass Pretreatment Solvents. *Green Chem.* **2020**, 22(10), 3132–3140. 10.1039/d0gc00058b.

[22] Meng, X.; Pu, Y.; Li, M.; Ragauskas, A. J. A Biomass Pretreatment Using Cellulose-Derived Solvent Cyrene. *Green Chem.* **2020**, 22(9), 2862–2872. 10.1039/d0gc00661k.

[23] Ragauskas, A. J.; Beckham, G. T.; Biddy, M. J.; Chandra, R.; Chen, F.; Davis, M. F.; Davison, B. H.; Dixon, R. A.; Gilna, P.; Keller, M.; Langan, P.; Naskar, A. K.; Saddler, J. N.; Tschaplinski, T. J.; Tuskan, G. A.; Wyman, C. E. Lignin Valorization: Improving Lignin Processing in the Biorefinery. *Science (80).* **2014**, 344(6185), 1246843. 10.1126/science.1246843.

[24] Ge, S.; Nai, P.; Yek, Y.; Wang, Y.; Xia, C.; Adibah, W.; Mahari, W.; Keey, R.; Peng, W.; Yuan, T.; Tabatabaei, M.; Aghbashlo, M.; Sonne, C.; Shiung, S. Progress in Microwave Pyrolysis Conversion of Agricultural Waste to Value-Added Biofuels: A Batch to Continuous Approach. *Renew. Sustain. Energy Rev.* **2021**, 135, 110148. 10.1016/j.rser.2020.110148.

[25] Wilson, A. N.; Dutta, A.; Black, B. A.; Mukarakate, C.; Magrini, K.; Schaidle, J. A.; Michener, W. E.; Beckham, G. T.; Nimlos, M. R. Valorization of Aqueous Waste Streams from Thermochemical Biorefineries. *Green Chem.* **2019**, 21(15), 4217–4230. 10.1039/c9gc00902g.

[26] Fiege, H.; Voges, H.-W.; Hamamoto, T.; Umemura, S.; Iwata, T.; Miki, H.; Fujita, Y.; Buysch, H.-J.; Garbe, D.; Paulus, W. Phenol Derivatives in *Ullmann's Encyclopedia of Industrial Chemistry*; Wiley-VCH Verlag GmbH & Co. KGaA: Weinheim, Germany, 2000; pp 503–519. 10.1002/14356007.a19_313.

[27] Zirbes, M.; Quadri, L. L.; Breiner, M.; Stenglein, A.; Bomm, A.; Schade, W.; Waldvogel, S. R. High-Temperature Electrolysis of Kraft Lignin for Selective Vanillin Formation. *ACS Sustain. Chem. Eng.* 2020. 10.1021/acssuschemeng.0c00162.

[28] Hayes, D. J. An Examination of Biorefining Processes, Catalysts and Challenges. *Catal. Today.* **2009**, 145(1–2), 138–151. 10.1016/j.cattod.2008.04.017.

[29] Andrić, P.; Meyer, A. S.; Jensen, P. A.; Dam-Johansen, K. Reactor Design for Minimizing Product Inhibition during Enzymatic Lignocellulose Hydrolysis: I. Significance and Mechanism of Cellobiose and Glucose Inhibition on Cellulolytic Enzymes. *Biotechnol. Adv.* **2010**, 28(3), 308–324. 10.1016/j.biotechadv.2010.01.003.

[30] Dey, P.; Pal, P.; Kevin, J. D.; Das, D. B. Lignocellulosic Bioethanol Production: Prospects of Emerging Membrane Technologies to Improve the Process – A Critical Review. *Rev. Chem. Eng.* **2020**, 36(3), 333–367. 10.1515/revce-2018-0014.

[31] Gan, Q.; Allen, S.; Taylor, G. Design and Operation of an Integrated Membrane Reactor for Enzymatic Cellulose Hydrolysis. *Biochem. Eng. J.* **2002**, 12(3), 223–229. 10.1016/S1369-703X(02) 00072-4.

[32] Gryta, M.; Morawski, A. W.; Tomaszewska, M. Ethanol Production in Membrane Distillation Bioreactor. *Catal. Today.* **2000**, 56(1–3), 159–165. 10.1016/S0920-5861(99)00272-2.

[33] Pal, P.; Kumar, R.; Ghosh, A. K. Analysis of Process Intensification and Performance Assessment for Fermentative Continuous Production of Bioethanol in a Multi-Staged Membrane-Integrated Bioreactor System. *Energy Convers. Manag.* **2018**, 171(June), 371–383. 10.1016/j. enconman.2018.05.099.

[34] Ahmetović, E.; Martín, M.; Grossmann, I. E. Optimization of Energy and Water Consumption in Corn-Based Ethanol Plants. *Ind. Eng. Chem. Res.* **2010**, 49(17), 7972–7982. 10.1021/ie1000955.

[35] Abels, C.; Carstensen, F.; Wessling, M. Membrane Processes in Biorefinery Applications. *J. Membr. Sci.* **2013**, 444, 285–317. 10.1016/j.memsci.2013.05.030.

[36] Onsekizoglu, P. Membrane Distillation: Principle, Advances, Limitations and Future Prospects in Food Industry. In Distillation – Advances from Modeling to Applications; InTech, 2012. 10.5772/ 37625.

[37] Huang, Y.; Baker, R. W.; Wijmans, J. G. Perfluoro–Coated Hydrophilic Membranes with Improved Selectivity. *Ind. Eng. Chem. Res.* **2013**, 52(3), 1141–1149. 10.1021/ie3020654.

[38] Nagy, E.; Mizsey, P.; Hancsók, J.; Boldyryev, S.; Varbanov, P. Analysis of Energy Saving by Combination of Distillation and Pervaporation for Biofuel Production. *Chem. Eng. Process. Process Intensif.* **2015**, 98, 86–94. 10.1016/j.cep.2015.10.010.

[39] 8/5/2004: U.S. EPA gives $69,000 to University of Nevada to find ways to improve alternative fuel https://archive.epa.gov/epapages/newsroom_archive/newsreleases/9e0feab118d98e95852570 d8005e166a.html (accessed Jun 26, 2020).

[40] Jaichandar, S.; Annamalai, K. The Status of Biodiesel as an Alternative Fuel for Diesel Engine – An Overview. *J. Sustain. Energy Environ.* **2011**, 2, 71–75.

[41] Sheehan, J.; Camobreco, V.; Duffield, J.; Graboski, M.; Shapouri, H. An Overview of Biodiesel and Petroleum Diesel Life Cycles. **1998**, 10.2172/1218368.

[42] Altin, R.; Çetinkaya, S.; Yücesu, H. S. Potential of Using Vegetable Oil Fuels as Fuel for Diesel Engines. *Energy Convers. Manag.* **2001**, 42(5), 529–538. 10.1016/S0196-8904(00)00080-7.

[43] Sharma, Y. C.; Singh, B. Development of Biodiesel: Current Scenario. *Renew. Sustain. Energy Rev.* **2009**, 13(6–7), 1646–1651. 10.1016/j.rser.2008.08.009.

[44] Rene, H.; Wijffels, J. M. B. An Outlook on Microalgal Biofuels. *Science.* **2010**, 329(August), 796–799.

[45] Mubarak, M.; Shaija, A.; Suchithra, T. V. A Review on the Extraction of Lipid from Microalgae for Biodiesel Production. *Algal Res.* **2015**, 7, 117–123. 10.1016/j.algal.2014.10.008.

[46] Kim, Y. H.; Choi, Y. K.; Park, J.; Lee, S.; Yang, Y. H.; Kim, H. J.; Park, T. J.; Hwan Kim, Y.; Lee, S. H. Ionic Liquid-Mediated Extraction of Lipids from Algal Biomass. *Bioresour. Technol.* **2012**, 109, 312–315. 10.1016/j.biortech.2011.04.064.

[47] Hegel, P. E.; Martín, L. A.; Popovich, C. A.; Damiani, C.; Leonardi, P. I. Biodiesel Production from Halamphora Coffeaeformis Microalga Oil by Supercritical Ethanol Transesterification. *Chem. Eng. Process. – Process Intensif.* **2019**, 145(August), 107670. 10.1016/j.cep.2019.107670.

[48] Karaosmanoğlu, F.; Cığızoğlu, K. B.; Tüter, M.; Ertekin, S. Investigation of the Refining Step of Biodiesel Production. *Energy Fuels.* **1996**, 10(4), 890–895. 10.1021/ef9502214.

[49] Demirbas, A.; Bafail, A.; Ahmad, W.; Sheikh, M. Biodiesel Production from Non-Edible Plant Oils. *Energy Explor. Exploit.* **2016**, 34(2), 290–318. 10.1177/0144598716630166.

[50] Rathore, V.; Madras, G. Synthesis of Biodiesel from Edible and Non-Edible Oils in Supercritical Alcohols and Enzymatic Synthesis in Supercritical Carbon Dioxide. *Fuel.* **2007**, 86(17–18), 2650–2659. 10.1016/j.fuel.2007.03.014.

[51] Van Kasteren, J. M. N.; Nisworo, A. P. A Process Model to Estimate the Cost of Industrial Scale Biodiesel Production from Waste Cooking Oil by Supercritical Transesterification. *Resour. Conserv. Recycl.* **2007**, 50(4), 442–458. 10.1016/j.resconrec.2006.07.005.

[52] Leung, D. Y. C.; Wu, X.; Leung, M. K. H. A Review on Biodiesel Production Using Catalyzed Transesterification. *Appl. Energy.* **2010**, 87(4), 1083–1095. 10.1016/j.apenergy.2009.10.006.

[53] Juan, J. C.; Kartika, D. A.; Wu, T. Y.; Hin, T.-Y. Y. Biodiesel Production from Jatropha Oil by Catalytic and Non-Catalytic Approaches: An Overview. *Bioresour. Technol.* **2011**, 102(2), 452–460. 10.1016/j.biortech.2010.09.093.

[54] Chauhan, D. S.; Goswami, G.; Dineshbabu, G.; Palabhanvi, B.; Das, D. Evaluation and Optimization of Feedstock Quality for Direct Conversion of Microalga Chlorella Sp. FC2 IITG into Biodiesel via Supercritical Methanol Transesterification. *Biomass Convers. Biorefinery.* **2020**, 10(2), 339–349. 10.1007/s13399-019-00432-2.

[55] Srivastava, G.; Paul, A. K.; Goud, V. V. Optimization of Non-Catalytic Transesterification of Microalgae Oil to Biodiesel under Supercritical Methanol Condition. *Energy Convers. Manag.* **2018**, 156, 269–278. 10.1016/j.enconman.2017.10.093.

[56] Gude, V. G.; Patil, P.; Deng, S. Microwave Energy Potential for Large Scale Biodiesel Production. *World Renew. Energy Forum, WREF 2012, Incl. World Renew. Energy Congr. XII Color. Renew. Energy Soc. Annu. Conf.* **2012**, 1(c), 751–759.

[57] Refaat, A. A.; Attia, N. K.; Sibak, H. A.; El Sheltawy, S. T.; El Diwani, G. I. Production Optimization and Quality Assessment of Biodiesel from Waste Vegetable Oil. *Int. J. Environ. Sci. Technol.* **2008**, 5(1), 75–82. 10.1007/BF03325999.

[58] Azcan, N.; Danisman, A. Microwave Assisted Transesterification of Rapeseed Oil. *Fuel.* **2008**, 87, 1781–1788. 10.1016/j.fuel.2007.12.004.

[59] Kumar, R.; Ghosh, A. K.; Pal, P. Sustainable Production of Biofuels through Membrane-Integrated Systems. *Sep. Purif. Rev.* **2020**, 49(3), 207–228. 10.1080/15422119.2018.1562942.

[60] Gomes, M. C. S.; Pereira, N. C.; De Barros, S. T. D. Separation of Biodiesel and Glycerol Using Ceramic Membranes. *J. Membr. Sci.* **2010**, 352(1–2), 271–276. 10.1016/j.memsci.2010.02.030.

[61] Saleh, J.; Tremblay, A. Y.; Dubé, M. A. Glycerol Removal from Biodiesel Using Membrane Separation Technology. *Fuel.* **2010**, 89(9), 2260–2266. 10.1016/j.fuel.2010.04.025.

[62] Alves, M. J.; Nascimento, S. M.; Pereira, I. G.; Martins, M. I.; Cardoso, V. L.; Reis, M. Biodiesel Purification Using Micro and Ultrafiltration Membranes. *Renew. Energy.* **2013**, 58, 15–20. 10.1016/j.renene.2013.02.035.

[63] Shuit, S. H.; Ong, Y. T.; Lee, K. T.; Subhash, B.; Tan, S. H. Membrane Technology as A Promising Alternative in Biodiesel Production: A Review. *Biotechnol. Adv.* **2012**, 30(6), 1364–1380. 10.1016/j.biotechadv.2012.02.009.

[64] Sharma, Y. C.; Singh, B.; Upadhyay, S. N. Advancements in Development and Characterization of Biodiesel: A Review. *Fuel.* **2008**, 87(12), 2355–2373. 10.1016/j.fuel.2008.01.014.

12 Green polymers and green building blocks

Sushil Kumar, Gyorgy Szekely

12.1 Introduction

Polymers have set their roots deep into almost every aspect of modern life. Polymers can be broadly classified into two groups based on their origin: natural polymers and synthetic polymers. Dwindling fossil resources and global warming have motivated the search for alternative green and sustainable sources for polymers and their building blocks. Biomass, for instance, represents an excellent source of biodegradable and compostable green polymers. This chapter reviews the utilization of biomass-derived building blocks as a sustainable feedstock for designing high-performance polymers.

Both natural and synthetic polymers are produced by the polymerization reaction of simple units called monomers. *Natural polymers* occur in our environment, while *synthetic polymers* are manufactured artificially. Polymers derived from biomass feedstock are called *bio-based polymers*, while polymers derived from petrochemicals and crude oil are known as *fossil-based polymers*. Figure 12.1 shows the relative demand for the different polymers in Europe [1].

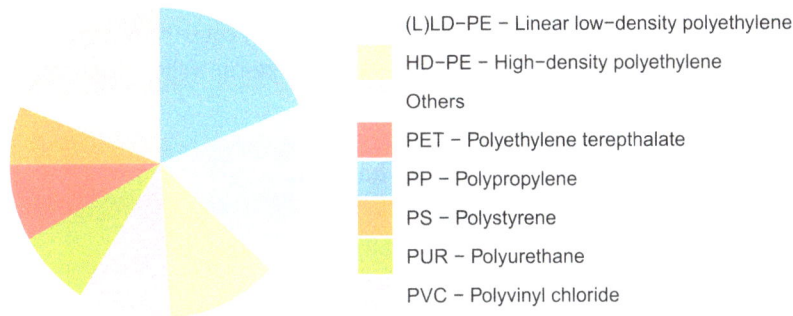

(L)LD–PE – Linear low–density polyethylene

HD–PE – High–density polyethylene

Others

PET – Polyethylene terepthalate

PP – Polypropylene

PS – Polystyrene

PUR – Polyurethane

PVC – Polyvinyl chloride

Figure 12.1: The relative demand for different polymers in Europe in 2018, based on data from reference [1].

The global production of fossil-based polymers in 2018 was estimated to be approx. 359 million metric tons, half of which consisted of single-use plastics for applications such as medical syringes and packaging [2]. The remainder was used for long-term infrastructure purposes such as water pipes, furniture, and vehicles. In 2015, approx. 6,300 million metric tons of plastic waste was produced, only 9% of which was recycled, 12% was burned for energy recovery, and the remaining 79% ended up in landfill [3]. The relentless rise of single-use fossil-based polymers has promoted the profligate consumption and uncontrolled accumulation of polymer waste and micro-

https://doi.org/10.1515/9783111028163-012

plastics, which persist in the environment for centuries. The amount of polymer waste that has entered the oceans from the land is approximated to be in the range of 4.8–12.7 million tons [4]. This plastic waste threatens not only the environment but also human health and the survival of other organisms. Table 12.1 shows the data for plastics produced, recycled, incinerated, and landfilled in different years in the USA [5].

Table 12.1: The total tons of plastics generated, recycled, composted, combusted for energy recovery, and landfilled in the USA from 1960 to 2017 [5]. The values are reported in municipal solid waste (MSW) expressed in thousands of US tons.

Management pathway	1960	1970	1980	1990	2000	2005	2010	2015	2016	2017
Generation	390	2,900	6,830	17,130	25,550	29,380	31,400	34,480	34,870	35,370
Recycled	–	–	20	370	1,480	1,780	2,500	3,120	3,240	2,960
Composted	–	–	–	–	–	–	–	–	–	–
Combustion with energy recovery	–	–	140	2,980	4,120	4,330	4,530	5,330	5,340	5,590
Landfilled	390	2,900	6,670	13,780	19,950	23,270	24,370	26,030	26,290	26,820

The consumption of petrochemical feedstock and energy for plastics manufacturing contributes to the overall industrial CO_2 emissions. The non-biodegradable nature or the slow degradation rate of fossil-based polymers are the main contributors to plastic waste accumulation. The generation of polymer waste can be controlled through the following measures:
- improved waste management systems;
- reuse of polymers for as long as they are still functional;
- recycling polymers into new products;
- reducing the production and use of single-use plastics;
- better final product design, and
- inherent polymer design based on polymerization–depolymerization cycles.

Advances have been made in the development, production, and consumption of bio-based and biodegradable polymers. The ingredients used to synthesize such polymers are greener, i.e., bio-based and biodegradable. Some of the bio-based polymers have properties, such as toughness, structural integrity, and flexibility, similar to fossil-based polymers. Many (but not all) bio-based polymers are biodegradable and thus represent a more practical choice to replace traditional fossil-based polymers in many applications. Bio-based resources can successfully replace fossil-based resources to reduce the environmental impact of polymers if their manufacturing technology is based on green and sustainable chemistry and engineering concepts [6, 7].

Developing efficient technologies for polymer manufacturing may be revolutionized by addressing three key areas: improved *reactor design* for better reactant delivery and control over reaction times and other reaction conditions; improved *online analysis* to extract real-time information of material synthesis, i.e., better control over reaction conditions; and increased *automation* through the engagement of machine learning-type algorithms with feedback loops, i.e., using artificial intelligence to deliver the desired material with less human interaction [8].

12.2 Polymers and the environment

The inefficient disposal of plastic products after use negatively affects the environment and society. Once released into the environment and exposed to the elements, plastic products often decompose into smaller plastic particles, which exacerbates the problem and complexity of plastic pollution. Plastic particles are classified according to their size as macro-, micro-, and nanoplastics. The presence of microplastic pollution particles in the environment was first recorded in 1972 [9]. Studies have shown several domestic products, such as shampoo, cosmetics, toothpaste, and packaged drinking water, are contaminated with small plastic fragments or fibers known as microplastics. These microplastics have a slow degradation rate [10] and, once reaching the ocean, can persist for decades in the marine environment. Additives such as fillers, colorants, stabilizers, pigments, lubricants, compatibilizers, and plasticizers are often introduced into commercial polymer products to enhance polymer functionality and performance. Without such additives, many polymeric products would be of limited use. However, some additive chemicals, often added to improve polymers' physical properties, are known to be harmful to human health [11, 12], such as bisphenol A, phthalates, bisphenone, organotins, and triclosan. Some of these unreacted precursors trapped inside the structure can leach from the polymer into the environment. Americans are believed to unwittingly consume at least 74,000 microplastic particles every year through eating, drinking, and breathing [13].

Plastic pollution is now present in sources of clean water and groundwater sources. Farmers have been using plastic coverings in agricultural practices for many years to balance the amount of water in the soil, maintain temperature, and prevent weed growth. This process is called mulching. However, these plastic coverings often leave some fibers or particles scattered across the agricultural land. These plastic coverings are typically made of polythene, which degrades slowly, and the plastic residue continues to accumulate in the fields.

In 2019, the European Union Scientific Advice Mechanism published a review in which they acknowledged that although we do not yet fully understand the adverse effects of plastic pollution, we should consider that microplastic contamination may become a global health issue if toxicological data reveal that microplastics are harmful.

One of the major causes of polymer waste accumulation is the linear economy approach (Figure 12.2). Polymer products are prepared from either renewable or non-renewable sources. The prepared polymer products are then transported, distributed, and consumed, eventually ending up either in landfill or incinerated [14]. A more sustainable route, known as the circular economy approach, requires polymers to be reusable without modification, reparable to prolong the polymer's lifetime, and recyclable into new products (Figure 12.2). Such materials circulate persistently in a closed-loop system of reuse, repair, and recycle, instead of being discarded after use. Follow the QR code on this page to learn about the initiatives of the Ellen MacArthur Foundation towards circular economy.

Circular Economy

Linear Economy

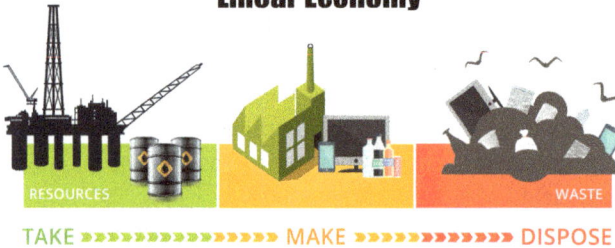

Figure 12.2: Linear and circular economy approaches. Reprinted with permission from Pro Carton.

The transformation of forest and agriculture biomass into building blocks and polymers derived from sustainable chemistry and engineering has gained increasing attention. Several innovations in bio-based materials have generated many new opportunities to reduce polymer waste and greenhouse gases. In 2019, approx. 0.79 million hectares of land – less than 0.02% of the global agricultural land – was required to grow the necessary renewable feedstock to produce bio-based polymers (Figure 12.3) [15]. Over the next five years, the land-use share for bio-based polymer production is predicted to remain at 0.02%, which indicates that the production of bio-based polymers will not compete with feedstock for food supply.

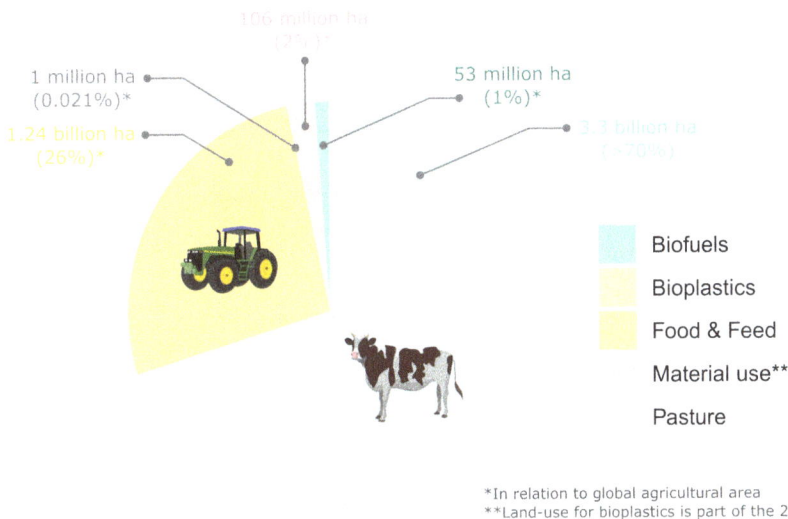

Figure 12.3: Land use estimation to grow biomass for bio-based polymers: 2019 to 2024 with data from reference [15].

The cellular biomass of prokaryotic microalgae may provide an alternative source for bio-based polymers derived from biomass, such as, for example, cyanobacteria for the production of polyhydroxyalkanoate (PHA) polymers [16]. Such microalgae use sunlight and atmospheric CO_2 as a source of energy to produce biomass. For example, the cyanobacterium *Spirulina* sp. LEB-18 can be cultured in a tank for up to 30 days under natural light. The biomass produced in the tank is separated by centrifugation. The obtained biomass is then washed with a hypochlorite solution, and the supernatant is decanted. The residual solid is treated with water and centrifuged. The supernatant is again discarded. The solid mass is treated with acetone at 45 °C for a few hours to precipitate the biopolymer. Another method involves the application of sonication on the biomass (suspended in water) followed by centrifugation. The obtained solid mass is then treated with hot chloroform, and the biopolymer is finally extracted by removing the solvent.

The Netherlands is a global leader in bio-based polymer technology as part of its aim to become one of the world's first CO_2 negative countries by 2050 [17]. They have been developing a system called Chemport Europe, which focuses on a more complete and cohesive production ecosystem based on green chemistry and engineering (Figure 12.4) [18]. In this system, different chemical companies collaborate to develop green processes through the exchange of feedstock. In the near future, they aim to close the loop of the circular economy. Their system has excluded any involvement from the petrochemical-based industry, making it the greenest chemical cluster in the world. Recently, in a similar direction, Saudi Arabia has proposed an international project called NEOM (from the word new future) to be built on the Red Sea. Follow the QR code on this page to browse the website of NEOM, and watch videos on building a new model for sustainable living, working, and prospering.

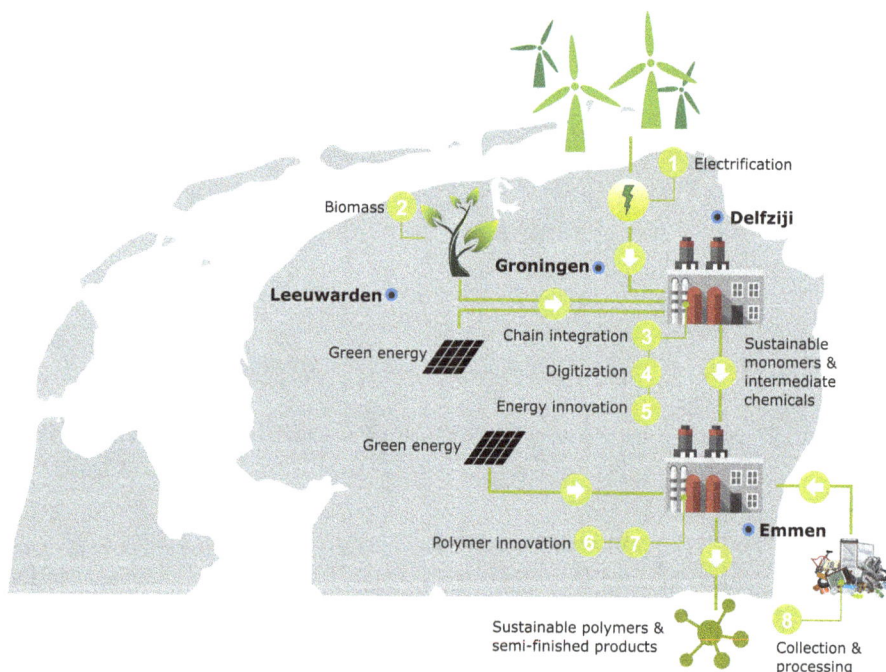

Figure 12.4: A coherent and innovative research, development, and production ecosystem in the Netherlands.

12.3 Plastic waste management: methods and limitations

The rate of recycling plastic waste in the European Union was 24% in 2005, and had increased to 42% by 2017 [19]. The plastic waste recycling process depends on the type of plastic to be recycled. Thermosetting polymers are thermally stable and cannot be recycled by standard melting procedures. On the other hand, thermoplastics are easy to recycle because they can undergo physical or chemical changes when heated. Plastic waste recycling is done in four ways, as shown in Figure 12.5 [20].

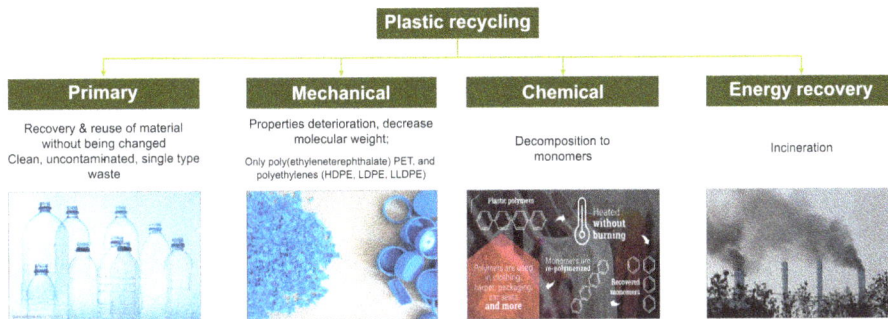

Figure 12.5: Types of processes used to recycle plastic waste.

The primary recycling process involves the recovery and reuse of polymeric materials without changing the product. This process is also called closed-loop recycling and ensures that recycling is simple and economical. However, only clean, uncontaminated, and single-use-type waste is eligible, which is the major limitation of this method.

The mechanical recycling process involves the physical breakdown of plastics into small pieces or pellets to be reprocessed into new plastic products. The plastic waste collected during this process is first separated from labels, stickers, or other contaminants. The polymer waste is further subdivided into small beads by melting and extrusion. The polymers' degradation results in the generation of volatile organic compounds such as phthalates, polycyclic aromatic hydrocarbons, furans, and dioxins. Their release into the environment is another source of pollution. This process is called downcycling because the product obtained after recycling is of a lower value and is inferior to its parental feedstock.

The chemical recycling process involves the breakdown of plastic waste into its smaller chemical units, namely, monomers, through chemical processes such as chemolysis, pyrolysis, fluid catalytic cracking, hydrothermal cracking, and gasification. This method suffers from several limitations, such as the requirement for high volumes to be cost-effective, and only works for condensation polymers. Furthermore, there is a loss of catalytic activity during the process, and the process is non-selective, therefore often producing a mixture of many products.

Incineration is a waste-to-energy recovery process that involves burning plastic waste in a controlled manner. The solid polymer waste can be reduced by up to one-third of its original volume, depending on the elements in the plastic waste [21]. Generally, the incineration of plastic waste is done to produce heat and power. For example, Sweden converts approximately 52% of its plastic waste into energy and recycles the rest. Incineration of plastic waste helps to curb unpleasant smells; methane gas is often generated in landfills when the waste decays. However, incineration produces toxic pollutants such as dioxins, smoke, heavy metals, and ash, leading to possible further environmental pollution [22].

Currently, the rate at which polymer products are produced, used, and disposed of is much higher than the recycling rate. It is important to harness bio-based materials at a rate at which they can be replaced naturally. Moreover, the rate of waste generation should always be lower than the rate of recycling. Manufacturers of sustainable bio-based materials must ensure that the supply of biomass remains reliable to avoid adversely affecting agriculture, water sources, and the environment in general.

12.4 Bioplastics

Polymer industries mainly produce two kinds of polymers to manufacture plastic products. *Addition polymers* are made up of saturated molecules, which generally make them unreactive and chemically inert. For example, polyethylene, polypropylene, and polystyrene are highly resistant to degradation and hydrolytic cleavage. This inertness also makes them highly durable and, unfortunately, non-biodegradable and difficult to recycle. When they are incinerated for energy recovery, toxic products such as dioxins are produced. On the other hand, *condensation polymers* are prepared through the condensation reaction of alcohol with amine or carboxyl functionalized molecules, such as polyamide or polyester. Condensation polymers are susceptible to degradation and hydrolytic cleavage, although it is a slow process.

Biodegradability describes how organic matter and materials degrade when subjected to natural conditions such as light, air, temperature, radiation, moisture, soil, and microorganisms in the presence of natural substances such as water or carbon dioxide. The term *bioplastic* refers either to a polymer's bio-based origin or to its biodegradable properties. These two properties are not synonymous. It is important to acknowledge that not all bio-based polymers are biodegradable, and not all biodegradable polymers are bio-based. Figure 12.6 shows some examples of polymers of fossil- and bio-based origin that are either biodegradable or non-biodegradable. Generally, the biodegradation of materials occurs in three stages:

1. Materials can undergo surface-level degradation when exposed to light, the environment, chemicals, and temperature, including changes in the materials' physical, mechanical, and chemical properties. This process of degradation is known as *bio-deterioration*.

2. The long polymer chains can be broken down into shorter oligomers and monomers through the action of microorganisms, known as *bio-fragmentation*.
3. The microorganisms absorb the monomers and oligomers to produce energy cur-
   rency (adenosine triphosphate) or other cell structure elements. The integration
   of these bio-fragmentation products into microbial cells is known as the *assimi-
   lation* process.

Fossil-based polymer products have held more than a 99% share of the global polymer
production market. The current global production capacity of bio-based polymers is
approx. 941,000 metric tons, which is estimated to grow by approx. 20% over the next
five years [23]. However, in 2019, 44.5% of all green polymers were bio-based but non-
biodegradable, 37.8% were both bio-based and bio-degradable, while 17.7% were petrochemical-based and biodegradable (Figure 12.7). Many bio-based polymers derived
from renewable biomass feedstock exhibit mechanical properties similar to their fos-
sil-counterparts, such as polyethylene (bio-PE), polystyrene (bio-PS), and polypropylene (bio-PP).

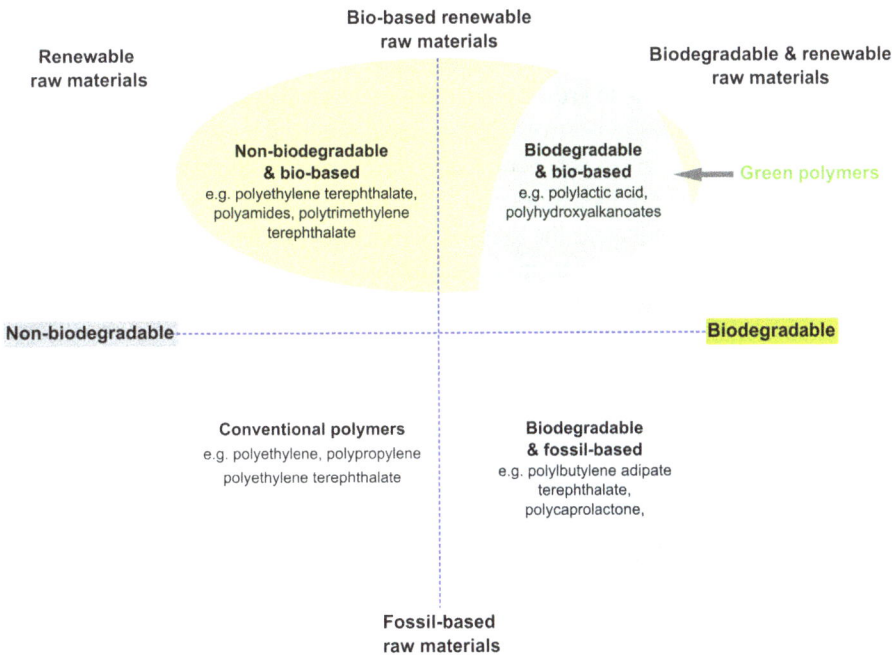

Figure 12.6: Classification of polymers on the basis of their origin and biodegradability. PA, polyamide;
PET, polyethyleneterephthalate; PTT, polytrimethylene terephthalate; PLA, polylactic acid; PHA,
polyhydroxyalkanoates; PBS, polybutylene succinate; PE, polyethylene; PP, polypropylene; PBAT,
polyethylene adipate terephthalate; PCI, polycaprolactone.

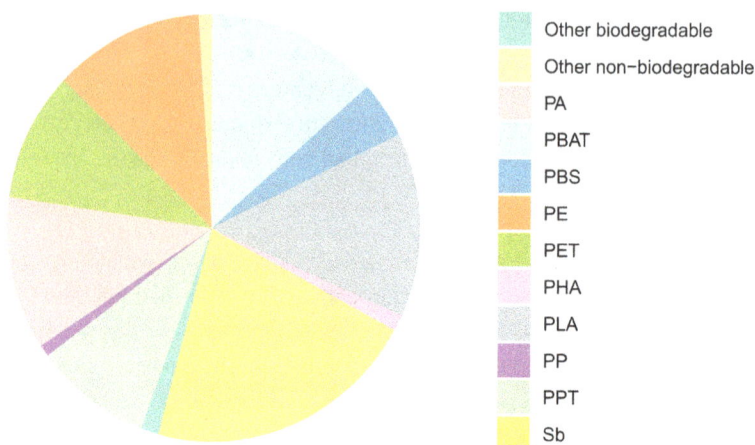

Figure 12.7: Global production of biodegradable and bio-based polymers.

12.5 Green polymers

Biomass production has witnessed surprisingly rapid growth in recent years, and it is expected that it will continue to grow considerably in the coming decades. Substantial biomass feedstock will have to be invested in material production every year to meet the global demand for new green polymers. Follow the QR code on this page to watch Hasso von Pogrell from European Bioplastics explain the future of bioplastics. The Dutch chemical sector has a vision of making the Netherlands the global leader in producing synthetic materials and chemicals from biomass and bio-based renewable raw materials. The chemical sector has targeted clean and sustainable production technologies to transform biomass into products suitable for a wide range of applications.

Biomass can be transformed into new chemicals and materials that are often impossible or difficult to produce from fossil-based raw materials, even after multistep reactions. For example, lactic acid is naturally produced by organisms as a chemical by-product of anaerobic respiration. Lactic acid is produced industrially via the bacterial fermentation of different biomass sources such as corn starch and sugarcane. As demonstrated in Figure 12.8, carbohydrates (cellulose, sugars, starch), chitin, terpenes, plant oil, and fats are some examples of naturally occurring macromolecules that are chemically amended in multiple ways to meet the demand for green polymers. Follow the QR code on this page to learn more about the biomass-derived sustainable polymers through the ACS Webinars.

Lignocellulose is generally considered the most abundant organic renewable resource for producing polymers, chemicals, and biofuels (such as ethanol, butanol, and biodiesel). Follow the QR code on this page to learn about getting more value from lignocellulose. Lignocellulosic feedstock comprises three major structural compo-

Plants fibres		Composites
Wood		Wood plastics
		Lignin
	Cellulose	Cellulose derivatives
		Polyesters PLA, PHB, PET
	Sugars	Monomers
Starch		Amylose thermoplastics
	Bioethanol	Polyolefins
		Butadiene polymers
		Biodiesel
Plant oil & fats	Glycerol	Epoxy resins
		Polyurethanes
Terpenes		Natural rubbers
Chitin		Chitosan

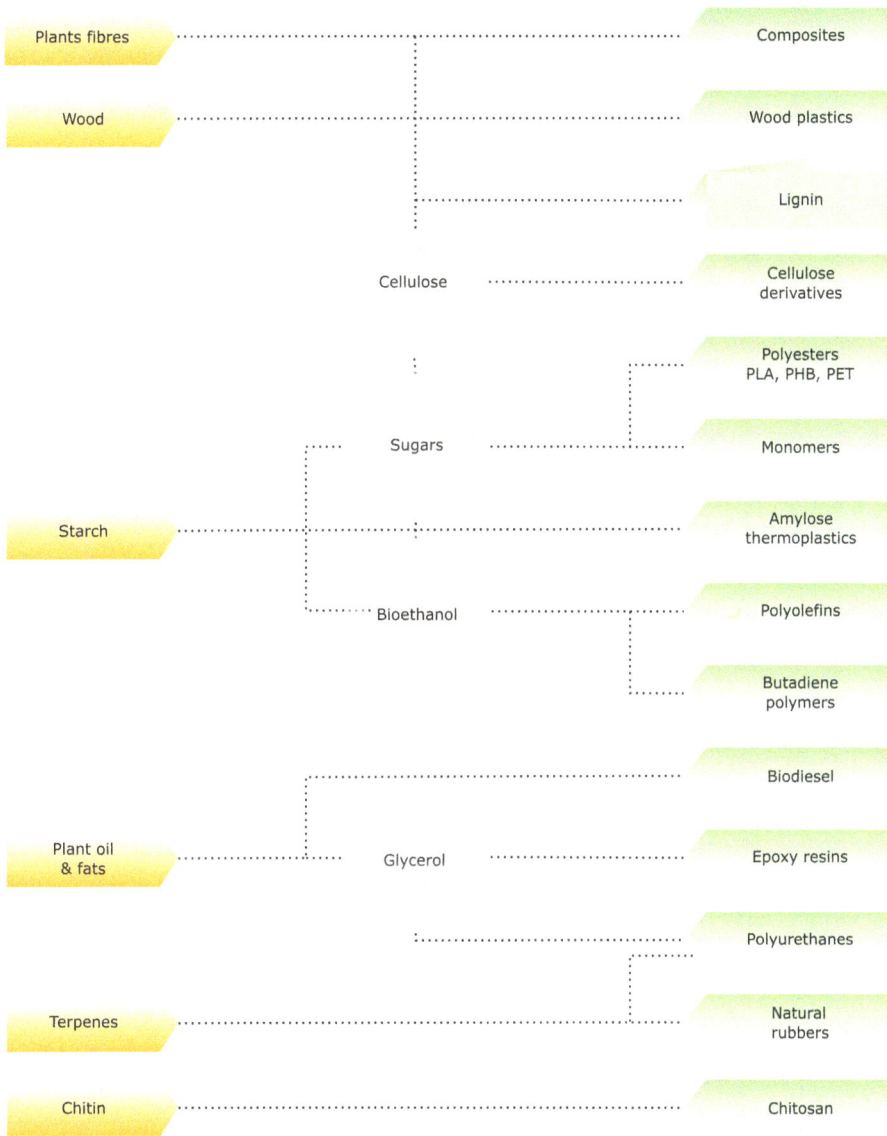

Figure 12.8: Natural polymer sources to different renewable bio-based green polymers [24].

nents: cellulose, hemicellulose, and lignin. *Cellulose* is a polysaccharide consisting of a linear chain of D-glucose units linked together by a glycosidic bond. Hemicellulose is a branched heteropolysaccharide comprised of diverse sugars (hexose and pentose) such as xylose, arabinose, mannose, galactose, and rhamnose. Cellulose and hemicellulose are often present together in plant cell walls, but both differ in their composition and structure. Cellulose is crystalline, strong, and resistant to hydrolysis. On the other hand, *hemicellulose* comprises a β-1,4-linked backbone and shorter side chains responsible for its amorphous structure. *Lignin* is a water-insoluble, long-chain natural polymer comprising three different phenylpropanoid monomer units: *p*-coumaryl alcohol, coniferyl alcohol, and sinapyl alcohol, which are often linked by ether bonds. Lignin is quite resistant to acid hydrolysis, is soluble in hot alkaline solutions, and is readily oxidized. Collectively, lignocellulose is naturally recalcitrant to hydrolysis. For the effective utilization of lignocellulose, it is pretreated in an acidic thermochemical process to make its components (cellulose, hemicellulose, and lignin) accessible by changing its physical or chemical structure.

Cellulose is the most abundant source of renewable polymers, and it serves as a sustainable carbon source for the green economy. *Depolymerization* of cellulose is usually done by acid or enzymatic hydrolysis, which breaks the glycosidic bond with glucose. Glucose further serves as a potential building block for preparing varieties of potential monomers such as ethanol, lactic acid, succinic acid, itaconic acid, glutamic acid, glucaric acid, 5-hydroxymethylfurfural, sorbitol, xylitol, erythritol, glycerol, propylene glycol, and ethylene glycol. These monomers can be further used to produce successive monomers, which could lead to different bio-based renewable green polymers through polymerization.

After acid or enzymatic hydrolysis of lignocellulose polysaccharides, the lignin is isolated as a solid side-product. The composition of lignin varies from plant to plant. Paper products turn yellowish with age due to the presence of lignin in the wood pulp. Lignin is amorphous and insoluble in inert solvents. Lignin is isolated as *lignosulfonates* during the sulfite pulping process. The presence of the sulfonate group makes lignin water-soluble. Lignosulfonates are used as dispersants, binders, emulsifying agents, and adhesives and in the manufacturing of vanillin and dyes. Lignin isolated from wood pulp serves as a feedstock for biofuel ethanol manufacturing and is burned for its fuel value. Refer to Chapter 10 to learn more about the sustainable manufacturing of biofuels.

Starch is a biodegradable, water-soluble natural polymer comprising a mixture of linear amylose and branched amylopectin fibers. Starch is commercially isolated from crops such as wheat, maize, potato, and other plants. Follow the QR code on this page to learn more about the extraction of starch. Some companies prepare starch-based biohybrids as a blend component with water-resistant non-biodegradable polyesters and polyolefins [25]. The solubility of starch in water has found application as an adhesive in the papermaking industry. Starch is fermented to produce bioethanol as a biofuel and feedstock for some useful monomers such as ethylene, acetaldehyde, and 1,3-butadiene. Starch is used as a thickener and stabilizer additive in different foods such as custard, salad, sauce, noodles, and pasta.

Chitin is the second most abundant natural polysaccharide. It is found in fungi, plankton, and the exoskeleton of arthropods such as crabs, lobsters, and shrimps. Follow the QR code on this page to learn about an economical method for the extraction of chitin developed by students of Imperial College London. Chitin is comprised of glucosamine linkages and is a rich source of *chitosan* polymer. Owing to their biodegradability, non-toxicity, biocompatibility, etc., chitin and chitosan are widely used in drug release, membrane separation, desalination, water purification, wound-dressing, and encapsulation. Chitin and chitosan are used to fabricate polymer scaffolds of nerves and blood vessels in tissue engineering. Chitin is also considered a starting material for preparing nitrogen-containing compounds, such as ethanolamine, pyrazine, pyridines, and other value-added chemicals such as furan and derivatives. The produced shell waste is often dumped into the sea or landfill in developing countries, which could instead be utilized and turned into value-added polymers and monomers.

Terpenes are major constituents of the essential oils and fats obtained from plants and flowers. Natural rubber consists of 97% polyterpene with a repeating unit of isoprene. They are valuable building blocks for manufacturing ionic and non-ionic surfactants in the chemical, perfumery and cosmetic, pharmaceutical, and biotechnology industries.

Biodiesel is derived from vegetable oil extracted from oilseed plantations such as rapeseed, sunflower, palm, or crops feedstock. Transesterification of vegetable oil with methanol in the presence of a basic catalyst yields methyl ester, which can be further used as biodiesel (Figure 11.8). Glycerol formed as a by-product is often used as feedstock to produce monomers such as 1,3-propanediol and epichlorohydrin to synthesize epoxy resins.

Proteins are classified as natural polymers, where amino acids are linked together through an amide linkage. Protein fibers such as wool and silk obtained from the pupae of the silk moth (*Bombyx mori*), or spider silk, have been used for weaving into textiles [26]. Protein-based fibers have found application in the fields of drug delivery, nanomedicine, and tissue engineering. Collagen protein fiber meshes are utilized for bone, cartilage, ligament, skin, muscle, and nerve regeneration. Casein protein isolated from cow milk is used as adhesives in labels, food additives (stabilizer) in processed food, and fiber for some modern fabrics.

12.6 Green monomers and building blocks

Figure 12.9 illustrates the transformation of cellulose into building blocks as a potential source of bulk chemicals and feedstocks. Simple combustion of glucose or cellulose is performed to prepare synthesis gas, which further can be transformed into biofuels. *Depolymerization* of cellulose by the hydrolysis of the β-1,4-glycosidic bond using mineral acids such as hydrochloric or sulfuric acid results in the formation of glucose and other polysaccharides. Follow the QR code on this page to watch a video explaining the chemistry of cellulose hydrolysis.

The petrochemical industry produces *ethylene* monomer via steam-cracking of naphtha or ethane (pathway A, Figure 12.10). The catalytic dehydration of bioethanol to produce ethylene is a sustainable alternative route. In fact, in the early twentieth

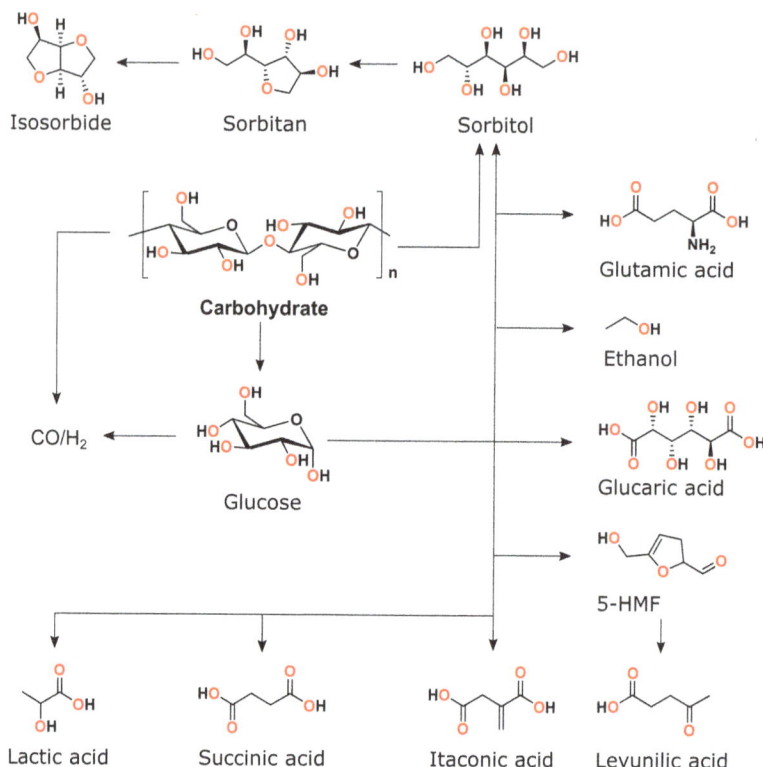

Figure 12.9: Transformation scheme of cellulose into green building blocks.

century, the mass production of ethylene was done by the dehydration of bio-based ethanol (pathway B). However, due to the high price of bioethanol, manufacturers continued to produce ethylene from fossil-based resources. In some countries such as Brazil and India, the cost of production of bioethylene is lower than in the United States and the European Union due to cheaper biomass feedstock. The companies Dow, Braskem, and Solvay are already engaged in the mass production of bioethylene from bioethanol. On the other hand, some plants and microorganisms are known to produce ethylene monomers. Ethylene is involved in controlling certain mechanisms during the life cycle of plants (pathway C). The production of ethylene from cyanobacteria is a promising method that is more sustainable and efficient.

Ethylene is one of the most important monomers for the synthesis of bulk polymers such as *polyethylene* (PE), polyethylene oxide (PEO), polyvinylchloride (PVC), and polystyrene (PS) (Figure 12.11). The manufacturing of PE is done by employing several conventional technologies such as high-pressure autoclave, high-pressure tubular, slurry, gas phase, and solution phase. Conventional PE manufacturing widely uses an initiator and high temperature and pressure (Table 12.2). However, these tech-

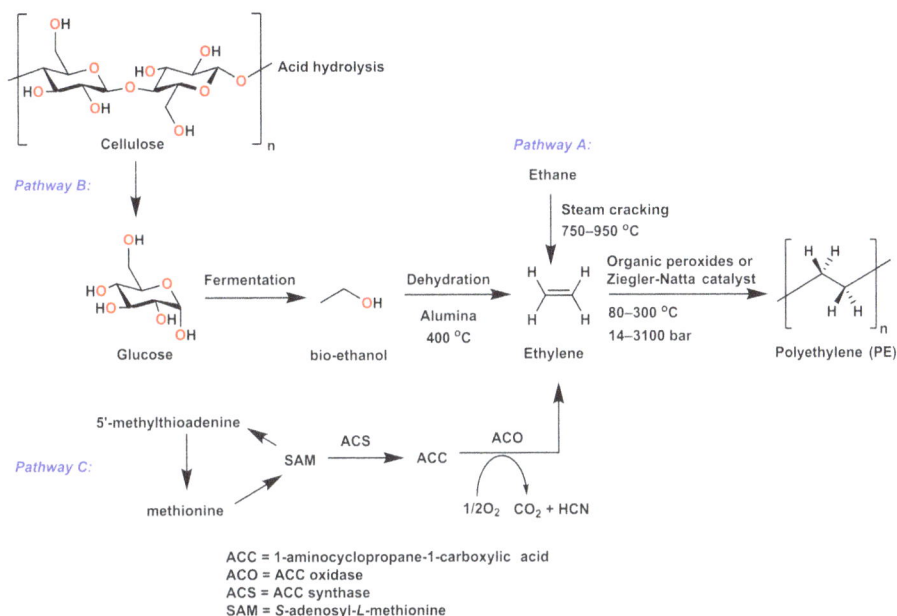

Figure 12.10: Different routes for synthesis of ethylene [27]. ACC, 1-aminocyclopropane-1-carboxylic acid; ACO, ACC oxidase; ACS, ACC synthase; SAM, S-adenosyl-L-methionine.

niques and operating conditions have drawbacks and limitations in terms of sustainable development.

The companies Braskem, Dow Chemicals, and Solvay are the major mass producers of bio-based polyethylene (bio-PE). Bio-PE has similar chemical, physical, and mechanical properties to its fossil-based counterparts. Due to its low price and high performance, bio-PE is widely used in agriculture, packaging, and other daily life applications. However, despite being bio-based, bio-PE is neither biodegradable nor compostable. Nonetheless, it can be recycled and processed into new polymer products. Bio-PE can be made degradable by introducing some breakpoints in the polymer structure. For example, adding CO during polyethylene formation leads to the introduction of a carbonyl group into the polymer structure. These carbonyl groups act as a convenient breakpoint so that light- or enzyme-mediated chemical reactions help the polymer to degrade.

The chlorination reaction of bioethanol-derived ethylene produces 1,2-dichlorethane, which, after dehydrochlorination, results in the formation of vinylchloride. The vinylchloride is polymerized into PVC with conventional processes. Another monomer that can be produced from bioethanol is 1,3-butadiene, which after polymerization forms polybutadiene, also known as synthetic rubber. As an additive, 1,3-butadiene improves the impact resistance of polymers such as polystyrene and acrylonitrile-butadiene-styrene. Styrene, a monomer of the widely used polystyrene, can be prepared from the dimerization of 1,3-butadiene.

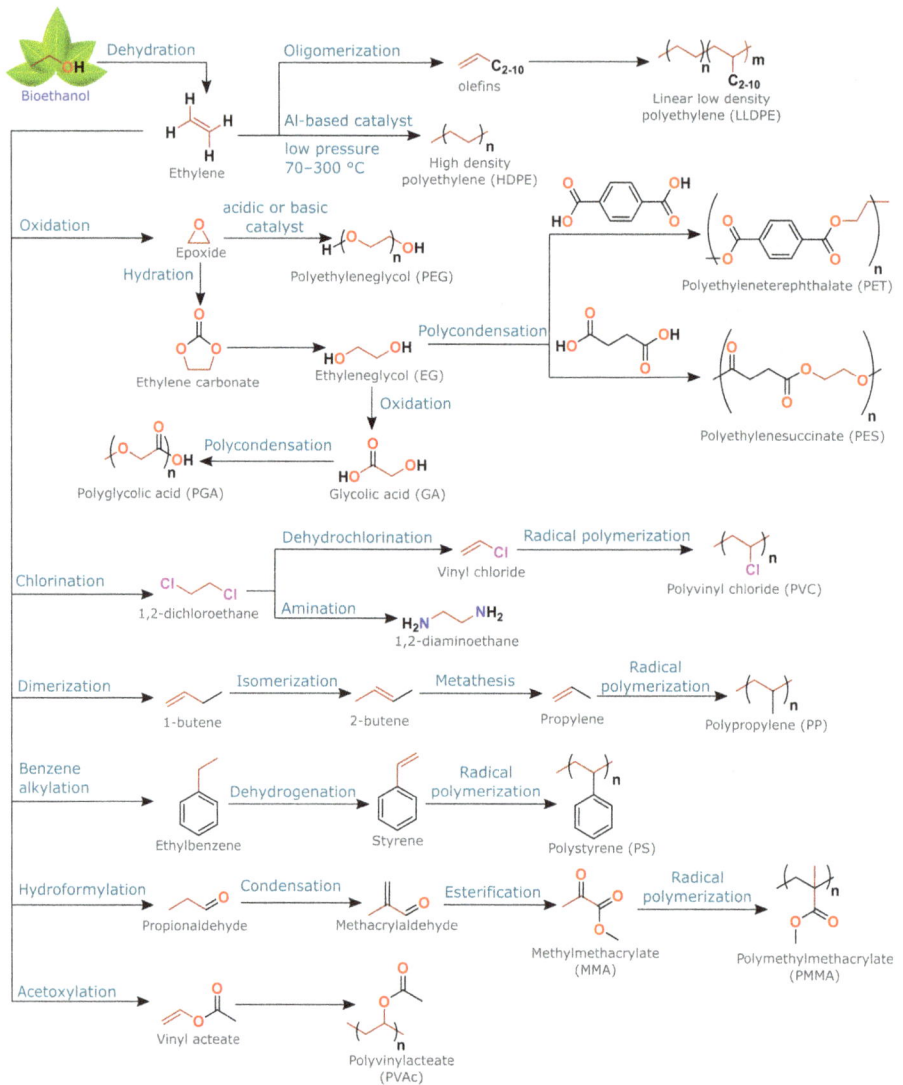

Figure 12.11: Bioethanol-derived bioethylene as a platform chemical for the synthesis of commodity monomers and their corresponding polymers [29].

Lactic acid has gained massive attention as a monomer for the production of green polymers [30]. Lactic acid is produced commercially via bacterial fermentation or chemo-catalytic conversion of polysaccharide (glucose) obtained from corn or wheat starch and sucrose in molasses. Lactic acid is naturally found inside organisms as a chemical by-product of anaerobic respiration. *Polylactic acid* (PLA) is a biodegradable polymer synthesized from the ring-opening polymerization reaction of lactides, a cyclic diester of lactic acid.

Table 12.2: Typical operating features of ethylene polymerization processes [28].

	High-pressure autoclave	High-pressure tubular	Slurry (suspension)	Gas phase	Solution
Temperature (°C)	180–300	150–300	80–110	80–110	160–220
Pressure (bar)	1,035–2,070	2,070–3,100	14–35	14–35	35–345
Catalyst	Organic peroxides	Organic peroxides	Cr/silica or Ziegler–Natta	Ziegler–Natta	Ziegler–Natta
Polymerization phase	Solution	Solution	Suspension	Fluidized beds	Solution

PLA is often employed in pharmaceutical and cosmetics for its disinfectant and kera-tolytic properties. It is also used in packaging materials, such as food and beverage packing, cups, bottles, and textiles. Recently, a PLA-based foam has been introduced as an alternative insulation material to the less sustainable polystyrene-based foam [31]. PLA is a bio-based polymer derived from renewable sources that is biodegradable as well as compostable. Owing to the compostability, low toxicity, environmentally benign characteristics, and low carbon footprint, PLA is an attractive green bio-based polymer. PLA does not have any toxic or carcinogenic effects, making it biocompatible and suit-able for biomedical applications.

PLA also serves as a potential renewable feedstock to manufacture acrylic acid monomer for polyacrylic acid and propylene glycol with application in polyesters, pol-ycarbonates, and polyurethanes. Acrylic acid and its amide derivatives serve as a commodity chemical in the production of acrylate polymers. Acrylic acid is prepared via the dehydration reaction of lactic acid. *Propylene glycol* (PG) is a green monomer utilized as a solvent medium to prepare unsaturated polyester resins, polycarbonate, and polyurethanes [32]. PG is synthesized via hydrogenation of lactic acid or lactates. The Food and Drug Administration (FDA) has classified PG as *generally recognized as safe* for indirect food additive use.

Fermentation is a well-established process used for the production of *succinic acid* from cellulose-derived sugars. Glycerol was found to be another source for the production of succinic acid in high yield compared to glucose-based sources [33]. The purification step of succinic acid involves separating by-products such as acetic acid, formic acid, lactic acid, and pyruvic acid formed during the fermentation process. Succinic acid is produced commercially from butane via malic anhydride intermedi-ates, but more sustainable fermentation processes are being developed. Succinic acid is an essential building block needed to form many important monomers.

1,4-Butanediol (BDO) is a crucial precursor for producing polymers such as polyur-ethanes, polyethers, polyamides, and the polyesters polybutylene terephthalate (which is an engineered polymer) and polybutylene succinate (which is a biodegradable polymer) [34]. At present, BDO is entirely derived from fossil-based feedstock. Although BDO is not naturally produced by organisms or plants, it is an important precursor to many other chemicals.

γ-Butyrolactone (GBL) is a simple C4 lactone generally synthesized by either dehydrocyclization of BDO or hydrogenation of fossil-derived maleic anhydride. It can also be prepared by the oxidation of furfural or derived from biomass feedstock; however, the low yield and harsh conditions limit this latter process's application. Li et al. [35] demonstrated the high-yield production of GBL from biomass-derived furfural in the presence of metal catalysts. Because it is prepared from biomass feedstock, GBL is considered a green solvent. As an important intermediate, GBL has found application as a solvent in the production of agrochemicals and pharmaceuticals [36]. Refer to Chapter 2 to learn about green solvents and their role in sustainable chemical manufacturing.

Treating GBL with ammonia or alkyl amine in the presence of some catalyst produces *2-pyrrolidone*, often used as a solvent due to its passive, non-toxic nature and high boiling point. It is used as a decolorizing agent for some hydrocarbons and film-forming agents. It also acts as a starting material for the production of *N*-vinylpyrrolidone (NVP), a monomer for polyvinylpyrrolidone (PVP). PVP is a water-soluble polymer widely used in medical products and cosmetics.

Itaconic acid (IA) or methylenesuccinic acid is a C5 unsaturated organic dicarboxylic acid produced by several organisms such as *Candida* sp., *Pseudozyma antartica*, and several other fungal species such as *Aspergillus*. The microorganisms *Aspergillus terreus* and *Ustilago maydis* are used in industrial processes to produce IA in fermentation processes using biomass-derived sugars. Owing to their chemical structure with one unsaturated double bond and two carboxylic groups, IA acts as a precursor in the synthesis of various high-value bio-based chemicals and materials. The presence of a double bond helps in the addition polymerization of IA. Thus, IA could replace fossil-based acrylic or methacrylic acid in polymer synthesis due to its trifunctional structure [37]. IA is employed as a building block to produce resins and synthetic fibers and as a plasticizer in certain polymer products, acrylic plastics, and antiscaling agents. Poly-IA has found potential applications in detergents and materials for water treatments and super-absorbent polymers owing to its superior calcium-binding tendency and biodegradability.

Glutamic acid (GA) is a non-essential amino acid found in living organisms, particularly in the proteins of plants and animals. GA is produced by the aerobic fermentation of polysaccharides such as glucose, fructose, sucrose, and molasses. Polyglutamic acid (PGA) is a polymer of glutamic acid naturally found in some bacteria. PGA comprises an enantiomeric mixture of D-GA and L-GA units linked via amide linkage between amino and carboxylic groups. PGA is an optically active polymer that has an active chiral center in every glutamate unit. *Bacillus* species are often used to produce PGA. Owing to its water solubility, biodegradability, and edible and non-toxic nature, PGA has found applications in medicine, foods, and pharmaceuticals. It is also well suited for producing thermoplastic fibers or membranes with transparent, high strength, and elastic properties.

The oxidation of glucose produces the high-value-added chemicals gluconic and glucaric acid. *Gluconic acid* and its derivatives are important materials in the chemical, food, and pharmaceutical industries. However, the slow reaction rate and complicated isolation procedure have raised concerns about developing efficient methods for its synthesis through glucose oxidation. *Glucaric acid*, also known as saccharic acid, is often prepared by the oxidation of glucose in the presence of nitric acid. Glucaric acid is a valuable C6 monomer for a range of hyper-branched polyesters and polyhydroxyl-polyamides. Glucaric acid has been identified as a top value-added chemical from biomass by the US Department of Energy for cholesterol reduction studies and cancer treatments [38].

The acid-catalyzed dehydration of hexose, such as fructose, is generally used to produce *5-hydroxymethylfurfural* (HMF). HMF can be synthesized not only from glucose but also directly from cellulose. The elimination of intermediate products addresses the 8th principle of green chemistry (Table 1.1). HMF serves as a building block in the manufacturing of polyesters, polyamides, and polyurethanes. HMF is utilized as an intermediate in the production of dimethylfuran liquid biofuel. Refer to Chapter 10 to learn more about the sustainable production of biofuels.

HMF is a building block for *levunilic acid* (LA), an additive used in polycarbonate synthesis. Because of the instability of HMF, LA is considered its degradation product. LA is prepared through acid hydrolysis of HMF. New methods for direct conversion of cellulose biomass into LA are currently intensively researched [39]. LA is a precursor of certain important chemicals and materials such as ketals, angelica lactone, plasticizers, pharmaceuticals, and cosmetics. Diphenolic acid derived from LA is a replacement for toxic bisphenol A that is often used as an additive in polycarbonate products. LA also serves as a potential building block for the green solvents γ-valerolactone and 2-methyl tetrahydrofuran. The company MaineBioproducts has developed a two-stage process called Biofine for LA's commercial production from lignocellulosic biomass feedstock in the presence of an acid such as sulfuric acid or hydrochloric acid [40].

Sorbitol is naturally found in some fruits such as apples, cherries, and apricots. Sorbitol is commercially prepared by the catalytic hydrogenation of polysaccharides. Fukuoka et al. [41] have reported the reduction of glucose to sorbitol using a Pt or Ru catalyst in aqueous media. Sorbitol is also prepared by fermenting polysaccharides, such as sucrose or a mixture of glucose and fructose, in the presence of the bacterium *Zymomonas mobilis* [42]. The chemical industry currently uses sorbitol as a dispensing agent and humectants in textiles, beauty care products, medicines, and surfactants. It is also added as a nutritive sweetener in diet drinks, cough syrups, and toothpaste. Upon dehydration, sorbitol is transformed into sorbitan, an intermediate in the conversion of sorbitol into isosorbide. The ester derivative of sorbitan is used as an emulsifying agent to prepare emulsions, medicinal creams, ointments, and beauty products. Sorbitan serves as a building block for several important polymers, such as polyesters. Sorbitan is also used as a plasticizer for PVC, PET, and PBT and is a chiral building block for non-linear optical polymers in the communication and photonic industries [43].

Moulijn et al. [44] demonstrated the isosorbide preparation by dissolving cellulose in molten salt hydrate Zn(II) chloride medium. Adding a co-catalyst, such as copper(II) chloride or nickel(II) chloride, or performing the hydrogenation at elevated temperatures results in the formation of isosorbide in high yield. *Isosorbide* is a precursor for the synthesis of polyethylene isosorbide terephthalate (PEIT) and dimethyl isosorbide green solvent. It is also used as an alternative plasticizer to toxic bisphenol A. The reactions of isosorbide with numerous aliphatic dicarboxylic acid dichlorides produce aliphatic polyesters. Fenouillot et al. [45] published a detailed review of polymers derived from renewable isosorbide.

Glycerol, ethylene glycol, propylene (including various mono- and polyols such as xylitol, erythritol, propylene glycol, and ethylene glycol), and even alkanes can be obtained via hydrogenolysis glucose. Ji et al. [46] used carbon-supported tungsten carbide as a catalyst for the hydrogenolysis of cellulose to produce high-yield ethylene glycol. *Propylene glycol* and ethylene glycol are high-value-added chemicals that could be potential alternative precursors to fossil-based polyolefin polymers. Ethylene glycol and propylene glycol are used as antifreeze depressants in water-cooling systems. Propylene glycol and ethylene glycol are transformed upon dehydration into propylene and ethylene, respectively. These alkenes could serve as a potential source of biomass-derived polymers polypropylene and polyethylene. *Glycerol* is a simple polyol widely used as a sweetener in the food industry. Due to its antiviral and antimicrobial activity, the FDA has approved glycerol for the treatment of burns and wounds.

12.7 Extraction methods

12.7.1 Mechano-catalytic depolymerization

Mechanical pretreatment of lignocellulose feedstock through ball milling is often described as mechanical-chemical treatment [47]. *Ball milling* involves efficiently breaking the chemical bond between lignin and hemicellulose to provide a high carbohydrate output, small particle size, high enzyme action, high specific surface area, and low crystallinity index. The process is energy-efficient and easy to scale up. The complete saccharification of the cellulosic component of biomass without forming any toxic side-products makes the process green. The mechano-catalytic depolymerization of lignocellulose on a kilogram scale has been successfully demonstrated (Figure 12.12) [48]. The α-cellulose, beechwood, and poplar chips are suspended in diethyl ether, followed by the addition of sulfuric acid. The cellulose microfibrils are embedded in the hemicellulose matrix, which is further protected by an additional layer of lignin. The lignin layer makes the lignocellulose recalcitrant to hydrolysis. The lignocellulose, hemicellulose, and lignin components are separated and converted into water-soluble products with the help of ball milling. The substrates are loaded into Simoloyer® ball mills, and depolymerization is performed at different rotor speeds, rotor torques, power consumption levels, and temperatures. After heating the suspension in water to 145 °C for approx. 1 h, the solid residue (lignin) is isolated from the sugar solution.

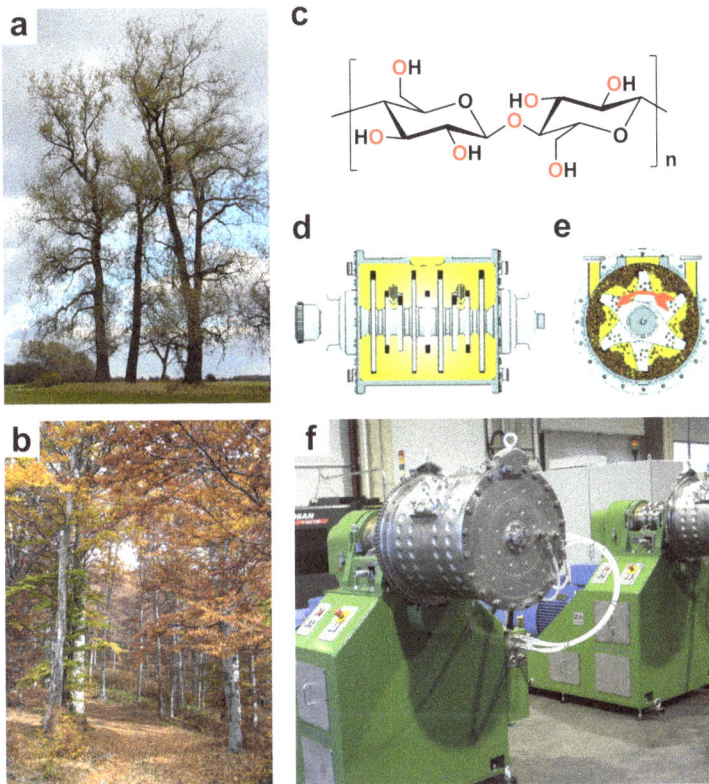

Figure 12.12: Mechano-catalytic depolymerization of lignocellulose using a Simoloyer® ball mill. (a) Poplar. (b) Beech. (c) α-Cellulose. (d and e) Ball mill cross-sections. (f) Photograph. Copyright, courtesy of Zoz GmbH and Deutsche Bundesstiftung Umwelt.

12.7.2 Integrated conversion

Hemicellulose and cellulose fractionation of lignocellulose biomass can be simultaneously converted into reactive furfural and levunilic acid in a single reactor (Figure 12.13) [49]. The low-boiling point product *furfural* is removed from the reactor by continuous distillation, which mitigates the reactive liquid's furfural degradation. An increase in the reactor's residence time allows the less reactive cellulose fraction to be converted into *levunilic acid* with a high yield. Catalytic conversion of the mixture (furfural and levunilic acid) into *GVL* is feasible without separating furfural from the reactive liquid. GVL acts as a solvent in the reaction and is one of the process's products, making the separation process more straightforward. The lignin and other degradation products present in the GVL solvent are precipitated in water and isolated by filtration.

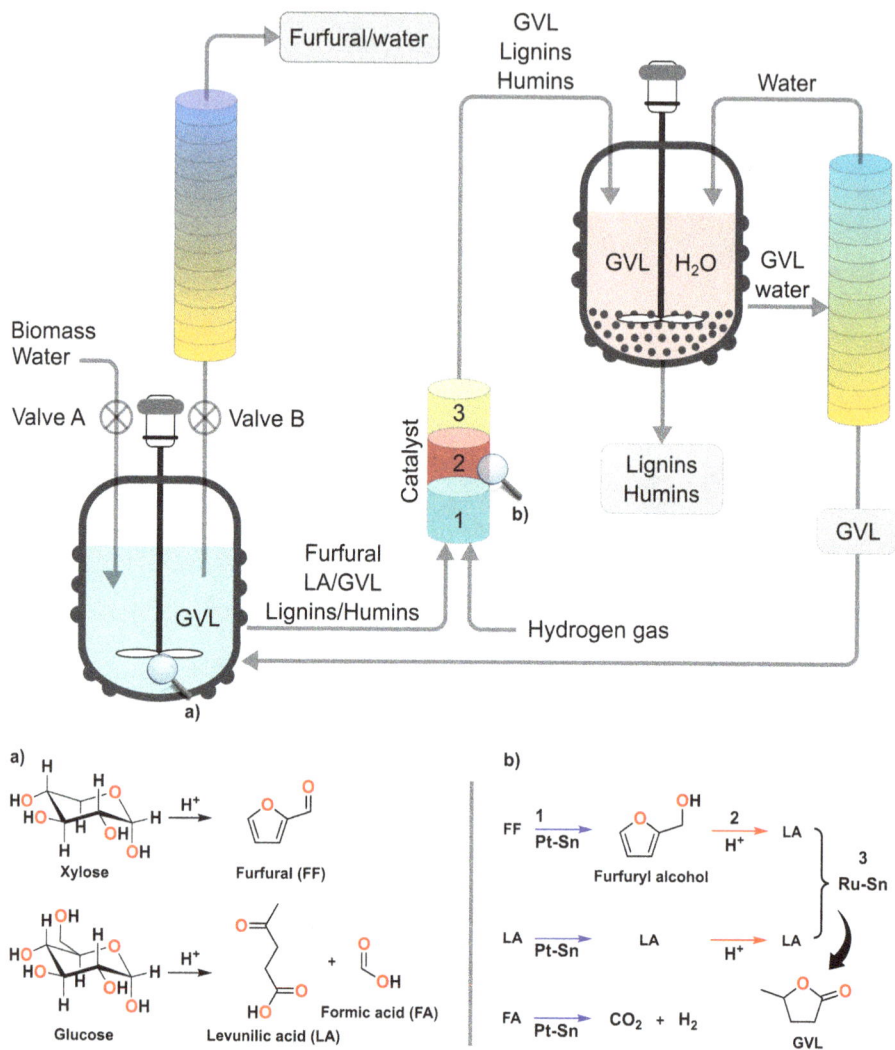

Figure 12.13: Schematic representation of the integrated conversion of hemicellulose and cellulose portions of lignocellulosic biomass (e.g., corn stover) to furfural and GVL, using a portion of the GVL as a solvent and the remainder for conversion to butene oligomers (C4) as a hydrocarbon.

12.7.3 Ultrasound-assisted radical depolymerization

A high-frequency ultrasound technique for the selective and complete depolymerization of cellulose into glucose without any catalyst, external heating source, or excess
pressure was developed (Figure 12.14) [50]. Micro-crystalline cellulose is first suspended in water, followed by exposure to ultrasonic irradiation at a high frequency
(525 kHz). Application of *high-frequency ultrasonic irradiation* (HFUS) induces the
nucleation, growth, and collapse of gas- and vapor-filled bubbles and the *in situ* formation of H· and ·OH radicals generated from the dissociation of water molecules.
Due to the bubbles' implosions, the radicals are released to either recombine or react
with the solutes, resulting in the selective and quasi-complete cleavage of glycosidic
bonds in cellulose. On the other hand, the low-frequency ultrasound irradiation (less
than 80 kHz) often induces physical effects such as shock waves, irregular water currents, and micro-jets. This energy induced from the low-frequency ultrasound is not
sufficient to break the glycosidic bond and necessitates the use of HFUS.

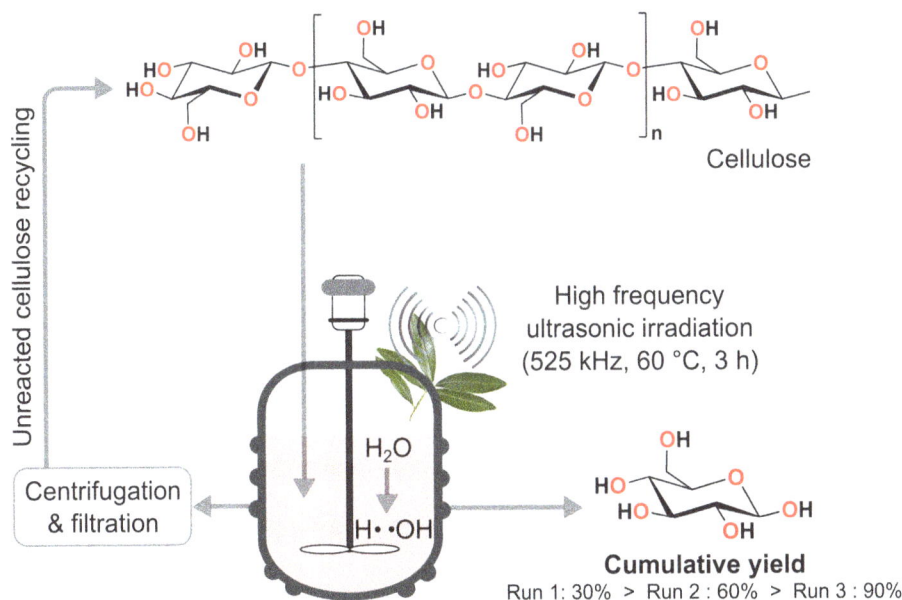

Figure 12.14: Depolymerization of cellulose to glucose induced by high-frequency ultrasound (HFUS)
irradiation coupled with filtration-based recycling of the unreacted cellulose.

Cellulose depolymerization performed under HFUS is more efficient in an H_2 environment than in air, O_2, argon, or a mixture of these gases, which suggests that the generated
H· radical transfers onto the surface of the cellulose particles and facilitates the cleavage
of the glycosidic bond. Despite the increasing cellulose loading from 0.5 to 5 wt%, when
the glucose concentration reaches approx. 17 mmol L^{-1}, the depolymerization of cellulose

terminates. The glucose released during the reaction's progress inhibits the reaction by preventing cellulose reduction induced by HFUS. This bottleneck can be overcome by recovering the unreacted cellulose by centrifugation and filtration, followed by resuspension in water and repeated HFUS. The recycling approach increased the yield of glucose from 30% (first run) to 90% (third run) (Figure 12.14).

12.7.4 Fermentation

Fermentation occurs when microorganisms, such as yeast or bacteria, convert carbohydrates into alcohol or organic acids under anaerobic conditions. Enzymes are biological molecules, typically proteins, that enhance the rate of reactions occurring inside cells. A specific temperature and pH range are usually required for efficient enzymatic action. Two separate unit operations, i.e., purification and fermentation, are commonly used to produce bioethanol from lignocellulosic biomass. However, these two unit operations can be performed simultaneously in a single process, often referred to as *simultaneous saccharification and fermentation* (SSF) [51]. In bio-based ethanol production, three main factors determine ethanol production costs:
1. the effective conversion of all polysaccharides to ethanol;
2. the concentration of ethanol in the fermentation broth; and
3. the number of unit operations involved.

Converting starch feedstock into ethanol is more straightforward than converting lignocellulose to monosaccharides. Lignocellulose feedstock is often cheaper than starch feedstock, and thus, lignocellulose feedstock is the better alternative for bioethanol production. Raw material costs contribute significantly to the final cost of ethanol. SSF of lignocellulose can significantly reduce the cost of bioethanol production. The SSF of glucose and xylose derived from corn-stover biomass is used to obtain high bioethanol concentrations from highly water-insoluble solids [51].

Some efficient methods for the production of value-added chemicals from lignocellulose biomass are compared in Table 12.3. However, these methods suffer from some limitations. The ball milling method has high carbohydrate production, a high enzyme action, scalability to hecto- and kilogram scales, and a high yield of polysaccharides and other high-value-added chemicals. It also produces sulfur-free lignin in a limited time. Substituting ether with a green solvent could make this method more sustainable. In the *integrated conversion* method, a γ-valerolactone and water mixture is used as a deep eutectic solvent to dissolve lignocellulosic biomass, which is a more sustainable choice that replaces harsh organic solvents. However, the removal of furfural through distillation is an energy-consuming process. Moreover, the use of expensive Pt–Sn and Ru–Sn catalysts is another limitation of this process. The SSF process is promising because it combines two different unit operations into a single process, ultimately reducing the costs involved in producing bioethanol. However, the high cost of enzymes and the long process times are some of this method's limitations, spurring a quest to develop a new strain of microbes that can convert biomass into ethanol more efficiently. Refer to Chapter 10 for in-depth discussions about the sustainable production of biofuels.

Table 12.3: Summary of the advantages and disadvantages of different methods producing valuable bio-based building blocks. SSF, simultaneous saccharification and fermentation; HFUS, high-frequency ultrasonic irradiation; SCFF, segmented continuous flow fractionation.

	Mechano-catalytic [48]	Integrated conversion [49]	Radical-based [50]	SSF [51]	SCFF [54]
Substrate	Cellulose, beech wood, poplar wood	Cellulose and hemicellulose	Cellulose	Corn stover	Lignin
Scalability	Hecto- to kilogram	Gram	Gram	Gram	Gram
Catalyst	Sulfuric acid (1%)	Sulfuric acid, Pt/Sn, and Ru/Sn	HFUS	Enzymes	–
Solvents	Diethyl ether	γ-valerolactone	Water	Water	Organic solvents
Yield	High	High	High	High	High
Products	Mixture of lignin oligomers, phenolic dimers, trimers	Furfural, γ-valerolactone	Glucose	Ethanol	

12.7.5 Segmented continuous flow fractionation

Lignin is a vital source of different value-added chemicals. Thus, tremendous effort has been invested in developing effective methods to extract lignins suitable for bulk-scale, high-value applications. Different methods developed to isolate carbohydrate streams alter the lignins' skeleton during processing and contribute to structural variations. High-value-added lignins can be prepared by fractionation methods. Traditional paper industries produce softwood kraft lignin (KL) via a lignoboost process. Wheat straw organosolv lignin (WL) is a by-product of the refinery process [52]. The KL process separates the acetone soluble low-molecular-weight fraction from bulk [53], whereas the WL method sequentially combines solvents such as acetone and methanol for fractionation. However, the implementation and upscaling of these processes in the continuous mode are cumbersome. For example, in membrane-based and chromatography-based continuous separation processes, clogging of the membrane pores or the column with fractionated solids is a common problem. This issue can be avoided by adopting a segmented flow approach, which is both continuous and scalable. This process is also known as segmented continuous flow fractionation (SCFF) of lignins.

An SCFF setup consists of a glass column, also called a lignin-sample holder, charged with dry solid lignin with a disk of porous PTFE frit at both ends to avoid any escape of solid lignin particles (Figure 12.15) [54]. The column is placed inside the column oven and is connected to the solvent flow system at one end. A photodiode array and refractive index (RI) detectors for the simple detection of the lignin fraction in the solvent are at the other end. At a suitable temperature, the desired solvent gradient is programmed and introduced in the setup. The detectors monitor the soluble lignin fraction. Owing to lignin's partial solubility in the solvent, the total volume of solid lignin inside the column continuously decreases during the process. This decrease in solid mass inside the column is compensated by moving the adjustable piece such that the solid lignin remains densely packed. Once the whole soluble part is fractionated, the fractions are concentrated, and the solvent is recycled.

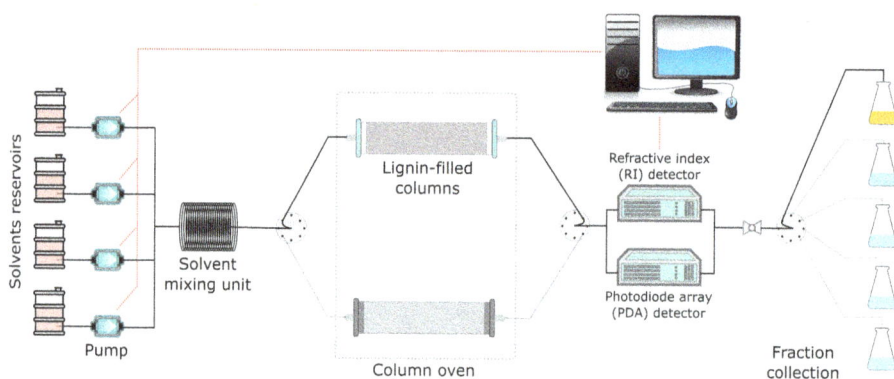

Figure 12.15: Schematic flow scheme of segmented continuous flow fractionation (SCFF) of lignin.

12.8 New design technology concepts for advanced polymer materials

Existing polymer material preparation technologies suffer from several limitations. These limitations include imprecise control over the polymer structure for precise application, inadequate green nature, unsustainable chemistry, high energy and time consumption, inefficiency on a large scale, etc., all of which present formidable challenges. Improving existing polymer synthesis techniques from a sustainable technology perspective can be achieved by further improving the reactor design, examining online, and automating the process. These improvements will also achieve increased accuracy and stability and make it easier to discover new polymer materials.

12.8.1 Reactor design configuration

Pressurized reactors: Laboratory-scale-based polymer synthetic routes often use glass vessels charged with reactants, such as monomers, exposed to external stimuli, usually heat, to initiate the reaction. Nowadays, reactors are typically commercial jacketed vessels attached to overhead stirrers for scale-up synthesis. For large-scale production, industries now prefer pressurized reactors. Pressurized reactors are also used to prepare ultrahigh-molecular weight polymers under high pressure via controlled radical polymerization methods.

Microwave reactors: Initiating the reaction using microwave radiation often results in an increased polymerization reaction rate without compromising the final polymeric material's quality.

Continuous flow reactors: In these reactors, a tube is subjected to a flow of reaction from inside and then exposed to heat or light to initiate the reaction. The integration of such reactors in polymerization routes provides reproducible synthetic conditions and simultaneously reduces human interference. In contrast to the batch process, continuous flow reactors offer several advantages, such as control over the molecular weight and conversion, scalability, and less operation time. Furthermore, integration with online monitoring/automation offers precise information on the conversion/molecular weight.

12.8.2 Online monitoring

Two factors are often involved in a polymerization reaction: (a) converting a monomer into a polymer and (b) the molecular weight and dispersity index of the final polymer. The latter factor is crucial for preparing a polymer with particular properties such as solubility, mechanical strength, and self-assembly. Current polymerization techniques often involve sample testing at intervals, which ultimately provides little information required to discover new polymers. With the help of online monitoring, detailed knowledge of a reaction's progress can be gathered, which could help smartly navigate the reaction with high efficiency. For example, the decomposition of a polymer with longer reaction times can be easily avoided. Real-time monitoring of the polymerization reaction can provide more precise information, such as introducing reactants at specific time intervals when preparing desired molecular weight polymers or conditions. Online monitoring in gathering real-time information involves different spectroscopic techniques such as UV-Vis, infrared, and Raman. The viscometry technique can also be used to extract real-time information during a polymerization reaction [55]. When finer details are required, the above-listed techniques are insufficient because they need to be calibrated with the system before use. Other techniques, such as nuclear magnetic resonance (NMR) and gel permeation chromatography (GPC), are often used for molecular weight distribution in polymers to provide a versatile platform. Unfortunately, neither technique provides real-time information because the sample preparation for analysis is a time-consuming process. An online GPC coupled with an automated sampler [56] or a high-throughput platform [57] or flow reactor [58] has been applied successfully for quickly extracting molecular weight information. A block copolymer's self-assembly during polymerization can be monitored using a small-angle X-ray scattering (SAXS) technique [59]. Automatic continuous online monitoring of polymerizations (ACOMP) technology is another powerful online analysis tool developed for comprehensive monitoring of polymerization by integrating a range of techniques [60].

12.8.3 Automation

The integration of computer-controlled and algorithm-based mechanical pumps, temperature controllers, samplers, etc., in a synthetic polymerization reactor setup can improve reproducibility and reduce human interactions in the process [61]. For example, a temperature controller can provide the necessary stable temperature to prepare reproducible products. With the help of a flow rate controller, a definite amount of monomer or other precursors can ultimately control the final polymer structure. Artificial intelligence (AI)-based automation can help predict the suitable conditions needed to prepare polymers with specific properties [62].

12.8.4 Membranes

Generally, lignin is derived from catalytic depolymerization of lignocellulosic feedstock comprised of phenolic dimers and lignin oligomers. Lignin products from the lignin phenolic stream are usually refined or extracted using distillation methods. However, distillation methods are often economically unfeasible for refinement or extraction of the lignin phenolic stream due to the high energy consumption, leading to high operating expenses. The difference in the molecular weight of the monophenols and oligomeric species in the lignin stream can be exploited in membrane-based separations. Non-thermal, *two-stage membrane fractionation* of catalytic upstream biorefining (CUB) liquor was realized for the separation of high-molecular-weight components from low-molecular-weight components (Figure 12.16) [63].

First, lignin biorefining is done in the presence of a heterogeneous hydrogenation catalyst, which leads to the formation of a lignin stream and a highly delignified pulp. A selective hydrodeoxygenation (HDO) process of lignin produces the lignin stream of a more uniform chemical nature than the other organosolv processes under similar conditions. Despite these advantages, the lignin streams are contaminated with high-molecular-weight lignin phenolic compounds (above 1 kDa), which generate pulp with a high degree of polymerization and high xylan content. Therefore, it is necessary to

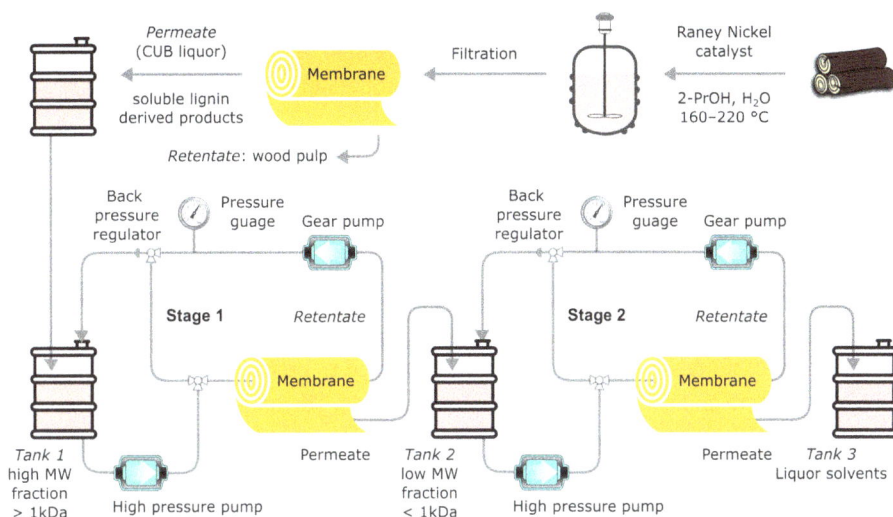

Figure 12.16: Schematic representation of a catalytic upstream biorefining (CUB) process coupled with a two-stage membrane fractionation of the CUB liquor.

separate the high-molecular-weight lignin-derived phenolic compounds from the lignin stream. For the separation process to be economically feasible, high-energy-consuming thermal processes such as distillation should be avoided.

The present method involves the fractionation of the CUB liquor into high- and low-molecular-weight lignin phenolic compounds and solvent recovery using membranes. The green solvents 2-propanol and water are used in the process. The process starts with the hydrogenation of poplar wood with a Raney nickel catalyst in a solvent mixture of 2-propanol (acting as a solvent and organic H donor for the selective HDO process) and water (7:3 v/v). The suspension is heated to 160–220 °C for 3 h. The catalyst is recovered from the pulp suspension using a magnet, and the CUB liquor is separated from the pulp suspension using a microfilter with a 4–7 μm pore diameter.

The obtained CUB liquor is then introduced to Tank 1 for stage 1 separation using GMT NC-1 from GMT Membrantechnik GmbH, Germany. In the first membrane stage, the high-molecular-weight phenolic compounds cannot permeate through the membrane assembly. Thus, they are separated and collected in Tank 1. The low-molecular-weight phenolic compounds can pass through the series of cross-flow membrane cells. The employed cross-flow membrane cells work in a recycle mode. The retentate and permeate streams are circulated back into feed Tank 1 to ensure that all the membrane cells are running with the same feed composition.

The permeate from stage 1 is introduced as a feed for stage 2. In the second stage, the Fimltec™ NF90 membrane enriches the low-molecular-weight lignin phenolic products in the retentate stream, which eventually accumulates in Tank 2. The liquor solvent is recovered in the permeate and stored in Tank 3. Due to the small pore size, the membrane permeance is lower than that observed in stage 1. However, in stage 2, a purity of 97% of low-molecular-weight lignin phenolic compounds is achieved. The applied pressures in stages 1 and 2 are in the ranges of 15–30 and 25–40 bar, respectively.

12.8.5 Membranes from chitosan and PLA

Membrane processes are widely recognized for their energy efficiency and sustainable nature. Membrane processes have emerged as one of the most promising sustainable processes in desalination, wastewater treatment, liquid, and gas purification. However, despite their usefulness, these methods are often used to prepare membrane materials such as polymers and solvents that are not sustainable. The membrane technology sector is currently evaluating bio-based polymers and green solvents with the aim of achieving a circular economy.

A thin-film composite membrane for forward osmosis was developed from chitosan (Figure 12.17) [64]. Utilizing glutaraldehyde as a crosslinker, crosslinked *chitosan* is prepared on an electrospun mat, followed by the synthesis of an amide-based polymer layer through interfacial polymerization. The resulting partially green membrane is then subjected to forward osmosis studies. The results indicate that the chitosan layer enhances salt rejection. Since *m*-phenylenediamine (MPD) and 1,3,5-trimesoyl chloride (TMC) are used to prepare the active layer, the overall membrane material is still far from green. MPD and TMC are highly reactive, toxic, and fossil-based chemicals. Hexane is also used during this polymerization, which is an undesired solvent, and sodium dodecyl benzenesulfonate (SDBS) is employed as a surfactant. Additives are undesired from a sustainability point of view, and thus, there is a need to develop an active layer from more sustainable building blocks using green solvents.

To address the sustainability challenges mentioned above, the active layer of thin-film composite membranes can be prepared from chitosan using TMC as a crosslinker. Although chitosan is a biobased polymer, the TMC reagent should be avoided. The membrane is fabricated on a sulfonated polyethersulfone-polyethersulfone (SPES-PES) support layer using the interfacial polymerization method (Figure 12.17b) [65]. These examples demonstrate that it is challenging to use only sustainable resources for materials fabrication. Nonetheless, in the forward osmosis studies, the membrane with higher chitosan concentrations exhibits higher salt rejection than the membrane composed of low chitosan concentrations. Maximizing the chitosan concentration not only improves the salt rejection but also makes the membrane material greener. The performance further improves when the membrane is treated with an aqueous NaOH solution.

The biopolymer *polylactic acid* (PLA) is also successfully utilized for membrane fabrication (Figure 12.17c) [66]. PLA is a bio-based biodegradable and compostable polymer. The process solvent for membrane preparation is ethyl lactate, a green solvent that is biodegradable, recyclable, bio-based, and non-toxic. Refer to Section 3.3 to

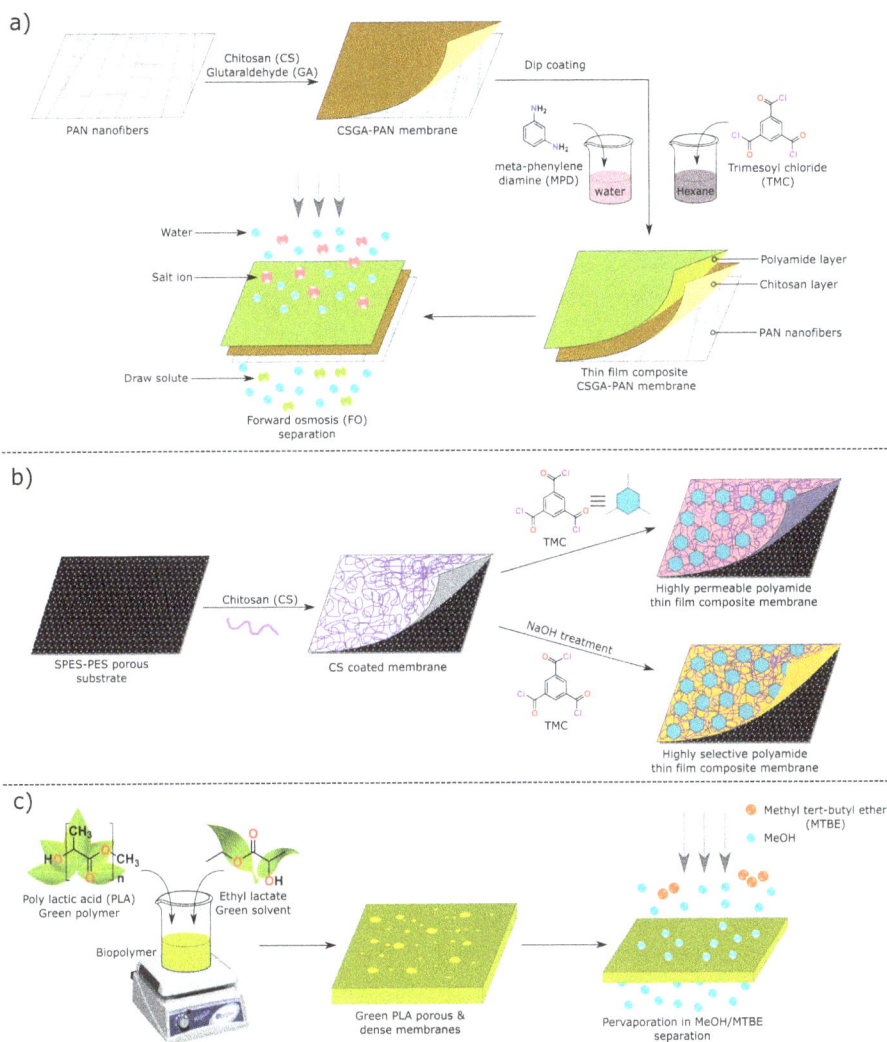

Figure 12.17: Schematic presentation of the use of biopolymers for membrane preparation. (a) A three-layered thin film composite membrane on a chitosan-based interlayer using SDBS surfactant. (b) A thin-film composite membrane with an active layer that incorporates chitosan. (c) A pervaporation membrane comprising of polylactic acid, made using ethyl lactate green solvent.

learn about green solvents. The obtained green integrally skinned asymmetric (ISA) membrane is used for the pervaporation of the azeotropic mixture of methanol and methyl *tert*-butyl ether. The membrane exhibits high permeation towards methanol with a selectivity value of higher than 75.

12.8.6 Production of bio-based polyethylene (bio-PE)

Polyethylene is the most common polymer used for packaging, plastic bags, containers, bottles, etc. Polyethylene accounts for 34% of the polymer market, producing 100 million tons of polyethylene resins per year [3]. Bio-PE is usually composed of different polymers of olefins. Bio-PE can be divided into three classes according to the polymer density:
1. high-density PE (HDPE), extruded at high pressure and temperature;
2. low-density PE (LDPE), extruded at low pressure and temperature; and
3. linear low-density PE (LLDPE), prepared by co-polymerization with longer (C4–C8) olefins.

Because the bio-based polymer market has only recently emerged, bio-based polyethylene is currently sold at high prices, around 30–40% higher than conventional PE. Bio-PE prepared from natural sources behaves similarly to its petroleum- or natural gas-derived counterparts. Like other synthetic polymers, it is not readily biodegradable and compostable and therefore accumulates in landfills. However, bio-based polymer products are usually thermoplastic and recyclable, and can be included in the waste plastic recycling system. An alternative to recycling is the production of oil from polymer waste through the *direct liquefaction pyrolysis* process performed in the presence [67] or absence of a catalyst [68]. Pyrolysis is a process where polymers are thermally cracked from longer hydrocarbon chains into smaller hydrocarbons. The pyrolysis process often leads to the generation of condensable products (wax/oil), non-condensable gases, and chars that are valuable to petrochemical industries and refineries [69]. In terms of energy production, the pyrolysis of polyethylene, polystyrene, and polypropylene yields approx. 89, 96, and 84 wt% fuel oil. Furthermore, the gaseous products obtained have high calorific value, making the process economical. Generally, in a pyrolysis process, catalysts such as bentonite clay, copper carbonate, CAT-2 zeolite, ZSM-5, and natural zeolite are added to enhance the hydrocarbon distribution in the pyrolysis liquid oil to obtain similar properties as conventional fuels like diesel or gasoline [67].

The pyrolysis of plastic waste for the catalyst-free generation of liquid fuels is a simple process. In general, plastic waste is first heated to a high temperature (450–500 °C) in a closed chamber under an inert nitrogen atmosphere (Figure 12.18) [70]. The resultant pyrolysis gas contains aliphatic hydrocarbons and acid gases such as hydrochloric acid or sulfur. The output gas is passed through an inline condenser, which quenches the pyrolysis reaction to predominantly form pyrolysis liquid hydrocarbon condensate. The formation of additional compounds such as wax and ash can block the passage, which can be easily controlled with a heating strip or restricted water flow. The concentration of gasoline in the pyrolysis liquid can be enhanced by passing the oil vapors through a condensate before they are vented out through a bubble scrubber. The bubble scrubber function is twofold: it scrubs the acid gases such as hydrochloric acid (the added sodium carbonate captures the chlorine released from the pyrolysis of PVC in the form of hydrochloric acid gas), and it averts the air backflow to the system.

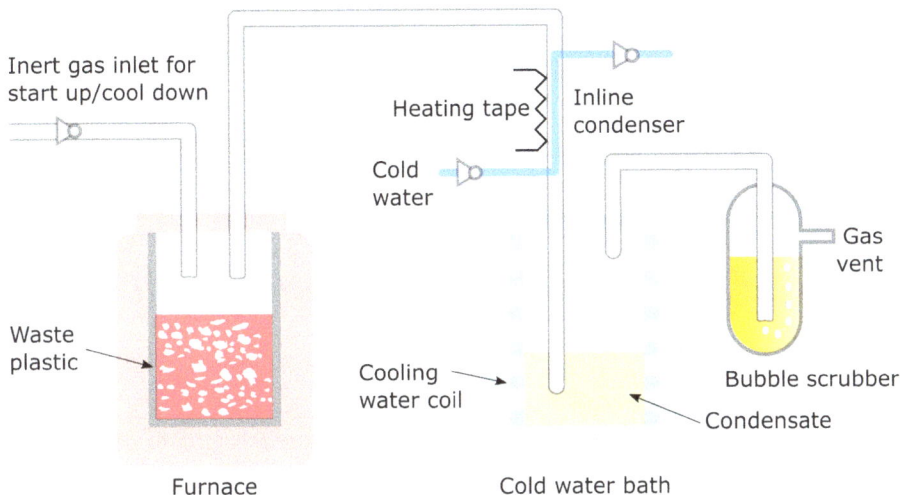

Figure 12.18: Schematic of pyrolysis conversion of waste plastic to oil.

In the synthesis of bio-based *ethylene* (Figure 12.19), the biomass is first converted to ethanol through a fermentation process [71]. Second, the produced ethanol is converted to ethylene through dehydration, which is finally converted to polyethylene through conventional polymerization processes. Bio-PE produced from biomass has the same chemical properties as bio-PE obtained from conventional ethylene.

Brazil and North America are the top producers of ethanol as *biofuel*, mainly from the yeast fermentation of sugar-rich and starch-rich biomass, such as sugarcane and maize, the so-called first-generation biomass. The USA produces 57×10^9 liters of ethanol from bio-resources. Ethanol's industrial production from lignocellulose biomass, also known as second-generation biomass, is rapidly developing. Refer to Chapter 10 to learn more about the conventional and sustainable production of biofuels.

Ethylene is an important monomer used to produce bulk polymers such as PE or PVC. The bio-based polyethylene field is growing rapidly. In conjunction with Crystalsev (a sugar and ethanol producer in Brazil), Dow Chemicals is already producing renewable LLDPE. In another venture with Mistsui Chemicals, Dow Chemicals has invested in constructing a production facility for making HDPPE and LLDPE, utilizing bio-based ethylene as raw material, with a production capacity of 350 kilotons per year. Braskem has developed a thermoplastic ethylene-based resin made from sugarcane ethanol, which helps to reduce greenhouse gas emissions. They have constructed a bio-based PE production facility with a capability of 400 kilotons per year. Looking at ethanol's global

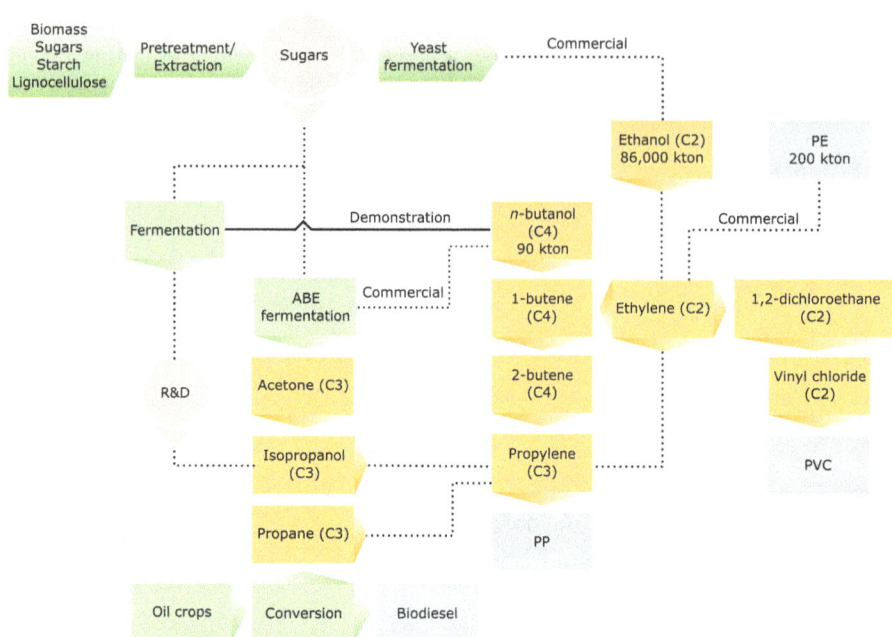

Figure 12.19: Synthetic route for conversion of biomass feedstock into renewable building blocks and biopolymers.

production, only 18% (16 megatons per year) of ethanol is currently used for non-fuel applications such as ethylene monomer production [72].

12.8.7 Bio-based 1,4-butanediol

The chemical industry widely uses 1,4-butanediol (BDO) as a solvent and building block to manufacture plastics, elastic fibers, and polyurethanes. Bio-BDO is an important commodity chemical used to synthesize some important chemicals. For example, under high temperatures and acidic conditions, it undergoes dehydration to form tetrahydrofuran [73]. Catalytic dehydrogenation of BDO in the presence of a ruthenium catalyst at high temperatures produces gamma-butyrolactone [74]. There are two different routes to prepare bio-BDO: the catalytic hydrogenation of succinic acid and sugar fermentation.

In a typical *catalytic hydrogenation* reaction, the reactor is charged with an aqueous solution of succinic acid and a monometallic catalyst (activated carbon-supported ruthenium [Ru] and palladium [Pd] catalyst) or bimetallic catalyst (Re-Pd/C and Re-Ru/C) under an inert argon atmosphere. Heating followed by the introduction of hydrogen gas initiates the hydrogenation reaction. Regarding the *sugar fermentation* route, no microbe can synthesize BDO directly from biomass. Consequently, a special strain of *Escherichia coli* has been developed for this purpose. A direct bio-catalytic route for the preparation of bio-based BDO from carbohydrate feedstock has been achieved [75]. In 2011, Italian and French legislation banned the sale of non-biodegradable light-weight plastic bags and packaging to mitigate polymer waste. With the integration of bio-BDO in their products, Novamont has paved the way to replace fossil-based raw materials with bio-based materials that meet biodegradability requirements.

12.8.8 BioFoam

BioFoam is a PLA-based foam to replace conventional foam used for technical and packaging applications such as domestic appliances, insulation, and surfboards [31]. BioFoam was designed to be interchangeable with conventional polymers such as polystyrene and polysterine [31]. Synbra Technology in the Netherlands manufactures BioFoam with an annual PLA production capacity of 6,000 tons. GMO-free sugar is used for the first-generation production of PLA raw material for BioFoam. The sugar is 100% bio-based feedstock extracted from cane and beets. The company is now developing second-generation PLA derived from thinning wood. BioFoam offers similar properties to conventional foam and is biodegradable, compostable, and can be remolded into new products. BioFoam production also uses CO_2 instead of pentane to blow the foam, leading to the encapsulation of CO_2 inside the foam structure. The bio-degradability, compostability, and CO_2 gas usage demonstrate the sustainability of BioFoam. It is highly durable and suitable for long-term use without a loss of quality. Follow the QR code on this page to watch Peter de Brujin explaining BioFoam production and products.

12.8.9 Desmodur eco N

Covestro AG, a German company, has designed a bio-based hardener (crosslinker) for quick polyurethane coatings commonly used to protect metal, wood, or polymeric surfaces [31]. Commercial applications include automotive, plastic, and wood coatings. Desmodur eco N is 70% bio-based, which means 70% of its carbon is derived from a bio-based source, and reduces the carbon footprint by 30% compared to fossil-based coating products. Desmodur eco N is produced from pentamethylenediamine, a raw material derived from the fermentation of corn starch feedstock (first generation). The company uses cellulosic and waste biomass as potential second-generation feedstock for sustainability in bio-based products. Under its current infrastructure, the company produces 20,000 tons of Desmodur eco N per year.

12.8.10 Rilsan HT and Rilsan Invent

Rilsan is a bio-based polyamide resin manufactured by Arkema, a France-based company [31]. Rilsan manufactures a high-performance resin derived from 100% bio-based feedstock castor oil for the automobile industry, sports equipment, and electrical appliances. Rilsan is manufactured in two steps: the first step involves the multistep organic transformation (transesterification, pyrolysis, hydrolysis, hydrobromination, and amination) of castor oil into amino acid. The side-products obtained from this step are used in cosmetics and lubricants. The second step involves the polymerization of the obtained amino acid precursor into a polymer. Mixing the amino acid building blocks with other monomers improves the temperature resistance of the prepared materials. Automobile parts manufactured by Rilsan are light-weight, reducing fuel consumption, and producing 30–40% less CO_2 emissions than petrochemical-based polymers.

12.8.11 Polycarbonates

Sustainable raw materials such as biomass and CO_2 are used for polymer synthesis to improve polymers' carbon footprint. CO_2 is a low-cost, non-toxic, and abundant raw material used to make green chemicals and materials, such as polycarbonate (PC). In PC, monomer units are linked by carbonate groups. Several methods have been developed for the preparation of PCs from several different precursors. The developed protocols include condensation reaction between bisphenol A and phosgene, metal-catalyzed epoxide and CO_2 polymerization, ring-opening of cyclic carbonates, transesterification, etc. A superbase-mediated reversible reaction of CO_2 with alcohols was performed to prepare α,ω-diene functionalized carbonate monomers [76]. After 1,4-di(hydroxymethyl)benzene and tetramethylguanidine are dissolved in DMF and treated with CO_2 gas, followed by the addition of allyl halide, α,ω-diene (yield 44%) and α,ω-mono-ene (yield 37%) functionalized carbonate monomers are produced. PCs are prepared via thiol-ene click [77] (yield 98%) and acyclic diene metathesis (yield 60%) protocols [78]. PC is a high-performance, amorphous, and thermoplastic polymer made using CO_2 as a raw material that can be used as an engineering plastic due to its high impact strength, thermal and chemical stability, and good electrical properties. Furthermore, PC is a biocompatible, biodegradable, recyclable, and compostable green polymer widely used in polymer blends such as PC-PET and PC-PMMA to improve material performance.

Bibliography

[1]   Plastics – the Facts www.plasticseurope.org/en/resources/publications/1804-plastics-facts-2019.

[2]   Global plastic production 1950–2018 | Statista https://www.statista.com/statistics/282732/global-production-of-plastics-since-1950/ (accessed Nov 12, 2020).

[3]   Geyer, R.; Jambeck, J. R.; Law, K. L. Production, Use, and Fate of All Plastics Ever Made. *Sci. Adv.* **2017**, 3(7), e1700782. 10.1126/sciadv.1700782.

[4]   Jambeck, J. R.; Geyer, R.; Wilcox, C.; Siegler, T. R.; Perryman, M.; Andrady, A.; Narayan, R.; Law, K. L. Plastic Waste Inputs from Land into the Ocean. *Science (80)*. **2015**, 347(6223), 768. 10.1126/science.1260352.

[5]   Advancing sustainable materials management: Facts and figures report.

[6]   Anastas, P. T.; Warner, J. C. Green Chemistry: Theory and Practice; Oxford univ. press New York, 2000.

[7]   Tsarevsky, N. V; Matyjaszewski, K. "Green" Atom Transfer Radical Polymerization: From Process Design to Preparation of Well-Defined Environmentally Friendly Polymeric Materials. *Chem. Rev.* **2007**, 107(6), 2270–2299. 10.1021/cr050947p.

[8]   Knox, S. T.; Warren, N. J. Enabling Technologies in Polymer Synthesis: Accessing a New Design Space for Advanced Polymer Materials. *React. Chem Eng. Royal Soc. Chem*. **March 1, 2020**, 405–423. 10.1039/c9re00474b.

[9]   Carpenter, E. J.; Smith, K. L. Plastics on the Sargasso Sea Surface. *Science (80)*. **1972**, 175(4027), 1240–1241. 10.1126/science.175.4027.1240.

[10] Roy, P. K.; Hakkarainen, M.; Varma, I. K.; Albertsson, A.-C. Degradable Polyethylene: Fantasy or Reality. *Environ. Sci. Technol*. **2011**, 45(10), 4217–4227. 10.1021/es104042f.

[11] Meeker, J. D.; Sathyanarayana, S.; Swan, S. H. Phthalates and Other Additives in Plastics: Human Exposure and Associated Health Outcomes. *Philos. Trans. – R. Soc., Biol. Sci.* **2009**, 364(1526), 2097–2113. 10.1098/rstb.2008.0268.

[12] Trasande, L.; Shaffer, R. M.; Sathyanarayana, S. Food Additives and Child Health. *Pediatrics.* **2018**, 142(2), e20181410. 10.1542/peds.2018-1410.

[13] Cox, K. D.; Covernton, G. A.; Davies, H. L.; Dower, J. F.; Juanes, F.; Dudas, S. E. Human Consumption of Microplastics. *Environ. Sci. Technol*. **2019**, 53(12), 7068–7074. 10.1021/acs.est.9b01517.

[14] Ignatyev, I. A.; Thielemans, W.; Vander Beke, B. Recycling of Polymers: A Review. *ChemSusChem.* **2014**, 7(6), 1579–1593. 10.1002/cssc.201300898.

[15] Bioplastics: a growing success – bioplastics MAGAZINE https://www.bioplasticsmagazine.com/en/news/meldungen/04122017-Bioplastics-a-growing-success.php (accessed Nov 12, 2020).

[16] Costa, S. S.; Miranda, A. L.; De Jesus Assis, D.; Souza, C. O.; De Morais, M. G.; Costa, J. A. V.; Druzian, J. I. Efficacy of Spirulina Sp. Polyhydroxyalkanoates Extraction Methods and Influence on Polymer Properties and Composition. *Algal Res.* **2018**, 33, 231–238. 10.1016/j.algal.2018.05.016.

[17] Climate Agenda: Resilient, Prosperous, and Green www.Government.Nl/Documents/Reports/2014/02/17/Climate-Agenda-Resilient-Prosperous-and-Green (Accessed Feb 17, 2014).

[18] Green Chemistry > ; TopDutch. A good place to be great. https://www.topdutch.com/invest/key-industries/green-chemistry (accessed Nov 12, 2020).

[19] How much plastic packaging waste do we recycle? – Product – Eurostat https://ec.europa.eu/eurostat/web/products-eurostat-news/-/DDN-20191105-2 (accessed Nov 12, 2020).

[20] Okan, M.; Aydin, H. M.; Barsbay, M. Current Approaches to Waste Polymer Utilization and Minimization: A Review. *J. Chem. Technol. Biotechnol*. **2019**, 94(1), 8–21. 10.1002/jctb.5778.

[21] Towards a greener waste cycle – Ramboll Group https://ramboll.com/ingenuity/towards-a-greener-waste-cycle (accessed Nov 12, 2020).

[22] Sharma, R.; Sharma, M.; Sharma, R.; Sharma, V. The Impact of Incinerators on Human Health and Environment. *Rev. Environ. Health*. **2013**, 28(1), 67–72. 10.1515/reveh-2012-0035.

[23] Bioplastic market data www.european-bioplastics.org/market (accessed Nov 12, 2020).

[24] Mülhaupt, R. Green Polymer Chemistry and Bio-Based Plastics: Dreams and Reality. *Macromol. Chem. Phys.* **2013**, 214(2), 159–174. 10.1002/macp.201200439.

[25] Rieger, B.; Künkel, A.; Coates, G. W.; Reichardt, R.; Dinjus, E.; Zevaco, T. A. Synthetic Biodegradable Polymers; Springer Science & Business Media, 2012; vol. *245*.

[26] Scheibel, T. Protein Fibers as Performance Proteins: New Technologies and Applications. *Curr. Opin. Biotechnol.* **2005**, 16(4), 427–433. 10.1016/j.copbio.2005.05.005.

[27] Mohsenzadeh, A.; Zamani, A.; Taherzadeh, M. J. Bioethylene Production from Ethanol: A Review and Techno-Economical Evaluation. *ChemBioEng Rev.* **2017**, 4(2), 75–91. 10.1002/cben.201600025.

[28] Malpass, D. B. Introduction to Industrial Polyethylene: Properties, Catalysts, and Processes; John Wiley & Sons, 2010; vol. *45*.

[29] Isikgor, F. H.; Becer, C. R. Lignocellulosic Biomass: A Sustainable Platform for the Production of Bio-Based Chemicals and Polymers. *Polym. Chem.* **2015**, 6(25), 4497–4559. 10.1039/C5PY00263J.

[30] Ren, J. In *Lactic Acid BT – Biodegradable Poly(Lactic Acid): Synthesis, Modification, Processing and Applications*; Ren, J., Ed.; Springer: Berlin Heidelberg: Berlin, Heidelberg, 2010; pp 4–14. 10.1007/978-3-642-17596-1_2.

[31] Bio-based products – from idea to market "15 EU success stories" www.op.europa.eu/en/publication-detail/-/publication/23ab58e0-3011-11e9-8d04-01aa75ed71a1 (accessed Nov 12, 2020).

[32] Zhang, J.; Zhao, Y.; Pan, M.; Feng, X.; Ji, W.; Au, C.-T. Efficient Acrylic Acid Production through Bio Lactic Acid Dehydration over NaY Zeolite Modified by Alkali Phosphates. *ACS Catal.* **2011**, 1(1), 32–41. 10.1021/cs100047p.

[33] Lee, P. C.; Lee, W. G.; Lee, S. Y.; Chang, H. N. Succinic Acid Production with Reduced By-Product Formation in the Fermentation of Anaerobiospirillum Succiniciproducens Using Glycerol as a Carbon Source. *Biotechnol. Bioeng.* **2001**, 72(1), 41–48. 10.1002/1097-0290(20010105)72:1<41::AID-BIT6>3.0.CO;2-N.

[34] Weissermel, K.; Arpe, H. J. Industrial Organic Chemistry. *Synthesis.* **2004**, 2004(07), 1127.

[35] Li, X.; Lan, X.; Wang, T. Highly Selective Catalytic Conversion of Furfural to γ-Butyrolactone. *Green Chem.* **2016**, 18(3), 638–642. 10.1039/C5GC02411K.

[36] Li, X.; Cui, Y.; Yang, X.; Dai, W.-L.; Fan, K. Highly Efficient and Stable Au/Mn$_2$O$_3$ Catalyst for Oxidative Cyclization of 1,4-Butanediol to γ-Butyrolactone. *Appl. Catal. A, Gen.* **2013**, 458, 63–70. 10.1016/j.apcata.2013.03.020.

[37] Robert, T.; Friebel, S. Itaconic Acid – A Versatile Building Block for Renewable Polyesters with Enhanced Functionality. *Green Chem.* **2016**, 18(10), 2922–2934. 10.1039/C6GC00605A.

[38] Sreeranjit, C. V. K.; Lal, J. J. GLUCOSE | Properties and Analysis. In Encyclopedia of Food Sciences and Nutrition, Second Edition; Caballero, B., Ed.; Academic Press: Oxford, 2003; pp 2898–2903. 10.1016/B0-12-227055-X/00557-5.

[39] Serrano-Ruiz, J. C.; West, R. M.; Dumesic, J. A. Catalytic Conversion of Renewable Biomass Resources to Fuels and Chemicals. *Annu. Rev. Chem. Biomol. Eng.* **2010**, 1, 79–100. 10.1146/annurev-chembioeng-073009-100935.

[40] Maine Bioproducts Business Pathways – Forest Bioproducts Research Institute – University of Maine https://forestbioproducts.umaine.edu/publication/maine-bioproducts-business-pathways/ (accessed Nov 12, 2020).

[41] Fukuoka, A.; Dhepe, P. L. Catalytic Conversion of Cellulose into Sugar Alcohols. *Angew. Chem., Int. Ed. Engl.* **2006**, 45(31), 5161–5163. 10.1002/anie.200601921.

[42] Jonas, R.; Silveira, M. M. Sorbitol Can Be Produced Not Only Chemically but Also Biotechnologically. *Appl. Biochem. Biotechnol.* **2004**, 118(1), 321–336. 10.1385/ABAB:118:1-3:321.

[43] Cho, M. J.; Choi, D. H.; Sullivan, P. A.; Akelaitis, A. J. P.; Dalton, L. R. Recent Progress in Second-Order Nonlinear Optical Polymers and Dendrimers. *Prog. Polym. Sci.* **2008**, 33(11), 1013–1058. 10.1016/j.progpolymsci.2008.07.007.

[44]   De Almeida, R. M.; Li, J.; Nederlof, C.; O'Connor, P.; Makkee, M.; Moulijn, J. A. Cellulose Conversion to Isosorbide in Molten Salt Hydrate Media. *ChemSusChem*. **2010**, 3(3), 325–328. 10.1002/cssc.200900260.

[45]   Fenouillot, F.; Rousseau, A.; Colomines, G.; Saint-Loup, R.; Pascault, J. P. Polymers from Renewable 1,4:3,6-Dianhydrohexitols (Isosorbide, Isomannide and Isoidide): A Review. *Prog. Polym. Sci*. **2010**, 35(5), 578–622. 10.1016/j.progpolymsci.2009.10.001.

[46]   Ji, N.; Zhang, T.; Zheng, M.; Wang, A.; Wang, H.; Wang, X.; Chen, J. G. Direct Catalytic Conversion of Cellulose into Ethylene Glycol Using Nickel-Promoted Tungsten Carbide Catalysts. *Angew. Chem., Int. Ed. Engl*. **2008**, 47(44), 8510–8513. 10.1002/anie.200803233.

[47]   Barakat, A.; Mayer-Laigle, C.; Solhy, A.; Arancon, R. A. D.; De Vries, H.; Luque, R. Mechanical Pretreatments of Lignocellulosic Biomass: Towards Facile and Environmentally Sound Technologies for Biofuels Production. *RSC Adv*. **2014**, 4(89), 48109–48127. 10.1039/C4RA07568D.

[48]   Kaufman Rechulski, M. D.; Käldström, M.; Richter, U.; Schüth, F.; Rinaldi, R. Mechanocatalytic Depolymerization of Lignocellulose Performed on Hectogram and Kilogram Scales. *Ind. Eng. Chem. Res*. **2015**, 54(16), 4581–4592. 10.1021/acs.iecr.5b00224.

[49]   Alonso, D. M.; Wettstein, S. G.; Mellmer, M. A.; Gurbuz, E. I.; Dumesic, J. A. Integrated Conversion of Hemicellulose and Cellulose from Lignocellulosic Biomass. *Energy Environ. Sci*. **2013**, 6(1), 76–80. 10.1039/C2EE23617F.

[50]   Haouache, S.; Karam, A.; Chave, T.; Clarhaut, J.; Amaniampong, P. N.; Garcia Fernandez, J. M.; De Oliveira Vigier, K.; Capron, I.; Jérôme, F. Selective Radical Depolymerization of Cellulose to Glucose Induced by High Frequency Ultrasound. *Chem. Sci*. **2020**, 11(10), 2664–2669. 10.1039/D0SC00020E.

[51]   Öhgren, K.; Bengtsson, O.; Gorwa-Grauslund, M. F.; Galbe, M.; Hahn-Hägerdal, B.; Zacchi, G. Simultaneous Saccharification and Co-Fermentation of Glucose and Xylose in Steam-Pretreated Corn Stover at High Fiber Content with Saccharomyces Cerevisiae TMB3400. *J. Biotechnol*. **2006**, 126(4), 488–498. 10.1016/j.jbiotec.2006.05.001.

[52]   Lange, H.; Schiffels, P.; Sette, M.; Sevastyanova, O.; Crestini, C. Fractional Precipitation of Wheat Straw Organosolv Lignin: Macroscopic Properties and Structural Insights. *ACS Sustain. Chem. Eng*. **2016**, 4(10), 5136–5151. 10.1021/acssuschemeng.6b01475.

[53]   Crestini, C.; Lange, H.; Sette, M.; Argyropoulos, D. S. On the Structure of Softwood Kraft Lignin. *Green Chem*. **2017**, 19(17), 4104–4121. 10.1039/C7GC01812F.

[54]   Majdar, R. E.; Crestini, C.; Lange, H. Lignin Fractionation in Segmented Continuous Flow. *ChemSusChem*. 2020, 10.1002/cssc.202001138.

[55]   Alb, A. M.; Reed, W. F. Fundamental Measurements in Online Polymerization Reaction Monitoring and Control with a Focus on ACOMP. *Macromol. React. Eng*. **2010**, 4(8), 470–485. 10.1002/mren.200900079.

[56]   Levere, M. E.; Willoughby, I.; O'Donohue, S.; De Cuendias, A.; Grice, A. J.; Fidge, C.; Becer, C. R.; Haddleton, D. M. Assessment of SET-LRP in DMSO Using Online Monitoring and Rapid GPC. *Polym. Chem*. **2010**, 1(7), 1086–1094. 10.1039/C0PY00113A.

[57]   Hoogenboom, R.; Fijten, M. W. M.; Abeln, C. H.; Schubert, U. S. High-Throughput Investigation of Polymerization Kinetics by Online Monitoring of GPC and GC. *Macromol. Rapid Commun*. **2004**, 25(1), 237–242. 10.1002/marc.200300218.

[58]   Rosenfeld, C.; Serra, C.; O'Donohue, S.; Hadziioannou, G. Continuous Online Rapid Size Exclusion Chromatography Monitoring of Polymerizations – CORSEMP. *Macromol. React. Eng*. **2007**, 1(5), 547–552. 10.1002/mren.200700024.

[59]   Alauhdin, M.; Bennett, T. M.; He, G.; Bassett, S. P.; Portale, G.; Bras, W.; Hermida-Merino, D.; Howdle, S. M. Monitoring Morphology Evolution within Block Copolymer Microparticles during Dispersion Polymerisation in Supercritical Carbon Dioxide: A High Pressure SAXS Study. *Polym. Chem*. **2019**, 10(7), 860–871. 10.1039/C8PY01578C.

[60] Reed, W. F., Alb, A. M. Monitoring Polymerization Reactions: From Fundamentals to Applications; Wiley Online Library, 2014.

[61] Leiza, J. R., Pinto, J. C. Polymer Reaction Engineering; Blackwell publishing limited: Oxford, 2007.

[62] Rubens, M.; Vrijsen, J. H.; Laun, J.; Junkers, T. Precise Polymer Synthesis by Autonomous Self-Optimizing Flow Reactors. *Angew. Chem., Int. Ed. Engl.* **2019**, 58(10), 3183–3187. 10.1002/anie.201810384.

[63] Sultan, Z.; Graça, I.; Li, Y.; Lima, S.; Peeva, L. G.; Kim, D.; Ebrahim, M. A.; Rinaldi, R.; Livingston, A. G. Membrane Fractionation of Liquors from Lignin-First Biorefining. *ChemSusChem.* **2019**, 12(6), 1203–1212. 10.1002/cssc.201802747.

[64] Chi, X.-Y.; Xia, B.-G.; Xu, Z.-L.; Zhang, M.-X. Impact of Cross-Linked Chitosan Sublayer Structure on the Performance of TFC FO PAN Nanofiber Membranes. *ACS Omega.* **2018**, 3(10), 13009–13019. 10.1021/acsomega.8b01201.

[65] Shakeri, A.; Salehi, H.; Rastgar, M. Chitosan-Based Thin Active Layer Membrane for Forward Osmosis Desalination. *Carbohydr. Polym.* **2017**, 174, 658–668. 10.1016/j.carbpol.2017.06.104.

[66] Galiano, F.; Ghanim, A. H.; Rashid, K. T.; Marino, T.; Simone, S.; Alsalhy, Q. F.; Figoli, A. Preparation and Characterization of Green Polylactic Acid (PLA) Membranes for Organic/Organic Separation by Pervaporation. *Clean Technol. Environ. Policy.* **2019**, 21(1), 109–120. 10.1007/s10098-018-1621-4.

[67] Budsaereechai, S.; Hunt, A. J.; Ngernyen, Y. Catalytic Pyrolysis of Plastic Waste for the Production of Liquid Fuels for Engines. *RSC Adv.* **2019**, 9(10), 5844–5857. 10.1039/C8RA10058F.

[68] Fivga, A.; Dimitriou, I. Pyrolysis of Plastic Waste for Production of Heavy Fuel Substitute: A Techno-Economic Assessment. *Energy.* **2018**, 149, 865–874. 10.1016/j.energy.2018.02.094.

[69] Faravelli, T.; Pinciroli, M.; Pisano, F.; Bozzano, G.; Dente, M.; Ranzi, E. Thermal Degradation of Polystyrene. *J. Anal. Appl. Pyrolysis.* **2001**, 60(1), 103–121. 10.1016/S0165-2370(00)00159-5.

[70] Miller, S. J.; Shah, N.; Huffman, G. P. Conversion of Waste Plastic to Lubricating Base Oil. *Energy Fuels.* **2005**, 19(4), 1580–1586. 10.1021/ef049696y.

[71] Harmsen, P. F. H.; Hackmann, M. M.; Bos, H. L. Green Building Blocks for Bio-Based Plastics. *Biofuels, Bioprod. Biorefining.* **2014**, 8(3), 306–324. 10.1002/bbb.1468.

[72] Guderjahn, L. Bioethanol Production in Europe-Sustainable and Efficient. 8th European Bioethanol Technology Meeting. Detmold 2012.

[73] Hunter, S. E.; Ehrenberger, C. E.; Savage, P. E. Kinetics and Mechanism of Tetrahydrofuran Synthesis via 1,4-Butanediol Dehydration in High-Temperature Water. *J. Org. Chem.* **2006**, 71(16), 6229–6239. 10.1021/jo0610170.

[74] Zhao, J.; Hartwig, J. F. Acceptorless, Neat, Ruthenium-Catalyzed Dehydrogenative Cyclization of Diols to Lactones. *Organometallics.* **2005**, 24(10), 2441–2446. 10.1021/om048983m.

[75] Yim, H.; Haselbeck, R.; Niu, W.; Pujol-Baxley, C.; Burgard, A.; Boldt, J.; Khandurina, J.; Trawick, J. D.; Osterhout, R. E.; Stephen, R.; Estadilla, J.; Teisan, S.; Schreyer, H. B.; Andrae, S.; Yang, T. H.; Lee, S. Y.; Burk, M. J.; Van Dien, S. Metabolic Engineering of Escherichia Coli for Direct Production of 1,4-Butanediol. *Nat. Chem. Biol.* **2011**, 7(7), 445–452. 10.1038/nchembio.580.

[76] Chai, Y.; Chen, Q.; Huang, C.; Zheng, Q.; North, M.; Xie, H. Introducing the Reversible Chemistry of CO_2 with Diols Mediated by Organic Superbases into Polycarbonate Synthesis. *Green Chem.* 2020, 10.1039/D0GC01197E.

[77] Hoyle, C. E.; Lowe, A. B.; Bowman, C. N. Thiol-Click Chemistry: A Multifaceted Toolbox for Small Molecule and Polymer Synthesis. *Chem. Soc. Rev.* **2010**, 39(4), 1355–1387. 10.1039/B901979K.

[78] Kreye, O.; Tóth, T.; Meier, M. A. R. Copolymers Derived from Rapeseed Derivatives via ADMET and Thiol-Ene Addition. *Eur. Polym. J.* **2011**, 47(9), 1804–1816. 10.1016/j.eurpolymj.2011.06.012.

13 Solar powered engineering

Gergo Ignacz, Edwing J. G. Gonzalez, Fabiyan A. Shamsudheen, Gyorgy Szekely

Excluding nuclear and geothermal energy, all energy sources on Earth are derived from the Sun (the energy from cosmic rays is practically negligible). Geothermal energy, for instance, is considered renewable and is harnessed from slow nuclear fission in the Earth's core. The total *solar energy* reaching the Earth's surface in a year is several thousand magnitudes (!) higher than the total annual energy consumption on Earth [1]. Solar energy is a *periodically renewable* energy source. Due to daily (and sometimes intermittent), seasonal, and regional changes in the exploitable solar radiation and the relatively energy-inefficient harvesting method, solar energy is usually not profitable. This chapter focuses on the different approaches available to capture and utilize solar energy. The fundamental aspects of light-harvesting by conventional solar cells and solar contractors are not covered here; the reader is referred to the literature for general concepts and ideas [2–4].

13.1 Water harvesting from air

Water vapor and the atmospheric water resource account for approx. 10% of the total freshwater reservoir [5]. The atmosphere contains approx. $1.4 \cdot 10^{17}$ L of water, which forms part of the natural hydrological cycle. Extracting this abundant water resource from the air could represent a sustainable solution where other freshwater sources are scarce or heavily polluted. Water vapor in the air can be considered renewable and – provided prevailing weather patterns do not change significantly in the future – a non-depleting source. Follow the QR code on this page to learn about RainMaker and the air-to-water application available therein. Although several ways exist to harvest dew and capture moisture and fog from the air [6, 7], practical, real-life applications have yet to be demonstrated. The main differences between fog collectors and dew-harvesting systems are summarized in Table 13.1. Both fog collectors and dew-harvesting systems require high humidity to operate, usually close to the saturation point. Due to the extra power supply required, the price of the produced water from dew-harvesting is 2–3 times higher than the price of produced water from fog collectors. However, water from fog collectors can be contaminated easily due to constant exposure to the surrounding environment. The schematic working principle and real-life pictures are presented in Figure 13.1.

For a water capture device to be viable, it must overcome its energy-intensiveness by using low-grade energy resources. For example, devices can be designed to generate electricity and store that energy during the day, while harvesting water at night when

https://doi.org/10.1515/9783111028163-013

Table 13.1: Differences between fog collectors and dew-harvesting systems. Adapted with permission from [6, 7].

Fog collectors	Dew-harvesting systems
Requires high humidity	Seasonal arid zones
Saturated air (1–30 μm)	80–90% RH
Mesh and storage tank	Extra power system (to reach dew point)
4.5 USD (m^{-3} water)	12 USD (m^{-3} water)

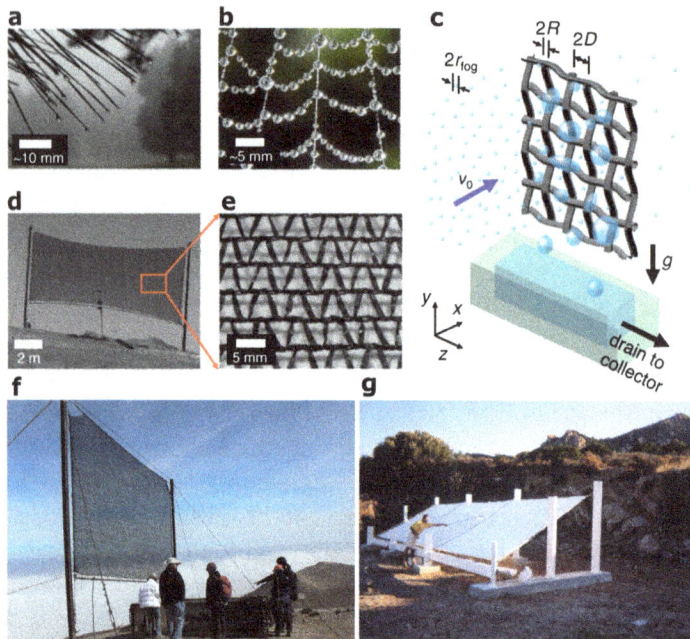

Figure 13.1: (a and b) Different fog and dew collectors in nature. (c) Schematic representation of the working principle of a fog collector. (d and e) Pilot-scale fog collector and close-up image. (f) Fog collector in use at a rural area (Atacama Desert, Chile). (g) Big OPUR Dew condenser in Corsica. (a–e) Reprinted with permission from [7]. (f and g) Reprinted under the CC-BY 3.0 license.

the relative humidity is at its highest.[1] Porous materials with high affinity towards water, such as certain zeolites and MOFs, may be ideal water-adsorbing materials. Regions with different levels of relative humidity will require different types of adsorbents with different water uptake capabilities as a function of the relative humidity. For

---

1 Usually the temperature at night is lower than during the day. The solubility of water vapor in cold air is significantly lower than in hot air. Thus, relative humidity is usually (!) higher at night.

example, in North Africa, where the relative humidity is low (20%), MOF-801 would be a suitable adsorbent, whereas in North India (40% relative humidity), UiO-66 would be more suitable (Figure 13.2a) [5]. A sharp increase in the water uptake capability with respect to relative humidity would enable removing the water from the MOF with only a small temperature change. This concept is visualized in Figure 13.2c; the device absorbs the water from the air with low relative humidity, which is released when the temperature rises, using the heat of the sun. Subsequently, when the device is heated, it is not exposed to fresh air because the water evaporates. Instead, the water – now in liquid form – is collected at the bottom of the device.

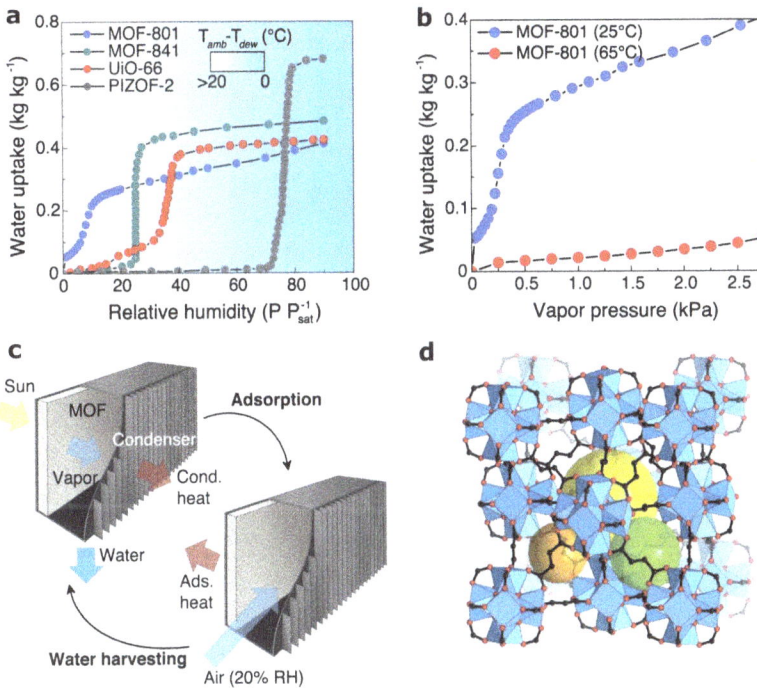

Figure 13.2: Water harvesting from air with MOFs. (a) Different porous materials share different water uptake characteristics based on different relative humidities. (b) Water uptake at different water vapor pressures at two temperatures. (c) Schematic representation of a water-harvesting device. (d) Structure of Y-based MOF-801. Reprinted with permission from [5].

Reusability, stability, complexity, and pricing are crucial factors for water-harvesting devices, primarily because the affected areas that need these devices are typically remote with limited transportation and operational expertise. In the case of MOF-801, the starting materials are cheap and readily available. Moreover, they have exceptional stability and can be recycled several times without loss of performance. The device is also capable of producing 2.8 kg of fresh water per kg of adsorbent without the need for any additional power supply, even at relative humidities as low as 20% [5].

13.2 Solar-driven membrane processes

Desalination technologies are separated into two groups, namely, membrane-based (reverse osmosis) and thermal processes. The leading thermal technologies include multistage flash (MFD), multieffect distillation (MED), and vapor compression (VC). Membrane-based technologies consist of reverse osmosis (RO) and electrodialysis (ED) [8]. The energy intensity of desalination processes is governed by fundamental thermodynamic limitations and thus, they are all energy-intensive processes. Reverse osmosis offers a cheaper and inherently safer option for water desalination than distillation processes; however, there are certain situations where distillation is still a viable option, such as when it is combined with a solar cell.

During conventional solar cell applications, only 10–20% of the solar power is utilized, and the rest of the heat is passively dissipated into the environment. The combination of distillation via excess heat utilization and electricity generation through solar cells could greatly improve the efficiency of the distillation process. Indeed, generating fresh water and electricity within the same unit has the dual benefit of providing two essential commodities while also reducing the operating space [9, 10]. In one configuration, a photovoltaic cell is introduced on top of a *multistage membrane distillation module* (MSMD). The layers of the membrane comprising the distillation units are separated by aluminum nitride for its high thermal conductivity (Figure 13.3) [10].

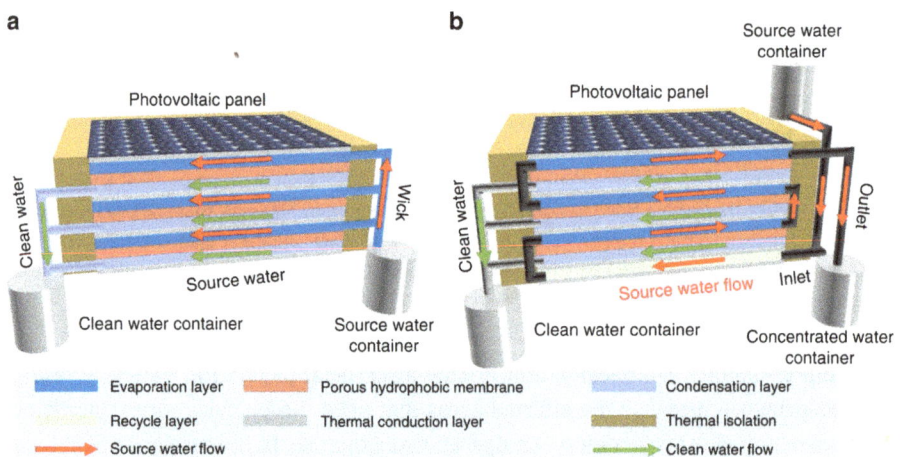

Figure 13.3: A multistage membrane distillation module (MSMD) with a photovoltaic panel on the top. (a) Dead-end filtration mode. (b) Flow-through filtration mode. Reprinted with permission from [10].

The photovoltaic cell is a commercially available polycrystalline silicon solar cell with approx. 11% efficiency. The *photovoltaic cell* is responsible for the relaxation of the short-wavelength light, while the long-wavelength sunlight undergoes photothermal conversion. This generated heat can be used as a resource instead of waste in general photovoltaic cells. Thus, the generated heat dissipates through the thermal conduction layers, eventually heating the water. The produced water vapor then passes through the hydrophilic membrane and condenses, producing potable water. The driving force behind this process is the difference in vapor pressure across the evaporation layers. The heat – released from the condensation process – is ultimately utilized in the next layer. Thus, the multistage process exploits all the available latent heat in the system. This device can generate water at a constant rate of 1.6 kg m^{-2} h^{-1}.

Passive modular distillers (without any moving parts such as pumps) can be constructed for seawater desalination (Figure 13.4). The absence of electrical moving parts is preferred for the lower cost and less maintenance [11]. The vapor pressure difference between the two hydrophilic layers is the driving force of the process. Reducing the distance between the evaporator and the condenser enables an evaporation–condensation cycle to take place at low temperature and ambient pressure. The seawater is supplied via capillary forces and enters the heated hydrophobic top layer. The thermal gradient between the top and bottom dictates the direction of the driving force and thus, the travel direction of the water vapor. The heat dissipation through several stages ensures efficient heat utilization. The spatially uneven salt content of the water in the system generates a different water flux across the device. Nevertheless, the net flux of water produced is steady when the energy supply is stable.

The non-linear relationship between each evaporator–condenser layer has a negative effect on the net water flux rate. For example, a three-stage configuration device has an approx. threefold higher specific mass flow rate compared to the mass flow rate of a single-stage distillation device. On the other hand, the mass flow rate of a ten-stage device is six-fold higher than that of a single-stage distillation device. The device can be operated with two different setups. In setup A, the system does not contain a membrane between the evaporator and condenser, but the gap is filled with air. In setup B, a hydrophobic membrane is placed between the evaporator and the condenser. Moreover, in setup B, the distance between the evaporator and the condenser is significantly smaller than in setup A. In general, the narrower the gap between the condenser and the evaporator, the higher the temperature gradient, and the higher the water flux. The thermal conductivity of the hydrophobic membrane is higher than the thermal conductivity of air, which results in a lower temperature gradient and lower water flux. These competing effects are in balance, and, interestingly, there is little difference in net water flux between setup A and setup B. The freshwater output can reach up to 3 L m^{-2} h^{-1} with *one sun illumination*.[2] These devices can be useful in emergencies due to their passive method. Also, floating gardens and other CO_2 sequestration applications could be envisaged where a freshwater supply is in constant demand.

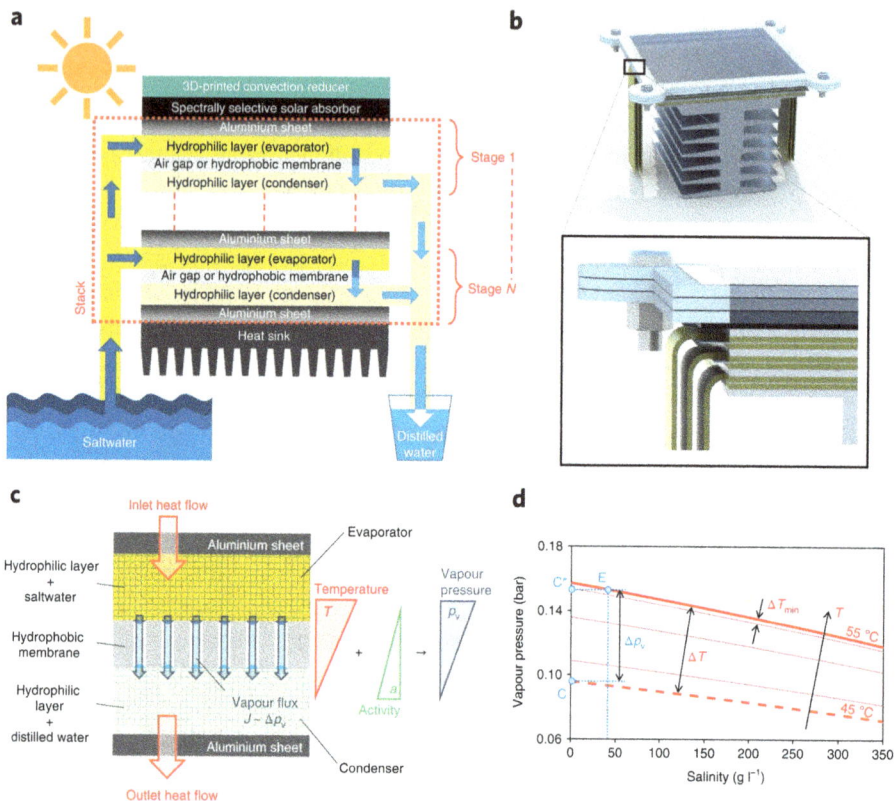

Figure 13.4: Schematic representation of the passive modular distiller. (a) Seawater travels to the system via capillary forces to the hydrophilic layers. Meanwhile, solar energy is converted into heat energy via solar absorbers. The heat is conducted towards the inside of the device using aluminum sheets. Evaporation takes place between the two hydrophilic layers. (b) Rendered image of the module. (c) Thermodynamic representation of the working principle: the vapor flux towards the driving force is proportional to the vapor pressure gradient across the hydrophobic layer. (d) Vapor pressure with respect to water salinity. Reprinted with permission from [11].

Direct solar-driven organic solvent purification has emerged recently. RO-type membrane processes are often cumbersome due to the harsh nature of the organic sol-

---

2 One sun illumination refers to the method where no additional solar reflectors are used (no mirrors are used to reflect the radiation).

vents involved. Although distillation processes offer a robust solution, they tend to be energy-intensive. The replacement of fossil fuel-based energy with solar power is a potentially viable option for the purification of organic solvents with low boiling points. The operating principle and the driving force for a passive evaporation device that exploits direct sunlight are similar to the freshwater distillatory devices described above. The electromagnetic waves heat the surface of a porous material immersed in the solvent. The solvent travels upwards via capillary forces where the vapor condenses, is collected, and is then reused. Direct evaporation of the organic solvents is achieved using sunlight and a coated porous material (Figure 13.5). The coating material is Prussian Blue ($Fe^{III}_4[Fe^{II}(CN)_6]_3$)-based MOF. This MOF has exceptional absorbance in the visible wavelength, is stable in a large variety of organic solvents (e.g., ketones, alcohols), and has a large surface area ($5.7\ m^2\ g^{-1}$). The device can achieve a $389.4\ mol\ m^{-2}\ h^{-1}$ vaporization rate for acetone and $7.6\ mol\ m^{-2}\ h^{-1}$ for DMF; these rates are well correlated with the change in their vaporization enthalpy [12]. This technology can compete with the already existing organic solvent nanofiltration technology, but its large-scale adoption and commercialization have yet to be realized.

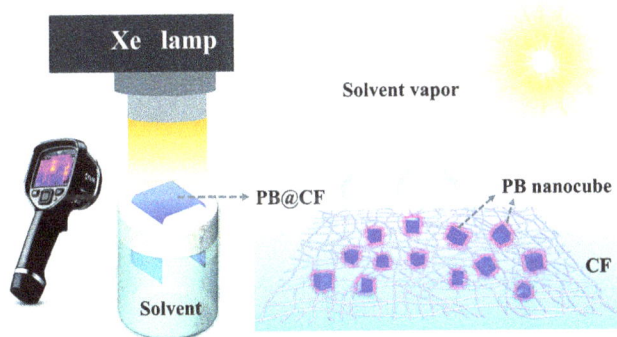

Figure 13.5: Solar-driven organic solvent purification through evaporation using Prussian Blue-coated porous materials. In this presented system, a xenon lamp heats the top layer of the material, which induces rapid evaporation from the surface. Capillary forces passively transfer the fluid to the surface. Reprinted with permission from [12].

13.3 Concentrated solar power

Concentrated solar power (CSP) devices generate heat or energy by projecting the sunlight using mirrors or lenses. The projected sunlight is converted to heat, which can then be directly used to heat a fluid to produce electricity or to perform chemical reactions [13]. Concentrated solar light can also be used in photochemical reactions [14]. Across the solar-belt region of the Earth (±40° latitudes where there is a low degree of

humidity and low levels of atmospheric aerosols), the solar radiosity is approximately 63 MW m^{-2}, which is ideal for heat generation. The Snell–Descartes law can be applied as the basic operational principle for solar concentrators.[3] Follow the QR codes on this page to watch videos about the CSP concept and its implementation.

Four different types of solar concentrators are shown in Figure 13.6. The first device is termed *parabolic trough* (PT). In this device, the curved mirrors reflect the light onto a tubing, which is filled with thermal fluid. The main advantage of this device is the simple design and the fact that it can be operated under continuous conditions. PTs and their analogs are used in large-scale CSP applications because they are easy to scale up at a reasonable cost. The second device uses *heliostats* (central receiver [CR]), which project the sunlight onto the surface of a receiver filled with thermal fluid, usually molten salts or molten metals. CRs are usually used for power generation. The third device is the *dish/engine*, where a parabolic reflector projects the sunlight onto a receiver engine. The main advantage of this configuration is that the parabolas can be controlled individually. The technology could be potentially used for water splitting or sustainable fuel production [15, 16]. The fourth device is the *linear Fresnel mirror solar concentrator*, which is a combination of PT and CR in a setup that allows a continuous flow way.

There are several feasible applications for commercial, industrial, and small-scale CSP generators. Domestic-grade devices usually generate low-grade heat for heating or small-scale biogas production. Commercial and industrial-scale solar plants are mostly used for grid-type electricity generation or to power different stand-alone or hybrid projects. Industrial process heat applications include boiling, melting, sterilizing, water desalination, and photochemical reactions. Electrical power generation using CSP is well established in the commercial sector and on the industrial scale [13]. Different types of CSPs have different average power outputs, operation temperatures, and efficiencies (Table 13.2) [16].

Electricity generation that utilizes heat is mainly achieved using steam engines. The two most well-known steam engine types are the Rankine cycle and the Brayton cycle. In the case of the *Rankine cycle*, water is pumped through a combustion chamber to generate steam. The steam (or vapor) is expanded in a turbine system, which is used for electricity generation. The low-pressure steam is then condensed and regenerated into the cycle. The low-grade heat from the condensation is used for heating [17]. On the other hand, the *Brayton cycle* uses a dual compressor-turbine system, where the compressor is physically connected to the turbine via a centerline. Compressed fresh air enters the chamber along with the fuel. After combustion, the exhaust gases enter the turbine, where they expand and actuate the turbine, before

---

3 The Snell–Descartes law, or simply refraction law, refers to the phenomenon when an incident beam is refracted upon entering a different medium. The incident beam from a material with refractive index n_1 entering another material with refractive index n_2 with angle ϑ_1 will be refracted at an angle ϑ_2, where $\frac{\sin \vartheta_1}{\sin \vartheta_2} = \frac{n_2}{n_1}$.

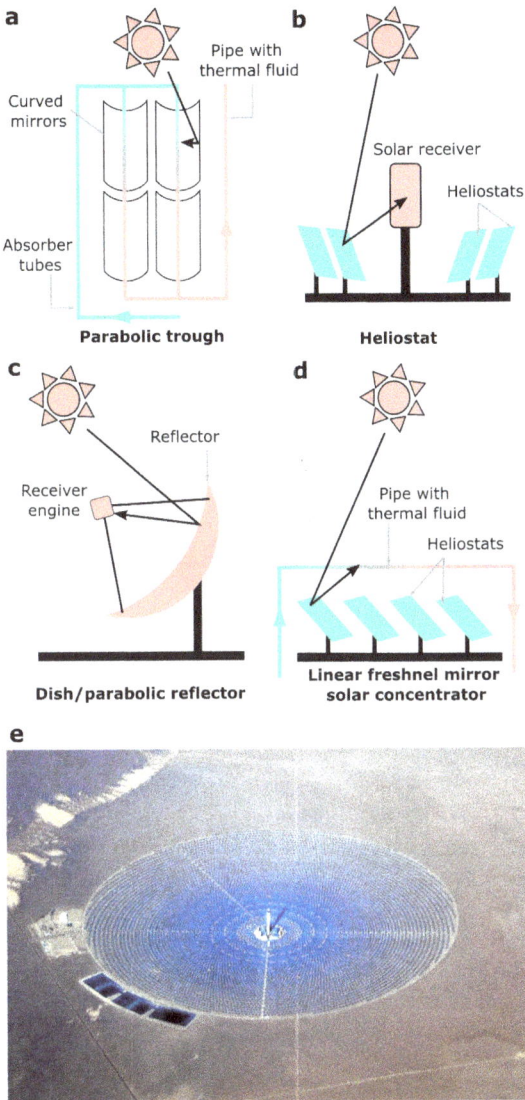

a

Curved mirrors

Pipe with thermal fluid

Absorber tubes

Parabolic trough

b

Solar receiver

Heliostats

Heliostat

c

Reflector

Receiver engine

Dish/parabolic reflector

d

Pipe with thermal fluid

Heliostats

Linear freshnel mirror solar concentrator

e

Figure 13.6: General configuration for (a) different solar concentrators and (b) a solar power tower with advanced molten salt energy storage technology concentrating light with 10,000 mirrored heliostats spanning across 1.2 km². The photo is licensed under the Creative Commons Attribution-Share Alike 4.0 International license.

exiting as exhaust [18]. The CSP-powered Rankine cycle for energy generation is nearly identical to the general Rankine cycle (Figure 13.7). Sunlight is converted into heat using a thermal fluid. The working fluid (water or organic solvent) is vaporized using heat exchangers and then expanded to generate electricity. The low-pressure vapor after the expander is subjected to heat exchange, and the low-grade heat can be used for heating.

Table 13.2: Concentrated solar power system characteristics.

	Parabolic troughs	Central receiver	Dish/engine
Power unit	30–80 MW[a]	10–200 MW[a]	5–25 kW
Temperature	390 °C	565 °C	750 °C
Annual capacity	23–50%[a]	20–77%[a]	25%
Peak efficiency	20%	23%	29.40%
Net annual efficiency	11–16%[a]	7–20%[a]	12–25%
Commercial status	Mature	Early projects	Prototypes/demos
Technology risk	Low	Medium	High
Thermal storage	Limited	Yes	Batteries
Hybrid schemes	Yes	Yes	Yes
Cost W installed			
USD per W	3.49–2.34[a]	3.83–2.16[a]	11.00–1.14[a]
USD per W_{peak}[b]	3.49–1.13[a]	2.09–0.78[a]	11.00–0.96[a]

[a]Data interval period of 2010–2025. [b]Without thermal storage. Data from [16].

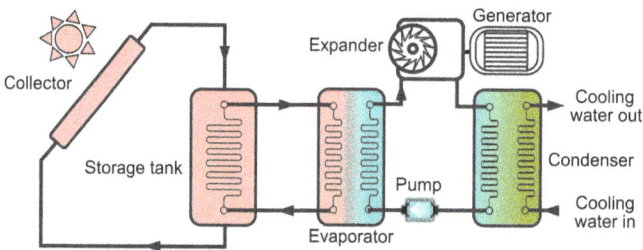

Figure 13.7: CSP-powered Rankine cycle. The general working principle is similar to those utilizing fossil fuel heat: the generated water vapor expands and powers a generator. Then the low-pressure vapor passes through a condenser generating low-grade heat. If the solar power is not enough to fill the storage tank, auxiliary power generators must be connected to the system.

The *fluctuation* problem regarding solar power is a challenging task to solve. The electromagnetic radiation of the sun can be considered steady in our timespan. On the other hand, due to the effect of the Earth's tilt, the effective radiation reaching any given location during the year varies spatially and periodically, depending on the season. Seasonal effects move slowly and change gradually. In contrast, daily changes due to the rotation of the Earth are significantly faster. Weather conditions can also result in a variation of the radiation energy, making the utilization of solar energy somewhat chaotic. These predictable and unpredictable changes make it challenging to implement a universal solar power-generating system across the globe. Another important issue is the uneven peak distribution. If peak energy demand hours occur before or after sunset, the collected solar power must be conserved and stored in large batteries. Further information about redox-flow batteries can be found following the QR code on this page [19]. Changing the general operational time for electrical-intensive industrial processes from nighttime to midday is an alternative solution to power storage.

Generally speaking, in power storage devices, the storage tank is filled during the day with high solar radiation and then emptied at night during low solar radiation. High-efficiency thermal storage can be achieved using molten salts. The main advantages of this technology include extremely low CO_2 generation and relatively high overall efficiency (55%), which is close to the regenerative Brayton cycle. Major drawbacks of this system include scale-up and integration with existing grid systems. As an example of a real-world case, consider a day when the weather conditions prevent the storage tank from filling fully and thus, the tank cannot supply enough electricity. In this instance, an alternative electrical generation system (renewable or otherwise) is needed immediately to meet energy demand. This not only makes the system more complicated but generates substantial idle time for the auxiliary power generator, which will eventually increase the cost.

13.4 Photochemistry and photocatalysis

Photochemistry refers to the process of chemical conversion induced by electromagnetic energy. Photocatalysis refers to the heterogeneous and homogeneous photolytic conversion of different compounds by electromagnetic energy. Photochemical reactions generate reactive species through the mild excitation of molecules. These highly *reactive intermediates* cannot be generated under mild conditions, and the use of conventional thermal methods results in low yield. Photochemical reactions are of utmost importance in nature.

Atoms and molecules can only absorb photons with a specific energy. This characteristic energy is called photon energy, which is proportional to the frequency (and inversely proportional to the wavelength) of the light. Higher-frequency light corresponds to higher energy. Upon absorption of a photon by an atom or molecule, the ground-state highest occupied molecular orbital (HOMO, S0) electron is excited to the lowest unoccupied molecular orbital (LUMO) or a higher MO while maintaining its spin (Figure 13.8). The excited electron on S2 quickly relaxes to a stable S1 via internal conversion (IC). Relaxation from S1 can take place in three different ways. First, through internal conversion back to S0. Second, the molecule or atom emits a photon, causing *fluorescence*. Internal conversion and fluorescent decay are both rapid processes, which is why fluorescent emitters fade quickly following excitation. The third way is called *intersystem crossing* (ISC), during which the electron on S1 undergoes a so-called *spin conversion*, causing the two SOMOs to share the same spin (T1). The T1 triplet state is relatively stable and has a long lifespan, but eventually radiates back to S0; this process is called *phosphorescence*. ISC is the basic concept of *photoredox catalysis*, which generates relatively stable triplet state radicals [20]. Electromagnetic radiation can also permit reactions to occur (called the Woodward–Hoffman rule), thus increasing the selectivity in some instances. For further reading on photochemistry, the reader is referred to [21].

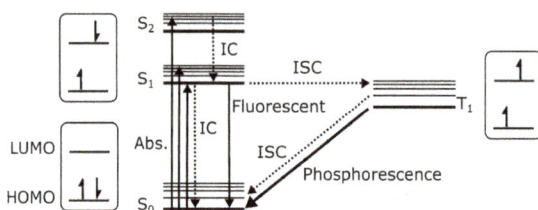

Figure 13.8: Jablonski diagram. The ground-state S0 electron absorbs light at a specific wavelength, reaching the S1 or S2 state based on the energy of the excitation. S2 converts back to S1 via radiationless de-excitation, called internal conversion. The unstable S1 state can undergo another IC either back to the ground state (fluorescence) or into the triplet T1 state via intersystem crossing (ISC). During ISC, the spin multiplicity of the LUMO electron changes. The relatively stable T1 can convert back to the ground state via phosphorescence.

Due to the renewable and non-polluting nature of solar energy, photochemistry is considered to be a green technology [22]. However, photochemical reactions are rarely conducted using direct sunlight due to fluctuations in sunlight intensity, broad wavelength distribution, and cumbersome technological implementations. Instead, artificial lighting is the most common method applied to perform photochemical reactions, which requires the use of LEDs, compact fluorescent lamps, or high-energy quartz UV lamps. A possible solution to replace artificial light with sunlight could be either the transformation of sunlight into useful wavelengths [23], or the inherent transformation of synthetic chemical methodologies to utilize sunlight directly (e.g., by designing catalysts that activate in sunlight). Since electromagnetic wave radiation generates a radical from the catalyst or the molecule, most photochemical reactions fall into the category of free radical transformations. The triplet state of the radical could be either oxidized or reduced; thus, photocatalysts can participate either in the reduction or oxidation part of the process.

13.4.1 Heterogeneous photocatalysis

The most common heterogeneous photocatalysts are transition metal oxides and semiconductors such as TiO_2, ZnO, CeO_2, CdS, and ZnS. Heterogeneous photocatalysis consists of a wide variety of reaction types that are capable of performing total or partial oxidations and reductions. Examples include dehydrogenation, hydrogen transfer, isotopic exchange, organic or inorganic pollutant removal, and metal deposition [24]. Heterogeneous photocatalytic reactions are similar to classical heterogeneous reactions in terms of the *process steps*, that is, the transfer of molecules near the surface, the adsorption of molecules to the surface, the type of reaction, the desorption of products and by-products, and the removal of molecules from the surface.

An *ideal heterogeneous photocatalyst* should have high light absorption efficiency in a broad spectrum and a strong interaction with chemical reactants, while meeting the energy requirements of a desired photochemical reaction and operating at ambient temperature and pressure. Moreover, an ideal photocatalyst should be easily regenerated and recycled. Regarding the *reactor design*, fixed bed and slurry batch photoreactors are preferred, although different designs and configurations are acceptable. The advantages and disadvantages of heterogeneous photocatalytic systems are shown in Table 13.3. Compared to homogenous catalysts, the main advantages of heterogeneous catalysts include higher stability towards photobleaching (i.e., degradation due to radiation) and their inert nature and low cost. However, they tend to have lower light-harvesting efficiency due to shadowing effects.

Table 13.3: Advantages and disadvantages of heterogeneous photocatalysis.

Advantages	Disadvantages
High stability	High recombination rate
High activity	Low harvesting efficiency
Non-toxic	
Chemically inert	Difficult isolation
Low cost	Poor electric adsorption
Easy to recover and recycle	

Catalytic activity and the rate of the reaction during heterogeneous photocatalysis are proportional to the mass of the catalyst and follow the Langmuir–Hinshelwood mechanism. Above a certain mass ratio threshold, the catalytic activity rapidly decreases, and the rate of the reaction becomes independent of the catalytic mass due to shadowing effects. Heterogeneous catalytic reactions do not require heating due to the radical photonic-type activation. Depending on the apparent activation energy, the optimal temperature of the reaction is usually between 20 °C and 80 °C. The rate of reaction depends on the radiant flux. The relationship between the reaction rate and the photon flux is called the quantum yield, Φ, which is calculated as follows:

$$\Phi = \frac{\text{molecules formed per second}}{\text{photons adsorbed per second}}. \tag{13.1}$$

The quantum yield is a kinetic definition, which is crucial in any photocatalytic system, and by definition, it cannot exceed 1. The quantum yield is wavelength-dependent; thus, monochromatic light sources at the excitation wavelength have far better quantum yield than, for example, white light. Most reactions have a quantum yield between 0.01

and 0.7, and common photosynthetic reactions have a quantum yield of approx. 0.1 [25]. Another important parameter of photocatalytic reactions is the photonic efficiency, ξ, which is calculated as follows:

$$\xi = \frac{\text{reaction rate}}{\text{photon flux}}.$$

(13.2)

The photonic efficiency in conventional batch reactors is usually between 0.004 and 0.009 and can reach up to 0.6 using a photomicroreactor [26]. Due to their efficiency, flow chemical photoreactors are preferred to batch reactors. Learn more about flow reactors and microreactors in Chapter 6.

Heterogeneous photocatalysis is frequently used for the removal of organic pollutants. However, the photocatalytic reaction rate for pollutant removal (photodegradation) is relatively low compared to other chemical removal methods. Photocatalytic reactions are often coupled with ozone generation to overcome these slow reaction times. The synergistic effect of photocatalytic and ozone treatments offers an efficient way for pollution removal. Heterogeneous catalysis has been successfully applied for dye photosensitization [27], semiconductor coupling [28], metal doping[4] [29], non-metal doping [30], and carbon dioxide sequestration.

In heterogeneous photocatalysis, the catalyst is either suspended or attached to the surface of a support material. For suspended catalysts, the reactor type is also referred to as a *slurry reactor*, where the catalyst particles are dispersed in the reaction matrix without any immobilization. The suspension results in a fully integrated system (the catalyst in the fluid system) and increases the photonic efficiency. The *immobilized catalyst reactor* design features a catalyst anchored to a fixed support dispersed on the stationary phase (catalyst support system). Figure 13.9 shows a reactor scheme of a fixed bed, more specifically a monolith-type reactor. The gas phase starting materials dissolve into the coated, interconnected monolithic channels and diffuse evenly. This setup improves light and mass distribution in the reactor, improving the overall efficiency. The products then desorb back into the gas stream and leave the reactor. The immobilized photocatalyst is commonly attached to activated carbon, fiberglass, glass beads, or zeolites, or onto inert polymer surfaces.

An example of a slurry-type reactor is shown in Figure 13.10. Slurry-type reactors are widely used in water treatment and have higher catalytic activity than fixed-bed reactors. One of the main disadvantages of slurry-type reactors is the difficulty in separating the catalyst from the reaction stream, in particular for catalysts below 0.1 μm in size. The immobilized catalyst beds have several drawbacks; for example, low surface area-to-volume ratio, large pressure drop, catalyst fouling, leaching, and cumber-

---

4 Doping in material science refers to a technique when a trace amount of impurities (doping agent) is added to the matrix, which alters the original compound's physical, electrical, and optical features. For example, silicon dopants change the Fermi level of silicon, leading to either *p*-or *n*-type semiconductors.

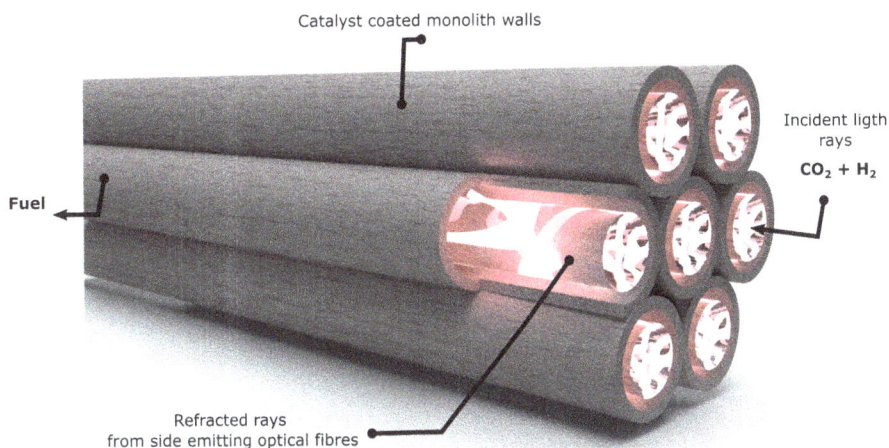

Figure 13.9: Single-channel honeycomb monolith with catalyst coating around an optical fiber. The light, CO_2, and water enter into the channel. The light is reflected several times in the optical fiber, which increases the quantum yield.

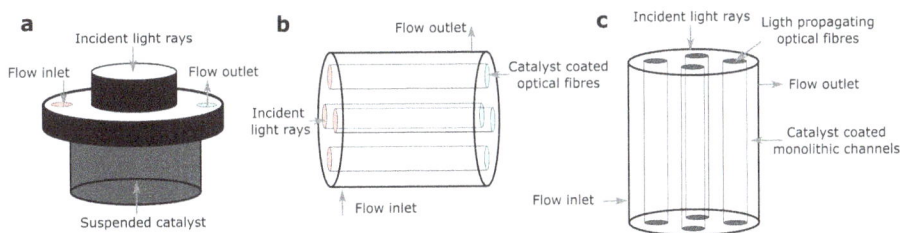

Figure 13.10: Schematic diagram of a photocatalytic slurry reactor. (a) Top illumination. (b) Optical fiber reactor design with side illumination. (c) Internal illumination.

some regeneration. The advantages and disadvantages of the slurry- and fixed-type reactors are detailed in Table 13.4.

13.4.2 Solar-driven advanced oxidation processes

Both heterogeneous and homogeneous systems have been developed for advanced solar oxidation processes (AOPs). Heterogeneous processes usually include $TiO_2/H_2O_2/$ UV or TiO_2/UV reaction matrix/catalyst compositions, and homogeneous systems usually utilize the Fenton process (UV/H_2O_2 or Fe/UV/H_2O_2) [31]. AOPs are radical chemical oxidation processes where the oxidizing agent is a hydroxyl radical (OH·). The hydroxyl radical redox potential is 2.8 V/(standard hydrogen electrode [SHE]). The presence of an unpaired radical makes the hydroxyl radicals very reactive, with half-lives of only a few nanoseconds. Generally, the hydroxyl radical destroys most living and

Table 13.4: Advantages and disadvantages of different reactor types in heterogeneous photocatalysis. Note that monolith and optical fiber reactors are subsets of fixed bed reactors.

Reactor type	Advantages	Disadvantages
Fluidized and slurry	Homogeneous temperature profile; Increased heat and mass transfer; High catalyst loading	Flooding; Corrosion and physical degradation; Low light utilization; Difficult catalyst separation; Additional filters and scrubbers; Restricted processing capabilities
Fixed bed	High surface area; Increased reaction rate; Low operating cost	Significant temperature gradient
Monolith	Low pressure drop; Easy to modify	Low quantum efficiency
Optical fiber	High surface area; High quantum yield	Low reactor volume utilization; Rapid degradation due to temperature discrepancies

non-living organic materials through excessive oxidation to CO_2, water, and other high-oxidation state molecules.[5] AOPs represent an efficient and environmentally friendly route to mitigate organic pollutants during wastewater treatment.

AOPs have been used for the treatment of wastewater in the textile industry [31]. For instance, in a pilot-scale study, dye-containing wastewater was oxidized using the solar TiO_2/H_2O_2 process in a compound parabolic collector. A similar parabolic collector is displayed in Figure 13.11. The reflectors help to increase the radiant flux and increase the temperature of the reaction. A typical pilot-scale solar-Fenton AOP can process around 30,000 L of textile wastewater in a day. Besides the UV/H_2O_2 system, the UV/chlorine process can achieve a similar or even better performance [32].

13.4.3 Hybrid advanced oxidation processes

The combination of photo- and sonoreactors presents an interesting *hybrid process intensification method*. Section 5.3.3 to learn about the exploitation of ultrasound in sustainable processes. This method simultaneously uses photo- and sono-irradiation to enhance the reaction rate and selectivity [35]. The catalyzed reaction is the selective oxi-

---

5 Interestingly, one of the liver's detoxifying process is based on a similar pathway. The liver removes unwanted by-products (and xenobiotics) from the body via oxidation. This oxidation increases the solubility of the degradation molecules in water, allowing them to be excreted in urine.

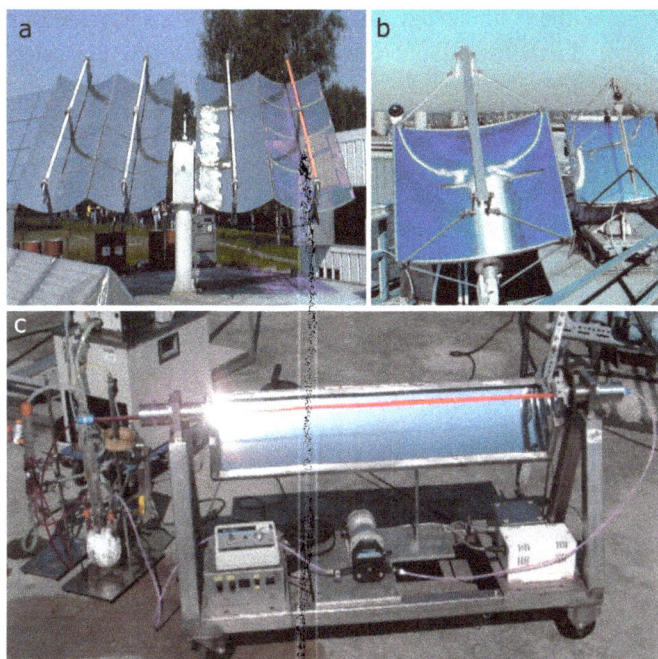

Figure 13.11: Different parabolic solar concentrators. (a) Parabolic Trough Facility for Organic Photochemical Synthesis (PROPHIS) in Cologne. (b) Solar line-focusing reactors on a roof. (c) Laboratory-scale horizontal parabolic trough reactor. Reprinted with permission from [33] and [34].

dation of benzyl alcohol to benzaldehyde using cobalt(III) oxide as a catalyst (Figure 13.12). The reaction can be performed in several solvents, for example, in water.

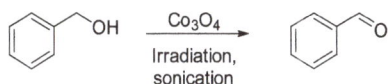

Figure 13.12: Selective oxidation of benzyl alcohol to benzaldehyde using irradiation and sonication.

The reactor contains the light source (Figure 13.13a, top illumination) and a cup horn, which provides the sonication and a cooling system to keep the temperature stable. Electromagnetic illumination is crucial to the system because, without light, the reaction cannot proceed. On the other hand, although sonication is not a crucial aspect, it nonetheless significantly improves the selectivity from about 50% to 80%. Moreover, at 365 nm, the overall conversion increases from 87% to 91% with the use of sonication. This system can be categorized as a slurry-type reactor; therefore, the smaller particles provide a larger surface area, which results in a higher conversion (higher quantum yield). Thus, the sonication increases the conversion by breaking down the catalyst particles and increasing the accessible surface area. Although this reaction was performed on a lab scale, the methodology can be scaled up or transferred to a flow reactor for increased productivity. Sonophotoreactors could be a potential selective alternative for the conversion of biomass.

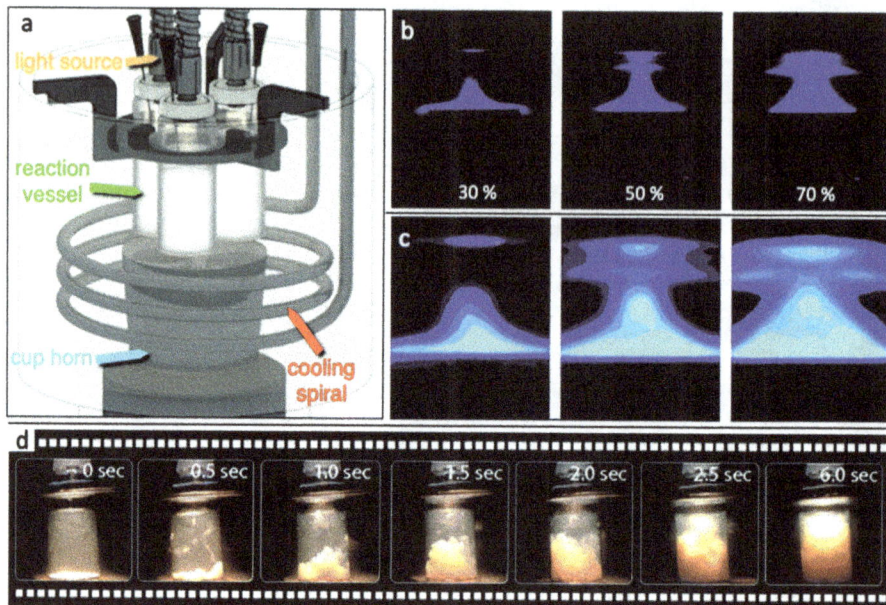

Figure 13.13: Hybrid process intensification through the synergistic combination of photocatalysis and sonochemistry. (a) Schematic representation of the sonophotoreactor (SoPhoReactor). (b) Real photograph during the testing phase (luminol test). (c) Power mapping of sonochemiluminescence to highlight the most intense regions of ultrasound waves derived from luminescence tests as a result of OH radical formation due to *acoustic cavitation*. (d) Actual photos of the reactor containing the catalyst at different times after starting the irradiations. Reprinted with permission from [35] with the Creative Commons Attribution-NonCommercial 3.0 Unported Licence.

13.4.4 Homogeneous photocatalysis

In homogenous photocatalysis, the light-absorbing catalytic units are dissolved in the reaction matrix [36]. The Lambert–Beer law (equation (6.3) in Chapter 6) describes the exponentially decaying electromagnetic radiation throughout the reactor.

The most well-known homogeneous photochemical approach is called photoredox catalysis. In photoredox catalysis, the catalyst undergoes a redox reaction in the excited state with the substrate before returning to its ground state. Organic photoredox catalysis is currently the most widely used photochemical conversion technique in the fine-chemical industry and is also a rapidly growing scientific field. Photoredox catalysis offers mild radical pathways and improved selectivity over classical radical reactions [37]. The general reaction mechanism for a photoredox cycle is shown in Figure 13.14. In oxidative quenching, the catalyst undergoes an oxidative cycle, while the substrate is reduced, and vice versa for reductive quenching. Every cycle also typically consumes a sacrificial reducer or oxidizer.

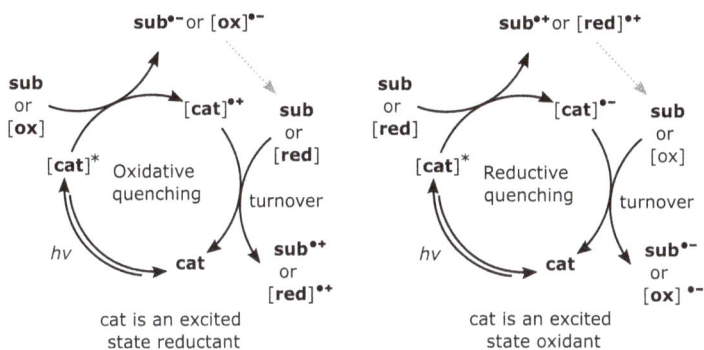

Figure 13.14: Photoredox pathways. (a) Oxidative quenching refers to the process when the catalyst undergoes reduction. (b) Reductive quenching refers to the process when the catalyst undergoes oxidation.

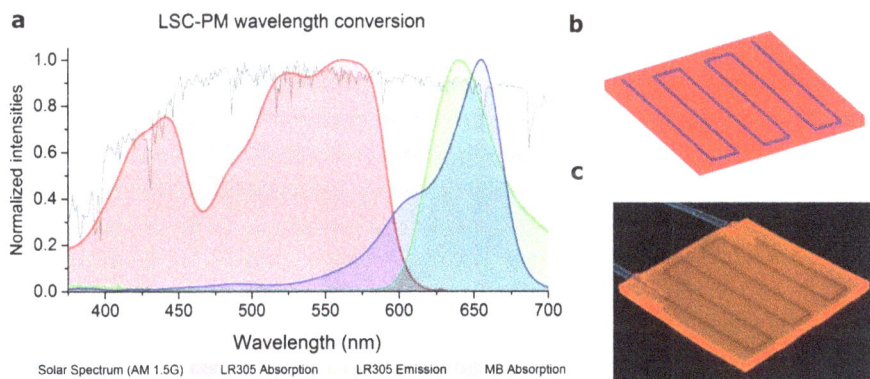

Figure 13.15: Core idea of luminescent solar concentrator reactors. (a) Different absorption and emission spectra of the LR305 device, the MB dye, and the solar spectrum. (b) Rendered image of a luminescent solar concentrator reactor. (c) Actual reactor. Reprinted with permission from Creative Commons CC BY [38] .

13.4.5 Luminescent solar concentrator reactors

Solar radiation (more precisely, extraterrestrial radiation) has a broad wavelength from 200 nm (UV region) to 1,000 nm and above (infrared region), as depicted in Figure 13.15. Since a specific catalytic reaction operates best at the absorption wavelength of the catalyst, the remaining energy from the solar radiation spectrum is wasted. Converting solar energy into different wavelengths could help to increase the overall efficiency of the system [38]. For example, the LR305 concentrator incorporates a patented dye with an absorption wavelength across the red region and an emitting spectrum in the blue region (Figure 13.15a). Thus, the LR305 device converts the energy (integral of the curve) from the red region and compresses it into the green region. This compression results in an increased intensity at approx. 630 nm. The reaction also exploits a special dye-type catalyst called Methylene Blue (MB), which has an absorption spectrum similar to the emission spectra of LR305. Note that the LR305 device has a patented dye inside, while the MB is dissolved in the reaction mixture. In the absence of the solar concentrator, the MB can only use the overlapping region with the solar spectrum (intersection of the gray and blue areas). When the LR305 solar concentrator is present, the red area is shifted and compressed into the green region (it also has a quantum efficiency). Therefore, the intersection of green and blue will be significantly higher than without a solar concentrator.

Nature-inspired chemical engineering is an approach that aims to provide innovative solutions to some of the twenty-first century's most significant challenges, in alignment with the UN's Sustainable Development Goals. Follow the QR code on this page to learn about the pioneering work in the field at CfNIE under the direction of Prof. Marc-Olivier Coppens. A nature-inspired (in particular leaf-inspired) self-optimizing reactor can overcome the typical variations in solar radiation during the day. The device (Figure 13.16c, placed on a rooftop) can adjust the flow rate of the reactants in real-time according to the actual direct radiation output using a light sensor. The device is semi-sustaining, meaning that it only requires the reactant vessels to be refilled. Such devices can facilitate the remote and cloud-controlled production of pharmaceuticals and other fine chemicals [39].

13.4.6 Cloud-inspired photochemical reactor

Atmospheric photochemical reactions occur in cloud water droplets in the air, which is how, for example, atmospheric ozone is generated from oxygen. Due to refraction and reflection caused by these atmospheric droplets, the average path length of light on a cloudy day is significantly higher than in clear weather (Figure 13.17a). This phenomenon results in an increased reaction rate when water droplets are present in the air. Nebulizer-type reactors mimic the conditions present in clouds by accelerating photooxidation using a pneumatically generated aerosol. However, the large-scale application of these useful and innovative reactors has yet to be demonstrated [40].

Another more robust and similarly efficient reactor has been constructed using a glass close-packed bed reactor (multiple scattering reactor). This reactor uses a high-refractive index contrast liquid–solid interface between the matrix of the reaction and the glass beads (Figure 13.17b). By using a light-scattering flow reactor, the absorption efficiency increases up to 0.19, which is three times higher than the absorp-

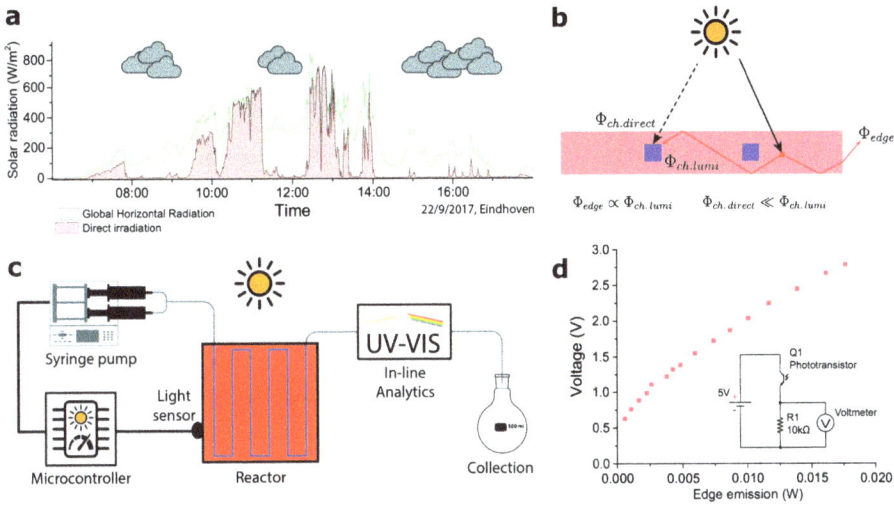

Figure 13.16: (a) Global radiation variation during a day in Eindhoven, the Netherlands. (b) Working principle of solar concentrators. (c) Microcontroller sensory-driven reactor directly utilizing sunlight. (d) Sensor voltage output with respect to edge emission. Reprinted with permission from Creative Commons CC BY [38].

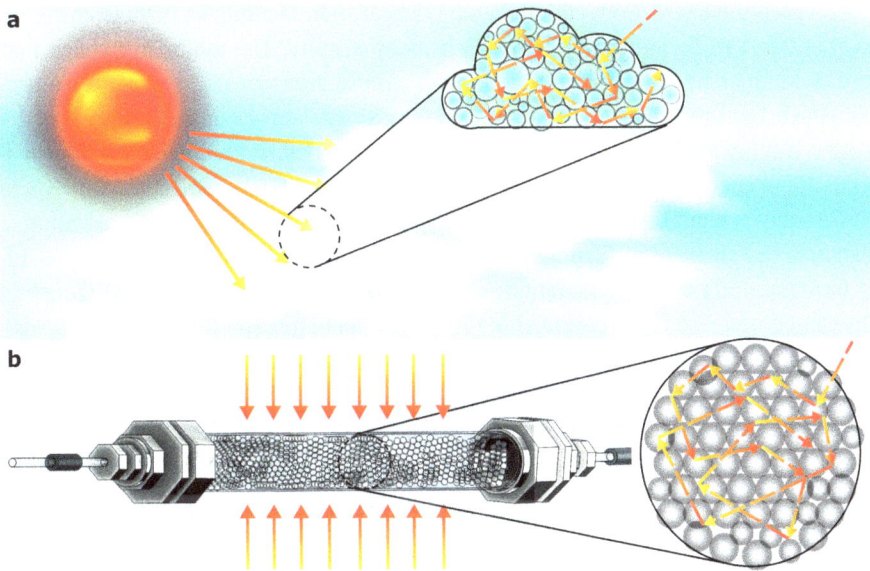

Figure 13.17: (a) Schematic representation of the refraction and reflection effects present in nature. (b) Biomimicking the clouds using a glass-packed bed reactor. Reprinted with the Creative Commons CC BY license from [41].

tion efficiency achievable using a similar setup without the scattering glass beads. Interestingly, doubling the diameter of the glass beads results in a 1.5-fold increase in light absorption efficiency [41]. The increased absorption induces a similar overall production rate increase in real-life testing. Using a 7-cm-long scattering reactor with a 10 mm bead, a production rate of 1.7 g day^{-1} can be achieved at the lab scale.

Even though photochemistry and photoredox chemical operations utilize electromagnetic radiation as a green and sustainable energy source, they still encounter a number of *challenges* to overcome. For example, the most common photoredox catalyst contains iridium and palladium metal ligands and due to photobleaching, these catalysts are usually not recoverable or reusable. Using organocatalytic photoredox catalysts (such as MB or the catalysts from Figure 6.11) could resolve this bottleneck. Furthermore, batch-type photocatalysis is not practical on a large scale. However, both the pharmaceutical industry and the fine-chemical industry are gradually switching to continuous flow reactors to resolve this issue (Chapter 6). Such continuous microreactors have excellent quantum efficiency and are easy to scale up.

13.4.7 Chiral separation using light

The separation of enantiomers[6] with the same physical features requires an external chiral-separating agent. Conventional enantioseparation techniques include chiral resolvation via salt formation and chiral chromatography. The major drawback that these techniques have in common is the need to use a *chiral agent* during the process, which requires an investment that increases costs and eventually decreases the E-factor. Additives such as chiral agents should be avoided, according to the 12 principles of green chemistry and the 12 principles of green engineering (Section 1.3).

The replacement of chiral agents by using light instead may provide a sustainable solution to the inherent difficulties encountered in enantioseparation. In Section 13.4.1, the fundamentals of electromagnetic radiation were introduced. Briefly, traveling light can be described in terms of three major components: the direction of propagation, the amplitude, and the electric component that can interact with matter. *Circular polarized light* (CPL) has chirality, which results in different interactions with enantiomer pairs. For example, naturally occurring sugars are chiral (e.g., glucose, sucrose). If CPL light shines through a solution of glucose, the light is diffracted, thus indicating a *chiral light–matter interaction* [42]. Note that there are two types of optical interactions between matter and light: achiral (not chiral), which does not differentiate between stereoisomers, and chiral, whose direction is opposite to that of the direction of the matter [43]. If chiral matter interacts with a monochromatic field (polar-

---

6 Enantiomers are mirror images of each other and are non-superposable (they are called stereoisomers).

ized light), the interaction forces can be calculated [43]. For two enantiomers, the force vectors point in the opposite direction. Thus, in principle, the separation of a racemic mixture (i.e., when both enantiomers are present in a 1:1 ratio) is thermodynamically feasible. The idea of separating racemic mixtures using light is a new concept, which has only been studied in theory and has not yet been experimentally validated (Figure 13.18). The importance of enantioseparation and the inherent chirality of light make it an appealing field of research. The realization of this concept and the subsequent technology is expected to result in a paradigm shift in enantioseparation and its sustainability.

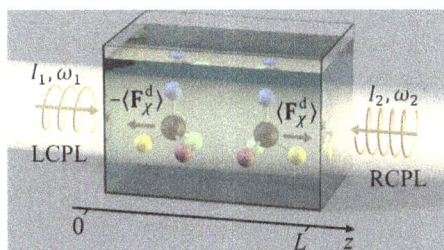

Figure 13.18: Separation of a racemic mixture using chiral light. The two enantiomers are traveling in the opposite direction due to the opposing forces generated by the left and right circular polarized light (LCPL and RCPL). Reprinted with permission from the Creative Commons CC BY license from [43].

Bibliography

[1] International – U.S. Energy Information Administration (EIA) https://www.eia.gov/international/data/world#/?tl_type=p&tl_id=44-A&pa=000000001&ct=0&ord=SA&c=ruvvvvvfvtvnv vlurvvvvfvvvvvvfvvvou20evvvvvvvvvvnvvuvo&f=A (accessed Jun 5, 2020).

[2] Available Solar Radiation. *Solar Engineering of Thermal Processes*. April 2013, pp 43–137. https://doi.org/10.1002/9781118671603.ch2.

[3] Solar Radiation. *Solar Engineering of Thermal Processes*. April 2013, pp 3–42. https://doi.org/10.1002/9781118671603.ch1.

[4] Cota, A.; Foster, R.; Ghassemi, M. *Solar Energy – Renewable Energy and the Environment*; 1st Edition, CRC Press, 2009. 10.1201/9781420075670.

[5] Kim, H.; Yang, S.; Rao, S. R.; Narayanan, S.; Kapustin, E. A.; Furukawa, H.; Umans, A. S.; Yaghi, O. M.; Wang, E. N. Water Harvesting from Air with Metal-Organic Frameworks Powered by Natural Sunlight. *Science (80)*. **2017**, *356* (6336), 430–434. 10.1126/science.aam8743.

[6] Lee, A.; Moon, M.-W.; Lim, H.; Kim, W.-D.; Kim, H.-Y. Water Harvest via Dewing. *Langmuir*. **2012**, *28* (27), 10183–10191. 10.1021/la3013987.

[7] Park, K.-C.; Chhatre, S. S.; Srinivasan, S.; Cohen, R. E.; McKinley, G. H. Optimal Design of Permeable Fiber Network Structures for Fog Harvesting. *Langmuir*. **2013**, *29* (43), 13269–13277. 10.1021/la402409f.

[8] Al-Karaghouli, A.; Kazmerski, L. L. Energy Consumption and Water Production Cost of Conventional and Renewable-Energy-Powered Desalination Processes. *Renew. Sustain. Energy Rev*. **2013**, *24*, 343–356. 10.1016/j.rser.2012.12.064.

[9]   Zaragoza, G.; Andrés-Mañas, J. A.; Ruiz-Aguirre, A. Commercial Scale Membrane Distillation for Solar
      Desalination. *Npj Clean Water*. **2018**, *1* (1), 20. 10.1038/s41545-018-0020-z.
[10] Wang, W.; Shi, Y.; Zhang, C.; Hong, S.; Shi, L.; Chang, J.; Li, R.; Jin, Y.; Ong, C.; Zhuo, S.; Wang,
      P. Simultaneous Production of Fresh Water and Electricity via Multistage Solar Photovoltaic
      Membrane Distillation. *Nat. Commun.* **2019**, *10* (1), 3012. 10.1038/s41467-019-10817-6.
[11] Chiavazzo, E.; Morciano, M.; Viglino, F.; Fasano, M.; Asinari, P. Passive Solar High-Yield Seawater
      Desalination by Modular and Low-Cost Distillation. *Nat. Sustain.* **2018**, *1* (12), 763–772. 10.1038/
      s41893-018-0186-x.
[12] Fang, Q.; Li, G.; Lin, H.; Liu, F. Solar-Driven Organic Solvent Purification Enabled by the Robust Cubic
      Prussian Blue. *J. Mater. Chem. A*. **2019**, *7* (15), 8960–8966. 10.1039/c9ta00798a.
[13] Sonawane, P. D.; Bupesh Raja, V. K. An Overview of Concentrated Solar Energy and Its Applications.
      Int. J. Ambient Energy. **2018**, *39* (8), 898–903. 10.1080/01430750.2017.1345009.
[14] Albarbar, A.; Arar, A. Performance Assessment and Improvement of Central Receivers Used for Solar
      Thermal Plants. *Energies*. **2019**, *12* (16). 10.3390/en12163079.
[15] Abanades, S.; Charvin, P.; Flamant, G.; Neveu, P. Screening of Water-Splitting Thermochemical
      Cycles Potentially Attractive for Hydrogen Production by Concentrated Solar Energy. *Energy*. **2006**,
      31 (14), 2805–2822. 10.1016/j.energy.2005.11.002.
[16] Romero, M.; Steinfeld, A. Concentrating Solar Thermal Power and Thermochemical Fuels. *Energy
      Environ. Sci.* **2012**, *5* (11), 9234–9245. 10.1039/C2EE21275G.
[17] Quoilin, S.; Van Den Broek, M.; Declaye, S.; Dewallef, P.; Lemort, V. Techno-Economic Survey of
      Organic Rankine Cycle (ORC) Systems. *Renew. Sustain. Energy Rev.* **2013**, *22*, 168–186. 10.1016/j.
      rser.2013.01.028.
[18] Viteri, F.; Anderson, R. E. Semi-Closed Brayton Cycle Gas Turbine Power Systems. **2003**, *2* (12).
[19] Soloveichik, G. L. Flow Batteries: Current Status and Trends. *Chem. Rev.* **2015**, *115* (20), 11533–11558.
      10.1021/cr500720t.
[20] Romero, N. A.; Nicewicz, D. A. Organic Photoredox Catalysis. *Chem. Rev.* **2016**, *116* (17), 10075–10166.
      10.1021/acs.chemrev.6b00057.
[21] Bates, R. *Photochemistry of Organic Synthesis Using Transition Metals*; 2009.
[22] Yoon, T. P.; Ischay, M. A.; Du, J. Visible Light Photocatalysis as a Greener Approach to Photochemical
      Synthesis. *Nat. Chem.* **2010**, *2* (7), 527–532. 10.1038/nchem.687.
[23] Cambié, D.; Zhao, F.; Hessel, V.; Debije, M. G.; Noël, T. A Leaf-Inspired Luminescent Solar
      Concentrator for Energy-Efficient Continuous-Flow Photochemistry. *Angew. Chem. Int. Ed. Engl.* **2017**,
      56 (4), 1050–1054. 10.1002/anie.201611101.
[24] Herrmann, J.-M. Heterogeneous Photocatalysis: Fundamentals and Applications to the Removal of
      Various Types of Aqueous Pollutants. *Catal. Today*. **1999**, *53* (1), 115–129. 10.1016/S0920-5861(99)
      00107-8.
[25] Tittor, J.; Oesterhelt, D. The Quantum Yield of Bacteriorhodopsin. *FEBS Lett*. **1990**, *263* (2), 269–273.
      10.1016/0014-5793(90)81390-A.
[26] Su, Y.; Hessel, V.; Noël, T. A Compact Photomicroreactor Design for Kinetic Studies of Gas-Liquid
      Photocatalytic Transformations. *AIChE J*. **2015**, *61* (7), 2215–2227. 10.1002/aic.14813.
[27] Qin, G.; Zhang, Y.; Ke, X.; Tong, X.; Sun, Z.; Liang, M.; Xue, S. Photocatalytic Reduction of Carbon
      Dioxide to Formic Acid, Formaldehyde, and Methanol Using Dye-Sensitized TiO2 Film. *Appl. Catal.
      B. Environ.* **2013**, *129*, 599–605. 10.1016/j.apcatb.2012.10.012.
[28] Marcì, G.; García-López, E. I.; Palmisano, L. Photocatalytic CO_2 Reduction in Gas–Solid Regime in the
      Presence of H_2O by Using GaP/TiO_2 Composite as Photocatalyst under Simulated Solar Light. *Catal.
      Commun.* **2014**, *53*, 38–41. 10.1016/j.catcom.2014.04.024.
[29] Hussain, M.; Akhter, P.; Saracco, G.; Russo, N. Nanostructured TiO_2/KIT-6 Catalysts for Improved
      Photocatalytic Reduction of CO_2 to Tunable Energy Products. *Appl. Catal. B. Environ.* **2015**, *170–171*,
      53–65. 10.1016/j.apcatb.2015.01.007.

[30] Phongamwong, T.; Chareonpanich, M.; Limtrakul, J. Role of Chlorophyll in Spirulina on Photocatalytic Activity of CO2 Reduction under Visible Light over Modified N-Doped TiO_2 Photocatalysts. *Appl. Catal. B Environ*. **2015**, *168–169*, 114–124. 10.1016/j.apcatb.2014.12.022.

[31] Vilar, V. J. P.; Pinho, L. X.; Pintor, A. M. A.; Boaventura, R. A. R. Treatment of Textile Wastewaters by Solar-Driven Advanced Oxidation Processes. *Sol. Energy*. **2011**, *85* (9), 1927–1934. 10.1016/j.solener.2011.04.033.

[32] Chan, P. Y.; Gamal El-Din, M.; Bolton, J. R. A Solar-Driven UV/Chlorine Advanced Oxidation Process. *Water Res*. **2012**, *46* (17), 5672–5682. 10.1016/j.watres.2012.07.047.

[33] Cambié, D.; Bottecchia, C.; Straathof, N. J. W.; Hessel, V.; Noël, T. Applications of Continuous-Flow Photochemistry in Organic Synthesis, Material Science, and Water Treatment. *Chem. Rev*. **2016**, *116* (17), 10276–10341. 10.1021/acs.chemrev.5b00707.

[34] Oelgemöller, M. Solar Photochemical Synthesis: From the Beginnings of Organic Photochemistry to the Solar Manufacturing of Commodity Chemicals. *Chem. Rev*. **2016**, *116* (17), 9664–9682. 10.1021/acs.chemrev.5b00720.

[35] Giannakoudakis, D. A.; Łomot, D.; Colmenares, J. C. When Sonochemistry Meets Heterogeneous Photocatalysis: Designing a Sonophotoreactor Towards Sustainable Selective Oxidation. *Green Chem*. **2020**, *22* (15), 4896–4905. 10.1039/d0gc00329h.

[36] Zhu, S.; Wang, D. Photocatalysis: Basic Principles, Diverse Forms of Implementations and Emerging Scientific Opportunities. *Adv. Energy Mater*. **2017**, *7* (23), 1700841. 10.1002/aenm.201700841.

[37] Jin, J.; MacMillan, D. W. C. Direct *α*-Arylation of Ethers through the Combination of Photoredox-Mediated C–H Functionalization and the Minisci Reaction. *Angew. Chem. Int. Ed. Engl*. **2015**, *54* (5), 1565–1569. 10.1002/anie.201410432.

[38] Cambié, D.; Noël, T. Solar Photochemistry in Flow. *Top Curr Chem*. **2018**, *376* (6), 45. 10.1007/s41061-018-0223-2.

[39] Skilton, R. A.; Bourne, R. A.; Amara, Z.; Horvath, R.; Jin, J.; Scully, M. J.; Streng, E.; Tang, S. L. Y.; Summers, P. A.; Wang, J.; Pérez, E.; Asfaw, N.; Aydos, G. L. P.; Dupont, J.; Comak, G.; George, M. W.; Poliakoff, M. Remote-Controlled Experiments with Cloud Chemistry. *Nat. Chem*. **2015**, *7* (1), 1–5. 10.1038/nchem.2143.

[40] Ioannou, G. I.; Montagnon, T.; Kalaitzakis, D.; Pergantis, S. A.; Vassilikogiannakis, G. A Novel Nebulizer-Based Continuous Flow Reactor: Introducing the Use of Pneumatically Generated Aerosols for Highly Productive Photooxidations. *ChemPhotoChem* **2017**, *1* (5), 173–177. 10.1002/cptc.201600054.

[41] Zheng, L.; Xue, H.; Wong, W. K.; Cao, H.; Wu, J.; Khan, S. A. Cloud-Inspired Multiple Scattering for Light Intensified Photochemical Flow Reactors. *React. Chem. Eng*. **2020**, *5* (6), 1058–1063. 10.1039/d0re00080a.

[42] Du, W.; Wen, X.; Gérard, D.; Qiu, C.-W.; Xiong, Q. Chiral Plasmonics and Enhanced Chiral Light-Matter Interactions. *Sci. China Phys. Mech. Astron*. **2019**, *63* (4), 244201. 10.1007/s11433-019-1436-4.

[43] Rukhlenko, I. D.; Tepliakov, N. V.; Baimuratov, A. S.; Andronaki, S. A.; Gun'Ko, Y. K.; Baranov, A. V.; Fedorov, A. V. Completely Chiral Optical Force for Enantioseparation. *Sci. Rep*. **2016**, *6*. 10.1002/anie.201410432.

14 Data-driven optimization of chemical processes

Andres Cardenas, Glenda T. Cuadrado, Kenia M. Martinez, Mohammed Ali,
Marwan Khayyat, Gyorgy Szekely

Chemical engineers and chemists aim to minimize the variability and maximize the efficiency of industrial processes. Process analytical technologies (PATs) are real-time analytical tools used to understand and control these processes effectively and safely. Refer to Chapter 9 to learn more about PAT.

Nevertheless, the next generation of process intensification is expected to achieve system-level transformations that use data-driven algorithms and *artificial intelligence* (AI) *techniques* to improve optimization, predictive control, and equipment design. *Machine learning* (ML) as an AI technique has enhanced intensified systems and processes in the next-generation approach and constitutes an essential component of *industry 4.0*. The latter concept was introduced in the 2010s, where the operational objective is the base of process systems within an interconnected framework.

Due to population growth and environmental concerns, chemical and biological processes have had to scale up and diversify considering the harm that could be caused to nature and the environment. Connecting these problems with the resources offered by data-driven technologies allows us to find a better answer to these troubling situations. Diverse approaches that use ML, AI, and Industry 4.0 are integrated to transform current industrial operations, optimize them, and achieve technological, economic, and sustainable innovations. One of these approaches incorporates autonomous processes, in particular, *autonomous experimentation platforms* (AEPs), which is a new research field that integrates "learned from data" algorithms with automated experimentations, on-the-fly data analysis, and decision-making algorithms. *Fault detection and diagnosis* (FDD) systems are another ML application, and they contribute to the intelligent diagnosis of the health status of equipment based on a model built with self-diagnostic information collected by specific devices (e.g., sensors). The data and insights gathered from the AEP- and FDD-based methods enable self-optimizing systems to continuously adjust and fine-tune their strategies for improving system performance based on real-time experimentation results. To achieve this, data-driven knowledge obtained using FDD approaches serves as input to inform decision-making algorithms that are designed to enhance and optimize chemical processes. Another analysis-based model is production scheduling, that includes the management of timing, sequencing decisions, and processing models that have technical, economic, and environmental constraints. *Refinery production scheduling* (RPS) integrates modelers to produce optimized production schedules that play a central role in refining business and the industry 4.0 environment.

Figure 14.1 illustrates the application of AEPs, FDD systems, and RPS, which addresses principles from *green chemistry* (prevention, atom economy, less hazardous chemical synthesis, designing for energy efficiency, real-time analysis for pollution, and safer chemistry for accident prevention) and *green chemical engineering* (integration of ma-

https://doi.org/10.1515/9783111028163-014

Green Chemistry
Safer Chemistry for Accident Prevention
Real-time Analysis for Pollution Prevention
Design for Degradation
Catalysis
Reduce Derivatives
Use of Renewable Feedstocks
Design for Energy Efficiency
Safer Solvents and Auxiliaries
Design Safer Chemicals
Less Hazardous Chemical Synthesis
Atom Economy
Prevention

Green Engineering
Inherent Rather Than Circumstantial
Prevention Instead of Treatment
Design for Separation
Maximize efficiency
Output pulled vs. input-pushed
Conserve Complexity
Durability Rather Than Immortality
Meet Need, Minimize Excess
Minimize Material Diversity
Integrate Material and Energy Flows
Design for Commercial Afterlife
Renewable Rather Than Depleting

Figure 14.1: Interconnectivity between the principles of green chemistry and green chemical engineering for the application of autonomous experimentation platforms (AEPs), fault detection and diagnosis (FDD) systems, and refinery production scheduling (RPS). The figure has been taken from Szekely (2021) and modified [1].

terial and energy flows, meeting needs, minimizing excess, maximizing efficiency, and designing for separation and prevention instead of treatment). AEPs meet safety and sustainability goals because they considerably reduce hazardous conditions for workers and/ or researchers during the undertaking of chemical processes. Moreover, the emission of greenhouse gases associated with process shutdowns and startups can be avoided and waste and carbon footprints can be reduced. The use of self-optimizing systems enables companies and laboratories to achieve these sustainability principles by reducing energy use, optimizing the reuse of resources and materials, preventing the release of greenhouse gases, and maximizing renewable fuel sources. FDD systems realize reliable and sustainable chemical industrial processes by preventing downtime operations, minimizing safety risks, reducing the unwanted disposal of hazardous materials, and optimizing resource utilization. Moreover, RPS may facilitate a decrease in energy consumption in long-term operations and the production of higher volumes with an environmental impact that is less detrimental. A more detailed study of the data-driven technologies, their applications, and practical case studies are discussed in the following sections.

14.1 Self-optimizing systems

As industry and manufacturing face a context of volatility, uncertainty, complexity, and ambiguity (VUCA), it is necessary to shift from the traditional research model to a new one in response to extreme VUCA conditions. The use of high-throughput screening platforms, such as ML, *deep learning* (DL), and/or AI, represents the path to enhancing the efficiency and reproducibility of chemical processes. Implementing an autonomous process can liberate researchers from routine tasks and, thus, accelerate industrial development to new levels of safety, sustainability, and profitability. *Automation* technology has been adopted in both academia and industry; for example, important applications include flow chemistry and the computer control of reactors [2]. Nevertheless, the enormous amount of data analysis, the decision-making still made by researchers or engineers, and the high dimensionality of problems with the process represent the main challenges to achieving a fully autonomous process [3]. An AEP is a new research paradigm that takes a closed-loop approach to integrate design, synthesis, characterization, testing, decision-making, and optimization into a more integrated pipeline with the use of "learned from data" algorithms and automation [3, 4]. Although efforts have been made to incorporate AI capabilities into existing operating technologies (follow the QR code on this page for Aspen Technology Inc.'s ongoing project), the AEP is currently being used for designing experiments (DoE) to be used by self-driving laboratories to achieve autonomous experimentation [3, 5]. An AEP has shown great potential for synthesizing chemical molecules and materials development [6–8]. Using the AI systems and strategies proposed in these works, insights into understanding the links between chemistry and information technology are provided. We expect that this report will serve as a guideline for understanding the principles used by AEPs and decision-making algorithms (such as *stable noisy optimization by branch and fit* (SNOBFIT)) for improving and optimizing chemical processes.

14.1.1 Autonomous experimentation platforms

Xie et al. gathered the essential components of an AEP. The components of a closed-loop self-driving lab are automated experiments, on-the-fly data analysis, and decision-making [3].

1. *Automated experimentation*: This refers to synthesis and processing platforms that are automated to have high efficiency by controllable and programmable equipment. PAT methods, such as for in situ and on-site characterization, can be integrated within these platforms [3]. PAT analyzers for spectroscopy (UV–Vis/IR, IR/PL, GC–MS, LC–MS, HPLC, NMR, EPR, etc.) and microscopy (AFM, SEM, TEM, SPM, STM, STEM, PFM, OM, digital imaging, etc.) provide a continuous stream of real-time data for characterizing the properties of materials and classifying the results of reactions. More information about PAT concepts and principles can be found in Chapter 9.
2. *On-the-fly data analysis*: Raw data can be automatically analyzed through ML/DL to qualify and quantify the outcome of the experiment and to provide real-time feedback that accelerates the development of discoveries.
3. *Decision-making*: Intelligent decision-making algorithms learn from previous results and suggest new parameters for the next experiment. They find the shortest path to the global optimum within the collected data, which is especially helpful for high-dimensional problems.

Among the many automated synthesis and processing platforms, the most commonly used are continuous flow reactors (CFRs) and desktop and mobile robots [8–11]. CFRs are a key technology for high-throughput experimentation. This reactor configuration consists of reactants being continuously fed into a tube that has a small diameter, which enables fast mass transport, efficient heat transfer, and the reduction of solvents and reactants [3]. Flow processes have also been demonstrated to increase the reaction selectivity and reproducibility of various chemistries [2], for instance, the utility of flow reactors in PET reversible addition–fragmentation chain transfer polymerization [12] or in photoredox-mediated metal-free ring-opening metathesis polymerization [13] improves the scalability and reliability of the processes. For more information, refer to Chapter 6.3, which discusses the green attributes of continuous flow processes and gives greater insight into their core features and how they impact flow chemistry [1].

Another way to perform automated experiments is with a desktop robot, widely used in life sciences and drug discovery. Desktop robots can handle, prepare, and process solid and liquid chemicals just before the reactions [3]. Compared to nonautonomous platforms, desktop robots can perform several operations in parallel, which increases the throughput of experimentation. For example, Grizou et al. developed a robotic platform called "Dropfactory" to study self-propelling, multicomponent oil-in-water protocell droplets [9]. They showed that the robotic platform could observe additional physicochemical parameters and perform highly reproducible experiments. Using the robotic platform, they were also able to increase the number of experiments per day from 50 to 300 compared to their previous nonautonomous platforms by performing the stages of preparation in parallel rather than sequentially. Two spinning wheels are used for mixing, droplet placing, recording, cleaning, and drying the oil and aqueous phases separately. Figure 14.2 illustrates the principles of Dropfactory. Follow the QR code on this page to watch Dropfactory in operation.

Mobile robots are another type of automated experimentation device that involves human-like operator researchers that can move freely between different components of a synthesis lab and can overcome the disadvantages of performing synthesis and characterizations in physically large and functionally complex equipment that require a special working environment (i.e., light-sensitive photochemical reactions) [3, 14]. They show superiority in consistency and efficiency by performing neat manipulations at various stations in the laboratory and by handling toxic and explosive chemicals [3]. Several advances in industrial robotics have been made by companies such as KUKA. Burger et al. used a mobile robot from these manufacturers to improve the synthesis of photocatalysts for producing hydrogen from water [14]. The robot locates its position in the laboratory using a combination of laser scanning and touch feedback, which allows it to operate in complete darkness. In addition, this robot operates for up to 22 h per day and requires no modifications to the laboratory space. The authors described five hypotheses for increasing the hydrogen evolution rate (HER) of their catalysts: (1) dye sensitization may improve light absorption and hence the HER; (2) pH may influ-

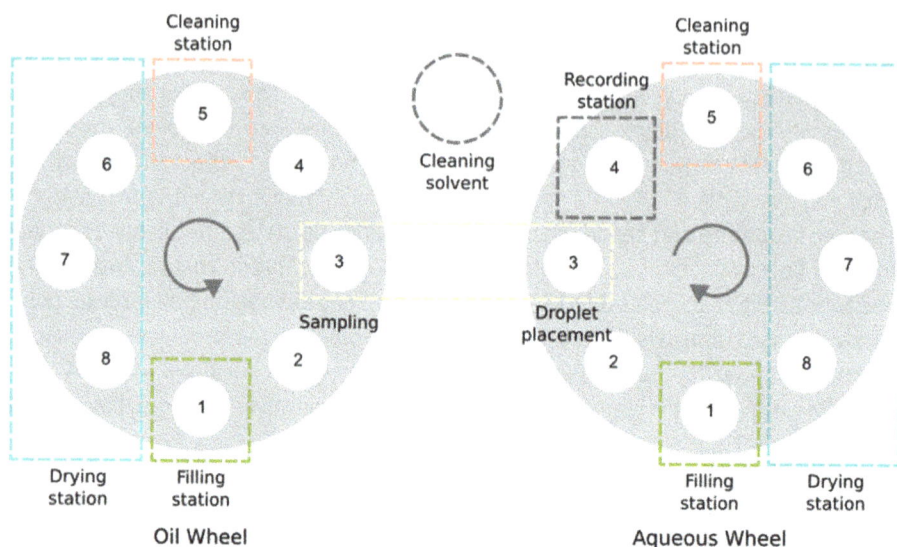

Figure 14.2: Conceptual design of Dropfactory. Taken from Grizou et al. [9] mentioned in the "Robotic platform: Dropfactory" section in the Supplementary Materials.

ence the catalytic activity; (3) ionic strength may also be important to the activity; (4) the addition of surfactants would affect the catalyst wettability; and (5) sodium disilicate may act as a hydrogen-bonding anchor. With the aid of the mobile robot, they were able to perform over 600 experiments to explore these five hypotheses, something a human researcher would have taken several months to complete. They claim that this autonomous workflow requires a half-day setup before running experiments 24/7 without any user input. As such, the manual method is only capable of performing two experiments per day, while the autonomous workflow is 500–1,000 times faster. Figure 14.3 shows a comparison of the time needed to perform 1,000 photocatalysis experiments using different methodologies, assuming a five-day working week.

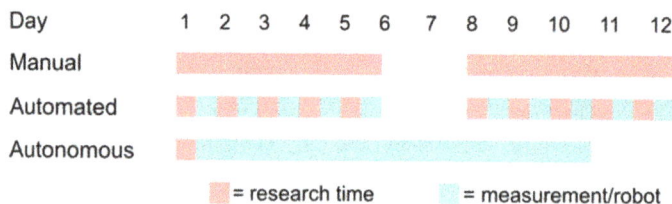

Figure 14.3: Comparison between manual, automated, and autonomous methods for photocatalysis experiments performed by Burger et al. [14] taken from the "Timescales for Steps in the Workflow" section in the Supplementary Information.

Although the use of automated experimentation devices can increase experimental throughput, the need for user-driven experiment design still limits their optimization [11]. This is because the methodology used for experiment design does not change with the rate at which experiments are performed on the said devices [9]. Characterization data must be analyzed to present and visualize the results for decision-making. Therefore, it is necessary to integrate ML/DL into the characterization instruments to achieve quick and efficient on-the-fly data analysis.

After the data are collected from the PAT analyzers, the next step in a typical workflow for constructing ML/DL models is to preprocess the data. This step involves processing the information related to the hypothesized problem and converting it to quantifiable data that can be read by a computer. For example, in the preprocessing of spectroscopic data, raw spectrum data should be converted to same-length vectors and normalized between 0 and 1. In addition, background noise should be removed by smoothing the spectra and correcting the baseline [3]. With the aid of ML/DL models, the classification of LC–MS spectral peaks [15], peak picking of GC–MS data [16], DL of accelerated NMR [17], validation of functional groups [18], and classification of XRD data sets [19] are possible. In contrast, DL models such as convolutional neural networks are widely used for preprocessing microscopic data because they perform better at image recognition, segmentation of areas of interest, reconstruction of hidden information, and retrofitting of new information from images [3]. Microscopic data preprocessing techniques can help in the identification of nanostructures [20], determination of crystal symmetry [21], characterization of the morphologies of nanoparticles [22], detection of chiral molecules [23], layer mapping of 2D materials [24], and perovskite crystallization [25]. Thus, on-the-fly data analysis can provide insightful treated information about the AEPs and generate new knowledge that can be used to achieve predefined targets for the following optimization step.

Automated experiments and on-the-fly analysis are the initial steps of the correct and complete optimization of chemical reactions. Unoptimized chemical reactions are considered costly and inefficient, regardless of the information obtained. Conventional approaches to achieving optimization include single-condition experiments or batch chemistry and exhaustive combinations of experimental conditions [26]. A new approach that uses *intelligent decision-making algorithms* has emerged as an efficient pathway to exploring and capturing global optima with reduced operational time and cost. Based on previous results and observations, a new set of optimal candidate parameters can be proposed for the subsequent experiment [3]. Coley et al. stated that there is no threshold between automation and autonomy. Nevertheless, autonomy involves some degree of decision-making for adapting to unexpected outcomes [27]. The field of decision-making algorithms is developing rapidly, and some algorithms have been implemented in chemical processes: *Bayesian optimization* (BO), *reinforcement learning* (RL), and *SNOBFIT* [3]. BO balances the use of *uncertainty* and *available information*, thereby obtaining high-quality configurations within fewer evaluations [28]. BO uses a surrogate model and an acquisition function, which can be correlated with an objective function and the desirability of sampling the next candidate, respectively. It offers many advantages, such as a reduced number of experiments to be tested, a great noise tolerance, and a balanced exploration between the best local optimum and a highly uncertain global optimum [3]. Figure 14.4 presents a graphical overview of the BO method.

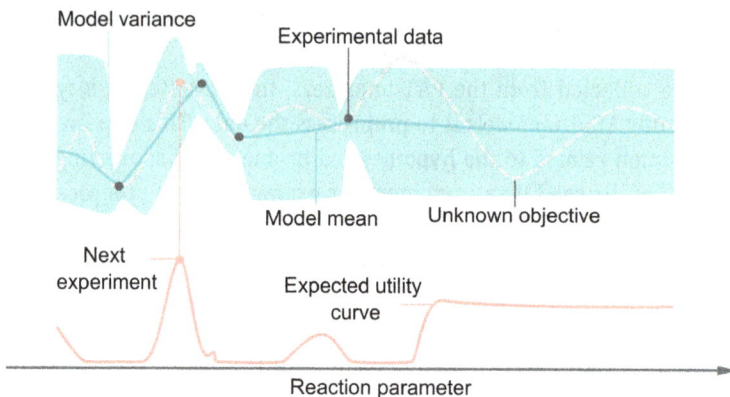

Figure 14.4: Graphical overview of Bayesian optimization for selecting the next experiment based on fitting a one-dimensional Gaussian process surrogate model. Taken and adapted from Shields et al. [28].

Shields et al. reported the fluorination of organic compounds in the drug industry as a perfect case for the application of BO [28]. They proposed the fluorination of an organic compound with a reaction space having ~312,500 possible configurations, defined by a set of fluorides, organic bases, solvents, and a grid of continuous parameters. Control reactions using conditions that were typical of commercial fabrication yielded an average yield of 36%. BO was applied to this process with a set of five experiments running in parallel per batch, with the initial experiments randomly chosen. It surpassed the benchmark result, with a 69% product yield under the identified reaction conditions. BO is widely used as an efficient tool for problems with few tunable parameters. However, the application of BO to processes with high dimensions and thousands of observations presents a challenge in finding the global optimum [29]. An alternative is the *multi-objective optimization* (MOO). This method does not consider a single point search space but rather a set of points, named a Pareto set, by transforming multiple objectives into one objective [3]. Diverse algorithms have been proposed for MOO objectives, such as TuRBO [29], Chimera [30], and TSEMO [31].

ML and AI have been implemented in the fields of chemistry and chemical engineering to obtain predictions based on experimental data and applied decision-making frameworks [26]. RL applies to dynamic problem-solving. A sequence of actions in a given domain is performed by the RL algorithm, which estimates the statistical relation between these actions and their possible outcomes to maximize a reward function that maps a specific experimental condition to a reaction result [3]. It is important to note that at every step, an action refers to all the changes that can be made to the experimental conditions; a state can include a full set of possible combinations of experimental parameters; a probability connects the transition from one state *s1* to another state *s2* with an action *a*; and the reward (R) is based on the state and action of the reaction [26].

Zhou et al. developed a method called deep reaction optimizer (DRO) with the help of RL to establish chemical reactions as the *environment*, the person or the algorithm as the *decision-maker*, the experimental conditions as the *actions*, and the yield of the reaction as the *reward* [26]. DRO uses recurrent neural networks to achieve optimal decision-making in a chemical process. The objective is to achieve the maximum reward (the optimal reaction condition) with the least number of steps. The authors claim that DRO allows the optimization for real microdroplet reactions within 30 min, using 71% fewer steps than other several optimization algorithms. Figure 14.5 visualizes the DRO model over the three steps proposed and developed by Zhou et al. Additional decision-making algorithms consider a different approach and are not based on many local optima searches. The *SNOBFIT* algorithm developed by Huyer et al. is considered a global optimization algorithm for many low-dimensional problems, and it uses constraints, fitting functions, and a set of experimental variables in a wide search space [3, 32]. One application of SNOBFIT is the construction of an automated feedback loop to generate the conditions required to continuously produce CdSe nanoparticles based on information obtained and controlled by a computer, such as flow rates, reactor temperatures, and online fluorimeter data [33].

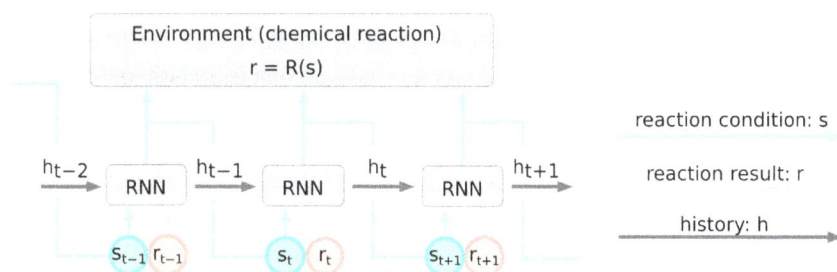

Figure 14.5: Visualization of the DRO model unrolled over three time steps. Taken from Zhou et al. [26].

14.2 Fault detection and diagnosis systems in industrial processes

In the era of big data, machine learning–based *FDD* systems have gained momentum and are a rapidly evolving research area [34, 35]. Instead of relying on expert knowledge, these systems automatically learn diagnostic information from collected data to build a model that links such data to the equipment's health status [36]. These methods, compared with other data-driven approaches such as multivariate statistical analysis and pure signal processing, excel in reducing manual intervention and enhancing the intelligence of machinery fault diagnosis [37].

FDD systems play a pivotal role in the reliability and safety of bench-scale and large-scale processes because they can identify abnormalities in the equipment/unit operations of each process for real-time correction via self-optimizing systems. Managing failures in any type of system, particularly in those with complex machinery, is paramount when serious problems such as costly downtime, economic losses, and even safety hazards occur [34]. Safe and reliable processes lead to risk minimization and efficient resource utilization, which in turn are indicators of process performance optimization and environmental footprint reduction [35]. The integration of FDD, AEP, and self-optimizing systems studied in Section 14.1 ensures that processes not only function efficiently but also respond effectively to faults and evolve to meet changing demands and conditions.

Within ML frameworks, *supervised learning methods* have been widely used for process monitoring, fault classification and identification, online operating mode localization designed for detection and separation of unknown faults from normal operating conditions and non-Gaussian distributed data [38], soft sensor modeling particularly useful for online classification of detected disturbances without the need to predict unknown (future) process behavior [39], quality prediction as the one applied in rotating machines whose potential faults were detected through a decision tree model in which the overfitting and complexity problems were avoided [40], and quality-related fault probability index which detected faults in a multiple operating conditions industrial hot strip mill process with the Gaussian mixture model [41].

This type of ML model employs datasets that contain input samples and their associated targets (labels) to help machines learn and make predictions. These input data are essentially labeled data samples, where discrete labels allow the classification of process data and continuous values enable the development of regression models for prediction or estimation of the equipment/process behavior [41, 42]. The following platforms have certain applications in industrial processes based on FDD:

1. *Shallow ML (SML)*: Fault diagnosis based on SML is essentially a pattern recognition process [43], which i) first collects operating data for equipment/unit operation using sensors installed on machines, ii) extracts fault features using time- and/or frequency-domain analyses while building feature vectors, and iii) finally detects the machinery/operation fault using ML-enabled models (Figure 14.6).

2. *DL*: In contrast to traditional ML models that depend on predefined algorithms and expert feature engineering, DL systems leverage multilayer neural networks to autonomously learn complex data features. This inherent capability marks a paradigm shift in how machines approach information processing and representation.

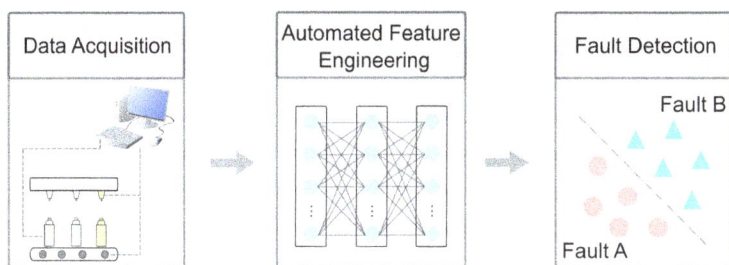

Figure 14.6: Shallow machine learning mechanism for fault detection. Adapted from Cen et al. [37].

3. *Transfer Learning* *(TL)*: TL empowers models to leverage knowledge across tasks and enhances adaptability and efficiency, particularly in scenarios that have few labeled data.

14.2.1 Shallow machine learning algorithms

Within the domain of SML, diverse models such as the *artificial neural network* (ANN), *support vector machine* (SVM), *K-nearest neighbor* (KNN), and *extreme learning machine* (ELM) are used to verify the health status of machinery. When it comes to simplicity and strong self-learning capabilities, ANNs are the best model type; however, they can become computationally intensive as their input data increases. SVMs excel in fitting small sample sizes and finding globally optimal solutions, although they may not be ideal for multitask classification. On the contrary, KNN models are easy to understand but can be computationally demanding when there are several samples, especially samples with imbalanced data (one data category has more data points than another category). ELMs are known for their fast training and generalization abilities; however, they lack high robustness due to their random generation of input weights and thresholds [37]. Refer to [36] for the detailed working mechanisms of these SML classifiers. Some examples are presented in the following paragraphs to demonstrate how SML algorithms contribute to the sustainability of processes in different industrial sectors.

Recently, Naimi et al. [44] developed and evaluated an ML-based FDD technique for sensors and actuators in the pressurized water reactors (PWRs) of nuclear power plants (NPPs). Typically, NPPs are associated with complex and unsafe systems, as exemplified by major accidents such as those at Three Mile Island, Chernobyl, and Fukushima [45]. Nonetheless, in the ongoing effort to demystify NNP operations, understanding the importance of continuously monitoring and promptly identifying faults in sensors and actuators is vital because they provide the data and control mechanisms needed to ensure safe, reliable, and efficient energy production. Faults were detected using neural networks to learn the normal behavior of the NPP systems and residual signals, which were generated by comparing the plant's output with the output predicted by the trained neural networks. This study found that KNN-based algorithms exhibited high accuracy for fault classification and required less computational time compared to the SVM and ensemble algorithms, which highlighted the suitability of this model for monitoring PWRs in current NPPs.

In the context of enhancing maintenance strategies and operational efficiency in buildings, particularly for safety and cost-effectiveness, the use of SML methods has proven essential. A compelling example is presented by Hosamo et al. [46] who developed a predictive maintenance system for air handling units (AHUs), considering that heating, ventilation, and air conditioning systems account for approximately half of the total energy consumed by buildings [43, 47]. Given that potentially 30% of the energy consumed by buildings could be saved through the implementation of FDD systems [48, 49], Hosamo et al. leveraged digital twin technology to assess AHU conditions over 2 years to avoid costly and disruptive reactive maintenance practices while mitigating operational carbon emissions at a university campus building in Norway [46].

This study followed a methodology (Figure 14.7) that integrates data between building information modeling (BIM) and facility management systems, integrates sensor data into BIM models, and includes a predictive maintenance process. The AHU condition was evaluated using expert rules based on the AHU performance assessment rules method (see Table 5 in the work by Hosamo et al. [46] for the description of detection rules). Metadata (nonphysical data for building components, such as their dimensions, materials, and installation year) are used to link data points with the diagnostic system and to store essential information related to AHU models for enhancing data interpretation. The final step of the model employs a multiclass classifier that uses ANN, SVM, and decision tree algorithms to predict faults in the AHU components. The suggested model was tested under real-world conditions by measuring the performance of AHUs using data from temperature, pressure, and flow rate sensors. The collected data allowed the authors of this paper to detect faults such as simultaneous heating and cooling, unexpected heating, and heat conservation during cooling when the supply air temperature exceeded the outside temperature.

Furthermore, the results demonstrated the superiority of ANN over SVM and decision trees in terms of prediction accuracy and error metrics. The ANN accurately classified all faults and exhibited a higher area under the curve value in the receiver operating characteristic analysis, which indicated that the model was more likely to correctly classify instances and make fewer incorrect classifications. This study emphasized the importance of maintenance planning to forecast future AHU conditions with the assistance of the trained ANN model; however, simultaneously, it recognized the need to look for new and more sophisticated approaches based on standardized data integration solutions for various types of sensors and application systems.

It is worth mentioning that works such as the one conducted by Hosamo et al. [46] pave the way to overcoming challenges that are related to the inaccuracy of results obtained from some FDD models that acquire data from laboratory environments. The fault feature extraction in laboratory scenarios does not capture the complexity and noise background that is inherent in the signal processing of industrial systems, which leads to misinformed and unrealistic decisions that have irreversible technical, economic, and environmental impacts.

Digital twin model

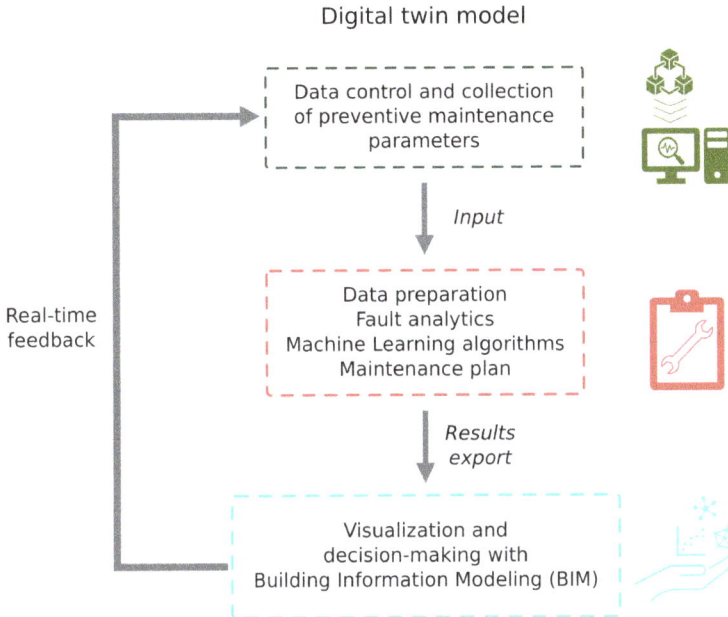

Real-time feedback

Figure 14.7: Digital twin predictive maintenance framework, proposed by Hosamo et al. [46], for detecting and diagnosing faults in air handling units (AHUs) of a campus building in Norway.

14.2.2 Deep learning

DL models, such as *deep belief networks, deep auto-encoders, convolutional neural networks* (CNNs), and *long short-term memory* (LSTM), can employ multilayer neural networks to automatically learn and extract intricate data features, thereby minimizing the need for manual feature selection. This enables the models to excel in fault identification within industrial processes [37]. The main benefits and shortcomings of each model are illustrated in Figure 14.8. Refer to [36] for detailed information about how each system works.

CNN is a well-established method for fault diagnosis in the realm of DL [37]. Ge et al. [51] developed a systematic FDD method with a focus on CNNs for local feature extraction and high-dimensional data processing in the production of formic acid through the hydrolysis of methyl formate via the reactive distillation (RD) process (see Chapter 14 of Doraiswamy et al. [52] for a detailed explanation of this process). The FDD process based on CNN was conducted using spatial datasets that represent process variables across differ-

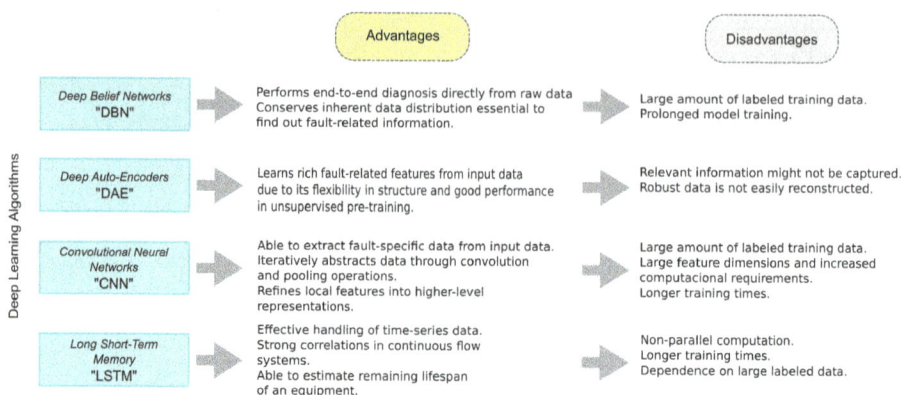

Figure 14.8: Main advantages and disadvantages of deep learning systems. Taken from Cen et al. [37] and Thoppil et al. [50].

ent locations in the RD process, while temporal characteristics reflect changes over time. Real-time data is preprocessed into two-dimensional matrices for training, with dimensions of m × n, where m is the time length and n is the number of process variables. Then, the CNN is able to extract relevant features from the real-time data streams leveraging convolutional layers [53]. As shown in Figure 14.9, the FDD procedure begins with data collection under normal and abnormal operating modes. Then, real-time data are preprocessed, labeled, and split into training, validation, and testing sets. The trained models are then tested and used for fault diagnosis, and the visualized results provide insights into the process conditions. Finally, if the accuracy of the model is satisfactory, the model can be used for online fault detection; otherwise, further modifications to the model may be needed. In this study, data were collected from normal conditions and 14 different fault conditions during dynamic simulations using Aspen Dynamics software. By considering both spatial and temporal features, the process data were transformed, even with measurement noise, into matrices for input to the CNN. Different CNN structures were trained, and the best-performing model successfully diagnosed most fault types with a *fault diagnosis rate* (FDR) that exceeded 92%. With average *false positive rates* (FPRs) (percentage of incorrect positive predictions made by a system) of 0%, 0.34%, and 0.68% and accuracies of 100%, 94.72%, and 91.31% on the training, validation, and test datasets, respectively, this CNN model demonstrated the promise of ML for fault detection in complex RD processes.

El-Shafeiy et al. [54] addressed the limitations of LSTM methods for conducting parallel computations by using a combined CNN–LSTM approach for the continuous measurement of water variables, including turbidity, specific conductance, and dissolved oxygen. Driven by the critical importance of effective real-time water quality monitoring for public health concerns and environmental preservation, these authors also wanted to overcome the shortcomings of existing methods, such as scalability problems, difficulties in handling sensor faults, and barriers to interpretability.

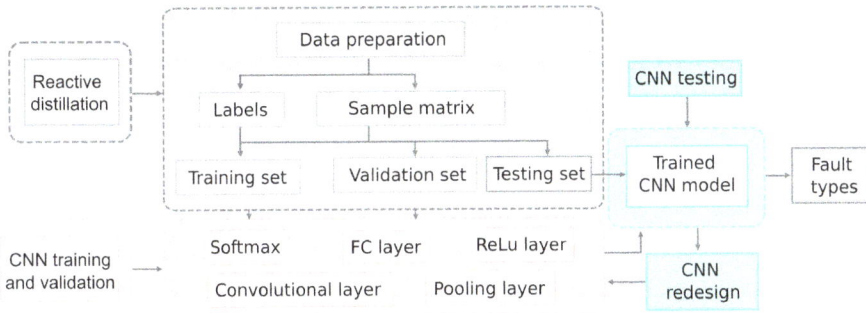

Figure 14.9: Convolutional neural network (CNN)-based model proposed by Ge et al. [51] to identify faults in the production of formic acid through the hydrolysis of methyl formate via reactive distillation (RD).

The results of their work presented CNN–LSTM as a suitable model for detecting real-time anomalies in water quality data with an accuracy of 92.3% and better performance than the SVM, ANN, CNN, or LSTM methods. With an F1 score of 0.93, which ensured effective anomaly detection with low FPRs (0.20) and high true positive rates (0.97), the real-world-tested CNN–LSTM model proved valuable for maintaining water quality as well as public and environmental health. To enhance the model, the developers plan to expand datasets, integrate them with IoT, and conduct comparative research, which would contribute to sustainable water resource management.

Despite the potential of DL models to achieve high-quality results, these models face challenges in the absence of large training data. Capturing new labeled data is often expensive and resource-intensive. In addition, these models tend to show poor recognition performance when they are used in scenarios beyond their training scope, such as in diagnosing faults in data related to diverse working conditions or machinery. This emphasizes the importance of adaptability and robustness in DL models [37, 41].

14.2.3 Transfer learning

In industrial settings, acquiring a substantial amount of labeled data with matching distributions for training and testing, as is required by traditional supervised and DL frameworks, can be quite challenging. However, TL offers a solution that addresses learning problems in target domains with limited or even no labeled samples. The core idea of TL is to enhance the learning of prediction functions in target domains by leveraging the knowledge obtained from source domains, as schematized in Figure 14.10 [55].

The process simulator built by the Tennessee Eastman (TE) company and proposed by Downs and Vogel [56] stands as a recognized benchmark in chemical engineering and serves as a complex simulation model that challenges researchers in the domain of process control and fault detection. The TE process simulator, with its

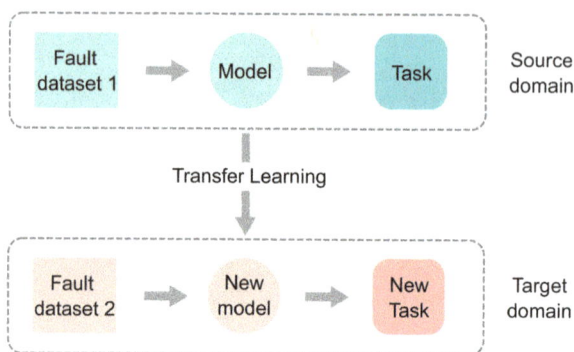

Figure 14.10: Schematic representation of the working principle of TL. Taken from Pan et al. [55].

many unit operations, variables, and operational modes (refer to [57] for a detailed explanation of the TE process), does not offer a robust FDD system, especially when transitioning to new operating modes, where labeled fault data may be scarce or absent. For this reason, Wu et al. [58] investigated the effectiveness of TL techniques in improving the adaptability of FDD models to diverse operating modes.

Figure 14.11 illustrates the multimode *FDD* framework. Historical data from TE processes train a *deep neural network* for fault detection. Real-time data are analyzed using trained detection networks for the identification of abnormalities, and a diagnostic network determines fault types. TL strategies, such as fine-tuning with labeled data and a joint adaptation network (JAN) with unlabeled data, enhance accuracy across different TE process modes. The fine-tuning method achieved an average FDR of 95.8% for labeled fault data in the target mode. Meanwhile, the JAN method reached a fault detection accuracy (FDR) of 93.4% for unlabeled fault data in the target mode, compared to 89.3% accuracy using labeled fault data in the source mode (where no fault data is available in the target mode). This multimode FDD framework addresses the challenges associated with FDD in chemical processes operating under different modes, which demonstrates its versatility and applicability. However, its limitations include its reliance on historically labeled fault data which leads to the challenge of not recognizing new fault types, and its assumption of consistent process variable dimensions across modes. Particularly, failure to recognize new fault types of constraints multi-mode operations by hindering the system's ability to adapt to emerging issues. This can give rise to operational disruptions, reduced efficiency, and increased downtime, undermining overall performance and reliability [37].

Current fault diagnostic methods based on TL often overlook the complexity of real industrial scenarios that may require data from multiple source domains. Federated TL, which addresses these concerns, aims to leverage data from diverse sources to achieve enhanced learning while maintaining privacy. This approach is a promising avenue for future advancements in fault diagnosis [37].

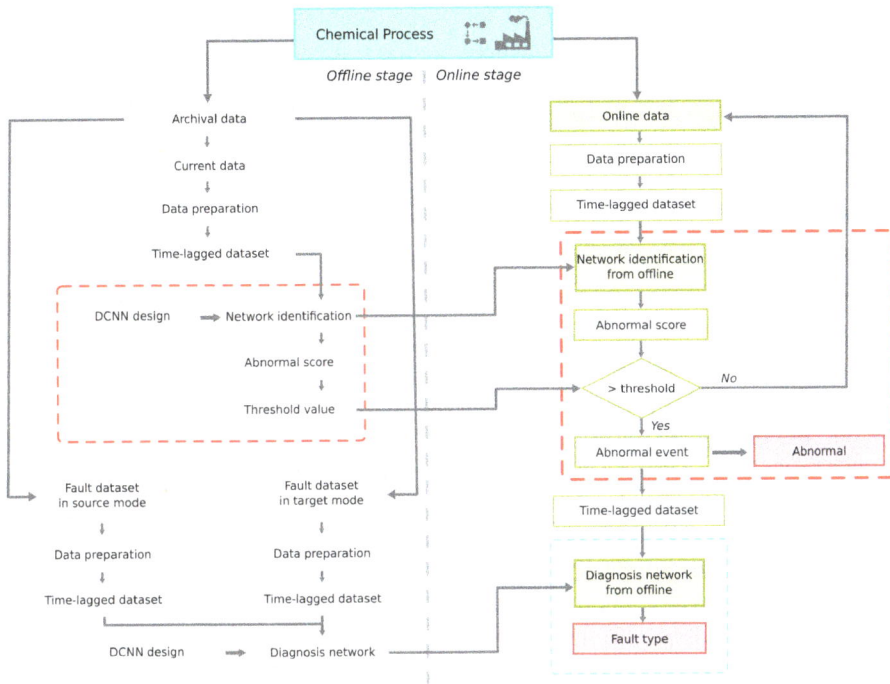

Figure 14.11: Transfer learning–based fault detection and diagnosis system proposed by Wu et al. [58] and implemented in the Tennessee Eastman process simulator.

14.2.4 Unsupervised machine learning algorithms

Although supervised ML methods have a strong presence in the FDD field, unsupervised ML methods have been broadly applied in processes for dimensionality reduction, outlier detection, process monitoring, and data visualization, thereby gaining momentum in the FDD area by leveraging tools such as *self-organizing maps* (SOMs) and *autoencoder* models [41]. For instance, Misbahulmunir et al. [59] addressed the critical importance of maintaining healthy and operational power transformers for the reliability and efficiency of electrical networks. Power transformer failures can lead to serious consequences, including the loss of human life. Dissolved gas analysis is a widely used method for monitoring and diagnosing incipient faults in power transformers. It focuses on gases dissolved in insulating oil as indicators. Nonetheless, the traditional gas ratio approach, as per the 60599 Standard set by the International Electrotechnical Commission, faces challenges in interpretation due to the complex nature of the oil degradation process.

In this study, the diagnostic process is outlined in two steps: training the SOM and validating the developed SOM feature map. The average diagnostic accuracy for various components (reactor, instrument transformer, bushing, and cable) was 84.3% for fault severity detection and 96.1% for fault type identification. These results validate the proposed power transformer model's accuracy. A comparison against the supervised SVM shows that 60% of the training data is adequate for SOM training.

Another example is related to monitoring the performance of ventilation systems to decrease the energy required to improve the indoor air quality (IAQ) of a subway station in South Korea. Motivated by the failure of sensors in the existing telemonitoring systems, which was attributed to the hostile conditions in underground environments, Loy-Benitez et al. [60] developed a novel IAQ monitoring approach using an AE model with a sensor self-validation framework to detect, identify, and reconstruct faulty IAQ sensor data. In addition, the authors integrated this model with a fault-tolerant ventilation control system to enhance IAQ and reduce energy consumption. The fault-tolerant control loop in the subway station in South Korea begins with obtaining data for normal sensor operation and normalizes the data using the mean and standard deviation. Then, for the online monitoring phase, an artificial faulty signal is introduced into the test dataset to be propagated through the AE model.

The distinctive feature of an AE model is its self-reconstruction capability. This capability is compared with that of *principal components analysis, independent components analysis*, and *dynamic components analysis* (DICA), which are known for the additional iterative steps for faulty sensor reconstruction. These steps make their algorithms computationally expensive. The AE method used in Loy-Benitez et al.'s study [60] detected between 44% and 100% of the total faulty signals for both low-bias ($18\ \mu g\ m^{-3}$) and high-bias ($180\ \mu g\ m^{-3}$) particulate matter concentration scenarios. Moreover, the AE method improved the reconstruction of faulty sensors by 31% compared to conventional dynamic components analysis. The proposed AE-based approach resulted in a 4% reduction in energy consumption and an 18.47% decrease in the average PM10 level. For confined spaces such as subway stations, this model allowed ventilation control systems to be fault-tolerant and achieved a sustainable balance between ventilation energy consumption and health risk level. Last but not least, the authors recognized the need to enhance AE architectures to improve the performance and accuracy of these models for representing the normal operating conditions of sensors.

14.3 Refinery production scheduling

RPS represents developing an analysis-based model that can overcome the production challenges of complex processes such as crude oil refining. The task involves overseeing decisions related to timing, sizing, allocation, and sequencing of refinery production scheduling in an interconnected and nonlinear refinery complex. This requires considering various factors, such as process unit optimization models and product blending prompts, which use linear programming that runs the refinery scheduling of their production. In addition, it is necessary to consider a range of technical, economic, environmental, and commercial limitations. RPS is currently recognized as a strategic tool that involves sophisticated proprietary mechanisms, a skilled workforce, an effective production strategy, and the seamless integration of data, systems, technologies, work processes, and personnel. It is responsible for implementing necessary changes within the company and plays a critical role in enhancing business performance during Industry 4.0. In the coming years, integrative modeling platforms and intelligent data management will become imperative to generating highly efficient production schedules. According to one source [61], schedulers are expected to transition from being software users/operators to becoming technically skilled modelers who play a role in developing innovative systems and integrated workflows within an Industry 4.0 setting. RPS is a lucrative endeavor that requires sophisticated tools, a skilled workforce, and a well-thought-out production strategy. RPS is an innovative activity that can bring about considerable organizational changes by incorporating data, systems, technologies, work processes, and individuals. Incorporating in-line blending operations is a prime example of how RPS can influence profitability. They refer to the operation where the scheduler blends streams of certain products like gasoline to meet the off-taker specification within production technologies. It is crucial to establish a collaborative partnership between industry and academia to advance the development of smart refineries.

A combination of discrete simulation techniques and rule-based heuristics marks a remarkable advancement in addressing practical challenges related to RPS. ORION, developed by AspenTech, is a groundbreaking commercial technology that has successfully tapped into the software market and has greatly impacted refinery schedulers. The introduction of this tool led to the replacement of intricate electronic spreadsheets using computer-aided decision-making tools that were standardized and easy for users to navigate. The initial commercial solutions, which third-party companies like Aspen or Aveva developed, are governed by Industry 3.0 and are still applied in most refineries. The simplicity of these systems, that is, Industry 3.0-based, catalyzed oil companies to pursue the development of their own RPS applications. This, in turn, prompted collaboration with academia in a quest for automated solutions. Optimization-based methods in the 2000s have enabled the resolution of clearly defined refinery subsystems. The following case studies exemplify the evolution of industry 3.0 and its transition to 4.0 in the oil and gas sector.

14.3.1 Optimizing production scheduling: industry 3.0 vs. industry 4.0 in oil refinery operations

Let us examine the crude oil section of an oil refinery, as depicted in Figure 14.12. In this setting, it is assumed that a failure will face the crude oil charging pump on the fifth day, which connects a crude oil tank to the crude oil distillation unit (CDU). The cause of this malfunction is attributed to notable deterioration in the rolling bearings

of the pump. The pump will undergo a 24-h maintenance period following the incident, which will render it temporarily unavailable.

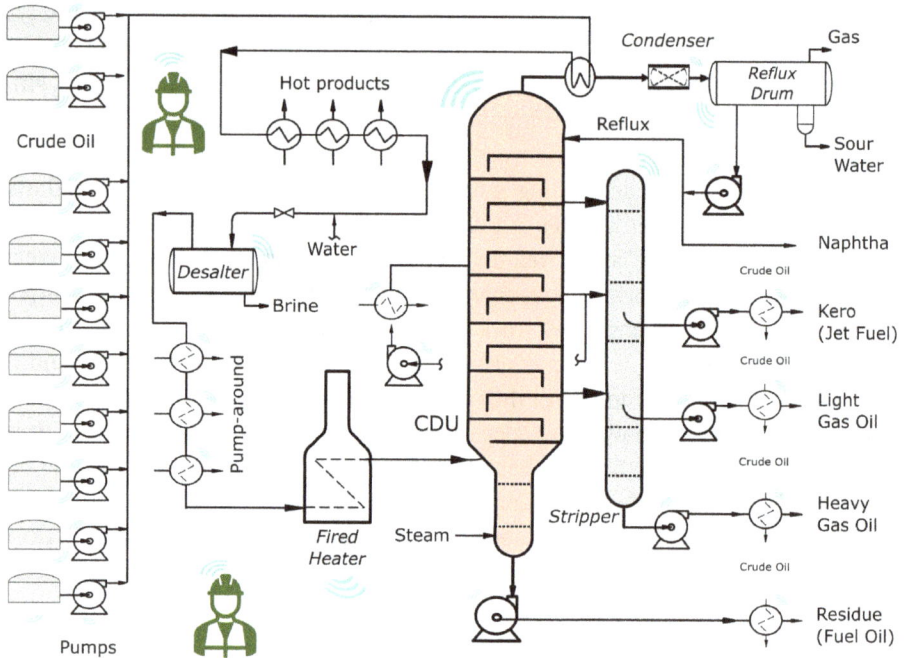

Figure 14.12: Scheme of the crude oil distillation unit. Taken from Joly et al. [61].

A 10-day production schedule will be assessed by considering two different technological contexts, *industry 3.0 (I3)* and *industry 4.0 (I4)*, as illustrated in Figure 14.13. In
the I3 scenario, the scheduler uses state-of-the-art decision-making technology. This
technology comprises an independent RPS solution that is seamlessly integrated with
corporate databases. The RPS application can autonomously retrieve the plant's initial
conditions from the databases and use them for offline operation. In the context of I4,
an industrial wireless network facilitates the independent execution and collaboration of various tasks using equipment, devices, workers, terminals, and other wireless
nodes. The utilization of a secure and exclusive cloud infrastructure allows for the
implementation of real-time scheduling. The RPS application in I4 operates in real-
time and uses supplementary data from the plant, which is akin to real-time traffic
updates. The data in question are drawn from various sources, including intelligent
entities such as sensors, machines, workers, and mobile devices. Additionally, distributed real-time optimizers and monitors are involved in the process. The RPS application enhances optimized trajectories by incorporating new data from these sources.

To maintain simplicity, our analysis primarily focuses on a specific element of I4,
namely, the self-diagnosis feature of the equipment. Therefore, this study takes a con-

servative approach when considering potential advantages. The process of determining the optimal operational schedule, whether I3 or I4, encompasses a range of interconnected decision-making procedures. The process involves making logical and logistical choices about the selection and order of resources over time, considering continuous variables (blending recipes, batch sizes, flow rates, and distillation cut points).

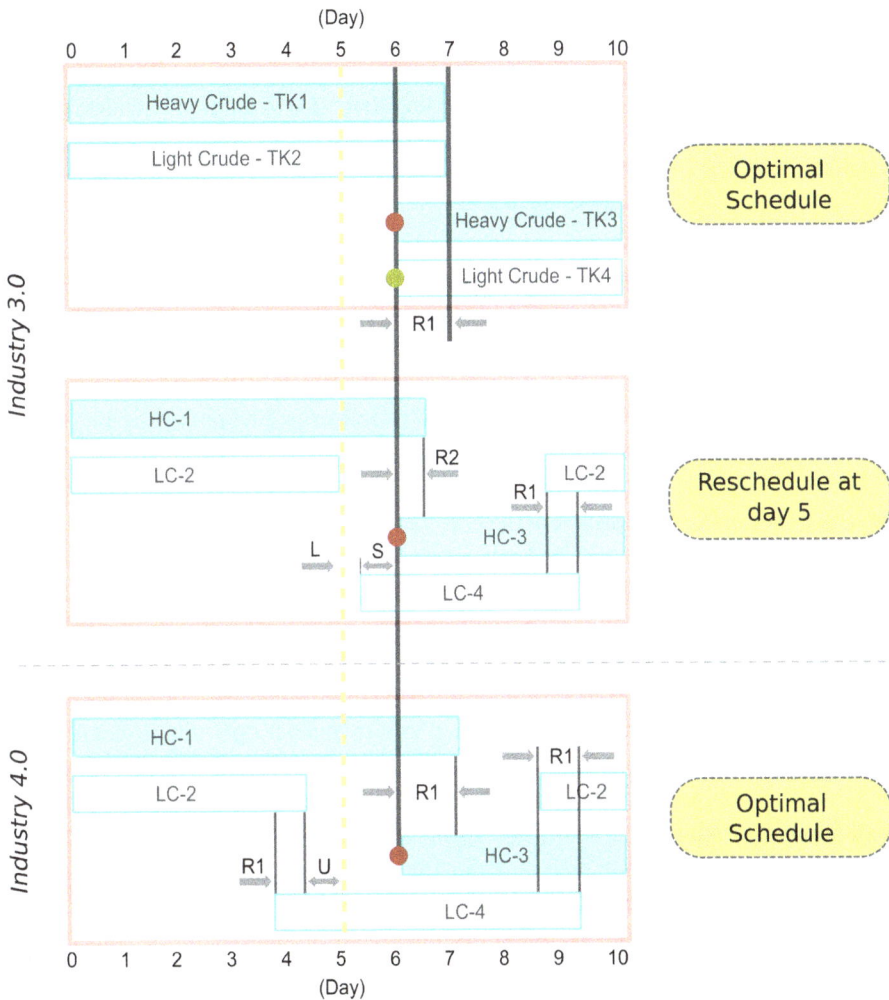

Figure 14.13: Industry 3.0 (I3) vs Industry 4.0 (I4) production scheduling theoretical technology backend configuration. Taken from Joly et al. [61].

The I3 scenario assumes that there will be no equipment failures. Therefore, to start, we look at previous data to select the most suitable data as a reference. The dataset encompasses a comprehensive collection of data regarding the plant's starting conditions, such as inventory levels and the quality of crude oil. Additionally, it contains known operational data for the future, including the schedule for crude oil supply, equipment maintenance, production targets, and more. A multiobjective optimization procedure is typically performed to meet operational and nonoperational objectives. Assuming that no unforeseen circumstances arise, the actual outcomes will closely align with the predetermined plan. The output of a skilled scheduler operating in the I3 scenario is depicted in the top panel of Figure 14.13. When evaluating a smooth operation without maloperation, the production schedule considers the necessary ramp times (R1) to ensure a seamless operation when switching between charging tanks in the CDU. Nevertheless, in the I3 realm, although no advanced correlation can automate the action, the scheduler needs to be able to predict upcoming operational upsets.

Suppose that on the fifth day, the light crude oil pump transfers oil from tank 2 (LC-2) to the CDU-phased series operation issues (as shown in the middle panel of Figure 14.13). Therefore, the amount of oil from LC-2 suddenly stops, which leads to disruptions in the functioning of the CDU despite the efforts made by the advanced process control (APC)/RTO applications to compensate for it. To overcome the loss of LC-2, these online applications regulate the feed flow rate of heavy crude oil from charging tank 1 (HC-1). As a result, the original deadline for completion has been moved up. Following a delay period (L) after the incident is identified (day 5), unplanned logistical activities are conducted in the tank farm until the completion of the current activities is projected by RPS. The injection of LC-4 into the crude oil blend mix starts without a seamless transition between tanks 2 and 4, which limits CDU operability until the loading from LC-4. The initial lack of content in tank 1, which results from the higher HC-1 flow rate instructed by the APC, results in a reduced ramp time (R2 < R1) between tanks 1 and 3. Regrettably, tank 3 will not be accessible until the start of the sixth day. Consequently, the CDU must contend with different feed compositions, which could lead to another operational disturbance. Hence, the scheduler must understand this problem and how long it will take to operate the CDU under this condition to update the RPS and prevent problems. On the contrary, in an I4 innovative industrial setting (as shown in the bottom panel of Figure 14.13), the RPS solver is given up-to-date data regarding the self-diagnosis of equipment. The data are acquired using digital twin models.

The differences between I3 and I4 are illustrated in this section by considering the particular scenario of tray damage inside the CDU. In such a scenario, the CDU still operates but the efficiency of the fractionation process within the column may be considerably affected. This damage leads to other performance issues, such as off-spec products or yield reduction. The challenge that the schedule needs to overcome is optimizing the RPS model for this operation until the next refinery operation turnaround (where the refinery stops operation for maintenance purposes and upgrades projects, if any). Unlike the previous case study, the scheduler must maintain the operation with the current situation and keep adjusting the parameters until the next turnaround and return to normal conditions. Therefore, it is difficult to predict similar issues and resolve those using I3 models. The manual action through I3 is critical when the control of the side-cut stream quality, where the tray is damaged and affects vapor–liquid equilibrium at that tar/product, is insufficient. As a result, this scenario's potential financial losses due to the damaged tray and I3 RPS implementation may be greater than those in the previous example. This is because schedulers may need to take measures to resolve problems in an I3 scenario.

However, the advanced tool in I4 that is equipped with self-diagnosis models detects undesired changes in CDU performance. Therefore, it is necessary to determine or estimate new crude oil assays using process simulators. These assays are constant parameters associated with distillation yields, and they depend on factors such as the type of crude oil and CDU operational mode. The RPS solution effectively integrates these newly established parameters to address the optimization problem in crude oil blending. When the integration system is implemented, APCs and RTOs can effectively perform their functions to determine the operating parameters for the system. Table 14.1 summarizes the distinctions between I3 and I4 in both cases.

Table 14.1: Differences between I3 and I4 related to equipment failure and dysfunction.

Example	Issue	Industry 3.0	Industry 4.0
Equipment failure (e.g., pump failure)	Rescheduling required?	Yes	No
	Stable operation affected?	Yes	No
	Production goals met?	No	Yes
	Profitability affected?	Yes	No
Equipment dysfunction (e.g., CDU tray failure)	Rescheduling required?	Yes	No
	Stable operation affected?	Yes	No
	Production goals met?	No	Yes
	Profitability affected?	Yes	No

14.3.2 Challenges

Refinery modeling. One of the primary difficulties is that an oil refinery functions as a system that remains open, constantly receiving and releasing materials (continuing in flow in and out). This perpetual cycle leads to the continuous accumulation and decomposition of its various constituents, such as refinery products, gasoline, and diesel. The concept of equifinality asserts that various starting conditions and routes can result in identical outcomes. According to Joly et al. [61] our comprehension of open systems is consistently incomplete and approximate. Moreover, the functioning of an oil refinery is an intricate process, and examining its components in isolation does not give a comprehensive understanding of the system as a whole. In addition, understanding the operating parameters and the mathematical assumptions for open systems is necessary to ensure good modeling representations. As a result, there is a tendency to overlook the potential for profit and performance. Likewise, more sophisticated controllers, such as multivariable model predictive control (MPC), which include DMCplus, RMPCT, and PACE/SMOCpro, do not directly consider the flowsheet. Instead, their primary focus is on the controlled (dependent) and manipulated (independent) variables. Another challenge is integrating real-world feedback and continuous modifications into the virtual setting, referred to as "parameter feedback." This particular point was addressed during the FOCAPO 2008 conference. To effectively implement a Rescheduling production scheduling (RPS) project in an Industry 4.0 context, gain and bias adjustments should be incorporated into the models. This is particularly important when the need for online scheduling capabilities is considered. In this situation, data resolution plays a critical role in accurately measuring the material balances, adjusting model parameters using plant feedback, and confirming the reliability of system measurements. Although state estimation, which is also known as Kalman filtering, is often seen as a potential solution, it may not effectively address nonlinearities. State estimation is a subset of data reconciliation that has been shown to be effective. Currently, there are industrial difficulties (reconfiguring all control systems at each unit and in each equipment and connecting them all under I4 RPS and IT, including cyber security and internet connectivity) related to the incorporation of real-time scheduling optimization alongside control system feedback and hybrid MPC as they are part of the I4 RPS of continues scheduling system. Another critical concern is the reduction in timeliness, which refers to the time between the anticipated availability of data and its actual usability. One example of the integration of the real-time scheduling and hybrid predictive model is implementing the "smart sweeping" application at a Canadian iron-ore processing facility operated by Rio Tinto. The system combines real-time scheduling and predictive control to regulate multiple feedstocks in the refinery. It works with a hold-up in surge tanks that accumulate the liquid before feeding the unit to ensure continuous and sustainable flow to the unit while monitoring the level using sophisticated transmitters. This example showcases the application of newly emerging technologies and concepts in the I4 landscape.

Refining business. Molecular management is widely used at present to increase operational profits [62]. One approach involves tracking crude oil composition to forecast the properties and yields of important distillate streams. Incorporating assay data and using feedback from the measurement system can lead to the achievement of this goal. The origins of this can be traced back to 1995 when Exxon Canada's Nanticoke refinery began recording considerable pipeline-to-tank and tank-to-CDU resources to create integrated solutions for real-time monitoring and control feedback obtained directly in the Honeywell TDC3000 DCS. Similarly, Petrobras in Brazil optimized refinery logistics using the GOMM project that they developed in-house. As the challenges of refinery scheduling continue after all these efforts from different companies, ExxonMobil and other oil companies have made notable investments in "molecular management," which refers to real-time optimization of refinery scheduling that tracks all molecules entering and leaving the refinery with minimized of the loss. This has led to implementing RPS projects as a preferred alternative to the AspenTech scheduling solution. The use of crude oil composition tracking results in their scheduling and distillation units' advanced controls remains to be determined. Numerous vendors specializing in APC need help managing distillation and fractionation units. This is primarily attributed to the dynamic nature of the crude oil diet or slate, which undergoes constant changes, particularly during crude-switching scenarios. A continuous process scheme makes running and optimizing molecular material balances difficult, especially for product tank blends. Monitoring crude oil compositions hourly is suggested, which would integrate data every minute. This approach would be particularly beneficial for the automated monitoring of running-gauge tanks. In addition, this method would offer insights into current transfers or movements involving crude oil tanks. Using other lab techniques to measure product specifications, such as cut-point temperature, it becomes feasible to characterize refinery product streams. The gain and bias model is one example that some refineries use to model the crude blend, especially if there is a laboratory measurement of crude specification, using historical routine operational data that can adjust the gain and bias each time, commonly called the past rolling horizon [63]. Nonetheless, one considerable obstacle to the improvement in the monitoring of crude oil feed composition lies in accurately recording real-time data on crude oil movements. This includes tracking deliveries from pipelines and marine vessels as well as conducting transfers between tanks, tank-to-blender operations, and tank-to-CDU processes. Hence, the objective of engineering in an Industry 4.0 context is to track crude oil composition using a "soft sensor." This application can be considered a noteworthy illustration of Industry 4.0 and Analytics 3.0. It uses a network model or a cyber-physical system, and it employs predictive analytics to enhance crude oil feed control, optimization, and scheduling. More specifically, it uses feedforward or anticipatory governance techniques.

14.3.3 Real case: Abqaiq Plants, a digital transformation success story

The digital transformation of Abqaiq Plants, which is recognized as the largest crude stabilization plant in the world, has been truly remarkable. For this plant, which is 70 years old, adopting 4IR technologies posed unique obstacles. Nonetheless, the significance of the establishment and its dedication to becoming a prominent digitalized energy corporation has resulted in international acclaim. The World Economic Forum recognized it as a global frontrunner for the 4IR in 2021 [64].

Digitally transforming aging oil and gas facilities, such as Abqaiq Plants, requires addressing infrastructure-level issues. This facility relies on legacy and decentralized systems, which makes data access and interpretation difficult. Employees have to manually analyze data and perform tasks using outdated processes and procedures. Overcoming these challenges is critical to achieving peak performance.

The process of digitizing Abqaiq Plants began with the development of a detailed plan to guarantee a seamless shift. Promising technologies were identified using a centralized technology management database. Collaboration with technology providers was facilitated to conduct pilot projects. Additionally, a dedicated center for 4IR technology was established to foster improved collaboration and knowledge-sharing. To enhance the value of the technologies, teams were encouraged to share data, which was made possible using an Industrial Internet of Things infrastructure. Furthermore, there was a strong emphasis on helping employees adopt technology by offering various skill-enhancement initiatives.

When Abqaiq Plants was built in the 1940s, the idea of industrial machines communicating with computers was inconceivable. However, 4IR technologies have revolutionized the facility. Through industrial Wi-Fi and advanced wireless communication, devices now share crucial performance data that enables AI systems to enhance efficiency and predict faults preemptively. State-of-the-art distributed control systems have replaced the outdated infrastructure and now offer improved interface and connectivity capabilities. This advanced connectivity has improved efficiency, reduced costs, and promoted safety.

Abqaiq Plants has revolutionized its manufacturing processes by leveraging big data and analytics. AI monitors equipment performance eliminates the need for manual inspections, and effectively prevents any production disruptions caused by equipment failure. The facility also employs ML models to enhance the performance of multiple crude stabilization units simultaneously, which results in improved efficiency. AI's ability to analyze data and adjust operating parameters surpasses that of humans in terms of speed and accuracy.

Abqaiq Plants acknowledges the increase in global energy demand and the importance of adopting approaches to sustainable production. The facility's efficiency is optimized using big data, advanced analytics, and automation techniques. Real-time data enables automated systems to make production adjustments promptly, which results in notable enhancements. The Abqaiq Plants have used data analytics to predict crude oil quality and to swiftly adapt to and execute different strategies, which has led to a considerable reduction of 31.8% in carbon intensity from 2019 to 2022.

The success of Abqaiq Plants is a prime example of how 4IR technology can play a crucial role in improving the effectiveness, output, and long-term viability of energy facilities. The facility's impressive transformation exemplifies the potential of embracing digitalization and demonstrates that even aging industrial facilities can benefit from a technological shift. Any industrial facility in the world has the potential to emulate the success of Abqaiq Plants and establish itself as a prominent leader in the adoption of 4IR technologies [64].

14.3.4 Enhancing heating control to increase refinery throughput

Integrating materials, energy, and processing tasks is one approach to realizing a sustainable process [65]. Separation is the most energy-intensive step of any chemical process, and crude refinery is no exception. More than 93 million barrels per day were produced in 2022 [66]. Crude oil is generally treated in a refinery to achieve higher-value products, mainly via separation. Hence, the ability to reduce energy consumption in any part of this industry is a win for the environment. A single refinery may contain multiple processes that include heat integration to ensure high performance and lower operational costs. Additionally, advanced controls optimize the process and maintain a smooth operation; smoother operations result in less frequent interruptions and lead to higher overall efficiency.

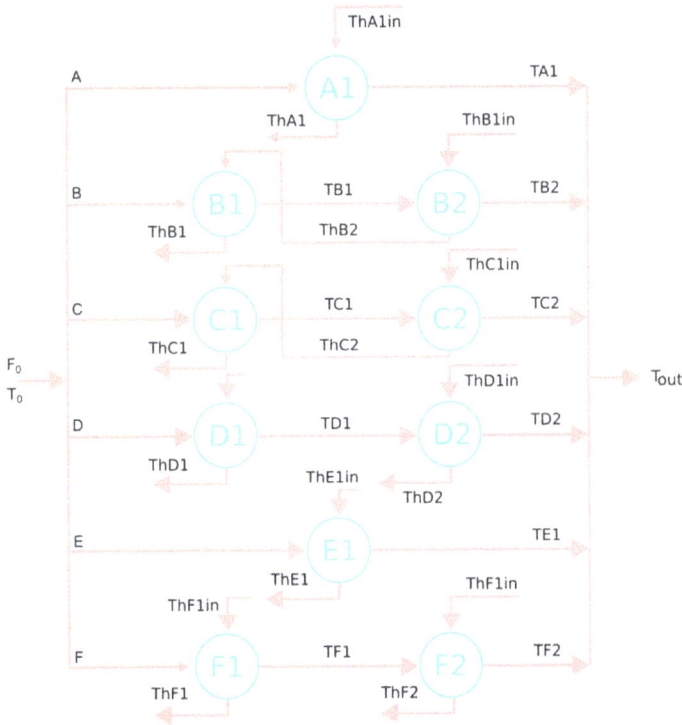

Figure 14.14: Preheating train heat exchanger network. Taken from Jäschke et al. [67].

Online optimization and offline model optimization are two of the standard methodologies used in process control. Instantaneous inputs are used in an online model, while the optimized output is calculated and applied in real-time such an approach furnishes excellent results given that an appropriate process model is maintained. However, the computation cost is relatively high despite the increased interest in this approach. On the contrary, offline model optimization allows the search of a near-optimal condition based on selected process variables [67]. Jäschke introduced a self-optimizing strategy based on the Jäschke temperature, where T is the branch temperature, T_0 is the cold stream inlet temperature, and T_h is the hot stream inlet temperature [67].

$$T_J = \frac{(T - T_0)^2}{T_h - T_0}. \tag{14.1}$$

This strategy allowed the implementation of a solution in the preheating train network at Mongstad Refinery. Optimal conditions are achieved when the maximum total heat transfer is maintained, which can be done by varying the split of the cold stream across the heat exchanger network. The control schema is applied by setting the difference between all Jäschke temperatures to zero. The cold crude feed is passed through a network of heat exchangers in which the fresh feed cools the products. This integration reduces the energy used to heat the feed before it enters the crude distillation unit (CDU). Maximum heat recovery is achieved by varying the split of the hot streams across the branches. A total of 10 heat exchangers are modeled in six parallel configurations, as shown in Figure 14.14. The objective function is defined as follows:

$$\max P = \sum_{i=A, ..., F} Q_i, \tag{14.2}$$

where P is the total heat transferred and Q is the heat transferred to the branch. This simple control schema enables near-optimal operation compared with model predictive control (MPC), which is more complex.

14.3.5 Model predictive control in scheduling a refinery

Unlike conventional controllers, in which the operation is based on action feedback, MPC is a control method in which a model is used to predict the response of the controlled variable, as shown in Figure 14.15.

The model is solved instantaneously as a constraint-based optimization problem in which a cost function is minimized [68]. Finding the optimal input values for the short term optimizes the long-term process. Because the control is carried out by following a trajectory of optimal values for the controlled variable, the MPC is highly suited to real-time processes. This is essential because the MPC algorithm depends on a predictive model and an optimization tool. MPC applications can be found in the process industry, which includes the oil and petrochemical industry [69], power electronics, building climate and energy, and manufacturing [68]. Oil is transported to the refinery by ship tankers, usually via docks. Later, oil is stored in storage tanks before being transferred to mixing (charging) tanks, where it is blended. The oil is discharged

a

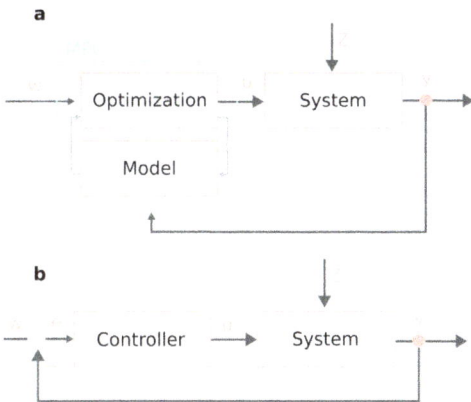

b

Figure 14.15: (a) MPC configuration and (b) conventional controllers. Taken from Schwenzer et al. [68].

to the CDU, where separation occurs. Planning and scheduling oil production is essential for increasing profit margins. The ability to minimize operating cost or maximize profit is the aim of schedule modeling [69]. Several assumptions must be made to formulate the problem correctly, such as assumptions about waiting and unloading times, flow rates, inventory levels, and critical component concentration.

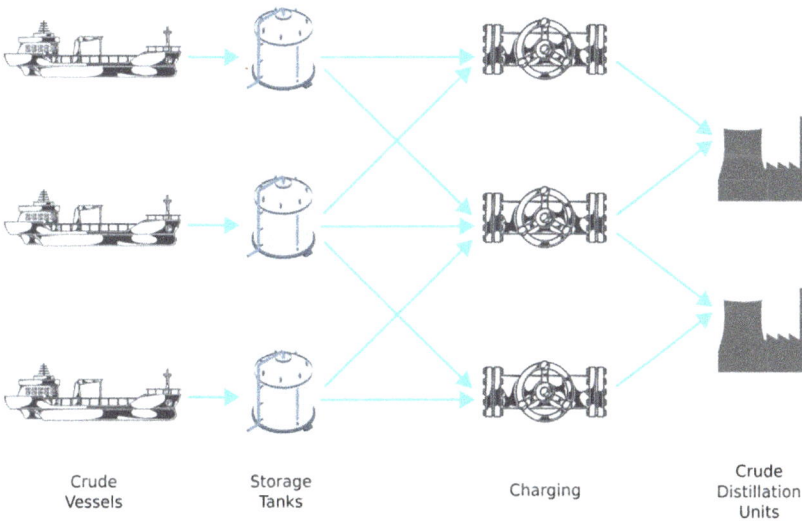

| Crude Vessels | Storage Tanks | Charging | Crude Distillation Units |

Figure 14.16: Overview of the oil scheduling problem network. Taken from Yüzgeç et al. [69].

The cost function consists of the sum of the sea waiting cost, unloading cost, storage inventory cost, and CDU changeover cost. The material balance is constructed around every stage. Figure 14.16 presents a system overview in which the parameters of the cost function are optimized. Multiple scenarios are assumed and assessed. The first case includes two vessels, two storage tanks, one charging tank, and one CDU; the resultant problem has 309 constraints, 288 variables, and 48 binary variables. The optimal solution reduces operating costs by more than $880,000 [69]. This reduction is achieved by allowing a higher volume of oil in storage tanks than in charging tanks. Further demonstration of the methodology's capability of this approach can be investigated when introducing a more complicated example. In this case, three types of crude are present, and three different mixtures are targeted. The generated model has 1,694 constraints, 1,200 variables, and 144 binary variables. The optimal cost reduction is estimated at around $900,000.

The vital role of data-driven processes in achieving sustainability cannot be overstated. In the contemporary landscape, harnessing the power of data is paramount to driving informed decision-making, optimizing resource utilization, and mitigating environmental impact. Self-optimizing systems are a representation of the beneficial fusion between AI-based methods and process intensification platforms. In this regard, AEPs with advanced decision-making algorithms represent a pivotal advancement in self-optimizing systems. These platforms, which are purpose-built for intelligent and efficient experimentation, greatly contribute to the acceleration of scientific development and bolster safety measures. They also enhance reliability and efficiency across industrial and laboratory domains. The integration of existing knowledge and theoretical models into decision-making processes through cutting-edge ML algorithms empowers AEPs to facilitate targeted and informed experimentation. This synergistic work between AI and traditional expertise not only streamlines processes but also sets the stage for a dynamic and transformative future in research and development. Moreover, machine learning-enabled FDD systems are recognized for their contribution to industrial process reliability and sustainability. By preventing downtime, optimizing resource utilization, and minimizing safety risks, these systems play a pivotal role in ensuring operational efficiency. Despite the challenges related to data constraints and adaptability, the potential of these systems remains promising. Integrating ML with self-optimizing systems is imperative for continuous improvement in industrial processes. Ongoing research endeavors are needed to refine the models, address sustainability concerns, and enhance the overall resilience of FDD systems and the dynamic landscape of industrial processes. In the oil and gas sector, RPS stands at the forefront of sustainable practices. These systems, when supported by MPC, can dynamically adapt to changing conditions, optimize resource allocation in real-time, and enhance overall operational efficiency. By leveraging data-driven decision-making, self-optimizing systems contribute to the reduction of energy consumption, emissions, cost, and environmental impact in refinery processes. This holistic approach resonates with the principles of green chemistry and engineering and promotes efficiency, safety, and sustainability in scientific experimentation and industrial operations.

Bibliography

[1]   Szekely, G. *Sustainable Process Engineering*; De Gruyter, 2021. 10.1515/9783110717136.

[2]   Knox, S. T.; Warren, N. J. Enabling Technologies in Polymer Synthesis: Accessing a New Design Space for Advanced Polymer Materials. *React. Chem. Eng.* **2020**, *5* (3). Royal Society of Chemistry, 405–423, 2020. 10.1039/c9re00474b.

[3]   Xie, Y.; Sattari, K.; Zhang, C.; Lin, J. Toward Autonomous Laboratories: Convergence of Artificial Intelligence and Experimental Automation. *Progress Mater. Sci.* **2023**, *132*. Elsevier Ltd, 2023. 10.1016/j.pmatsci.2022.101043.

[4]   Flores-Leonar, M. M. *et al.* Materials Acceleration Platforms: On the Way to Autonomous Experimentation. *Curr. Opin. Green Sustain. Chem.* **2020**, *25*, 100370. 10.1016/j.cogsc.2020.100370.

[5]   Häse, F.; Roch, L. M.; Aspuru-Guzik, A. Next-Generation Experimentation with Self-Driving Laboratories. *Trends Chem.* **2019**, *1*, (3), 282–291. 10.1016/j.trechm.2019.02.007.

[6]   Li, J. et al. Autonomous Discovery of Optically Active Chiral Inorganic Perovskite Nanocrystals through an Intelligent Cloud Lab. *Nat. Commun.* **2020**, *11*, (1), 2046. 10.1038/s41467-020-15728-5.

[7]   Cortés-Borda, D. et al. An Autonomous Self-Optimizing Flow Reactor for the Synthesis of Natural Product Carpanone. *J. Org. Chem.* **2018**, *83*, (23), 14286–14289. 10.1021/acs.joc.8b01821.

[8]   MacLeod, B. P. et al. Self-driving Laboratory for Accelerated Discovery of Thin-film Materials. *Sci. Adv.* **2020**, *6*, (20). 10.1126/sciadv.aaz8867.

[9]   Grizou, J.; Points, L. J.; Sharma, A.; Cronin, L. A Curious Formulation Robot Enables the Discovery of A Novel Protocell Behaviour, *Sci. Adv.* **2020**, *6*, (5). 10.1126/sciadv.aay4237.

[10]   Gongora, A. E. et al. A Bayesian Experimental Autonomous Researcher for Mechanical Design. *Sci. Adv.* **2020**, *6*, (15). 10.1126/sciadv.aaz1708.

[11]   Epps, R. W. *et al.* Artificial Chemist: An Autonomous Quantum Dot Synthesis Bot. *Adv. Mater.* **2020**, *32*, (30). 10.1002/adma.202001626.

[12]   Rong, L.; Caldona, E. B.; Advincula, R. C. PET-RAFT Polymerization under Flow Chemistry and Surface-initiated Reactions. *Polym. Int.* **2023**, *72*, (2), 145–157. 10.1002/pi.6475.

[13]   Rigoglioso, V. P.; Boydston, A. J. Flow Optimization of Photoredox-Mediated Metal-Free Ring-Opening Metathesis Polymerization. *ACS Macro. Lett.* **2023**, *12*, (11), 1479–1485. 10.1021/acsmacrolett.3c00545.

[14]   Burger, B. *et al.* A Mobile Robotic Chemist. *Nature.* **2020**, *583*, (7815), 237–241. 10.1038/s41586-020-2442-2.

[15]   Kantz, E. D.; Tiwari, S.; Watrous, J. D.; Cheng, S.; Jain, M. Deep Neural Networks for Classification of LC-MS Spectral Peaks. *Anal Chem.* **2019**, *91*, (19), 12407–12413. 10.1021/acs.analchem.9b02983.

[16]   Borgsmüller. *et al.* WiPP: Workflow for Improved Peak Picking for Gas Chromatography-Mass Spectrometry (GC-MS) Data. *Metabolites.* **2019**, *9*, (9), 171. 10.3390/metabo9090171.

[17]   Qu, X. *et al.* Accelerated Nuclear Magnetic Resonance Spectroscopy with Deep Learning, *Angewandte Chemie International Edition.* **2020**, *59*, (26), 10297–10300. 10.1002/anie.201908162.

[18]   Fine, J. A.; Rajasekar, A. A.; Jethava, K. P.; Chopra, G. Spectral Deep Learning for Prediction and Prospective Validation of Functional Groups. *Chem. Sci.* **2020**, *11*, (18), 4618–4630. 10.1039/C9SC06240H.

[19]   Oviedo, F. et al. Fast and Interpretable Classification of Small X-ray Diffraction Datasets Using Data Augmentation and Deep Neural Networks. *Npj Comput. Mater.* **2019**, *5*, (1), 60. 10.1038/s41524-019-0196-x.

[20]   Gordon, O. M.; Hodgkinson, J. E. A.; Farley, S. M.; Hunsicker, E. L.; Moriarty, P. J. Automated Searching and Identification of Self-Organized Nanostructures. *Nano Lett.* **2020**, *20*, (10), 7688–7693. 10.1021/acs.nanolett.0c03213.

[21]   Kaufmann, K. *et al.* Crystal Symmetry Determination in Electron Diffraction Using Machine Learning. *Sci (1979).* **2020**, *367*, (6477), 564–568. 10.1126/science.aay3062.

[22] Lee, B. *et al.* Statistical Characterization of the Morphologies of Nanoparticles through Machine Learning Based Electron Microscopy Image Analysis. *ACS Nano.* **2020**, *14*, (12), 17125–17133. 10.1021/acsnano.0c06809.

[23] Li, J. *et al.* Machine Vision Automated Chiral Molecule Detection and Classification in Molecular Imaging. *J. Am. Chem. Soc.* **2021**, *143*, (27), 10177–10188. 10.1021/jacs.1c03091.

[24] Dong, X. *et al.* 3D Deep Learning Enables Accurate Layer Mapping of 2D Materials. *ACS Nano.* **2021**, *15*, (2), 3139–3151, 10.1021/acsnano.0c09685.

[25] Kirman, J. *et al.* Machine-Learning-Accelerated Perovskite Crystallization. *Matter,* **2020**, *2*, (4), 938–947. 10.1016/j.matt.2020.02.012.

[26] Zhou, Z.; Li, X.; Zare, R. N. Optimizing Chemical Reactions with Deep Reinforcement Learning. *ACS Cent. Sci.* **2017**, *3*, (12), 1337–1344. 10.1021/acscentsci.7b00492.

[27] Coley, C. W.; Eyke, N. S.; Jensen, K. F. Autonomous Discovery in the Chemical Sciences Part I: Progress. *Angew. Chem. Int. Ed.* **2020**, *59*, (51), 22858–22893. 10.1002/anie.201909987.

[28] Shields, B. J. et al. Bayesian Reaction Optimization as a Tool for Chemical Synthesis. *Nature,* **2021**, *590*, (7844), 89–96. 10.1038/s41586-021-03213-y.

[29] Eriksson, D.; Pearce, M.; Gardner, J.; Turner, R. D.; Poloczek, M. Scalable Global Optimization via Local Bayesian Optimization. *Adv. Neural Inf. Process. Syst.* **2019**, 32.

[30] Häse, F.; Roch, L. M.; Aspuru-Guzik, A. Chimera: Enabling Hierarchy Based Multi-objective Optimization for Self-driving Laboratories. *Chem. Sci.* **2018**, *9*, (39), 7642–7655. 10.1039/C8SC02239A.

[31] Bradford, E.; Schweidtmann, A. M.; Lapkin, A. Efficient Multiobjective Optimization Employing Gaussian Processes, Spectral Sampling and a Genetic Algorithm. *J. Global Optim.* **2018**, *71*, (2), 407–438. 10.1007/s10898-018-0609-2.

[32] Huyer, W.; Neumaier, A. SNOBFIT – Stable Noisy Optimization by Branch and Fit. *ACM Trans. Math. Software.* **2008**, *35*, (2), 1–25. 10.1145/1377612.1377613.

[33] Krishnadasan, S.; Brown, R. J. C.; deMello, A. J.; deMello, J. C. Intelligent Routes to the Controlled Synthesis of Nanoparticles. *Lab Chip.* **2007**, *7*, (11), 1434. 10.1039/b711412e.

[34] Lei, Y.; Yang, B.; Jiang, X.; Jia, F.; Li, N.; Nandi, A. K. Applications of Machine Learning to Machine Fault Diagnosis: A Review and Roadmap. *Mech. Syst. Signal Process.* **2020**, *138*, 106587. 10.1016/j.ymssp.2019.106587.

[35] Liu, R.; Yang, B.; Zio, E.; Chen, X. Artificial Intelligence for Fault Diagnosis of Rotating Machinery: A Review. *Mech. Syst. Signal Process.* **2018**, *108*, 33–47. 10.1016/j.ymssp.2018.02.016.

[36] Zhao, R.; Yan, R.; Chen, Z.; Mao, K.; Wang, P.; Gao, R. X. Deep Learning and Its Applications to Machine Health Monitoring. *Mech. Syst. Signal Process.* **2019**, *115*, 213–237. 10.1016/j.ymssp.2018.05.050.

[37] Cen, J.; Yang, Z.; Liu, X.; Xiong, J.; Chen, H. A Review of Data-Driven Machinery Fault Diagnosis Using Machine Learning Algorithms. *J. Vib. Eng. Technol.* **2022**, *10*, (7), 2481–2507. 10.1007/s42417-022-00498-9.

[38] Huang, -C.-C.; Chen, T.; Yao, Y. Mixture Discriminant Monitoring: A Hybrid Method for Statistical Process Monitoring and Fault Diagnosis/Isolation. *Ind. Eng. Chem. Res.* **2013**, *52*, (31), 10720–10731. 10.1021/ie400418c.

[39] Gins, G.; Van den Kerkhof, P.; Vanlaer, J.; Van Impe, J. F. M. Improving Classification-based Diagnosis of Batch Processes through Data Selection and Appropriate Pretreatment. *J. Process Control.* **2015**, *26*, 90–101. 10.1016/j.jprocont.2015.01.006.

[40] Karabadji, N. E. I.; Seridi, H.; Khelf, I.; Azizi, N.; Boulkroune, R. Improved Decision Tree Construction Based on Attribute Selection and Data Sampling for Fault Diagnosis in Rotating Machines. *Eng. Appl. Artif. Intell.* **2014**, *35*, 71–83. 10.1016/j.engappai.2014.06.010.

[41] López-Guajardo, E. A.; Delgado-Licona, F.; Álvarez, A. J.; Nigam, K. D. P.; Montesinos-Castellanos, A.; Morales-Menendez, R. Process Intensification 4.0: A New Approach for Attaining New, Sustainable

and Circular Processes Enabled by Machine Learning. *Chem. Eng. Process – Process Intensif.* **2022**, *180*, 108671. 10.1016/j.cep.2021.108671.

[42] Peng, K.; Zhang, K.; You, B.; Dong, J. Quality-related Prediction and Monitoring of Multi-mode Processes Using Multiple PLS with Application to an Industrial Hot Strip Mill. *Neurocomputing.* **2015**, *168*, 1094–1103, 10.1016/j.neucom.2015.05.014.

[43] Xiao, F.; Wang, S. Progress and Methodologies of Lifecycle Commissioning of HVAC Systems to Enhance Building Sustainability. *Renewable Sustain. Energy Rev.* **2009**, *13*, (5), 1144–1149. 10.1016/j.rser.2008.03.006.

[44] Naimi, A.; Deng, J.; Doney, P.; Sheikh-Akbari, A.; Shimjith, S. R.; Arul, A. J. Machine Learning-Based Fault Diagnosis for a PWR Nuclear Power Plant. *IEEE Access.* **2022**, *10*, 126001–126010. 10.1109/ACCESS.2022.3225966.

[45] Rehm, T. E. Advanced Nuclear Energy: The Safest and Most Renewable Clean Energy. *Curr. Opin. Chem. Eng.* **2023**, *39*, 100878. 10.1016/j.coche.2022.100878.

[46] Hosamo, H. H.; Svennevig, P. R.; Svidt, K.; Han, D.; Nielsen, H. K. A Digital Twin Predictive Maintenance Framework of Air Handling Units Based on Automatic Fault Detection and Diagnostics. *Energy Build.* **2022**, *261*, 111988. 10.1016/j.enbuild.2022.111988.

[47] Pérez-Lombard, L.; Ortiz, J.; Pout, C. A Review on Buildings Energy Consumption Information, *Energy Build.* **2008**, *40*, (3), 394–398. 10.1016/j.enbuild.2007.03.007.

[48] Fernandez, N. E. P.; Katipamula, S.; Wang, W.; Xie, Y.; Zhao, M.; Corbin, C. D. *Impacts of Commercial Building Controls on Energy Savings and Peak Load Reduction*; Richland, WA (United States), 2017. 10.2172/1400347.

[49] Piette, M. A.; Kinney, S. K.; Haves, P. Analysis of an Information Monitoring and Diagnostic System to Improve Building Operations. *Energy Build.* **2001**, *33*, (8), 783–791. 10.1016/S0378-7788(01)00068-8.

[50] Thoppil, N. M.; Vasu, V.; Rao, C. S. P. Deep Learning Algorithms for Machinery Health Prognostics Using Time-Series Data: A Review. *J. Vib. Eng. Technol.* **2021**, *9*, (6), 1123–1145. 10.1007/s42417-021-00286-x.

[51] Ge, X.; Wang, B.; Yang, X.; Pan, Y.; Liu, B.; Liu, B. Fault Detection and Diagnosis for Reactive Distillation Based on Convolutional Neural Network. *Comput. Chem. Eng.* **2021**, *145*, 107172. 10.1016/j.compchemeng.2020.107172.

[52] Doraiswamy, L. K.; Uner, D. *Chemical Reaction Engineering*; CRC Press, 2013. 10.1201/b14951.

[53] Ge, X.; Wang, B.; Yang, X.; Pan, Y.; Liu, B.; Liu, B. Fault Detection and Diagnosis for Reactive Distillation Based on Convolutional Neural Network. *Comput. Chem. Eng.* **2021**, *145*, 107172. 10.1016/j.compchemeng.2020.107172.

[54] El-Shafeiy, E.; Alsabaan, M.; Ibrahem, M. I.; Elwahsh, H. Real-Time Anomaly Detection for Water Quality Sensor Monitoring Based on Multivariate Deep Learning Technique. *Sensors.* **2023**, *23*, (20), 8613. 10.3390/s23208613.

[55] Pan, S. J.; Yang, Q. A Survey on Transfer Learning. *IEEE Trans. Knowl. Data Eng.* **2010**, *22*, (10), 1345–1359. 10.1109/TKDE.2009.191.

[56] Downs, J. J.; Vogel, E. F. A Plant-wide Industrial Process Control Problem. *Comput. Chem. Eng.* **1993**, *17*, (3), 245–255. 10.1016/0098-1354(93)80018-I.

[57] Ricker, N. L. Optimal Steady-state Operation of the Tennessee Eastman Challenge Process, *Comput. Chem. Eng.* **1995**, *19*, (9), 949–959. 10.1016/0098-1354(94)00043-N.

[58] Wu, H.; Zhao, J. Fault Detection and Diagnosis Based on Transfer Learning for Multimode Chemical Processes. *Comput. Chem. Eng.* **2020**, *135*, 106731. 10.1016/j.compchemeng.2020.106731.

[59] Misbahulmunir, S.; Ramachandaramurthy, V. K.; Thayoob, Y. H. M. D. Improved Self-Organizing Map Clustering of Power Transformer Dissolved Gas Analysis Using Inputs Pre-Processing. *IEEE Access.* **2020**, *8*, 71798–71811. 10.1109/ACCESS.2020.2986726.

[60] Loy-Benitez, J.; Li, Q.; Nam, K.; Yoo, C. Sustainable Subway Indoor Air Quality Monitoring and Fault-tolerant Ventilation Control Using a Sparse Autoencoder-driven Sensor Self-validation. *Sustain. Cities Soc.* **2020**, *52*, 101847. 10.1016/j.scs.2019.101847.

[61] Joly, M.; Odloak, D.; Miyake, M. Y., Menezes, B. C.; Kelly, J. D. Refinery Production Scheduling toward Industry 4.0. *Front. Eng. Manage.* **2018**, *0* (0), 0. 10.15302/J-FEM-2017024.

[62] Wu, Y.; Zhang, N. Molecular Characterization of Gasoline and Diesel Streams. *Ind. Eng. Chem. Res.* **2010**, *49*, (24), 12773–12782. 10.1021/ie101647d.

[63] Kelly, J. D.; Zyngier, D. Unit-operation Nonlinear Modeling for Planning and Scheduling Applications. *Optim. Eng.* **2017**, *18*, (1), 133–154. 10.1007/s11081-016-9312-7.

[64] Davy, M. How 70-year-old Abqaiq Plants Transformed to Become a 4IR Leader. *Elements Magazine*, Jul. 31, 2023. Accessed: Nov. 22, 2023. [Online]. Available: https://www.aramco.com/en/news-media/elements-magazine/2023/how-70-year-old-abqaiq-plants-transformed-to-become-a-4ir-leader

[65] Lima, F. V.; Li, S.; Mirlekar, G. V.; Sridhar, L. N.; Ruiz-Mercado, G. Modeling and Advanced Control for Sustainable Process Systems. In *Sustainability in the Design, Synthesis and Analysis of Chemical Engineering Processes*; Elsevier, 2016; 115–139. 10.1016/B978-0-12-802032-6.00005-0.

[66] KPMG and Kearney. *Oil Production Worldwide From 1998 to 2022*; Energy Institute.

[67] Jäschke, J.; Skogestad, S. A Self-Optimizing Strategy for Optimal Operation of A Preheating Train for A Crude Oil Unit. 2014, 607–612. 10.1016/B978-0-444-63456-6.50102-2.

[68] Schwenzer, M.; Ay, M.; Bergs, T.; Abel, D. Review on Model Predictive Control: An Engineering Perspective. *Int. J. Adv. Manuf. Technol.* **2021**, *117*, (5–6), 1327–1349. 10.1007/s00170-021-07682-3.

[69] Yüzgeç, U.; Palazoglu, A.; Romagnoli, J. A. Refinery Scheduling of Crude Oil Unloading, Storage and Processing Using a Model Predictive Control Strategy. *Comput. Chem. Eng.* **2010**, *34*, (10), 1671–1686. 10.1016/j.compchemeng.2010.01.009

15 Worked examples

Gergo Ignacz, Martin Gede, Viktor Toth, Gyorgy Szekely

The following examples are created to help students to practice and better understand green metrics calculations. The green metrics used in the worked examples are introduced and explained in depth in Chapter 1. The examples are the modified versions of real transformations and reactions, but they have been modified to highlight different concepts.

15.1 Example 1 – Green metrics analysis for hazardous chemistry scale-up and decision-making

You are working in a pilot plant of a pharmaceutical company as a scale-up engineer, and you are responsible for the stage 1 scale-up, that is, from laboratory scale to pilot scale. The organic chemist is persuading you to scale up the reaction. A simplified reaction scheme is provided to assist with the decision (Figure 15.1). For simplicity, assume that the reaction yields 100% product using a stoichiometric amount of reagent, a trace amount of catalyst, and no solvent.

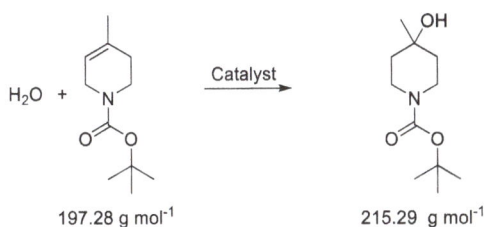

Figure 15.1: Schematic representation of the reaction.

15.1.1 Part A problem statements

(a) Is the reaction in Figure 15.1 complete? Suggest a pathway and catalyst to obtain the final product from the given starting material.
(b) Is the information in the reaction scheme sufficient to decide whether or not to start the scale-up experiments? Justify your answer.

15.1.2 Part B problem statements

The lab-scale procedure is as follows: "A 5 L autoclave was charged with 690.5 g starting material, 1115.4 g Hg(OAc)$_2$, 130 mL water (d = 1.00 g cm^{-3}), and 2.5 L THF (d = 0.889 g cm^{-3}),

https://doi.org/10.1515/9783111028163-015

followed by heating to 95 °C and left for stirring overnight. Then 132.4 g sodium borohy-
dride (NaBH₄) and 400 g of sodium hydroxide (NaOH) were added and left for stirring for
4 h. The organic solvent was removed using distillation, then water (500 mL) and chloro-
form (500 mL, d = 1.49 g cm⁻³) were added for the subsequent extraction. After the extrac-
tion and separation (repeated three times), the organic layer was evaporated using
distillation. The aqueous phase was discharged into the organic waste container due to its
acetate content. The final product was obtained as 749.2 g white powder with 100% purity."

(a) Point out the mistake that requires immediate action regardless of scale-up
considerations.
(b) Identify the reagents, catalyst, solvent, intermediate, product, and side-product(s).
What is missing from the reaction scheme?
(c) Based on the information above, calculate as many green metrics of the system as
possible. Would you recommend the process to be scaled up? Justify your answer.
(d) What are the chemical engineering tools that can mitigate the identified drawbacks?
(e) Is there any reasonable justification to proceed with the hazardous process?

15.1.3 Part A solutions

(a) No, we need more information! One possible solution is depicted in Figure 15.2.
The reaction comprises two steps: (i) the addition of water and mercury acetate
$Hg(OAc)_2$, followed by (ii) the reductive elimination of the mercury.

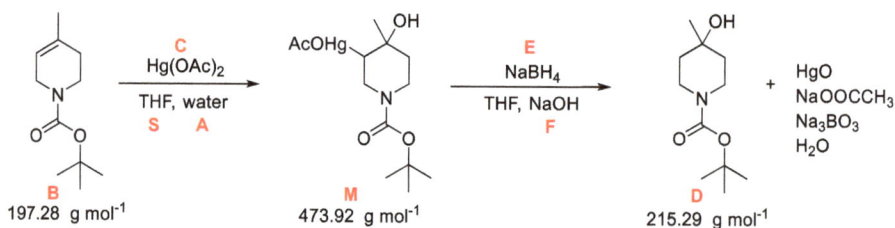

Figure 15.2: Schematic representation of the detailed reaction path for Example 1.

(b) Without any further information, only AE and CE can be calculated:

$$AE = \frac{215.29}{18.01 + 197.28} \cdot 100 = 100\%, \tag{15.1}$$

$$CE = \frac{11}{11} \cdot 100 = 100\%. \tag{15.2}$$

These metrics are sufficient to give an initial idea of a reaction's sustainability, strictly
in theory only. However, your colleague's initial information was superficial, and the

real process is more complicated than it seems. Therefore, these factors are insufficient to make a final decision.

In light of the new information, the calculation of both AE and CE gives different
results. Addition reactions (Step 1) usually have high AE and CE since there is no leaving
group. However, the second reaction is an elimination, which has a low AE value.

$$\text{Step 1: AE} = \frac{\text{MW}_{\text{product}}}{\sum n \times \text{MW}_{\text{reagents}}} \times 100\% = \frac{473.92}{18.01 + 197.28 + 318.67} \cdot 100 = 89\%, \qquad (15.3)$$

$$\text{Step 1: CE} = \frac{\text{Amount of carbon in product}}{\text{Total carbon present in reactants}} \times 100\% = \frac{13}{11+4} \cdot 100 = 87\%, \qquad (15.4)$$

$$\text{Step 2: AE} = \frac{215.29}{473.92} \cdot 100 = 45\%, \qquad (15.5)$$

$$\text{Step 2: CE} = \frac{11}{13} \cdot 100 = 85\%. \qquad (15.6)$$

Besides the overall low AE and CE values, the reaction employs a highly toxic mercury
derivative in stoichiometric amounts. This example demonstrates that a detailed understanding of the reaction is essential to make an appropriate decision. An experimental protocol is necessary to calculate more green metrics.

15.1.4 Part B solutions

(a) The incorrect disposal of the waste containing mercury requires immediate attention. Disposing of mercury requires dedicated waste containers and special treatment.
It is forbidden to dispose of mercury along with organic waste. Waste management
always starts in the lab! It is advisable not to scale up the process based on these poor
metrics and the employed heavily toxic reagent (mercury). Although THF, NaOH, and
$Hg(OAc)_2$ are all considered hazardous, the hazard level is not the same. NaOH is a
commonly used household material and can be used by everyone (using proper per-
sonal protective equipment). THF is frequently used in laboratories but THF may be
carcinogenic and flammable. However, the bottleneck is $Hg(OAc)_2$ because its discharge into the environment has serious consequences that affect all living creatures;
its disposal and transportation are expensive and cumbersome.

(b) The reagents are A, B, C, E, and F. There is no catalyst in the reaction. The solvent is
S (tetrahydrofuran, THF), and the product is D. The side-products are HgO, sodium acetate, water, and boric acid or sodium borate, depending on the amount of sodium hydroxide used. The side-products play an important role in decision-making.

(c) The calculation of green metrics is as follows:

$$RME = \left(\frac{MW_C}{MW_A + (MW_B \times molar\,ratio\,B/A)} \right) \times yield$$

$$= \frac{749.2}{690.5 + 1115.4 + 130 + 132.4 + 400} = 30\%, \tag{15.7}$$

$$EMY = \frac{Mass\,of\,product}{Mass\,of\,non\text{-}benign\,reagents} \times 100\%$$

$$= \frac{749.2}{690.5 + 1115.4 + 132.4 + 400} = 32\%. \tag{15.8}$$

For a comprehensive list of what is considered dangerous or hazardous, follow the QR code on this page.

$$MP = MI^{-1} \times 100\% = \frac{Mass\,of\,product}{Total\,mass\,(excl.H_2O)} \times 100\%$$

$$= \frac{1}{PMI\text{-}m(workup)} = \frac{749.2}{690.5 + 1115.4 + 132.4 + 130 + 400 + 2222.5}$$

$$= 16\%, \tag{15.9}$$

$$sEF = \frac{\sum m_{raw\,materials} + \sum m_{reagents} - m_{product}}{m_{product}}$$

$$= \frac{1115.4 + 132.4 + 400}{749.2} = 2.2\,kg\,kg^{-1}, \tag{15.10}$$

$$cEF = \frac{\sum m_{raw\,materials} + \sum m_{reagents} + \sum m_{solvents} + \sum m_{water} - m_{product}}{m_{product}}$$

$$= \frac{1115.4 + 132.4 + 400 + 3 \cdot 500 \cdot 1.49 + 2500 \cdot 0.889}{749.2} = 8.15\,kg\,kg^{-1}. \tag{15.11}$$

The differences between the simple E-factor (sEF) and the complete E-factor (cEF) are nearly fourfold. As previously mentioned, sEF does take the used solvents into account (in this case, THF and chloroform). Solvent-intensive reactions and purifications significantly increase the EF.

$$MI = \frac{Total\,mass\,in\,process\,(excl.H_2O)}{Mass\,of\,product}$$

$$= \frac{690.5 + 1115.4 + 132.4 + 400 + 3 \cdot 500 \cdot 1.49 + 2500 \cdot 0.889}{749.2}$$

$$= 9.07\,kg\,kg^{-1}, \tag{15.12}$$

$$\text{WWI} = \frac{\text{Mass of total waste water generated}}{\text{Mass of product}}$$

$$= \frac{\left(130 - \dfrac{690.5}{197.28} \cdot 18.01\right) + 3 \cdot 500}{749.2} = 2.09 \text{ kg kg}^{-1}, \tag{15.13}$$

$$\text{SI} = \frac{\text{Mass of solvents (excl. water)}}{\text{Mass of product}}$$

$$= \frac{3 \cdot 500 + 3 \cdot 500 \cdot 1.49 + 2222.5}{749.2} = 7.95 \text{ kg kg}^{-1}, \tag{15.14}$$

$$\text{PMI} = \frac{\text{Total mass in process (incl. } H_2O)}{\text{Mass of product}}$$

$$= \frac{690.5 + 1115.4 + 130 + 2222.5 + 132.4 + 400 + 1500 + 2235}{749.2}$$

$$= 11.25 \text{ kg kg}^{-1}. \tag{15.15}$$

(d) There are several ways to improve the green assessment of the reaction. The conventional THF could be replaced with a green solvent, such as MeTHF, which is a more sustainable choice than THF because it is derived from sugars (through furfural). Also, MeTHF has a higher boiling point, which is appealing for reactions that occur at a higher temperature (although it is more energy-intensive to recover). Compared to THF, MeTHF is non-miscible with water; thus, it could also be used in liquid–liquid extraction processes. Sections 3.3 and 3.4 provide an overview of green solvents and how to select green solvents for a specific process. The solvent and the reagent can be recycled *in situ*, significantly reducing the environmental burden associated with the reaction. $NaBH_4$ and NaOH are already considered green reagents; therefore, their replacement is not mandatory. Changing the reaction from batch to microflow would result in better processing, heat transfer, and selectivity and easier integration of analytical technologies (Chapters 6 and 8, respectively). Read more about solvent recycling processes in Chapter 8.

(e) The reaction can be scaled up if this reaction (Figure 15.2) is the only way to produce D and if D happens to be an extraordinarily valuable product. Since final products can usually be synthesized in many different ways, the above-mentioned criteria are highly unlikely, and a more environmentally friendly and safer reaction route should be designed.

Example 1 illustrates that using green metrics only can be misleading, and one must have high conciseness when designing reaction pathways to be scaled up.

15.2 Example 2 – Green metric analysis of catalytic synthesis and purification of a pharmaceutical building block

Provide a green metric analysis for a reaction with the following experimental procedure. A 1 m³ reactor was filled with dimethylformamide (DMF, 300 kg) as a solvent. 1-Iodo-4-methylbenzene (A, reagent, 32.7 kg), tetrahydrofuran-2-carboxylic acid (B, reagent, 19.16 kg), and a catalyst (C, 1.27 kg, 850 g mol⁻¹ molecular weight) were added, and the reaction mixture was stirred for 30 min to reach completion at room temperature (Figure 15.3). Product D was formed with an average conversion of 93%.

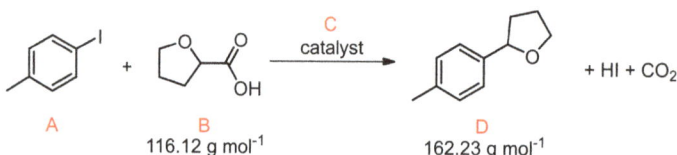

Figure 15.3: Schematic representation of the reaction route for Example 2.

15.2.1 Part A problem statements

(a) Assume that D is 100% pure. Calculate the kernel mass of waste produced in this reaction.
(b) Calculate E_{aux}, E_{excess}, and cEF.

15.2.2 Part B problem statements

"The analytical report revealed the purity of the product to be 94.6%, and the remaining 5.4% is side-product (structurally related to the product). The initial purification method is flash chromatography in which 105 kg silica gel and 720 L eluent mixture (ethyl acetate–heptane, $d = 0.8$ g cm⁻³) were used. The product purity increased to 99.8% after purification."

(a) Calculate $EF_{chromatography}$ and cEF after purification. Compare the previous cEF (without purification) with the new cEF (with purification).
(b) Calculate EF_{kernel} and EF_{aux} after purification.
(c) Which environmental factor has the largest impact on the sustainability of the reaction, and why? Is there any way to reduce the amount of waste?
(d) Your task is to reduce the cEF by 75%. How much weight do you need to recover? What is the easiest material to recover?
(e) Assess the effect of solvent recovery on the EF. Plot your results.
(f) Is it possible to reach 100% solvent recovery?

15.2.3 Part A solutions

(a) MW of the reactants:

$$218.04 + 116.12 = 334.16 \, \text{g mol}^{-1}. \tag{15.16}$$

MW of the products:

$$M(\text{product}) + M(CO_2) + M(HI) = 162.23 + 44.01 + 127.92 = 334.16 \, \text{g mol}^{-1}. \tag{15.17}$$

The limiting reagent is A. The by-products of the reaction are carbon dioxide and hydrogen iodide. The observed number of moles of product D is

$$\frac{22.7}{162.23} = 0.14 \, \text{kmol}. \tag{15.18}$$

Since the balanced chemical equation shows that one mole of HI and CO_2 is produced for every mole of product, the observed mass of CO_2 is

$$0.14 \cdot 44.01 = 6.16 \, \text{kg}. \tag{15.19}$$

and for HI it is

$$0.14 \cdot 127.92 = 17.9 \, \text{kg}. \tag{15.20}$$

Assuming that no side reactions have occurred, the number of moles and masses of unreacted stoichiometric reagents can be summarized in Table 15.1.

Table 15.1: Summary of the reaction in Example 2 with the obtained data for stoichiometric reagents.

Metric	Reagent A	Reagent B
Molecular weight (g mol^{-1})	218.04	116.12
Mass (kg)	32.7	19.16
Moles (kmol)	0.15	0.165
Stoichiometric moles (kmol)	0.15	0.15
Reacted mole (kmol)	0.14	0.14
Unreacted stoichiometric moles (kmol)	0.01	0.01
Unreacted stoichiometric mass (kg)	2.18	1.16
Unreacted mass (kg)	2.18	2.9

The total mass of stoichiometric unreacted reagents is 3.34 kg; therefore, the kernel mass of waste for this reaction is

$$6.16 + 17.9 + 3.34 = 27.4 \, \text{kg}. \tag{15.21}$$

The total mass of input materials is 353.13 kg. The mass of the product collected is 22.71 kg. Therefore, the mass of waste produced in this reaction is

$$353.13 - 22.71 = 330.42 \text{ kg}. \tag{15.22}$$

The sources of waste in this reaction are the reaction solvent, catalyst, unreacted starting materials, and by-product.

(b) The only auxiliary input material used in the reaction is the reaction solvent (DMF). Hence,

$$E_{aux} = \frac{300}{22.71} = 13.2 \text{ kg kg}^{-1}. \tag{15.23}$$

The total mass of excess reagents is 1.74 kg. Hence,

$$E_{excess} = \frac{1.74}{22.71} = 0.077 \text{ kg kg}^{-1}. \tag{15.24}$$

The total mass of waste produced is 330.43 kg. Hence,

$$cEF = \frac{330.43}{22.71} = 14.55 \text{ kg kg}^{-1}. \tag{15.25}$$

15.2.4 Part B solutions

(a) $EF_{chromatography}$ (or any other type of purification) is calculated by the amount of mass of chemical used in the purification compared to the product's mass. In our example, the cEF needs to be updated because the total mass of waste is different after the purification step. The used solvent, the silica gel, and the loss of mass from the final product contribute to the waste's total mass.

$$\text{Total mass of waste} = 330.42 + 576 + 105 + 1.22 \cong 1012.6 \text{ kg}. \tag{15.26}$$

Using the updated total mass of waste, we can calculate the new cEF:

$$cEF = \frac{1012.6}{21.5} = 47.1 \text{ kg kg}^{-1}. \tag{15.27}$$

From the cEF, the chromatography contribution can also be calculated:

$$EF_{chromatography} = \frac{105 + 576}{21.5} = 31.7 \text{ kg kg}^{-1}. \tag{15.28}$$

The difference between equation (15.27) and equation (15.28) is the reaction cEF calculated previously (equation (15.25)).

(b) First, we must update all EFs because the mass of the product has been reduced during the crystallization. The kernel mass of waste is 27.4 kg. Hence,

$$EF'_{kernel} = \frac{27.4}{21.5} = 1.27 \text{ kg kg}^{-1}, \tag{15.29}$$

$$EF'_{aux} = \frac{300 + 105 + 576 + 1.27}{21.5} = 45.7 \text{ kg kg}^{-1}, \tag{15.30}$$

$$EF'_{excess} = \frac{1.74}{21.5} = 0.08 \text{ kg kg}^{-1}, \tag{15.31}$$

$$EF'_{solvent} = \frac{300 + 576}{21.5} = 40.74 \text{ kg kg}^{-1}. \tag{15.32}$$

(c) $EF_{chromatography}$ has the highest impact on sustainability because of the large amount of solvent and silica gel used. Using distillation or membrane filtration would be two examples to regenerate or recycle the solvent. This includes the recycling of DMF and the ethyl acetate–heptane eluent mixture as well. Recovery of high-boiling point solvents is energy-intensive, and thus, in the case of DMF (boiling point: 153 °C), membrane-based recovery is advised. The boiling points of ethyl acetate and heptane are respectively approx. 77 °C and 98 °C, and therefore, they can be separated using distillation.

(d) The 75% reduction means that the new environmental factor is reduced to $cEF'_{new} = 11.78$ by simply taking the 25% of the old $cEF = 47.1$. We can calculate the amount of waste to be reused by taking 75% of the total mass of waste, which is around 760 kg. For example, in some circumstances, silica gel can be reused several times without significant compromise. The ethyl acetate–heptane mixture can be recovered using distillation, and the DMF can be recovered via membrane separation. To do this, we need to plot the cEF as a function of the material recovery rate. Recycling only the silica gel,

$$cEF = \frac{330.42 + 576 + 1.22}{21.5} = 42 \text{ kg kg}^{-1}. \tag{15.33}$$

Recovering the DMF postreaction,

$$cEF = \frac{30.42 + 576 + 1.22}{21.5} = 28 \text{ kg kg}^{-1}. \tag{15.34}$$

Recovering the ethyl acetate–heptane mixture after purification,

$$cEF = \frac{30.42 + 1.22}{21.5} \cong 1.5 \text{ kg kg}^{-1}. \tag{15.35}$$

The regeneration of silica gel (equation (15.33)) is a step function (either reused or not). In contrast, the recycling of DMF and the ethyl acetate–heptane mixture is a linear function of the material recovery (Figure 15.4).

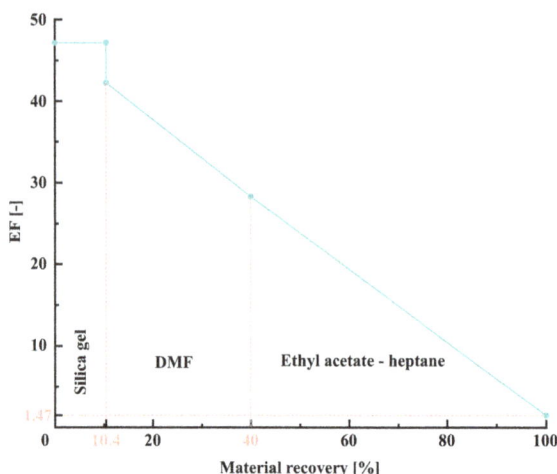

Figure 15.4: EF as a function of material recovery. Note that silica gel is not a continuous function in this case because of its complete reuse.

(e) Although it is possible, it is not advised to reach 100% recovery because other processes' collateral costs will increase the cEF overall.

15.3 Example 3 – Comparison of batch and microflow processes in diazomethane-based chemistry

In the following example, we examine 3-hydroxyquinolin-2(1H)-one synthesis using two different approaches. The first approach is a conventional batch synthesis, and the second approach is a microflow chemical synthesis of 3-hydroxyquinoline-2(1H)-one. This example highlights the differences between batch and microflow procedures, calculates the different green metrics, and proposes a sustainable solution.

15.3.1 Part A problem statements

A 1.5 m^3 reactor was charged with 29.4 kg of 1, followed by the slow addition of 200 kg of ethanol. A solution of 27.4 kg 2 in 89 kg ethanol was slowly added to the reaction mixture, followed by the slow addition of 4.57 kg catalyst (1,8-diazabicyclo[5.4.0] undec-7-ene [DBU] MW: 152.24 g mol^{-1}). After completion of the reaction (3 h) at room temperature, 0.884 kg of catalyst 2 ($Rh_2(OAc)_4$) was added to the mixture, and the reaction was left to reach completion (1 h). The intermediate product 4 was formed as white crystals in 81.9% yield. After particle filtration and air drying, the crystals were transferred to a second reactor. The second reactor was charged with 100 L water, followed by slow addition of aqueous sodium hydroxide (16.2 kg NaOH in 223 kg water), heated to reflux, and left for stirring until reaction completion (7 h). After completion, the reaction was left to cool down to room temperature, and aqueous hydrogen chloride was added to set the pH to 1–2 (225 kg, 2 M, $d = 1.082$ k gm^{-3}), which resulted in the precipitation of the final product 6 in 90% yield. To reach GMP purity, the product must be recrystallized using an ethanol–water mixture (210 kg in total) with a 95% yield. We ignore the heating in this example (Figure 15.5).

(a) Calculate AE, CE, and RME for the two separate reactors and overall.
(b) Calculate cEF, EMY, PMI, WWI, and SI for the two separate reactors and overall excluding crystallization.
(c) What is the effect of including the crystallization in the analysis? Which green metrics are affected? Calculate cEF, PMI, and SI after crystallization for the whole process.

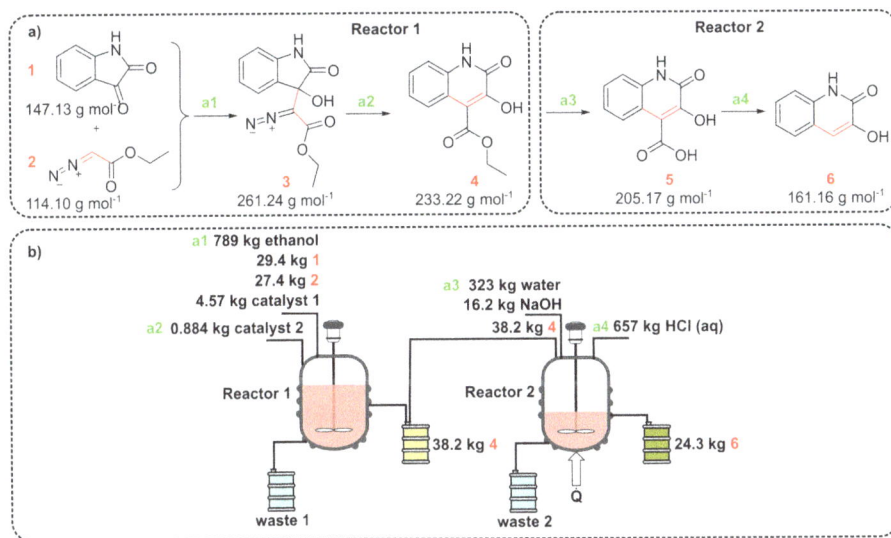

Figure 15.5: Reaction pathways for Reactors 1 and 2 (a) and the schematic representation of Reactors 1 and 2 with the input and output materials (b).

15.3.2 Part B problem statements

The product can be synthesized under continuous flow conditions. A suitable pump provides NaOH solution (0.6 M) in water–ethanol (1:1 v/v, $d = 0.98$ g cm^{-3}) mixture with a flow rate of 0.4 mL min^{-1}. Another pump provides Diazald solution (0.6 M) in water–ethanol (1:1 v/v, $d = 0.92$ g cm^{-3}) mixture in 0.4 mL min^{-1}. The two streams are combined (the inner tubing is made of a special semi-permeable membrane that allows quick gas diffusion but nothing else), and the diazomethane forms. The conversion of Diazald is considered 100%. The starting material 1 in dry THF (0.4 M, d = 0.9 g cm^{-3}) is supplied via continuous pumping (0.4 mL min^{-1}) in the outer tubing. As discussed in Section 6.3.2, the diazomethane diffuses into the outer tubing, reacting with 1. The final product 6 and the waste stream are collected at a 0.4 mL min^{-1} flow rate. Owing to the high selectivity and separated reaction mixtures, the product needs no further purification steps. Assume that p-toluenesulfonic acid sodium salt (NaPTSA) at the end of the reaction is considered a waste. A schematic process diagram is presented in Figure 15.6.

(a) Calculate AE, CE, and RME for the flow reactor considering the system as a whole.
(b) Calculate cEF, EMY, PMI, WWI, and SI for the flow reactor considering the system as a whole.
(c) Compare the processes in Part A and Part B, and discuss the differences and their limitations.

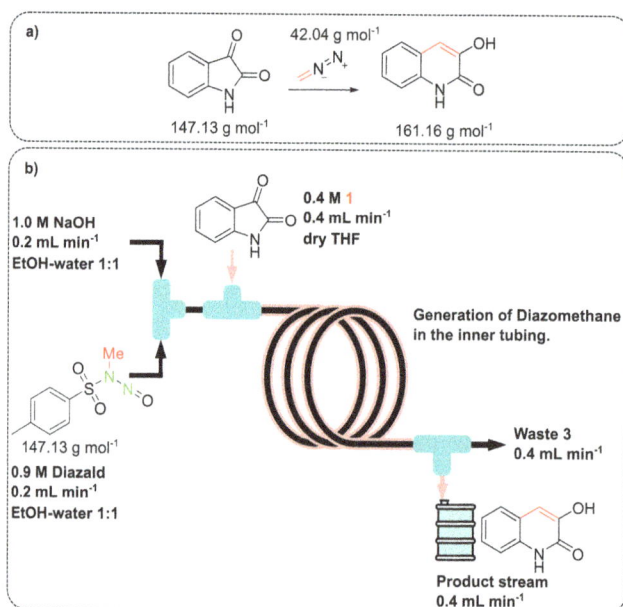

Figure 15.6: Reaction pathway for the flow reactor (a), and the schematic representation of the flow reactor with the input and output materials (b).

15.3.3 Part A solutions

(a) Since everything is given, we only need to calculate the required green metrics. Starting with an overall atom economy,

$$AE_1(\%) = \frac{MW_{product}}{\sum n \times MW_{reagents}} \times 100\% = \frac{233.22}{147.13 + 114.1} \cdot 100 \cong 89\%, \tag{15.36}$$

$$AE_2(\%) = \frac{161.16}{233.22} \cdot 100 \cong 69\%, \tag{15.37}$$

$$AE_{total}(\%) = \frac{161.16}{147.13 + 114.1} \cdot 100 \cong 62\%,$$

$$AE_{total} = AE_1 \cdot AE_2 = 0.892 \cdot 0.692 \cong 62. \tag{15.38}$$

Equation (15.36) reveals that the AE is relatively low due to the triple eliminations after the first addition reaction. Moreover, there is one carbon lost as CO_2 and two carbons are lost as ethanol, which lowers the carbon efficiency:

$$CE_1(\%) = \frac{\text{Amount of carbon in product}}{\text{Total carbon present in reactants}} \times 100\% = \frac{12}{8+4} \cdot 100 = 100\%, \tag{15.39}$$

$$CE_2(\%) = \frac{9}{12} \cdot 100 = 75\%, \tag{15.40}$$

$$CE_{total}(\%) = \frac{9}{8+4} \cdot 100 = 75\%, \tag{15.41}$$

Calculating the mass reaction efficiency for the first reactor,

$$RME_1(\%) = \left(\frac{MW_4}{MW_1 + (MW_2 \times \text{molar ratio } B/A)} \right) \times \text{yield}$$

$$= \frac{233.22}{147.13 + 1.2 \cdot 114.1} \cdot 81.9 = 67\%, \tag{15.42}$$

$$RME_2(\%) = \left(\frac{MW_6}{MW_4} \right) \times \text{yield} = \frac{161.16}{233.22} \cdot 90 = 62\% \tag{15.43}$$

$$RME_{total}(\%) = \left(\frac{MW_6}{MW_1 + (MW_2 \times \text{molar ratio } B/A)} \right) \times \text{yield}$$

$$= \frac{161.16}{147.13 + 1.2 \cdot 114.1} \cdot (0.9 \cdot 81.9) = 41\%. \tag{15.44}$$

By definition, RME_{total} is the product of the two previous processes: $RME_{total} = RME_1 \cdot RME_2$. Note that the catalyst is not included in the RME since they are not incorporated into the final product. Although NaOH and HCl are not catalysts, they are also not included in the RME calculation for the same reason.

(b) The AE, CE, and RME metrics calculated above do not truly focus on the process itself but on the reaction. They are only useful for preliminary calculations and assumptions. Therefore, we need to calculate additional metrics to be able to compare the two different processes.

Since our process mixes water with an organic solvent, the water itself is always contaminated to some degree; therefore, it is advised to use cEF instead of sEF because sEF does not consider water a waste. Thus, using sEF in a mixed organic–water system can lead to data misrepresentation and result in environmental issues upon water release.

$$
\begin{aligned}
\text{cEF}_1 &= \frac{\sum m_{\text{raw materials}} + \sum m_{\text{reagents}} + \sum m_{\text{solvents}} + \sum m_{\text{water}} - m_{\text{product}}}{m_{\text{product}}} \\
&= \frac{289 + 29.4 + 27.4 + 4.57 + 0.884 - 38.2}{38.2} = 8.2\,\text{kg kg}^{-1},
\end{aligned}
\tag{15.45}
$$

$$
\text{cEF}_2 = \frac{38.2 + 323 + 16.2 + 225 - 24.3}{24.3} = 23.8\,\text{kg kg}^{-1},
\tag{15.46}
$$

$$
\text{cEF}_{\text{total}} = \text{cEF}_1 + \text{cEF}_2 = 31.0\,\text{kg kg}^{-1}.
\tag{15.47}
$$

Calculation of the effective mass yield gives

$$
\begin{aligned}
\text{EMY}_1(\%) &= \frac{\text{Mass of product}}{\text{Mass of non-benign reagents}} \times 100\% \\
&= \frac{38.2}{29.4 + 27.4 + 4.57 + 0.884} \cdot 100 = 61\%.
\end{aligned}
\tag{15.48}
$$

Note that we consider ethanol as a benign material. Ethanol is probably the greenest solvent. It is vastly abundant in nature and can be supplied from a sustainable source. It is slightly toxic ($LD_{50} = 7$ g kg^{-1}),[1] although most living creatures can metabolize it. It is also easy to recover, with a moderately low boiling point. Ethanol is probably often contaminated with reagents and cannot be discarded to nature; thus, it must be treated to avoid uncontrolled waste disposal.

$$
\text{EMY}_2(\%) = \frac{24.3}{38.2 + 323 + 26.2 + 25} \cdot 100 = 4\%.
\tag{15.49}
$$

It does not matter that a benign product (NaCl) will be one of the products at the end; the corresponding reagents (NaOH and HCl) are still considered non-benign.

$$
\text{EMY}_{\text{total}}(\%) = \text{EMY}_1 \cdot \text{EMY}_2 = 2.4\%.
\tag{15.50}
$$

Calculation of the PMI gives

$$
\begin{aligned}
\text{PMI}_1 &= \frac{\text{Total mass in reactor 1 (incl. H}_2\text{O)}}{\text{Mass of product 4}} \\
&= \frac{289 + 29.4 + 27.4 + 4.57 + 0.884}{38.2} = 9.19\,\text{kg kg}^{-1},
\end{aligned}
\tag{15.51}
$$

---

1 LD_{50} is the lethal dose with a 50% mortality rate. Therefore, in our example, approximately 700 mL of pure ethanol would result in a 50% mortality rate in a population of humans weighing on average 80 kg.

$$PMI_2 = \frac{\text{Total mass in reactor 2 (incl. } H_2O)}{\text{Mass of product 6}}$$

$$= \frac{38.2 + 323 + 16.2 + 225}{24.3} = 24.8 \text{ kg kg}^{-1}, \tag{15.52}$$

$$PMI_{total} = \frac{\text{Total mass in process (incl. } H_2O)}{\text{Mass of product 6}}$$

$$= \frac{289 + 29.4 + 27.4 + 4.57 + 0.884 + 323 + 16.2 + 225}{24.3}$$

$$= 38 \text{ kg kg}^{-1}. \tag{15.53}$$

With respect to the calculation of wastewater intensity, since water has not been used nor formed in the first reactor, WWI is zero:

$$WWI_1 = \frac{\text{Mass of total waste water generated}}{\text{Mass of product}} = 0 \text{ kg kg}^{-1}. \tag{15.54}$$

Compared to this, water has been heavily used in the second reactor, and the overall WWI will be

$$WWI_2 = WWI_{total} = \frac{323 + 210}{24.3} = 22 \text{ kg kg}^{-1}. \tag{15.55}$$

Calculation of the solvent intensity gives

$$SI_1 = \frac{\text{Mass of solvents (excl. water)}}{\text{Mass of product}} = \frac{289}{38.2} = 7.56 \text{ kg kg}^{-1}, \tag{15.56}$$

$$SI_2 = \frac{0}{24.3} = 0 \text{ kg kg}^{-1}, \tag{15.57}$$

$$SI_{total} = \frac{289}{24.3} = 11.89 \text{ kg kg}^{-1}. \tag{15.58}$$

(c) Crystallization does not involve chemical transformation; thus, the reaction-based metrics are not subject to change. For example, AE, CE, and RME will not change, and EMY and WWI will slightly change due to the loss of product after the crystallization. If we use a non-benign solvent or water for recrystallization, EMY and WWI will change significantly. To update the cEF, we need to implement the solvent weight used for recrystallization.

$$cEF_{crystallization} = \frac{289 + 29.4 + 27.4 + 4.57 + 0.884 + 323 + 16.2 + 225 + 200 - 23.1}{23.1}$$

$$= 47.3 \text{ kg kg}^{-1}, \tag{15.59}$$

$$\begin{aligned} \text{PMI}_{\text{total}} &= \frac{\text{Total mass in process (incl. H}_2\text{O})}{\text{Mass of product 6} \times \text{yield}} \\ &= \frac{289 + 29.4 + 27.4 + 4.57 + 0.884 + 323 + 16.2 + 225 + 200}{24.3 \cdot 0.95} \\ &= 48.3 \, \text{kg kg} -1, \end{aligned} \tag{15.60}$$

$$\text{SI}_{\text{crystallization}} = \frac{289 + 200}{23.1} = 21.2 \, \text{kg kg}^{-1}. \tag{15.61}$$

15.3.4 Part B solutions

The continuous flow configuration does not change how we calculate the metrics because all metrics are *unitless*. This results in a reasonably straightforward calculation process considering that the reaction in Part B is less complicated compared to Part A.

(a) The reaction partners are the three starting materials: Diazald, NaOH, and 1.

$$\text{AE}_{\text{flow}}(\%) = \frac{161.16}{147.13 + 212.24 + 40.01} \cdot 100 = 40\%, \tag{15.62}$$

$$\text{CE}_2(\%) = \frac{9}{8+8} \cdot 100 = 56\%. \tag{15.63}$$

The ratio between diazomethane and the substrate is 0.45/0.4 due to the 100% conversion of Diazald. The excess amount of NaOH compared to 1 is 0.5/0.4.

$$\text{RME}_{\text{flow}}(\%) = \frac{161.16}{147.13 + 1.125 \cdot 214.24 + 1.25 \cdot 40.01} \cdot (0.9 \cdot 81.9) = 27\%. \tag{15.64}$$

(b) The calculation of cEF is simple; we need to follow the definition and simply add the different input material flow rates of the streams. To calculate the mass of the product mass flow rate,

$$\begin{aligned} \dot{m}_{\text{product}} &= 0.4 \, \text{mL min}^{-1} \cdot 0.4 \, \text{mol L}^{-1} \cdot 161.16 \, \text{g mol}^{-1} \\ &= 25.8 \, \text{mg min}^{-1}, \end{aligned} \tag{15.65}$$

$$\begin{aligned} \text{cEF}_{\text{flow}} &= \frac{0.2 \, \text{mL min}^{-1} \cdot 0.98 \, \text{g cm}^{-3} + 0.2 \, \text{mL min}^{-1} \cdot 0.92 \, \text{g cm}^{-3}}{25.8 \, \text{mg min}^{-1}} \\ &\quad + \frac{0.4 \, \text{mL min}^{-1} \cdot 0.9 \, \text{g cm}^{-3} - (25.8 + 35.0 \, \text{mg min}^{-1})}{25.8 \, \text{mg min}^{-1}} \\ &= 27.7. \end{aligned} \tag{15.66}$$

Based on the reasoning in Part A, we consider the ethanol–water mixture as benign even though it is probably contaminated with the reagents. Although Diazald is non-benign, the resulting *p*-toluene sulfonic acid sodium salt is a non-toxic material; thus, the waste stream could be considered benign. It must be mentioned that even if the waste stream is benign, it does not mean that it can be released into nature. Proper waste management is crucial in downstream processing!

$$EMY = \frac{0.2\,\text{mL min}^{-1} \cdot 0.92\,\text{g cm}^{-3}}{25.8\,\text{mg min}^{-1}} \cdot 100 = 7100\%, \tag{15.67}$$

$$PMI_{flow} = \frac{0.2\,\text{mL min}^{-1} \cdot 0.98\,\text{g cm}^{-3} + 0.2\,\text{mL min}^{-1} \cdot 0.92\,\text{g cm}^{-3}}{25.8\,\text{mg min}^{-1}}$$

$$+ \frac{0.2\,\text{mL min}^{-1} \cdot 0.9\,\text{g cm}^{-3}}{25.8\,\text{mg min}^{-1}}$$

$$= 28.6\,\text{kg kg}^{-1}. \tag{15.68}$$

The density of the water is approx. $1\,\text{g cm}^{-3}$, thus for WWI,

$$WWI_{flow} \approx \frac{0.2\,\text{g min}^{-1} \cdot 1.00\,\text{g cm}^{-3}}{25.8\,\text{mg min}^{-1}} \approx 7.75\,\text{kg kg}^{-1}. \tag{15.69}$$

The density of ethanol is $0.789\,\text{g cm}^{-3}$ (from Part A). Using this, for the SI

$$SI_{flow} \approx \frac{0.2\,\text{g min}^{-1} \cdot 0.789\,\text{g cm}^{-3}}{25.8\,\text{mg min}^{-1}} \approx 6.60\,\text{kg kg}^{-1}. \tag{15.70}$$

Equations (15.69) and (15.70) are only close approximations due to the liquids' density changes after mixing.

(c) The best way to compare the processes is to create a table from our results. Table 15.2 shows a comparison between Part A and Part B.

   In our example, Parts A and B score the same in the calculated metrics. Interestingly, if we exclude the crystallization from Part A, the favor would shift towards the batch reactor instead of the flow reactor. Reagent 2 (Part A) is a *sacrificial reagent* due to the CO_2 loss at the end, meaning the only usable material from 2 is the leftover ethanol in a small quantity. Notice that this ethanol is different from the solvent! Compared to this, p-toluene sulfonic acid sodium salt (the end product of Part B) is reusable (p-toluenesulfonic acid is a widely used reagent) and, more importantly, Diazald can be made from p-toluenesulfonic acid. Thus – in theory – the waste stream could not just be utilized but regenerated, meeting the circular economy principles. The last statement indicates that the p-toluene sulfonic acid is better not considered a waste rather than another useful product. The last column in Table 15.2 shows how the metrics change if we consider the NaPTSA as a valuable side-product and not a waste.

Table 15.2: Comparison between the metrics from Part A and Part B. The bordered area denotes the final results. PTSAS means that the p-toluenesulfonic acid sodium salt is considered a product and not a waste (it can be used as a catalyst itself or can be converted to other useful chemicals as well).

Metric	Part A			Part B		
	Reactor 1	Reactor 2	Reactors 1 & 2	After crystallization	Flow reactor	With NaPTSA
AE (%)	89	69	62	62	40	100
CE (%)	100	75	75	75	56	100
RME (%)	67	62	41	41	37	56
cEF (kg kg^{-1})	8.2	23.8	31	47.3	27.7	11.2
EMY (kg kg^{-1})	61	4	2.4	2.4	7.10	3.02
PMI (kg kg^{-1})	9.19	24.8	38	48.3	28.6	9.22
WWI (kg kg^{-1})	0	22	22	22	7.5	3.29
SI (kg kg^{-1})	7.56	0	11.89	21.2	6.60	2.59

15.4 Example 4 – Bioethanol production: conventional batch fermentation versus continuous membrane bioreactor

15.4.1 Part A problem statements

In this example, bioethanol was produced via a conventional batch process and purified with extractive distillation, as illustrated in Figure 15.7. During the fermentation, glucose was converted into ethanol by yeast cells in an aqueous solution, producing CO_2 as a co-product. The reaction medium consisted of 120 g L^{-1} glucose, 5 g L^{-1} yeast extract, 3 g L^{-1} peptone, 3.5 g L^{-1} KH$_2$PO$_4$, and 2 g L^{-1} MgSO$_4$. The fermentation was performed for 24 h, which yielded 6.5 vol% ethanol with a residual sugar concentration of 4 g L^{-1}. After the biomass was filtered out, the ethanol–water mixture was separated by distillation. To overcome the azeotropism, extractive distillation was used, with the addition of ethylene glycol and calcium chloride as entrainer agents. For each kilogram of ethanol produced, 0.4 kg ethylene glycol and 35 g calcium chloride were consumed. The overall energy consumption of the purification process was 1425 kJ kg^{-1} of ethanol. Refer to Chapter 11 to learn about the sustainable production of biofuels.

(a) What are the bottlenecks of this production process? Recommend some possible solutions to maximize the energy, material, and economic efficiency of this production.

(b) Green metrics are important for comparing the production process with alternative methods and devising which one is more suitable from an environmental perspective. Calculate the AE, RME, CE, EMY, E-factors (and the corresponding EF$_{kernel}$, EF$_{excess}$, and EF$_{aux}$ values, the complete E-factor (cEF), and E$^+$), MI, PMI, MP, and WWI.

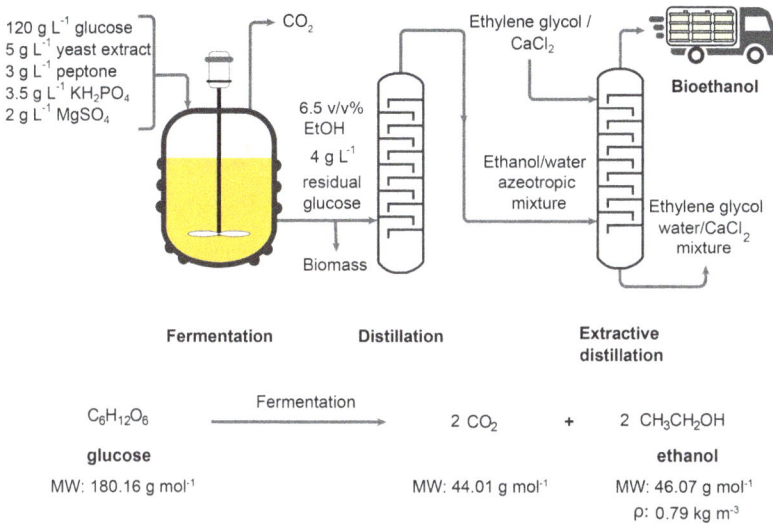

Figure 15.7: Schematic representation of the conventional production of bioethanol. The glucose, yeast, and additional substances are fed into the reactor, which eventually yields 6.5 vol% ethanol. After the biomass is separated, the ethanol–water mixture is distilled until the azeotropic point, followed by an extractive distillation to purify the ethanol further and make it suitable for biofuel.

15.4.2 Part A solutions

(a) The possible bottlenecks of this production process originate from the fermentation process itself: ethanol production from glucose yields a significant amount of CO_2, which means that approx. half of the glucose mass is lost. It would be beneficial to capture and utilize the CO_2, for example, to produce carbonated beverages. The biomass may also represent a source of income; for example, it can be sold as a fertilizer. The fermentation process requires external cooling; therefore, this heat surplus can be utilized for purification purposes because distillation requires considerable energy. Extractive distillation also generates waste in the form of aqueous ethylene glycol; hence, the entrainer agent's regeneration is advisable.

(b)
$$AE(\%) = \frac{\text{Molecular Weight}_{product}}{\text{Molecular Weight}_{all\ reagents}} \times 100\%$$
$$= \frac{\text{Molecular Weight}_{ethanol}}{\text{Molecular Weight}_{glucose}} \times 100\%$$
$$= \frac{46.07}{0.5 \cdot 180.16} \cdot 100\% = 51.14\%. \tag{15.71}$$

The fermentation starts with 120 g L^{-1} glucose, and a final ethanol concentration is 6.5 vol%. This means that 1 L of fermentation broth contains 65 mL of ethanol, and by multiplying it with the density (0.79 g mL^{-1}), we get the ethanol concentration of

51.35 g L^{-1}. According to the fermentation equation (illustrated in Figure 15.7), ideally, 120 g L^{-1} glucose (0.67 mol) would yield 1.33 mol of ethanol per L, which is 61.37 g L^{-1}. The yield can be calculated as follows:

$$\text{Yield}(\%) = \frac{\text{Weight of product}}{\text{Theoretical weight}} \times 100\% = \frac{51.35}{61.37} \cdot 100\% = 83.67\%, \tag{15.72}$$

$$\begin{aligned}\text{RME}(\%) &= \frac{\text{mass of desired product}}{\text{mass of reactant}} \times 100\% \\ &= \frac{\text{mass of ethanol}}{\text{mass of glucose}} \times 100\% = \frac{51.35}{120} \cdot 100\% = 42.79\%.\end{aligned} \tag{15.73}$$

$$\begin{aligned}\text{CE}(\%) &= \frac{\text{Amount of carbon in product}}{\text{Total carbon present in reactants}} \times 100\% \\ &= \frac{\text{moles of ethanol} \times \text{carbons in ethanol}}{\text{moles of glucose} \times \text{carbons in glucose}} \times 100\% \\ &= \frac{2 \cdot 2}{1 \cdot 6} \cdot 100\% = 66.67\%.\end{aligned} \tag{15.74}$$

The EMY is the percentage of the desired product's mass divided by the total mass of all non-benign materials used in the synthesis. Since the fermentation takes place in a diluted aqueous medium and the glucose and the added salts are non-toxic, only the KH_2PO_4 can be considered a potentially non-benign substance.

$$\begin{aligned}\text{EMY}(\%) &= \frac{\text{Mass of product}}{\text{Mass of non-benign reagents}} \times 100\% = \frac{51.35}{3.5} \cdot 100\% \\ &= 1467.14\%.\end{aligned} \tag{15.75}$$

EF$_{total}$ can be calculated as follows:

$$\text{EF}_{total} = \text{EF}_{kernel} + \text{EF}_{excess} + \text{EF}_{aux}. \tag{15.76}$$

EF$_{kernel}$ originates from the core chemical reaction; EF$_{excess}$ consists of all the un-reacted material that does not contribute to the product, even though it is vital for the yeast cells' functioning. EF$_{aux}$ includes the waste generated during the extractive dis-tillation, involving ethylene glycol and calcium chloride. The starting glucose concen-tration is 120 g L^{-1}, and the final fermentation broth contains 51.35 g L^{-1} ethanol; thus, the EF$_{kernel}$ can be calculated following equation (15.77), where the total mass waste involves the carbon dioxide co-product, the possible by-products (different alcohols), and the residual glucose.

$$\begin{aligned}\text{EF}_{kernel} &= \frac{\text{Total mass of waste (excl. } H_2O)}{\text{Mass of final product}} = \frac{120 - 51.35}{51.35} = 1.34 \, g\, g^{-1} \\ &= 1.34 \, kg\, kg^{-1}.\end{aligned} \tag{15.77}$$

EF_{excess} comprises the nitrogen, potassium, and phosphorus sources:

$$EF_{excess} = \frac{\text{Total mass of waste (excl. } H_2O)}{\text{Mass of final product}} = \frac{5 + 3 + 3.5 + 2}{51.35} = 0.26 \text{ g g}^{-1}$$

$$= 0.26 \text{ kg kg}^{-1}, \tag{15.78}$$

$$EF_{aux} = \frac{\text{Total mass of waste (excl. } H_2O)}{\text{Mass of final product}} = \frac{400 + 35}{1000} = 0.435 \text{ g g}^{-1}$$

$$= 0.435 \text{ kg kg}^{-1}. \tag{15.79}$$

Therefore, EF_{total} is expressed as follows:

$$EF_{total} = EF_{kernel} + EF_{excess} + EF_{aux} = 1.34 + 0.26 + 0.435 = 2.035 \text{ kg kg}^{-1}. \tag{15.80}$$

cEF takes into account the solvent, which is water. The ethanol concentration of the final broth (51.35 g L^{-1}) equals 6.5 vol%; therefore, the amount of water is 935 mL. Keep in mind that the total waste generated during the purification is given for 1 kg of ethanol; therefore, it has to be corrected for 51.35 g ethanol. The cEF for 1 L of fermentation broth can be calculated as follows:

$$cEF = \frac{\sum m_{\text{raw materials}} + \sum m_{\text{reagents}} + \sum m_{\text{solvents}} + \sum m_{\text{water}}}{m_{\text{product}}}$$

$$+ \frac{\sum m_{\text{purification}} - m_{\text{product}}}{m_{\text{product}}}$$

$$= \frac{m_{\text{glucose}} + m_{\text{reagents}} + m_{\text{water}} + \sum m_{\text{purification}} - m_{\text{product}}}{m_{\text{product}}} \tag{15.81}$$

$$= \frac{120 + 13.5 + 935 + \left(435 \cdot \frac{51.35}{1000}\right) - 51.35}{51.35} = 20.24 \text{ g g}^{-1}$$

$$= 20.24 \text{ kg kg}^{-1}.$$

The energy consumption of the purification process is 1,425 kJ kg^{-1}, which has to be converted into kWh by dividing it by 3,600 s, and then it can be inserted into the equation:

$$E^+ = \frac{\sum m_{\text{waste}}}{m_{\text{product}}} \left[\text{kg kg}^{-1}\right] + \frac{W[\text{kWh}] \times CO_2EF\left[\text{kgCO}_2\text{kWh}^{-1}\right]}{m_{\text{product}}[\text{kg}]}$$

$$= 1.41 \text{kg kg}^{-1} + \frac{\frac{1425\text{kJ}}{3600\text{s}} \cdot 0.30720}{1} = 1.53 \text{kg kg}^{-1}. \tag{15.82}$$

$$MI = \frac{\text{Total mass in process (excl. } H_2O)}{\text{Mass of product}}$$

$$= \frac{120 + 13.5 + \left(435 \cdot \frac{51.35}{1000}\right)}{51.35} = 3.03 \,\text{g g}^{-1} = 3.03 \,\text{kg kg}^{-1}. \tag{15.83}$$

$$PMI = \frac{\text{Total mass in process (incl. } H_2O)}{\text{Mass of product}}$$

$$= \frac{120 + 13.5 + \left(435 \cdot \frac{51.35}{1000}\right) + 935}{51.35} = 21.24 \,\text{g g}^{-1} = 21.24 \,\text{kg kg}^{-1}. \tag{15.84}$$

$$MP = MI^{-1} \cdot 100\% = \frac{\text{Mass of product}}{\text{Total mass (excl. } H_2O)} \cdot 100\%$$

$$= 3.03^{-1} \cdot 100\% = 33.0\%. \tag{15.85}$$

$$WWI = \frac{\text{Mass of total waste water generated}}{\text{Mass of product}} = \frac{935}{51.35}$$

$$= 18.21 \,\text{g g}^{-1} = 18.21 \,\text{kg kg}^{-1}. \tag{15.86}$$

15.4.3 Part B problem statements

Part A described the batch-to-batch production of bioethanol and its purification with extractive distillation. This method's main bottlenecks are the low ethanol yield due to the unconverted sugar content resulting from product inhibition and the energy-intensive distillation process. A greener alternative is offered by continuous production and the implementation of less energy-demanding purification processes. In Part B, ethanol is separated in a multistage membrane filtration process, as illustrated in Figure 15.8. The continuous separation of ethanol from the fermentation broth is achieved by applying direct contact membrane distillation. The solvent (water), residual sugars, and yeast cells are recycled back into the fermentation. After steady state is achieved, ethanol's productivity is 14.6 g L^{-1} h^{-1}, and the ethanol flux is 29.5 kg m^{-2} 24 h^{-1} with a total membrane surface of 900 cm^2. The dilution rate is 0.15 h^{-1}, and the feed composition is the following: 200 g L^{-1} glucose, 1.0 g L^{-1} urea, 1.5 g L^{-1} $(NH_4)_2SO_4$, 1.75 g L^{-1} KH_2PO_4, and 0.75 g L^{-1} $MgSO_4$. The pumps' energy consumption is 3.2 kWh m^{-3}, while the membrane distillation consumes 7 kWh m^{-3}.
(a) Calculate the E-factors (and the corresponding EF_{kernel}, EF_{excess}, and EF_{aux} values, the complete E-factor (cEF), and E^+), MI, PMI, MP, WWI. Note that the dilution rate is given; however, we do not know the reaction volume. First, we have to calculate the permeate flow rate of ethanol and then get the reaction volume and the corresponding feed volume.

(b) Compare the results with the conventional production described in Part A. What are the key differences between the two processes?

(c) The wastewater intensity, in this case, is close to zero since the production is done in a closed-loop fashion, where water is recycled back to the fermentor. However, the system requires a certain amount of maintenance and cleaning, which results in the disposal of the fermentation broth, thus generating wastewater. Calculate the wastewater intensity if maintenance occurs every 24 h, every 72 h, every week, or every three months.

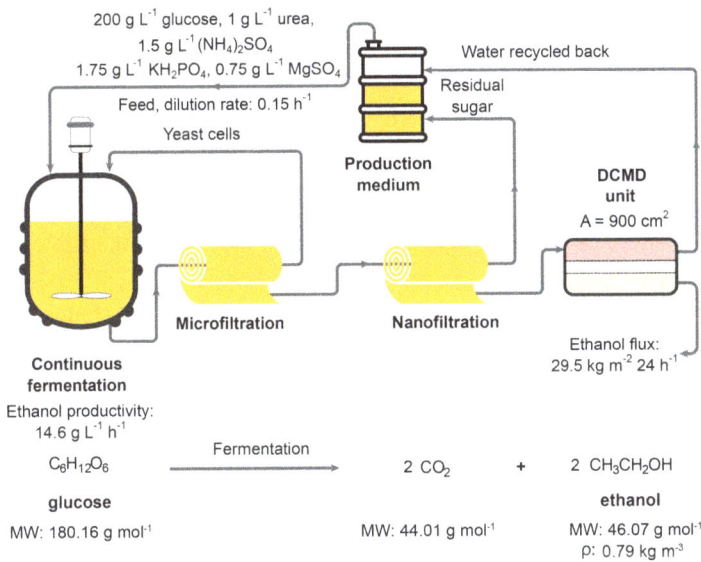

Figure 15.8: Schematic representation of a continuous fermentation coupled with a multistage membrane filtration process. Bioethanol is continuously separated from the fermentation broth, with the implementation of direct contact membrane distillation (DCMD). Simultaneously, the yeast cells, residual sugar, and water are recycled back into the fermentor.

15.4.4 Part B solutions

(a) The permeate flow rate in the membrane distillation unit, which is the pure ethanol, can be calculated using the flux and the total membrane surface area. Since the ethanol flux is given for a day, it has to be divided by 24 h.

$$Q_{permeate} = F \times A = \frac{29.5}{24} \cdot \frac{900}{10^4} = 1.23 \cdot 0.09 = 0.11 \, \text{kg h}^{-1}. \tag{15.87}$$

The productivity of the fermentation unit is 14.6 g $L^{-1} h^{-1}$; therefore, the volume of the reaction is:

$$V_{reaction} = \frac{Q_{permeate}}{Productivity} = \frac{0.11 \frac{kg}{h}}{0.0146 \frac{kg}{Lh}} = 7.53 \text{ L.} \tag{15.88}$$

The feed rate can be calculated as the multiplication of the dilution rate and $V_{reaction}$:

$$\text{Feed} = V_{reaction} \times \text{Dilution rate} = 7.53 \text{ L} \cdot 0.15 \frac{1}{h} = 1.13 \text{ Lh}^{-1}. \tag{15.89}$$

The feed consumption for each kilogram of ethanol can be calculated as follows:

$$\text{Feed consumption}_{1 \text{ kg ethanol}} = \frac{1 \text{ kg ethanol}}{Q_{permeate}} \times \text{Feed} = \frac{1}{0.11} \cdot 1.13$$

$$= 10.27 \text{ L.} \tag{15.90}$$

The composition of the feed is given in Figure 15.8. The following green metrics can be calculated accordingly:

Since the fermentation process is the same, the atom economy does not change:

$$\begin{aligned} AE(\%) &= \frac{\text{Molecular Weight}_{product}}{\text{Molecular Weight}_{all \text{ reagents}}} \times 100\% \\ &= \frac{\text{Molecular Weight}_{ethanol}}{\text{Molecular Weight}_{glucose}} \times 100\% \\ &= \frac{46.07}{0.5 \cdot 180.16} \cdot 100\% = 51.14\%. \end{aligned} \tag{15.91}$$

The permeate flow rate of ethanol is 0.11 kg h^{-1}, and the reaction feed rate is 1.13 L h^{-1}. This means that 1 kg of ethanol production requires 10.27 L feed volume, which has a sugar concentration of 200 g L^{-1}. According to the fermentation equation illustrated in Figure 15.8, 1 mol glucose produces 2 moles of ethanol; thus, the yield is

$$\begin{aligned} \text{Yield}(\%) &= \frac{\text{Weight of product}}{\text{Theoretical weight}} \times 100\% \\ &= \frac{1000 \text{ g}}{\frac{(10.27 \cdot 200)}{180.16} \cdot 2 \cdot 46.07} \cdot 100\% = 95.19\%, \end{aligned} \tag{15.92}$$

$$\begin{aligned} \text{RME}(\%) &= \frac{\text{mass of desired product}}{\text{mass of reactant}} \times 100\% \\ &= \frac{\text{mass of ethanol}}{\text{mass of glucose}} \times 100\% = \frac{1000 \text{ g}}{10.27 \cdot 200} \cdot 100\% = 48.69\%. \end{aligned} \tag{15.93}$$

Carbon efficiency remains the same as in Part A:

$$CE(\%) = \frac{\text{Amount of carbon in product}}{\text{Total carbon present in reactants}} \times 100\%$$

$$= \frac{\text{moles of ethanol} \times \text{carbons in ethanol}}{\text{moles of glucose} \times \text{carbons in glucose}} \times 100\%$$

$$= \frac{2 \cdot 2}{1 \cdot 6} \cdot 100\% = 66.67\%. \tag{15.94}$$

The fermentation is similar to the process in Part A. The reaction takes place in an aqueous medium. Only KH_2PO_4 can be considered as potentially non-benign. For 1 kg of product, 10.27 L feed is required; thus, the EMY is

$$EMY(\%) = \frac{\text{Mass of product}}{\text{Mass of non-benign reagents}} \times 100\% = \frac{1000}{10.27 \cdot 1.75} \cdot 100\%$$

$$= 5564\%. \tag{15.95}$$

$$EF_{total} = EF_{kernel} + EF_{excess} + EF_{aux}. \tag{15.96}$$

One major change compared to the previous batch process (Part A) is that during the separation, waste generation does not occur; therefore, the EF_{aux} is zero.

$$EF_{kernel} = \frac{\text{Total mass of waste (excl. }H_2O)}{\text{Mass of final product}} = \frac{10.27 \cdot 200 - 1000}{1000}$$

$$= 1.054\ g\,g^{-1} = 1.054\ kg\,kg^{-1}, \tag{15.97}$$

$$EF_{excess} = \frac{\text{Total mass of waste (excl. }H_2O)}{\text{Mass of final product}}$$

$$= \frac{10.27 \cdot (1 + 1.5 + 1.75 + 0.75)}{1000\ g} = 0.0514\ g\,g^{-1}$$

$$= 0.072\ kg\,kg^{-1}, \tag{15.98}$$

$$EF_{aux} = \frac{\text{Total mass of waste (excl. }H_2O)}{\text{Mass of final product}} = 0.0\ kg\,kg^{-1}, \tag{15.99}$$

$$EF_{total} = EF_{kernel} + EF_{excess} + EF_{aux} = 1.054 + 0.051 + 0.0$$

$$= 1.105\ kg\,kg^{-1}. \tag{15.100}$$

cEF considers the solvent, which is water. The production takes place in a closed-loop system, and water is recycled back to the fermentation broth and for the production medium's preparation. Therefore, it can be left out of the equation. The membrane distillation unit does not generate additional waste; thus, the purification mass is also zero.

$$cEF = \frac{\sum m_{\text{raw materials}} + \sum m_{\text{reagents}} + \sum m_{\text{solvents}} + \sum m_{\text{water}}}{m_{\text{product}}}$$

$$+ \frac{\sum m_{\text{purification}} - m_{\text{product}}}{m_{\text{product}}}$$

$$= \frac{10.27 \cdot (200 + 1 + 1.5 + 1.75 + 0.75) - 1000}{1000} = 1.105 \, \text{g g}^{-1}$$

$$= 1.105 \, \text{kg kg}^{-1}. \tag{15.101}$$

The energy consumption of the pumps and membrane distillation unit is altogether 10.3 kWh m^{-3}, which has to be calculated for 1 kg of ethanol:

$$E^+ = \frac{\sum m_{\text{waste}}}{m_{\text{product}}} \left[\text{kg kg}^{-1}\right] + \frac{W[\text{kWh}] \times CO_2EF \left[\text{kg } CO_2 \text{ kWh}^{-1}\right]}{m_{\text{product}}[\text{kg}]}$$

$$= 1.105 + \frac{\frac{3.7 + 7}{1000 \times 0.79} \cdot 0.30720}{1} = 1.105 \, \text{kg kg}^{-1} + 0.004 \, \text{kg kg}^{-1}$$

$$= 1.109 \, \text{kg kg}^{-1}. \tag{15.102}$$

The MI is:

$$MI = \frac{\text{Total mass in process (excl. } H_2O)}{\text{Mass of product}}$$

$$= \frac{10.27 \, \text{L} \cdot (200 + 1 + 1.5 + 1.75 + 0.75)}{1000 \, \text{g}}$$

$$= 2.105 \, \text{g g}^{-1} = 2.105 \, \text{kg kg}^{-1}. \tag{15.103}$$

The process mass intensity includes the water consumption of the process. However, water is continuously recycled back to the fermentor (for the preparation of the production medium), and therefore no wastewater is generated. Hence,

$$PMI = \frac{\text{Total mass in process (incl. } H_2O)}{\text{Mass of product}}$$

$$= \frac{10.27 \, \text{L} \cdot (200 + 1 + 1.5 + 1.75 + 0.75)}{1000}$$

$$= 2.105 \, \text{g g}^{-1} = 2.105 \, \text{kg kg}^{-1}. \tag{15.104}$$

$$MP = MI^{-1} \times 100\% = \frac{\text{Mass of product}}{\text{Total mass (excl. } H_2O)} \times 100\%$$

$$= 2.105^{-1} \cdot 100\% = 47.51\%. \tag{15.105}$$

$$WWI = \frac{\text{Mass of total waste water generated}}{\text{Mass of product}} = 0 \, \text{kg kg}^{-1}. \tag{15.106}$$

(b) Comparison of the continuous process (Part B) over the batch (Part A) proves that the main advantages of the continuous process are the better glucose conversion and higher ethanol yield due to continuous removal of the product and residual sugar recycling. Membrane distillation does not require additional entrainer ingredients and is more energy-efficient than the extractive distillation. Production in a closed-loop system means that the solvent is recycled and only required for operation purposes; therefore, there is a remarkable difference between the process mass intensity and wastewater intensity (Table 15.3).

Table 15.3: Green metric comparison: conventional batch fermentation coupled with distillation versus continuous fermentation coupled with membrane separation.

Green metric	Batch fermentation and distillation	Continuous fermentation and membrane separation
AE (%)	51.14	51.14
RME (%)	42.79	48.69
CE (%)	66.67	66.67
EMY (%)	1467	5564
EF_{total} (kg kg^{-1})	2.03	1.105
EF_{kernel} (kg kg^{-1})	1.34	1.05
EF_{excess} (kg kg^{-1})	0.26	0.05
EF_{aux} (kg kg^{-1})	0.435	0
cEF (kg kg^{-1})	20.24	1.105
E^{+}-factor (kg kg^{-1})	1.53	1.109
MI (kg kg^{-1})	3.03	2.105
PMI (kg kg^{-1})	21.24	2.105
MP (%)	33	47.51
WWI (kg kg^{-1})	18.21	0

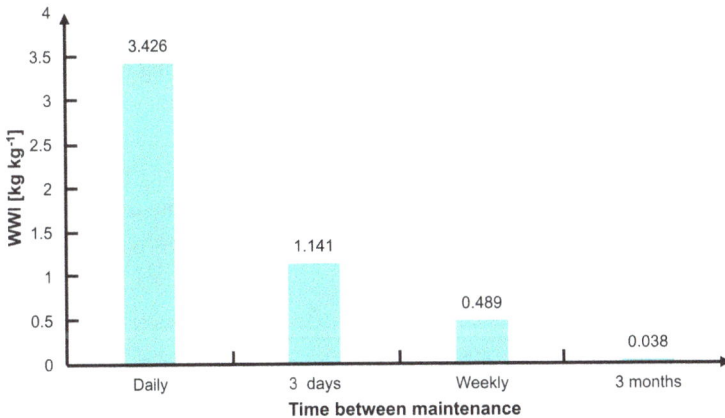

Figure 15.9: Wastewater intensity as a function of downtime for maintenance purposes.

(c) During maintenance, the fermentation broth is disposed of, with a total volume of 7.53 L (reaction volume). Although we do not know the exact mass of this fermentation broth, we can estimate the density to be roughly 1.2 kg L^{-1} by considering the 200 g L^{-1} glucose. The product's mass can be calculated as the operation time multiplied by ethanol's permeate flow rate (0.11 kg h^{-1}), assuming unchanged conditions. One can predict that having longer continuous production times, less wastewater will be generated per kg of ethanol, as illustrated in Figure 15.9.

$$WWI_{24h} = \frac{\text{Mass of total waste water generated}}{\text{Mass of product}}$$

$$= \frac{7.53 \cdot 1.2}{24 \cdot 0.11} = 3.423 \text{ kg kg}^{-1}, \tag{15.107}$$

$$WWI_{72h} = \frac{7.53 \cdot 1.2}{72 \cdot 0.11} = 1.141 \text{ kg kg}^{-1}, \tag{15.108}$$

$$WWI_{weekly} = \frac{7.53 \cdot 1.2}{168 \cdot 0.11} = 0.489 \text{ kg kg}^{-1}, \tag{15.109}$$

$$WWI_{3 \text{ months}} = \frac{7.53 \cdot 1.2}{90 \cdot 24 \cdot 0.11} = 0.038 \text{ kg kg}^{-1}. \tag{15.110}$$

15.5 Example 5 – Application of process analytical technologies in continuous catalytic hydrogenation

15.5.1 Problem statements

Consider the reaction depicted in Figure 15.10, where the desired product can be achieved by catalytic hydrogenation. After the starting material is converted into the desired product, the hydrogenation is terminated; otherwise, an overreduction impurity is formed with the unwanted halogen group. Therefore, monitoring the reaction with the application of process analytical technology (PAT) enables us to determine the reaction end-point and eliminate by-product formation effectively. This will result in a much cleaner crude product, and the time- and energy-consuming purification steps can be minimized or omitted. Refer to Chapter 9 to learn about the use of PAT in sustainable processing.

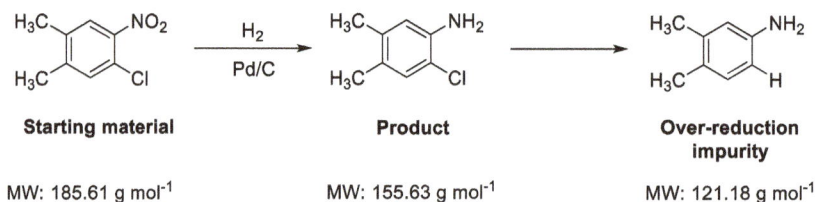

Figure 15.10: Catalytic hydrogenation of an aromatic compound, with possible formation of an overreduction impurity.

A purity of 92% can be achieved with offline sampling, and further purification of the crude material is carried out with recrystallization from toluene. The crystallization and filtration process consumes 7 kg of toluene for each kg of product. With PAT's application, the desired purity (98%) can be achieved at once, without the need for further purification.

(a) Calculate the EF_{kernel} with and without the implementation of PAT and the EF_{aux} of the process.
(b) The implementation of PAT into a manufacturing process is capital-intensive. Assuming that the instruments and the method development cost 105,000 USD, how many product batches are required to return this investment if each batch produces 55 kg of product, regardless of purity? The product's selling price is 150 USD kg^{-1} for 92% purity and 425 USD kg^{-1} for 98% purity.
(c) There are some safety risks associated with hydrogen; it is highly flammable over a wide range of concentrations (4–74 vol%) in air. It is advisable to switch from batch production to small-scale continuous manufacturing. Calculate how much hydrogen is present at the beginning of a batch reaction if the starting material's amount is 60 kg. Then, compare the result with the amount of hydrogen present in a continuous flow reactor if the feed solution's concentration is 1.5 mol L^{-1} and the volume of the reactor is 5 mL.
(d) How many continuous flow reactors have to be run in parallel if the volumetric flow rate is 1 L h^{-1}, and we wish to produce the same amount of product in under two days, as in the case of the batch process (assuming 100% yield)?

15.5.2 Solutions

(a)

$$EF_{kernel, without PAT} = \frac{\text{Total mass of waste (excl. } H_2O)}{\text{Mass of final product}}$$

$$= \frac{185.61 - 155.63 \cdot 0.92}{155.63 \cdot 0.92} = 0.296 \text{ g g}^{-1} = 0.296 \text{ kg kg}^{-1}, \quad (15.111)$$

$$EF_{kernel, with PAT} = \frac{\text{Total mass of waste (excl. } H_2O)}{\text{Mass of final product}}$$

$$= \frac{185.61 - 155.63 \cdot 0.98}{155.63 \cdot 0.98} = 0.217 \text{ g g}^{-1} = 0.217 \text{ kg kg}^{-1}, \quad (15.112)$$

$$EF_{aux} = \frac{\text{Total mass of waste (excl. } H_2O)}{\text{Mass of final product}} = \frac{7}{1} = 7 \text{ kg kg}^{-1}. \quad (15.113)$$

(b) The price difference between 92% and 98% purity is 275 USD kg^{-1}; therefore, each batch done with PAT implementation produces 15,125 USD additional income. If the instrumentation and method development cost 105,000 USD, the investment will return after seven cycles of production.

(c) Reduction of the nitro group requires 3 mol hydrogen; therefore, in the case of 60 kg starting material with a molar mass of 185.61 g mol^{-1},

$$m_{hydrogen} = \frac{60 \cdot 1000}{185.61} \cdot 3 \cdot 2 = 1939.6\,g. \tag{15.114}$$

This amount of hydrogen has a volume of 23.76 m^3 at standard conditions.

In the case of a continuous flow reactor, the amount of hydrogen present can be calculated as follows:

$$m_{hydrogen} = V_{reactor} \times \text{concentration} \times \text{amount of hydrogen}$$
$$\times MW_{hydrogen}$$
$$= 0.005 \cdot 1.5 \cdot 3 \cdot 2 = 0.045\,g = 45\ mg. \tag{15.115}$$

This is a very tiny amount, which will not raise any safety concerns in the event of a runaway reaction.

(d) If we want to produce the same amount as in the case of the batch process, we have to compare the productivity of the two processes:

$$\text{Productivity}_{batch} = \frac{60 \cdot 1000}{185.61} = 323.26\ mol\,batch^{-1}, \tag{15.116}$$

$$\text{Productivity}_{continuous\text{-}flow\ reactor} = 1.5 \cdot 1 \cdot 48 = 72\ mol\ 48\ hours^{-1}, \tag{15.117}$$

$$\frac{\text{Productivity}_{batch}}{\text{Productivity}_{continuous\text{-}flow\ reactor}} = \frac{323.26}{72} = 4.49. \tag{15.118}$$

The results indicate that five continuous flow reactors have to run in parallel to achieve the same productivity as the batch process.

15.6 Example 6 – Green metrics analysis for hazardous chemistry and purification optimization

You are working in a pharmaceutical company that is focused mainly on the synthesis of Oseltamivir which is an antivirotic used for treating influenza A and influenza B. However, there is a problem with the supply of the starting material, and you have only access to the racemic starting materials. To try to cut the cost, the company has decided to do a diastereomeric Diels–Alder reaction in which compound **C** is the intermediate that you must synthesize and purify (Figure 15.11). Assume that the yield of the compound **C** is 80% and the rest is compound **D** (20%), without any side products. Assume that the masses listed below are the masses after the reaction is complete.

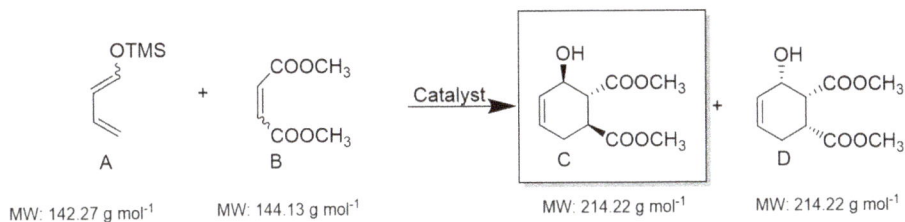

Figure 15.11: Diels–Alder reaction of 1-(trimethoxysilyl)-1,3-butadiene (**A**) with maleic acid (**B**).

15.6.1 Part A problem statements

(a) Suggest a catalyst from Table 15.4 below for the Diels–Alder reaction. Take into consideration every aspect of the available catalysts such as price, efficiency of the reaction and the toxicity of each catalyst.

Table 15.4: Information about the Diels–Alder catalysts.

Catalyst	Pictograms	Yield	Cost
Methyllithium (MeLi)		50% (81% ee)	1 kg/915.12€
Cesium fluoride (CsF)		95% (95% ee)	1 kg/1700.00€
Barium isopropoxide [Ba(O^{i}Pr)$_2$]		90% (95% ee)	1 kg/48,500€

Enantiomeric excess (ee) is the difference between the relative abundance of two enantiomers. For example, if a mixture contains 95% of one enantiomer the enantiomeric excess is 90%.

$$Ee\ (\%) = \text{major enantiomer}\ \% - \text{minor enantiomer}\ \% \qquad (15.119)$$

(b) Calculate the atom economy AE (%) and the carbon economy CE (%) from the information available above.

15.6.2 Part A solutions

(a) Diels–Alder reactions (DAR) are mostly endothermic when there is no catalyst added to the reactions. That means that the reaction must be heated, which in most cases results in poor enantiomer- and diastereoselectivity. To circumvent the selectivity issues organometallic catalysts are widely used. In this case, we can compare methyl lithium (MeLi), cesium fluoride (CsF) and barium isopropoxide (Ba(O^{i}Pr)$_2$) catalysts for this reaction. since the reaction with methyl lithium produced only 81% ee, it is not a suitable catalyst for this reaction. Enantiomeric excess is a crucial property in the synthesis of pharmaceuticals, methyl lithium is not suitable as a catalyst for this reaction. Moreover, the reaction yield is only 50% Barium isopropoxide gives a reaction yield of 90% with 95% ee. While the reaction catalyzed by CsF gives a 95% yield with a 95% ee. Unfortunately, cesium fluoride is the most toxic chemical from the three catalysts. Cesium fluoride is a corrosive chemical, that causes skin burns, and severe eye damage, toxic if swallowed or inhaled. Cesium fluoride can cause serious health problems, however, it is 28 times cheaper than barium isopropoxide. On the other hand, barium isopropoxide is flammable and can be harmful only when ingested or directly inhaled which makes it much safer to handle compared to cesium fluoride. When all things are considered, the health risks involved in working with cesium fluoride outweigh the benefits of lower price and better yield for the reaction. Therefore, barium isopropoxide would be the more suitable catalyst for this reaction, because of the lower health issues associated with the compound. As stated in the principles of green chemistry highly toxic chemicals should be avoided completely and replaced with less toxic ones.

Green metrics for this specific reaction would improve significantly if the starting material would only be a single enantiomer, so the product will be the desired diastereomer. The reaction with a single enantiomer looks the following:

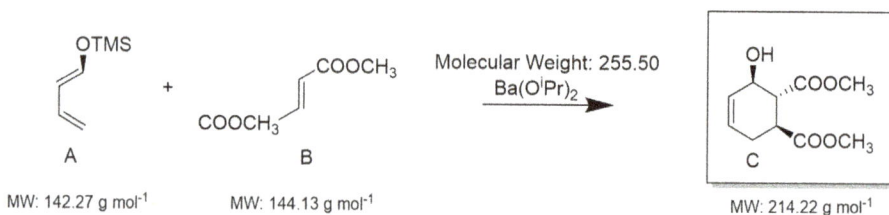

Figure 15.12: Diastereoselective Diels–Alder reaction catalyzed with barium isopropoxide.

(b)

$$AE\,(\%) = \frac{214.22}{142.27 + 144.13} \times 100 = 100\%, \qquad (15.120)$$

$$CE\,(\%) = \frac{8}{11} \times 100 = 73\%. \qquad (15.121)$$

15.6.3 Part B problem statements

Let's assume that you are asked to prepare 10 kg of compound **C** from the Part A problem statement Figure 1.13.

The laboratory procedure is as follows: *"An oven dried nitrogen filled flask is charged with (E)-(buta-1,3-dien-1-yloxy)trimethylsilane (20.20 kg, 140 mmol) followed by dimethyl fumarate (6.65 kg, 47 mmol). The starting material is dissolved in 30 L THF (anh., 0.89 g mL^{-1}). A solution of barium isopropoxide (1.19 kg, 4.7 mmol) is added to the reaction mixture in 2 L of THF. The reaction mixture is stirred for 96 h at 50 °C. After the reaction mixture was cooled to room temperature, it was quenched by adding 1 M HCl (2 L, 1.2 g mL^{-1}). The reaction mixture was extracted with 3 × 15 L of ethyl acetate (0.902 g mL^{-1}) and 20 L of water. The organic layers were combined and dried over Na$_2$SO$_4$ (500 g). The product was purified by silica gel (20 kg) column chromatography with a mixture of ethyl acetate (0.902 g mL^{-1}) and hexane hexane (0.661 g mL^{-1}) (4:1, 50 L). The product was isolated as a pale-yellow oil with a yield of 100% (10 kg, 47 mmol)."*

(a) Based on the reaction scheme from Figure 15.12 identify reactants, catalysts, solvents, side products.

(b) Calculate the E_{factor}, $E_{column\text{-}chromatography}$, E_{excess}, solvent intensity.

15.6.4 Part B solutions

(a) The reactants are **A** and **B**, the catalyst is Ba(O^{i}Pr)$_2$. Based on the reaction scheme from Figure 15.12, THF as a solvent is missing and should be listed below the reaction arrow. There is a missing side product which is trimethyl silyl chloride (**D**).

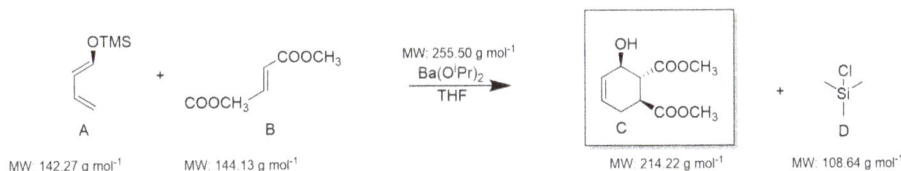

Figure 15.13: Diastereoselective Diels–Alder reaction catalyzed by barium isopropoxide with trimethyl silyl chloride side product (**D**).

(b)

$$E = \frac{\text{Total mass of waste (excl. water)}}{\text{Mass of product}}$$

$$= \frac{\text{mass of THF, ethyl acetate, hexane, HCl, silica gel, sodium sulphate}}{\text{Mass of product}}$$

$$= \frac{\left(\begin{array}{c} 32L \times 0.89 \, kg \, L^{-1}(THF) + 85L \times 0.902 \, kg \, L^{-1}(EA) + 10L \times 0.661 \, kgL^{-1}(Hex) \\ + 2L \times 1.2 \, kgL^{-1}(HCl) + 20 \, kg(silica) + 0.5 \, kg(Na2So4) \end{array}\right)}{10kg}$$

$$= 13.47 \, kg \, kg^{-1}$$

(15.122)

$$SI = \frac{\text{Total mass of waste (excl. water)}}{\text{Mass of product}}$$

$$= \frac{\text{mass of THF, ethyl acetate, hexane, HCl}}{\text{Mass of product}}$$

$$= \frac{\left(\begin{array}{c} 32 \, L \times 0.89 \, kgL^{-1}(THF) + 85 \, L \times 0.902 \, kgL^{-1}(ethylacetate) \\ + 10 \, L \times 0.661 \, kgL^{-1}(hexane) + 2 \, L \times 1.2 \, kgL^{-1}(HCl) \end{array}\right)}{10kg} = 11.42 \, kg \, kg^{-1}$$

(15.123)

$$E_{\text{column chromatography}} = \frac{m_{\text{silica gel}} + m_{\text{eluent}}}{m_{\text{product}}}$$

$$= \frac{20 \, kg + 59.48 \, kg}{10 \, kg} = 7.95 \, kg \, kg^{-1} \qquad (15.124)$$

$$E_{\text{excess}} = \frac{m_{\text{excess reagent}}}{m_{\text{product}}} = \frac{13.47 \, kg}{10 \, kg} = 1.35 \, kg \, kg^{-1} \qquad (15.125)$$

15.7 Example 7 – Green metrics analysis and reaction optimization

Imagine that you are working in a chemical company mainly focused on the nucleophilic substitution and additions using aniline. Aniline is a toxic compound, however, there is an order pending for 1,4-quinone derivatives. The standard reaction is displayed in Figure 15.14 where the yield of the reaction is 90%, which is of a high standard. However, the management has decided to shift more towards green chemistry. The task was to make the quinone derivative with a greener method. The production of the product requires 30 L ethanol (0.789 kg L^{-1}) to be used as solvent.

This reaction, however, works even better when the solvent used is water. With the same reaction conditions the yield of the reaction is 100% and the reaction is also faster (Figure 15.15).

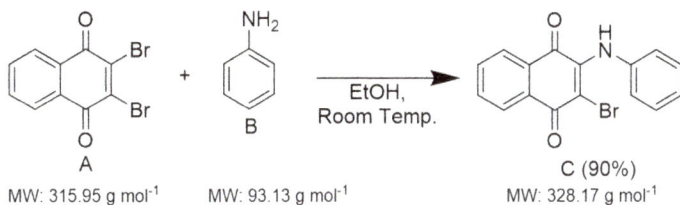

Figure 15.14: Addition of 2,3-Dibromo-1,4-naphtoquinone to Aniline in ethanol.

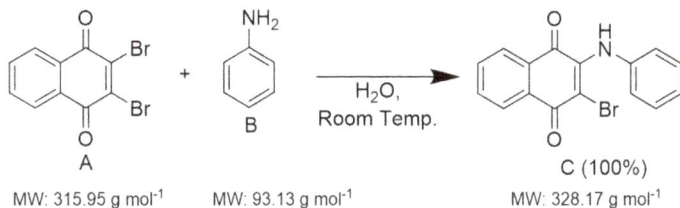

Figure 15.15: Addition of 2,3-Dibromo-1,4-naphtoquinone to aniline in water.

15.7.1 Part A problem statements

(a) From the available information calculate the E_{kernel} for the reaction.
(b) Calculate the E-aux for the reaction where it is feasible.
(c) Calculate the atom economy of the reaction.

15.7.2 Part A solutions

(a)

$$E_{kernel} = \frac{\text{Total mass of waste (excl. } H_2O)}{\text{Mass of final product}}$$

$$= \frac{(315.95 + 93.13) - (328.17 \times 0.90)}{328.17 \times 0.90} = 0.385 \text{ g g}^{-1} = 0.385 \text{ kg kg}^{-1} \qquad (15.126)$$

(b)

$$EF_{aux\,(EtOH)} = \frac{\text{Total mass of waste (excl. } H_2O)}{\text{Mass of final product}} = \frac{(30 \text{ L} \times 0.789 \ kg \ L^{-1})}{10 \ kg} = 2.367 \text{ kg kg}^{-1}$$

$$(15.127)$$

(c)

$$AE(\%) = \frac{\text{Molar weight of product} \times 100}{\text{Sum of molar weights of reactants}}$$

$$= \frac{328.17}{(315.95 + 93.13)} \times 100 = 80.2\% \qquad (15.128)$$

15.7.3 Part B problem statements

The laboratory procedure for the reaction in EtOH is as follows:

"*A reactor was charged with 2,3-dibromonaphthalene-1,4-dione (15.000 kg, 1 Eq, 47.476 mol) dissolved in 45 L of ethanol (0.789 kg L⁻¹). Aniline (4.421 kg, 1 Eq, 47.476 mol) was added to the reaction mixture and the reaction was stirred at room temperature for an hour. When the reaction was complete, the excess solvent was evaporated under reduced pressure. The product is a red crystalline compound which was isolated in 90% yield (15.000 kg).*"

The laboratory procedure for the reaction in water is as follows:

"*A reactor was charged with 2,3-dibromonaphthalene-1,4-dione (15.000 kg, 1 Eq, 47.476 mol) dissolved in 30 L of H₂O (0.997 kg L⁻¹). Aniline (4.421 kg, 1 Eq, 47.476 mol) was added to the reaction mixture and the reaction was stirred at room temperature for 50 min. After the reaction, the formed crystals were filtered off and washed with 10 L of H₂O (0.997 kg L⁻¹). The product is a red crystalline compound which was isolated in 100% yield (16.246 kg).*"

Based on the reaction procedures calculate the following green metrics: Mass intensity (MI), Solvent intensity (SI), Mass productivity (MP) and reaction mass efficiency (RME) for the reactions using ethanol and water.

15.7.4 Part B solutions

Mass intensity$_{(EtOH)}$:

$$MI(EtOH) = \frac{\text{Total mass in process (exl. Water)}}{\text{Mass of product}}$$

$$= \frac{\text{mass of (ethanol + reactant A + reactant B)}}{\text{mass of product}}$$

$$= \frac{(45 \text{ L} \times 0.789 \text{ kg L}^{-1}) + 15.000 \text{ kg} + 4.421 \text{ kg}}{15 \text{ kg}} = 3.66 \text{ kg kg}^{-1} \qquad (15.129)$$

Mass intensity$_{(H_2O)}$:

$$MI\,(H_2O) = \frac{\text{Total mass in process (exl. Water)}}{\text{Mass of product}} = \frac{\text{mass of reactant A + reactant B}}{\text{mass of product}}$$

$$= \frac{(30\,L \times 0.997\ kg\ L^{-1}) + 15.000\ kg + 4.421\ kg}{16.246\ kg} = 1.195\ kg\,kg^{-1} = 3.04\ kg\,kg^{-1}$$

$$(15.130)$$

Solvent intensity$_{(EtOH)}$:

$$SI(EtOH) = \frac{\text{Mass of solvents}}{\text{Mass of product}} = \frac{\text{mass of ethanol}}{\text{mass of product}} = \frac{(45\,L \times 0.789\ kg\ L^{-1})}{15\ kg} = 2.367\ kg\,kg^{-1}$$

$$(15.131)$$

Solvent intensity$_{(H_2O)}$:

$$SI(H_2O) = \frac{\text{Mass of solvents}}{\text{Mass of product}} = \frac{\text{mass of H2O}}{\text{mass of product}} = \frac{(40\,L \times 0.997\ kg\ L^{-1})}{16.246\ kg} = 2.455\ kg\,kg^{-1}$$

$$(15.132)$$

Mass productivity$_{(EtOH)}$:

$$MP(EtOH)(\%) = \frac{\text{Mass of product}}{\text{Total mass (incl. solvents)}}$$

$$= \frac{\text{mass of product}}{\text{mass of (ethanol + reactant A + reactant B)}}$$

$$= \frac{15\ kg}{(45\,L \times 0.789\ kg\ L^{-1}) + 15.000\ kg + 4.421\ kg} x\ 100 = 27.3\% \quad (15.133)$$

Mass productivity $_{(H_2O)}$:

$$MP(H_2O)(\%) = \frac{\text{Mass of product}}{\text{Total mass (incl. solvents)}}$$

$$= \frac{\text{mass of product}}{\text{mass of (H2O + reactant A + reactant B)}}$$

$$= \frac{16.246\ kg}{(40\,L \times 0.997\ kg\ L^{-1}) + 15.000\ kg + 4.421\ kg} x\ 100 = 27.4\% \quad (15.134)$$

Reaction mass efficiency$_{(EtOH)}$:

$$RME(EtOH)(\%) = \frac{\text{Mass of product}}{\text{Total mass of reactants}} = \frac{\text{mass of product}}{\text{mass of reactant A + reactant B}}$$

$$= \frac{15\ kg}{15.000\ kg + 4.421\ kg} \times 100 = 77.2\% \quad (15.135)$$

Reaction mass efficiency $_{(H_2O)}$:

$$RME(H_2O)(\%) = \frac{\text{Mass of product}}{\text{Total mass of reactants}} = \frac{\text{mass of product}}{\text{mass of reactant A + reactant B}}$$

$$= \frac{16.246 \ kg}{15.000 \ kg + 4.421 \ kg} \times 100 = 83.6\% \qquad (15.136)$$

15.7.5 Part C problem statements

The synthesis of quinone requires aniline in excess, which is difficult to remove from the reaction mixture. To remove the residual aniline a greener separation method was selected using the diafiltration membrane technique. The initial volume of the ethanol solution is 18 L. The mass of the quinone derivative is 1,650 g (MW: 328.17 g mol^{-1}) with a rejection value of 0.99 and the mass of the residual aniline is 230 g (MW: 93.13 g mol^{-1}) with a rejection value of 0.32. Calculate the required amount of solvent needed for the diafiltration process if 95%, 99%, and 99.99% purity is required for the product and compare the solvent consumption differences.

15.7.6 Part C solutions

First, we need to calculate the concentrations of the product and the impurity:

$$c_{A0} = \frac{n}{V_0} = \frac{m_q}{M_q \times V_0} = \frac{1650 \ g}{328.17 \ g \ mol^{-1} \times 18 \ L} = 0.279 \ g \ L^{-1} \qquad (15.137)$$

$$c_{B0} = \frac{n}{V_0} = \frac{m_{an}}{M_{an} \times V_0} = \frac{230 \ g}{93.13 \ g \ mol^{-1} \times 18 \ L} = 0.137 \ g \ L^{-1} \qquad (15.138)$$

After we have the concentrations, we can calculate the diafiltration factor as follows (note that the P in the equation is a constant for the relative purity, c_{B0} is the concentration of the impurity, c_{A0} is the concentration of the product, R_A is the rejection of the product and R_B is a rejection of the impurity):

$$D = \frac{\ln\left(P\frac{c_{B0}}{c_{A0}}\right)}{(R_A - 1) - (R_B - 1)} = \frac{\ln\left(P\frac{0.137 \ g \ L^{-1}}{0.279 \ g \ L^{-1}}\right)}{(0.99 - 1) - (0.32 - 1)} \qquad (15.139)$$

Calculated diafiltration factors based on the purity of the product:

$$D_{95\%} = 10.21, D_{99\%} 12.67, D_{99\%} 16.12 \qquad (15.140)$$

After we have the diafiltration factor calculated, we can calculate the diafiltration volume that is needed to reach the desired purities of the product.

$$D = \frac{V_K}{V_0} \qquad (15.140)$$

$$V_{K_{95\%}} = D \times V_0 = 10.21 \times 18 \ L = 183.78 \ L \qquad (15.141)$$

$$V_{K_{99\%}} = D \times V_0 = 12.67 \times 18 \ L = 228.06 \ L \qquad (15.142)$$

$$V_{K_{99.99\%}} = D \times V_0 = 16.12 \times 18 \ L = 290.16 \ L \qquad (15.143)$$

From the calculated diafiltration volumes we can see that with increasing purity of the product, the solvent consumption increases significantly. A product with a purity of 99.99% consumes 62.1 L more solvent than the product with the purity of 99%. In the fine chemical production, the waste generation is remarkably higher with a lot higher E_F values as well when compared to other chemical sectors.

15.7.7 Part D problem statements

Now that every information is available about the reaction and the optimization, how would you present to the management team the improvements that the solvent change caused? Write at least three examples that prove that the reaction is "greener".

15.7.8 Part D solutions

1. Water is the greenest solvent that can be used in organic synthesis. [1–3] Issues that can emerge when trying to use water as a solvent include poor solubility and a difficult extraction of some compounds from water.
2. The solvent change from EtOH to water improved the yield of the reaction from 90% to 100%. This change has reduced the waste production of the reaction. Mass intensity decreased from 3.66 kg kg^{-1} in EtOH to 3.04 kg kg^{-1} in water.
3. The reaction mass efficiency increased from 77.2% in ethanol to 83.6% in water.
4. Overall, the production of waste in this reaction was decreased with increased yield and efficiency.

Bibliography

[1]   Lajoie, L.; Fabiano-Tixier, A.-S.; Chemat, F. Water as Green Solvent: Methods of Solubilisation and Extraction of Natural Products – Past, Present and Future Solutions. *Pharmaceuticals*. **2022**, *15* (12), 1507. https://doi.org/10.3390/ph15121507

[2]   Castro-Puyana, M.; Marina, L. M.; Plaza, M. Water as Green Extraction Solvent: Principles and Reasons for Its Use. *Curr. Opin. Green Sustain. Chem*. **2017**, *5*, 31–36. https://doi.org/10.1016/j.cogsc.2017.03.009.

[3]   Hartonen, K.; Riekkola, M. -J. Water as the First Choice Green Solvent. *Appl. Green Solvents Sep. Process*. **2017**, 19–55.https://doi.org/10.1016/B978-0-12-805297-6.00002-4

Index

accident 109, 112
acetic anhydride 13–14, 16, 183
acetylsalicylic acid 13–14, 183–184
activated carbon 138, 153, 262, 282
activation energy 189, 191, 222, 281
active site 152, 202, 212
actual yield 13, 65
adsorbent 137, 140, 152–153, 155, 197–201, 223,
    271–272
adsorption 84, 122, 137–140, 147, 153–155, 159,
    198–202, 281
advanced oxidation process 283–284
advanced solar oxidation process 283
air-to-water application 269
alcohol 104, 157–158, 208, 210, 218, 220, 222–224,
    236, 240, 252, 285
algae 208, 218
alternative energy resource 76
alternative solvent 125
amidation 49
area normalized flow rate 125
aromatic aldehyde 104–105
artificial intelligence 116, 141, 150, 168, 231, 294
AspenTech 311, 317
aspirin 13–21
Asterix and Obelix 98
atom economy 6, 13, 56, 69, 217, 338, 350
automation 122, 168, 172, 231, 255–256
autonomous experimentation platforms 294–295
auxiliary component 16
auxiliary material 17
awareness 37, 72, 83, 207

ball milling 248, 252
BASF 3
batch reactor 69, 77, 79–80, 98–102, 104, 107, 112,
    115, 132, 282, 343
big data 172
bioaccumulative chemical 27
bio-based polymers 229–230, 233, 236–238, 258
biodegradability 39, 42, 217, 237, 241, 246, 263
bio-demineralization 202, 204
biodiesel 90–91, 207–208, 217–224, 238, 241
bioethanol 207–210, 214–216, 240–241, 243, 252,
    344–345, 348
biofuel 83, 207–208, 210, 240, 247, 261, 345

biomass 42, 83, 87–89, 98, 153, 192, 207, 209–213,
    217, 229, 233, 236–238, 242, 246–250, 252,
    261–264, 286, 344–345
biomass recalcitrance 210
biomass valorization 87–89
bio-mineralization 202
bio-oxidation 202
bioplastics 238
bioreactor 213–215, 344
bio-reduction 202–204
bio-sorption 202
boiling point 6, 43, 48, 55–56, 148–150, 221, 246,
    249, 275, 331, 335, 340
brivanib 180–181
building block 51–53, 87–88, 100, 105, 138, 212,
    229, 233, 240–242, 245–247, 253, 258, 262,
    264, 331
butadiene 49–50, 240, 243
by-product 14–15, 17, 48, 50, 54, 56, 69, 73, 79, 85,
    102, 121, 165, 167, 182, 207, 217, 333, 346, 354

caffeine 123
capacity factor 192–193
capillary microflow reactor 110
carbon dioxide 22, 25, 112, 123, 125, 131, 208, 236,
    282, 333
carbon efficiency 57, 339
carbon emission 23, 25, 151
carbon footprint 22–24, 25, 42, 128–130, 140–141,
    147, 149, 151, 155–157, 245, 263–264
carbon intensity 25–26, 56, 69
case study 39, 48, 106, 129, 179, 182, 215
catalysis 111, 210, 212, 220–221, 282, 286
catalyst recovery 127
catalytic activity 235, 281–282
cation 200, 202
cellulose 16, 67, 88–89, 91, 200, 210–213, 215, 223,
    238–242, 245, 247–251
centrifugal partition chromatography 136–137
charcoal 153, 155, 207
CHEM-21 43
chemical engineering 1, 10, 72–73, 288, 328
chemical sector 10–11, 17, 238
chemical waste 58, 60
chemical yield 13
chemometric techniques 171, 174

https://doi.org/10.1515/9783111028163-016

chilling 58
chiral 246–247, 290–291
chiral agent 290
chiral light-matter interaction 290
chitin 200, 238, 240
chitosan 200, 241, 258–259
chlorinated solvent 36
chromatography 50, 54, 116, 121, 123, 135–136, 155, 165, 253, 255, 290, 332, 334
circular economy 48, 51, 232, 234, 258, 344
circular polarized light 291
clean energy 192
coal ash 196
combustion 25, 58, 83, 123, 189–190, 208, 210, 218, 241, 276
commercialization 5, 275
complexity 7, 9, 51, 56–57, 121, 126, 231, 271
concentrated solar power 276
concentration polarization 125, 128, 131
concession step 56–57
conductive heating 89
constant-volume diafiltration (CVD) 155
construction 35, 42, 56–57, 194
construction step 56–57
contamination 37, 122–123, 132, 157, 201–202, 231
continuous flow reactor 78, 98–100, 110, 128, 255, 290
continuous membrane reactor 224
continuous oscillatory baffled crystallizer 133–134
continuous plug flow crystallizer 133
continuous segmented flow crystallizer 133–134
continuous separation process 253
control 4, 7, 69, 77–78, 79, 100, 102, 106, 122, 134, 147, 165, 167–168, 170, 172–174, 178, 182–183, 185, 192, 216, 231, 254–256
conventional solvent 37–39, 42–43, 150
conversion factor 22
conversion rate 101, 110, 181, 213
cooling water consumption 58
co-product 14–17, 344, 346
core chemical reaction 17, 346
cost 29, 38, 100, 102, 115, 131–132, 149–151, 155–156, 159–160, 175, 195, 198–202, 212, 217, 224, 235, 242, 252–253, 264, 273, 276, 279, 281, 284, 354–355
counter-current type liquid-liquid adsorption process 135
cross-flow filtration 125
crude oil 25–26, 217, 229

crude oil distillation unit 311–312
crystal formation 134
crystallization 83, 122–123, 129, 132–133, 135, 334, 337, 341, 343, 354
cyclic anhydride 50
cyrene 150

dead-end filtration 272
decision-making 28, 48, 76, 327, 329
deep eutectic solvent 104, 223, 252
deep learning 295, 306
degradation 7, 92, 111, 154, 182, 230–231, 235–236, 247, 249, 281, 284
dehydration 153, 159, 210, 214, 241, 245, 247–248, 261–262
demobilization 202–203
de-oxygenation 112
depolymerization 212, 230, 248–249, 251, 256
deposition 85, 123–124, 183–184, 218, 280
desalination 241, 258, 272, 276
dew-harvesting 269–270
diafiltration 153, 156
diazomethane 106–109, 336–337, 342
digital transformation 317
dioxane 104
dipolar rotation 89–90
direct contact membrane distillation 214, 216, 223, 348–349
disconnection 51
dish/engine 276
disposal 5, 17, 23, 36–37, 43, 121, 147, 183, 201, 231, 329, 348
dissolve 9, 32, 34, 121, 128, 152, 165, 178, 183, 195, 197, 203, 212, 220, 252, 264, 282, 286, 288
distillation 39, 53, 58, 82–83, 86–87, 122, 125, 129, 131–132, 147, 149–152, 154–155, 157, 159–160, 210, 214, 217, 249, 252, 256–257, 272–273, 275, 328, 335, 344–345, 348, 352–353
downstream processing 102, 106, 121, 137–138, 182, 342
dynamic viscosity 123

eco-friendly 48, 91
economics 1, 208
effective mass yield 16, 67–68, 340, 344, 346
electromagnetic spectrum 89
electrospinning 183–185
emission conversion factor 19, 22
emulsification 124

enantioseparation 290

energy consumption 5, 9, 18–19, 25, 36, 58, 60, 72–73, 80, 91, 104, 121, 125, 130, 134, 152, 155, 157, 159, 168, 211–212, 214, 256, 269, 344, 347–348, 351

energy efficiency 7, 26, 85, 109, 150, 258

enrichment 152, 195

entrainer 150, 210, 344–345, 352

environment 2–3, 5–7, 10, 27, 36–37, 42–43, 82–84, 194, 200, 202, 207–208, 220, 229–231, 235–236, 251, 269, 272, 329

environmental effect 4, 7, 43

enzyme 212, 243, 248, 252

equipment 10, 58, 72–82, 112, 123, 134–135, 212, 264, 329

ester 51, 53, 181, 215, 219, 223, 241, 247

esterification 49, 180, 220

ethyl 2-diazoacetate 107

ethyl acetate 34, 66, 123, 139, 141, 155, 177, 184, 332, 335

ethyl nicotinate 105

ethyl piperidine-3-carboxylate 105

evacetrapib 102

evaluation of sustainability 13

evaporation 86, 122, 129, 154, 217, 221, 273, 275

excess reagent 7, 17, 334

extraction 53, 58, 66, 83, 88, 102, 121–124, 136, 138, 141, 155, 158–159, 189, 195, 197–200, 219–221, 240, 256, 328, 331

extractive distillation 147, 149–150, 214, 344–346, 348, 352

extraterrestrial radiation 287

extreme condition 112–113, 220

extruder machine 103

Fault detection and diagnosis 294, 301

FDA 165, 245, 248

feature extraction 304–305

feedstock 7, 10, 51, 83, 207–208, 221, 223, 229–230, 233–235, 237–238, 240–242, 245–248, 252, 256, 262–264

fermentation 89, 175, 207, 209–211, 213, 215–217, 238, 244–246, 252–253, 261–263, 344–353

filtration 13, 17–18, 125, 127, 131, 141, 154, 178, 250–252, 272, 335–336, 348–349, 354

fission 189–192, 269

flammability 38, 43

flaring intensity 26

flow chemistry 102

flow photoreactor 110

fluctuation 182, 278

fluorescence 175, 279–280

flux 98, 110, 125–126, 154, 159, 215, 217, 223, 273–274, 281, 284, 348–349

fog collectors 269–270

fouling 85–86, 134, 159, 216–218, 223, 282

fractional distillation 149

free fatty acid 220

freshwater reservoir 269

furfural 211, 246, 249–250, 252, 331

fusion 189–191

geosteering 27

geothermal energy 269

glucose 212, 239–241, 244–248, 251–252, 290, 344–346, 348, 350, 352

glycerol 39, 149, 212, 218–219, 221, 223–224, 240, 248

green chemistry 3–4, 5, 6, 7, 10, 17, 20, 27, 29, 38, 52, 54, 74, 76, 102–103, 109, 167–168, 170, 217, 234, 247, 290

green engineering 3, 5, 9–10, 38, 74, 76, 102–103, 109, 167–168

green metric 7, 13–14, 16, 19, 21, 25, 48, 60, 69, 140, 155, 327–329, 331, 336–338, 350

green motion 29

green reagent 104, 331

green solvent 33, 35, 37–38, 40–42, 48–49, 51, 141, 178, 211, 223, 246–248, 252, 257–259, 331

greenhouse gases (GHG) 22

Grignard reaction 79–80

grinding 88, 210–211

hazard 4, 10, 27–28, 53, 66, 329

health and safety hazard 27, 66

heat transfer 78–79, 90, 98, 100, 102, 109, 134, 150, 222, 331

heating 1, 18, 24, 39, 45, 58, 60, 87, 89–92, 103–104, 114, 129, 134, 137, 207, 217, 221–222, 249, 251, 260, 273, 276, 281, 328, 337

Heck reaction 106

heteroatom 51, 56

heterogeneous catalyst 111, 281

heterogeneous photocatalysis 281–282, 284

Higee (High-Gravity) Technology 82

high-risk process 102

HOMO 279

homogenous catalyst 130, 281

homogenous photocatalysis 286
homologation agent 107
hybrid process intensification method 284
hybrid separation 76, 83, 86
hydrocyanation 49
hydroformylation 130–131
hydrogel 200
hydrogen 33, 64, 88, 102, 105, 112–113, 123, 131,
    138, 165, 180, 190, 202, 207, 212, 220, 222, 262,
    280, 333, 336, 355
hydrogenation 44, 102, 105, 112, 136, 167, 177,
    179–181, 245–248, 256–257, 262, 354
hydrolysis 13, 49, 88, 180, 209–213, 218, 239–241,
    247–248, 264

ideality 56–57
immobilized catalyst reactor 282
impurity 54–55, 56, 155–156, 166, 170, 178–180,
    182–183, 354
in situ leaching 195–196
incineration 43, 58, 147, 149, 151, 236
industry 4.0 294, 311–312
infrared spectroscopy 173–174
in-line gas mixing 112
in-process control (IPC) 170
integrated conversion 250, 252
integration 9, 72–73, 76, 83, 100–101, 106, 109, 122,
    125, 128, 157, 175, 215, 221, 237, 255–256, 263,
    279, 331
intersystem crossing 279–280
ion competition 201
ionic conduction 89–90
ionic liquid 38–39, 80, 104, 211, 219, 223
irradiation 88, 221–222, 251, 253, 284–285
isomer 6, 50
isotropic membrane 126

kinetics 36, 73, 101, 134, 173, 180–181
Knudsen diffusion 127

Labtrix 115
Lambert–Beer 109, 286
large-scale 104, 107, 200–201, 212, 255,
    275–276, 288
life cycle assessment 42, 212
lignin 88, 153, 211–212, 239–240, 248, 250,
    252–254, 256–257
lignocellulose 91–92, 240, 248–249, 252, 261

linear economy 232
linear programming 310
liquid polymer 38–39
liquid-liquid separation 126
lithium diisopropylamide (LDA) 53
low-grade energy resource 269
LUMO 279–280

machine learning 161, 231, 294, 301, 303, 309, 322
market 6, 33, 35, 37, 116, 237, 260
mass intensity 19
mass-transfer 76, 78, 84
material cost 29, 104, 252
materials 1, 3, 5–6, 7, 9, 14, 16–20, 27–28, 34,
    37–38, 42, 51, 54, 57, 61, 63, 66, 68–69, 78–79,
    81, 88, 90–92, 103, 106, 121–123, 127, 137–139,
    194–195, 199–200, 209, 220, 232–233,
    235–236, 238, 245–247, 254, 258, 263–264,
    270–271, 282, 284, 333, 337–338, 342
membrane 48, 83–86, 108, 112, 121, 125–130, 132,
    139–140, 147, 151, 153–155, 158–159, 213–217,
    220, 223–225, 241, 253, 256–259, 272–274,
    335, 337, 344, 348–349, 351, 353
membrane cascade 153–155
membrane cross-section 126
membrane distillation 83, 86–87, 225,
    348–349, 351
membrane reactor 83–84, 129, 224
methane 22, 26, 138, 236
methane leak detection 26
method 28, 51, 67, 72, 83, 86, 88, 90, 103–104, 112,
    134, 149–150, 166, 168, 171, 173, 177–178, 196,
    199, 212, 219, 221–222, 225, 233, 235, 240, 242,
    252–253, 257–258, 269, 274, 280, 284, 332,
    348, 354–355
methyl 5-(dimethylamino)-2-methyl-5-
    oxopentanoate 52
methyl methacrylate (MMA) 51
Michael addition 51, 53, 128
microalgae 208, 218, 233
microflow chemistry 70, 98
microfluidic device 98, 101
microfluidic regime 98
microreactor 78, 80, 111, 113
microwave 84, 89–92, 220–222, 255
microwave reactor 91, 221
minimization 17, 72, 106
mining 194–195

mixer 78–79, 81–82
mixing 34, 45, 73–74, 77–79, 81, 99, 102, 112, 128, 133–134, 172, 224, 343
mobilization 202–204
molecular sieve 153, 210
molecular weight 13, 60, 127, 157–158, 213, 253, 255–257, 331
molecularly imprinted polymer 139, 153
morphology 133, 153
multifunctional reactor 76, 83
multiple scattering reactor 288
multi-stage membrane distillation module 272

N,N-dimethylacetamide (DMAc) 51
nanofiltration 127–129, 155, 157, 214, 216–217
natural polymers 200, 229, 241
non-benign material 16, 67, 346
nuclear binding energy 189
nuclear energy 189, 191–193, 200, 204
nuclear fuel 193–195, 201–202
nuclear magnetic resonance 165, 255
nuclear waste 193, 201–202, 204
nucleation 134, 251

offline sampling 171, 178
oleuropein 138–140
on-demand 106
online monitoring 255
optimization 3, 69, 79, 102, 106, 150–151, 155, 176
organic pollutant 27, 282, 284
organic solvent 33–37, 80, 88, 127, 131, 147–148, 150, 153–154, 157–159, 168, 177, 212, 252, 275, 277, 328, 339
organic solvent forward osmosis 154, 159
organic solvent nanofiltration 127, 131, 147, 154, 157–158
organic solvent pressure assisted osmosis 154
oxalic acid 54
oxidation 44, 51, 83, 88, 109, 189, 203–204, 246–247, 280, 283–285, 287
oxidative delignification 88
oxidative desulfurization 87–88

packed bed reactor 82, 102, 112, 128, 288–289
parabolic trough 276, 285
parallelization 101
particle 33, 100–101, 123, 133–134, 191, 210, 248, 336
particle formation 101, 123

particle size distribution 134
passive modular distiller 274
penalty point system 29
people 2
percent yield 65, 336
performance 9, 37, 56, 63, 69, 83, 109, 115–116, 127, 165, 198, 201, 215, 218, 229, 231, 243, 258, 264, 271, 284
periodically operated dynamic process 138
periodically renewable 269
peristaltic pump 116
permeability 84, 126, 128, 217
permeance 125–126, 258
permeate stream 128, 131, 139, 159, 216, 224, 257
pervaporation 147, 154, 159–160, 214–215, 224–225, 259
petrochemical 17, 25, 33–34, 35, 37, 124, 230, 234, 237, 241, 260, 264, 320
Pfizer 42–43, 150–151
pharmaceutical industry 15, 17, 35, 57, 77, 105, 130, 132, 138, 150, 172, 184
phenol 212
phosphate deposit 196
phosphorescence 279–280
photocatalysis 109, 279–280, 282, 286, 290
photonic efficiency 282
photoredox catalysis 110, 279, 286
photovoltaic cell 272–273
phthalic anhydride 112–113
pictogram 28
pK_a 69
pKa 52–53
planet 2, 22, 188
plastic recycling 260
plastic waste 229–230, 235–236, 260
plasticizer 246–248
plug-flow 111
polar aprotic solvent 33, 35, 48–49, 157
polarity 33, 39, 48, 121
pollution 5, 36, 88, 111, 167, 231, 235–236, 282
polyethylene 9, 38–41, 198, 224, 236–237, 242–243, 248, 260–261
polyethylene glycol (PEG) 39
polylactic acid 237, 258–259
polymers 39, 103, 198–199, 229–233, 235–248, 255–256, 258, 260–261, 263–264
polypropylene glycol (PPG) 39
pore flow model 127
porous material 123–124, 138, 153, 200, 271, 275

potassium tert-butoxide (t-BuOK) 53
precursor 106, 123–124, 180, 245–248, 264
predictive control 294, 316, 320
pressure 7, 26, 38, 48, 79, 86, 88–89, 92, 99, 105–106,
    108, 112–114, 116, 121, 123, 125–128, 130–131,
    136–138, 147–151, 154–155, 157, 159, 165, 170,
    178–179, 183, 214, 220, 222, 242, 245, 251, 255,
    260, 271, 273–274, 276, 278, 281–282, 284
pressure difference 125–127, 159, 214, 273
pressure swing adsorption 137–138
pressure swing distillation 150
pretreatment 87–88, 211–212, 218, 220–221, 223, 248
prevention 17, 103, 109, 111–112, 172
process analytical technology 165–168, 354
process efficiency 29, 36, 72, 222
process integration 152
process mass intensity 20, 352
process safety 42, 79
process systems engineering 74, 76
product inhibition 212, 214–215, 348
profit margin 5, 17, 103
profitability 4, 72–73
pump 103, 113, 128, 130–131, 337
purification 9, 16–17, 19, 21, 35, 50, 58, 66, 87, 100,
    102, 104, 106, 110, 114, 127, 130, 137–138, 147,
    153, 166, 209, 215, 219–221, 223, 241, 245, 252,
    258, 275, 331–332, 334–335, 338, 344–345,
    347–348, 351, 354
purity 21, 25, 35, 54–55, 56, 69, 100, 102, 104, 106,
    111, 133, 150, 153, 155–157, 184, 212, 258, 328,
    332, 337, 355
pyrolysis 83, 91–92, 212, 235, 260–261, 264

QR code 6, 28, 38, 191–192, 196–197, 330
quantitative analysis 174–175, 177
quantity of activity 22
quantum yield 281, 283–285

radiative forcing 188
radioactive waste 201–202
RainMaker 269
Raman spectroscopy 175, 183–184
Rankine cycle 276, 278
raw material 7, 9, 11, 18, 23–24, 28, 33, 58, 83, 174,
    209, 220–221, 238, 261, 263–264
reaction monitoring 173, 177–178, 182–183
reaction-separation module 136
reactive extrusion 83, 104
reactive intermediate 279

reactor configurations 115
reactor design 213, 231, 254, 281, 283
real-time feedback 111
recrystallization 13, 17–18, 125, 341, 354
recycling 9, 18, 23, 42–43, 85, 128–129, 132, 141,
    147, 149, 152, 155, 230, 235–236, 251–252, 260,
    335, 352
reduction 8, 17, 26, 33, 44, 51, 74, 80, 88, 104–105,
    129, 134, 141, 151, 159, 165, 177, 180, 198, 202,
    204, 208, 222, 247, 252, 280, 287, 335, 354
refinery 1, 26, 253, 295, 310–311, 314–316,
    319–320, 322
Refinery production scheduling 294, 310
regioisomer 50, 54, 80
rejection 86, 126, 155–157, 258
renewable 7, 10, 25, 29, 38–39, 42, 45, 49, 51, 53,
    83, 106, 153, 192, 194, 199–200, 212, 217, 224,
    232–233, 237–240, 245, 248, 261–262, 269,
    279–280
renewable resource 53, 153, 238
renewable-based solvent 38, 45
residence time distribution 73, 100
retentate stream 125, 128, 131, 224, 257
retrosynthesis 51
ring-opening reaction 112
robot 117, 297

saccharification 211, 248, 252–253
safety 4, 7, 29, 36, 42–43, 48, 53, 56, 69, 76, 78, 92,
    102, 147, 172, 178, 193, 222–223, 355–356
safety controls 4
sampling 165, 168, 171, 173, 178–179, 183, 354
sampling methods 171, 179
Sanofi 42–43
Saudi Aramco 25–26
scale-up 1, 77, 86, 90, 101, 115, 125, 167, 181, 255,
    279, 327–329
scouting reaction 106
seawater desalination 273
secure energy 192
selectivity 65–66, 69, 74, 76–77, 79–80, 85, 100,
    102, 106, 110, 128, 153, 159, 172, 197, 200, 215,
    217, 259, 280, 284–286, 331, 337
sensor self-validation 310
separation process 5, 82–84, 121–122, 137, 148,
    150, 153–155, 159, 215, 223, 250, 257
simple distillation 149–150, 210
slurry reactor 282–283
slurry-type reactor 282, 285

society 10, 207, 231
sodium borohydride 104
sodium hydroxide 104, 328–329, 336
solar energy 109, 192, 269, 274, 278, 280, 288
solar radiation 269, 279, 288
solar-driven organic solvent purification 274
solid drug formulation 183
solubility 38, 42, 104, 122–123, 125, 128, 153, 178,
    240, 246, 254–255, 270, 284
solubility problem 104
solute 32, 121–122, 126, 128, 132
solution diffusion model 127–128
solvent 6, 13, 16, 20–21, 32–39, 42–43, 45, 48–49, 53,
    66, 68–69, 88, 90, 102–106, 111, 121, 123–124,
    126–132, 135–136, 139–141, 147, 149–160, 168,
    174, 181, 183, 212, 219, 221, 223, 233, 245–246,
    249–250, 254, 257–258, 262, 275, 327–329,
    331–335, 340–341, 343, 347–348, 351–352
solvent exchange 155
solvent intensity 68, 155, 341
solvent recovery 121, 127, 130, 140–141, 147,
    149–157, 159, 223, 257, 332
solvent recycling 140, 147, 149–150, 152, 331
solvent selection guide 38, 42–43, 158
solvent-free system 103
sonocrystallization 134
Sonogashira reaction 80
space-time yield 101–102
spin conversion 279
spinning disk reactor 78, 81
spiral wound membrane module 125–126
stability 38, 42–43, 48, 179, 204, 211, 218, 254, 264,
    271, 281
static mixer 78–79
stationary phase 121, 135, 282
steam distillation 149–150
stirring 58, 328, 336
stoichiometric 7, 13, 60, 327, 329, 333
supercritical carbon dioxide 113
supercritical condition 112, 123, 222
supercritical fluid 38, 84, 112, 123, 220
supersaturation 135, 173
sustainability 2–3, 5, 7, 11, 13, 20, 23–24, 27, 29, 36,
    42, 48, 51, 58, 66, 77, 102, 111, 129, 138, 140,
    149, 160, 166, 168–170, 172, 192, 195, 201, 207,
    217, 223–224, 258, 263, 291, 328, 332, 335
sustainability assessment 13, 23
sustainable chemistry 4, 230, 233

sustainable development 1, 3, 5, 10, 200, 243
switchable solvent 38–39
synthesis route 6, 13, 55–57, 59, 63, 65, 68
synthetic equivalent 51–52
synthetic polymers 199, 229, 260
synthetic route 13–14, 18, 48–51, 54, 56, 58, 60,
    66–68, 255
synthon 51
syringe pump 116
Syrris 115

take-make-dispose 103
telescoping 101, 115
temperature swing adsorption 138, 140–141
tetrahydrofuran 39, 49, 104, 109, 159–160, 247,
    262, 329, 331
thalidomide 6
theoretical yield 65
thermostat 129, 140
time-lag 165, 171
toolbox 76, 151
torrefaction 83
total generated wastewater 21
total waste stream analysis 18
toxicity 6, 16, 27, 36–38, 43, 48–49, 131, 204, 211,
    241, 245
training data 307, 310
transesterification 207, 218–221, 223–224,
    264
Transfer Learning 303
tube-in-tube reactor 107–109

ultrasound 76, 84, 87–89, 134–135, 139, 251,
    284, 286
uranium 189, 192–193, 195–204
UV-cutoff 177

vacuum 13, 33, 58, 138, 149, 159
vacuum distillation 58, 149
vacuum filtration 13
vacuum pressure adsorption 138
valorization 212
vanadium 201
vaporization energy 123
Vapourtec 115
vegetable oil 208, 217–219, 241
volatility 38–39, 43, 148, 150
volume-time-output (VTO) 29

waste disposal 42, 155, 194, 202, 204, 340
waste generation 5, 14, 17, 23, 25, 36, 54, 58,
  72–73, 103, 130, 150, 155, 211–212, 236, 351
waste management 202, 230, 235, 342
wastewater 21, 208, 219–220, 258, 284, 341,
  348, 352
wastewater intensity 341, 348, 352
water 9, 13, 16, 18–19, 20, 21, 24–25, 27, 33, 38–39,
  42, 48, 55, 58, 66–68, 84, 87–88, 90–91,
  104–105, 108, 112–113, 125, 136, 153, 158, 170,
  174, 177, 184, 192, 197–198, 200–202, 204,
  207–210, 214–218, 220–221, 223–224, 229, 231,

233, 236, 240–241, 246, 248, 250–252, 257,
  260, 269–274, 276, 278, 282–285, 288,
  327–329, 331, 336–337, 339, 341–345,
  347–349, 351–352
water harvesting 271
water vapor 270–271, 273, 278
water-soluble 104, 202, 204, 240, 246, 248

yield 13, 15, 19, 21, 29, 36, 56, 65, 68–70, 76, 79–80,
  83, 85, 89, 91, 101–104, 106, 108, 121, 134, 166,
  209, 220–221, 245–246, 248–249, 252, 264,
  279, 281, 337, 345, 348, 350, 352, 355

www.ingramcontent.com/pod-product-compliance
Lightning Source LLC
Chambersburg PA
CBHW080706220326
41598CB00033B/5320